NOBEL LECTURES

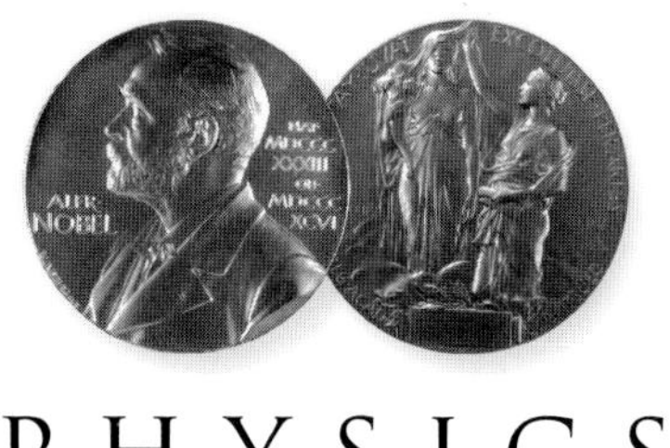

PHYSICS

1996–2000

NOBEL LECTURES

INCLUDING PRESENTATION SPEECHES
AND LAUREATES' BIOGRAPHIES

PHYSICS
CHEMISTRY
PHYSIOLOGY OR MEDICINE
LITERATURE
PEACE
ECONOMIC SCIENCES

NOBEL LECTURES

INCLUDING PRESENTATION SPEECHES AND LAUREATES' BIOGRAPHIES

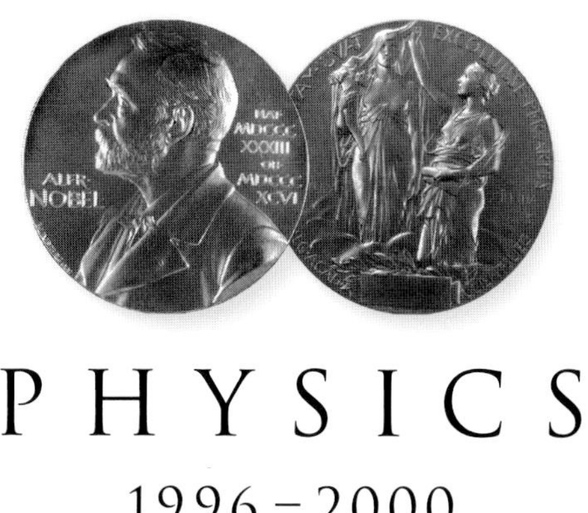

PHYSICS

1996–2000

EDITOR

Gösta Ekspong

Department of Physics

University of Stockholm

Published for the Nobel Foundation in 2002 by

World Scientific Publishing Co. Pte. Ltd.
5 Toh Tuck Link, Singapore 596224
USA office: Suite 202, 1060 Main Street, River Edge, NJ 07661
UK office: 57 Shelton Street, Covent Garden, London WC2H 9HE

NOBEL LECTURES IN PHYSICS (1996–2000)

ISBN 981-238-003-5
ISBN 981-238-004-3 (pbk)

Printed by FuIsland Offset Printing (S) Pte Ltd, Singapore

FOREWORD

Since 1901 the Nobel Foundation has published annually "Les Prix Nobel" with reports from the Nobel award ceremonies in Stockholm and Oslo as well as the biographies and Nobel lectures of the laureates. In order to make the lectures available for people with special interests in the different prize fields the Foundation gave Elsevier Publishing Company the right to publish in English the lectures for 1901–1970, which were published in 1964–1972 through the following volumes:

Physics 1901–1970	4 vols.
Chemistry 1901–1970	4 vols.
Physiology or Medicine 1901–1970	4 vols.
Literature 1901–1967	1 vol.
Peace 1901–1970	3 vols.

Thereafter, and until the year 2000, the Nobel Foundation has given World Scientific Publishing Company the right to bring the series up to date and also publish the Prize lectures in Economics from the year 1969. The Nobel Foundation is very pleased that the intellectural and spiritual message to the world laid down in the laureates' lectures, thanks to the efforts of World Scientific, will reach new readers all over the world.

Bengt Samuelsson
Chairman of the Board

Michael Sohlman
Executive Director

Stockholm, August 2001

PREFACE

The Nobel price for Physics like those for Chemistry, for Medicine or Physiology and for Literature is awarded every year at a solemn ceremony in Stockholm on December 10th, the day of the donor Alfred Nobel's death in San Remo, Italy. The presentation speech given by a member of the Nobel Committee for physics is included in this volume. This is quite short to suit the occasion and the audience. More information may be found on the website www.nobel.se.

The main content of the present volume is comprised of the full texts of the Nobel lectures given by the Laureates for the years 1996–2000. They are here reprinted unabridged and represent documents of historical value. Only minor changes have been made such as to correct printing errors. The reader may enjoy here excellent and penetrating papers of great significance.

The range of subjects is well described by the wordings of the citations carefully chosen by the Royal Swedish Academy of Sciences as follows:

"for their discovery of superfluidity in helium-3" (1996) to David Lee, Douglas Osheroff and Robert Richardson

"for development of methods to cool and trap atoms with laser light" (1997) to Steven Chu, Claude Cohen-Tannoudji and William Phillips

"for their discovery of a new form of quantum fluid with fractionally charged excitations" (1998) to Robert Laughlin, Horst Störmer and Daniel Tsui

"for elucidating the quantum structure of electroweak interactions in physics" (1999) to Gerardus 't Hooft and Martinus Veltman

"for developing semiconductor heterostructures used in high-speed and opto-electronics" (2000) to Zhores Alferov and Herbert Kroemer (one-half of the prize)

"for his part in the invention of the integrated circuit" (2000) to Jack Kilby (one-half).

Included also are the autobiographies of the fourteen Laureates. These texts ensure highly interesting reading and describe a large variety of family backgrounds and paths to scientific achievements and successes.

Gösta Ekspong
Professor Emeritus
Member of the Royal Swedish Academy of Sciences

CONTENTS

Physics 1996

DAVID M. LEE, DOUGLAS D. OSHEROFF and ROBERT C. RICHARDSON

"for their discovery of superfluidity in helium-3"

THE NOBEL PRIZE IN PHYSICS

Speech by Professor Carl Nordling of the Royal Swedish Academy of Sciences. Translation of the Swedish text.

Your Majesties, Your Royal Highness, Ladies and Gentlemen,

The air that we breathe not only contains oxygen and nitrogen, but also small quantities of other gases. One of these–often mentioned in connection with the global greenhouse effect–is carbon dioxide. Another is the noble gas helium, which makes up five parts per million of our atmosphere.

This chemical element exists in two forms, or isotopes: a heavier one called helium-4 and a lighter one called helium-3. The heavier variety accounts for nearly all helium. Helium-3 makes up only one millionth of the total quantity of helium, which is already very insignificant to begin with. Yet this year's Nobel Prize in Physics is all about helium-3.

David Lee, Douglas Osheroff and Robert Richardson used a few cubic centimeters of liquid helium-3 to perform the experiments that would lead to the discovery being rewarded with this year's Nobel Prize. They changed its pressure, temperature and volume, carefully monitoring the mutual dependence of these variables.

In the resulting diagram, they observed two small jogs in the curve, a few thousandths of a degree above absolute zero. Many researchers would probably have shrugged their shoulders at these deviations, dismissing them as minor imperfections in their measuring apparatus.

Not these three researchers. Might new magnetic states be manifesting themselves in this way? What the three were actually looking for was magnetism in solid helium-3. At first, they also believed that this was what they had seen. But there was not a perfect correspondence with the measurement data. By means of measurements employing the same technique that has since come to be used in hospital magnetic resonance imaging systems, Lee, Osheroff and Richardson were able to show that the phenomenon was occurring in liquid helium-3, not in solid helium-3.

In other words they had discovered two new forms of liquid helium-3, both superfluid. The team eventually discovered three superfluid phases. And as is so often the case in basic research, they had discovered something other than what they were looking for!

Superfluidity is a remarkable and unusual property that had previously been observed only in helium-4. It manifests itself in different ways: A superfluid liquid lacks viscosity and cannot be stored in an unglazed ceramic vessel, because it seeps out through the microscopic pores in the ceramic. If an empty beaker is lowered part way into the liquid, the liquid flows upward, over the edge and down into the beaker.

To describe the phenomenon of superfluidity at a fundamental, atomic level, we usually say that the atoms have undergone a Bose-Einstein condensation. This means that all the atoms join in a common quantum state. Such a condensation is only possible in a category of particles called bosons. In the category called fermions, this type of condensation is not possible.

As it happens, helium-3 atoms are fermions and should thus not be capable of undergoing a Bose-Einstein condensation and form a superfluid liquid. Yet they can. The explanation is that the atoms form pairs, where the atoms orbit each other. Such a pair behaves like a boson and–presto–Bose-Einstein condensation can now occur and the liquid becomes superfluid!

Lee's, Osheroff's and Richardson's discovery triggered intensive research activity in all the low-temperature laboratories of the world. The phase transitions to superfluidity in liquid helium-3 showed that the quantum laws of microphysics sometimes also govern the behavior of macroscopic quantities of matter. They are being used to define the temperature scale at very low temperatures. They are contributing to our growing understanding of "warm" superconductors, and they were recently used as the model for how "cosmic strings" may have been formed in the universe.

Professor Lee, Professor Osheroff, Professor Richardson,

You have been awarded the 1996 Nobel Prize in Physics for your discovery of superfluidity in helium-3. Your discovery has greatly enlarged our knowledge of the possible states of condensed matter. On behalf of the Royal Swedish Academy of Sciences, I wish to congratulate you on this achievement, and I now ask you to step forward and receive your Prize from the hands of His Majesty the King.

David M. Lee

DAVID M. LEE

My parents were born and brought up in New York City. My father was trained as an electrical engineer and my mother was an elementary school teacher. They were the children of Jewish immigrants who had come to the United States from England and Lithuania in the late 1800's. One of my great grandfathers had actually settled in the United States considerably earlier. When I was born on January 20, 1931, my parents lived in a small suburban town, Rye, New York, just outside New York City. My father commuted by train to his job at a small but growing electrical manufacturing company in the city. During the great economic depression of the early 1930's we moved to the city for a few years to save money, but eventually moved back to Rye, where I received my early education. As time went on our family circumstances improved as my father advanced in his company, which was expanding rapidly, and eventually became its president.

As a child I was fascinated by living things in the fields and along the coast line near our home. I was constantly roaming around collecting frogs, fish, salamanders, snakes and worms. Starting at the age of six, I spent every summer away from home at various children's camps in New England, giving me further opportunities to explore this interest.

My other childhood passion was railways. I managed to accumulate an extensive collection of railway timetables covering the entire U.S.A. and became a young travel expert. When I was a very young child my father gave me a set of spring operated "wind up" trains. The first thing I did was to insert the tracks into the electric socket in our kitchen. A shower of sparks flew all over the room. Fortunately my parents were indulgent and everyone laughed about the incident.

As a young teenager I became very interested in meteorology. I kept my own weather records and subscribed to the daily weather map issued by the U.S. weather bureau. One day I asked my father about a book in his library entitled "The Mysterious Universe" by Sir James Jeans. He indicated that no one really understood what was in the book. I immediately picked up the book and began to read it. There was a beautiful discussion of the cosmology known at that time, which I found totally fascinating. I think that this book really sparked my interest in physics.

The high school in Rye had an excellent program. There was emphasis on acquiring the necessary basic skills in writing and mathematics through extensive exercises but we were also taught to think for ourselves. I owe a considerable debt of gratitude to my teachers. Of course most young boys during that time wanted to be sports heroes and I was no exception. I was a reasonably good short distance runner and so was active on our school track team, as well as a participant in our high school football program, but there was no chance that I would ever be a sports hero.

Following graduation from high school in 1948, I attended Harvard University where I became a physics major. Having grown up in a small town, I found Harvard to be an enormously enriching experience. Students in my class came from all walks of life and from a great variety of geographical locations. I still stay in touch with many of my college friends. At one time during my college years I considered the possibility of a career in medicine. With this in mind I took some of the pre-medical courses in addition to my physics major. I especially enjoyed the course in organic chemistry, but in spite of my early interests, I did not find the biological sciences fascinating. Therefore I gave up the idea of a career in medicine and continued with my studies of physics. My main extracurricular activity was the Harvard Yacht Club. In June 1950 a group of us sailed in the Bermuda race from Newport, Rhode Island, to Hamilton, Bermuda. It was a wonderful adventure.

After 3 1/2 years at Harvard, I had enough credits to graduate in January 1952. In April 1952, I entered the U.S. Army for 22 months and served at various posts in the continental United States during the final stages of the Korean War. One night during this period I was serving as corporal of the guard. One of the guards was a young soldier named Herbert Fried. It turned out that he had been a graduate student at the University of Connecticut with Professor Paul Zilsel who specialized in the theory of superfluidity. We had a wonderful discussion about superfluid helium 4. Later on Herbert Fried became a Professor of Theoretical Physics at Brown University.

Following my honorable discharge from the army, I entered the University of Connecticut in February 1954 partly as a result of my discussion with Herbert Fried and partly because my parents had moved to Connecticut, so it was now my home state. The one and one half year stay at the University of Connecticut was extremely beneficial. It gave me the chance to study physics in a relatively relaxed setting and to learn about experimental physics. My first project was to build an ionization gauge control circuit for Professor Edgar Everhart's Cockcroft–Walton accelerator. In those days vacuum tubes were the active components in electronic circuits. I can still recall the warm orange glow of the vacuum tube filaments and the cool blue glow of the thyratron tubes. In assembling and trouble shooting my circuit, I can also still remember all the 300 volt electric shocks from the vacuum tube power supply.

While at the University of Connecticut, I met my life-long friend John Reppy who was later to become my colleague in our Cornell low temperature group. John was doing experimental research on superfluid liquid helium with Professor Charles Reynolds. It was Professor Reynolds who really excited my interest in superfluidity and low temperature physics.

In addition to John Reppy's prowess as an experimental physicist, he was a rock climber and mountaineer, par excellence. He somehow persuaded me to overcome my natural fear of heights and took me on some wonderful climbs in the Grand Tetons of Wyoming and the Black Hills of South Dakota in the American west. I still enjoy hiking in the mountains.

Eventually I completed my requirements for the Master of Science degree at the University of Connecticut, after which I enrolled in the Ph.D. program in physics at Yale University in the summer of 1955. My summer project at Yale was to build a mercury jet stripper for the Heavy Ion Linear Accelerator then under construction. By removing more electrons from an ion, one could increase its net charge and thus accelerate it to higher energies. Electrons from the ions were removed rather efficiently when the ions were passed through a supersonic jet of mercury atoms. Also during my first summer at Yale I met Russell Donnelly who was finishing his Ph.D. thesis on rotating superfluid helium in the Yale low temperature group with Professor Cecil T. Lane. Russ was a talented experimentalist with tremendous enthusiasm for physics. He has had a distinguished career and is now a Professor at the University of Oregon. In addition to my work on the accelerator, I enjoyed helping Russ with his experiments that summer. In a very short time, I learned a great deal about experimental low temperature physics and the life of an experimental physicist. As time went on my growing fascination with low temperature physics led me to the decision that this would be my area of specialization in graduate school. Fortunately, Professor Henry A. Fairbank of the Yale low temperature group had a position for me. Henry was an excellent mentor and a helpful and understanding thesis adviser. At that time, the isotope ^{3}He was first becoming available. My thesis topic involved research on liquid ^{3}He and is discussed in my Nobel lecture. I look back upon graduate school as being a very happy period in my life. The chance to be thoroughly immersed in physics and to be surrounded by friends pursuing similar goals was a marvelous experience. It was totally rewarding to observe exciting new effects in an apparatus that I had designed and constructed with my own hands.

In January 1959, I completed my research at Yale and joined the Cornell University faculty. My responsibilities were to set up a research laboratory in low temperature physics and to teach courses in the physics department. I was also responsible for the operation of our helium liquifier. Shortly after arriving at Cornell I met my wife, Dana, who was a Ph.D. student in nutrition and biochemistry. She was born and raised in Thailand. Her father originally came from Copenhagen and her mother was a native Thai. For more than 36 years she has been a wonderful companion. Without her loving support my career would certainly have been far less successful. We now have two grown sons who, with their wives, joined us at the Nobel celebration in Stockholm. Over the years I worked my way up through the ranks to the position of Professor in the Cornell physics department. Meanwhile our low temperature group increased in size with the addition in the 1960's of Professors John D. Reppy who had also been a graduate student at Connecticut and later at Yale and Robert C. Richardson who joined us from Duke University. More recently Professor Jeevak Parpia has joined our group. Over the years our program has been very successful.

Highlights, in addition to the work on superfluid ^{3}He, include the discovery of the tricritical point on the phase separation curve of liquid ^{3}He–^{4}He mixtures by graduate student Erlend Graf, John D. Reppy and myself, the discovery of the antiferromagnetic ordering in solid ^{3}He by graduate student William P. Halperin, Robert C. Richardson and their associates, and the discovery of nuclear spin waves in spin polarized atomic hydrogen gas as part of a collaboration between myself and Jack H. Freed of our chemistry department. In addition, John Reppy and his students conducted extensive investigations of persistent currents in superfluid ^{4}He and ^{3}He. His experiment with graduate student David Bishop provided a striking example of the Kosterlitz-Thouless transition in superfluid ^{4}He films. For this work John was awarded the 1981 Fritz London Memorial Prize. Jeevak Parpia has recently performed some very exciting studies of superfluid ^{3}He in confined geometries.

Other prizes awarded to members of the group include the 1976 Sir Francis Simon Memorial Prize of the British Institute of Physics and 1981 Oliver Buckley Prize of the American Physical Society. Both of these prizes were awarded to Douglas D. Osheroff, Robert C. Richardson and myself for the discovery of superfluid ^{3}He. In addition, Robert Richardson, John Reppy and myself have been elected to the National Academy of Sciences and the American Academy of Arts and Sciences.

One of the most rewarding aspects of an academic career is the opportunity to work with graduate students, and to watch them develop after leaving graduate school. My fellow laureate, Doug Osheroff, is a prime example of a scientist who was extremely successful as a graduate student but who later had a distinguished career at AT & T Bell Laboratories and at Stanford University. Most of our other students have had very responsible and rewarding careers in science and technology. It is a special pleasure to thank my students and my colleagues for their role in our success.

THE EXTRAORDINARY PHASES OF LIQUID ^{3}He

Nobel Lecture, December 7, 1996

by

DAVID M. LEE

Laboratory of Atomic and Solid State Physics, Dept. of Physics,
Cornell University, Ithaca, N.Y. 14853, USA

INTRODUCTION

Modern low temperature physics began with the liquefaction of helium[1] by Kamerlingh Onnes and the discovery of superconductivity[2] at the University of Leiden in the early part of the 20th century. There were really two surprises that came out of this early work. One was that essentially all of the electrical resistance of metals like mercury, lead, and tin abruptly vanished at definite transition temperatures. This was the first evidence for superconductivity. The other surprise was that, in contrast to other known liquids, liquid helium never solidified under its own vapor pressure. Helium is an inert gas so that the interactions between the helium atoms are very weak; thus the liquid phase itself is very weakly bound and the normal boiling point (4.2 K) is very low. The small atomic masses and the weak interaction lead to large amplitude quantum mechanical zero point vibrations which do not permit the liquid to freeze into the crystalline state. Only if a pressure of at least 25 atmospheres is applied will liquid ^{4}He solidify.[3] It is thus possible, in principle, to study liquid ^{4}He all the way down to the neighborhood of absolute zero.

Quantum mechanics is of great importance in determining the macroscopic properties of liquid ^{4}He. Indeed, liquid helium belongs to a class of fluids known as quantum fluids, as distinct from classical fluids. In a quantum fluid the thermal de Broglie wavelength $\lambda_T = h(2\pi mkT)^{-\frac{1}{2}}$ is comparable to or greater than the mean interparticle distance. There is then a strong overlap between the wave functions of adjacent atoms, so quantum statistics will have important consequences. ^{4}He atoms contain even numbers of elementary particles and thus obey Bose-Einstein statistics, which means that any number of atoms can aggregate in a single quantum state in the non-interacting particle approximation. In fact macroscopic numbers of atoms in a quantum fluid can fall into the lowest energy state even at finite temperatures. This phenomenon is called Bose-Einstein condensation. On the other hand ^{3}He atoms, each of which contains an odd number of elementary particles, must obey Fermi-Dirac statistics: only one atom can occupy a given quantum state. Therefore one should expect a very large difference between the behavior of liquid ^{4}He and that of liquid ^{3}He for low temperatures where the thermal de Broglie wavelength becomes greater than the mean interparticle distance.

A remarkable phase transition was discovered in liquid ^{4}He under saturated vapor pressure at 2.17 K. As the liquid cooled through this temperature, all boiling ceased and the liquid became perfectly quiescent.[4] We now know that this effect occurs because the liquid helium becomes an enormously good heat conductor so that thermal inhomogeneities which can give rise to bubble nucleation are absent. The specific heat vs. temperature curve of liquid ^{4}He was shaped like the Greek letter lambda, characteristic of a second order phase transition at 2.17 K. This temperature is called the lambda point.[4,5] Below this temperature, liquid ^{4}He was found to possess remarkable flow properties as well as the "super" heat transport mentioned above. If a small test tube containing the liquid was raised above the surrounding helium bath, a mobile film would form, allowing the liquid to be transported up the inner walls, over the top and down the outer walls, eventually dripping back into the bath and thereby emptying the test tube.[6] Furthermore, liquid ^{4}He could flow freely through the tiniest pores and cracks as shown by Kapitza,[7] who performed a number of ingenious experiments involving flow properties of superfluid helium. Perhaps the most dramatic manifestation of anomalous flow behavior was the so-called fountain effect discovered by Allen and Jones.[8] If a glass tube packed tightly with a powder such as jeweler's rouge was partially immersed in a ^{4}He bath and then heated, a fountain of helium rising high above the level of the surrounding helium bath was produced. A model called the Two Fluid Model to describe these phenomena was developed by Landau[9] and Tisza.[10] According to this model, liquid ^{4}He below T_λ can be thought of as two interpenetrating fluids, the normal and the superfluid components. The latter component is involved in superflow through pores and cracks and does not carry entropy. Furthermore, it does not interact with the walls of a vessel containing the fluid in a dissipative fashion. Superimposed on this background superfluid component is the normal component which transports heat efficiently and exhibits viscosity, allowing transfer of energy between the liquid and the walls. This latter effect was the basis for the ingenious experiment by Andronikashvili[11] who actually measured the normal fluid density as a function of temperature by studying the damping of a torsional pendulum, which interacted only with the normal fluid. It was found that the normal fluid density decreased with decreasing temperature and consequently, the superfluid density increased, becoming dominant at the lowest temperatures. The normal fluid carries heat away from the heat source and is replaced by the superfluid component, so we have a countercurrent heat flow. The flow of the superfluid component toward a source of heat is spectacularly manifested in the fountain effect, mentioned above.

According to Landau the normal fluid consists of a gas of quantized thermal excitations which include the ordinary longitudinal sound waves (phonons) and short wavelength compact excitations which he named the rotons. On the basis of the two fluid model, it was predicted that heat transport would obey a wave equation which describes the compressions and rarefactions in the phonon/roton "gas". Such a wave phenomenon was indeed discovered experimentally[12] and was named second sound.

The nature of the superfluid background still needed to be characterized. Fritz London's great contribution[13] was to note that superfluidity could be viewed as quantum behavior on a macroscopic scale associated with the Bose-Einstein (BE) condensation. As the temperature is reduced through the transition temperature, the occupancy of the one particle ground state becomes macroscopic and can be thought of as the BE condensate. The superfluid component in the two fluid picture could be roughly identified with this condensate, although strong interactions between the atoms in the liquid modify this picture. In this scheme, the superfluid atoms are governed by a wave-function-like entity called the order parameter, as introduced by Ginzburg and Landau[14] for the case of superconductivity. The order parameter ψ for superfluid ^{4}He is characterized by a phase ϕ and an amplitude ψ_0 and is given by $\psi = \psi_0 e^{i\phi}$ where ψ_0 can be roughly thought of as the square root of the density of the superfluid component. The fact that the macroscopic order parameter is also described by a definite phase is called broken gauge symmetry. It has been shown that the superfluid velocity is proportional to the gradient of the phase. It is this macroscopic order parameter picture which describes how the helium atoms march in "lock-step" during superfluid flow. One beautiful consequence is the existence of quantized vortices in superfluid ^{4}He. This is a generalized phenomenon seen in all superfluids including superfluid ^{3}He and superconductivity, where a quantized current vortex must enclose a quantum of flux.

Superfluidity in liquid ^{4}He is thought to be a manifestation of BE condensation. What about the electrons in a superconducting metal, which obey Fermi-Dirac (FD) statistics? The theory behind superconductivity remained a mystery for about half a century. There were tantalizing clues such as the isotope effect,[15,16] which showed that the superconducting transition temperature of a particular metal is dependent on the atomic mass of the isotope comprising that metal sample, thus connecting the superconductivity of the electrons to the dynamical behavior of the crystalline lattice of metallic ions.

The major breakthrough in our understanding of superconductivity occurred in the late 1950's when Bardeen, Cooper and Schrieffer (BCS)[17] proposed their theory of superconductivity. This theory resulted in a vast revolution in the field of superconductivity. As mentioned earlier, BE particles (bosons) can congregate in their ground state at finite temperatures as a result of BE condensation. This provides the basis for the establishment of a superfluid order parameter. (The situation is really more complicated and requires that interactions be taken into account.) For a simple model involving non-interacting electrons, the conduction electrons in a metal form a sea of FD particles (fermions). At $T = 0$ all the lowest states are occupied up to the Fermi energy. Because of the Pauli exclusion principle, only one electron is allowed in each quantum state, so that macroscopic congregation in the ground state is not permissible. The BCS theory overcame this difficulty by showing that, when a metal became superconducting, the electrons in the metal formed pairs (now known as Cooper pairs),[18] which had some of the properties of bosons. These pairs could thus congregate into a single ground state (in a loose analogy to BE condensation) described by an order parameter which does not violate the Pauli principle but which leads to a conducting superfluid of electrons. The wave function describing this ground state was devised by Robert Schrieffer.[17] The partners in a Cooper pair consist of two electrons whose motion is correlated even though they may be separated by distances much larger than the interparticle spacing. In other words, the pairs do not behave like Bose condensed discrete diatomic molecules. The difference may be understood in terms of modern rock and roll dancing vs. ballroom dancing, according to a marvellous analogy invented by Schrieffer and discussed by him in a number of public lectures. In ballroom dancing, the partners hold tightly to one another in analogy with diatomic molecules. The Cooper pair, on the other hand, would consist of two rock and roll dancers whose gyrations are closely related in spite of their distant separation. In between the partners of a pair, members of other pairs may pass by. The strong correlation of the pairs demonstrated by BCS leads to the pairs marching in lock step, in a fashion similar to the bosons in superfluid ^{4}He.

Why do electrons form pairs? Leon Cooper[18] in fact showed that at low enough temperatures electrons form pairs as long as there is a *net* attractive force, even a very weak one. We know that the electrons all have negative charges which result in strong Coulomb repulsion but this can be balanced out and even reversed by the dynamic response to the electrons of the positive ions forming a crystal. The results of the isotope effect experiments mentioned earlier provided the key to this insight. As an electron moves through the lattice it attracts the positive ions, forming a region with a higher density of positive ions, which can in turn *attract* other electrons. This role of the massive positive ions explains the isotope effect. The density fluctuations of the ions that are associated with the passage of an electron can be described in terms of quantized lattice waves called phonons, and the attraction is thus associated with electron-phonon interactions.

The temperature T_c at which a metal becomes superconducting is the temperature for which it becomes energetically favorable to form pairs. The temperature T_c is typically 1000 times smaller than the Fermi degeneracy temperature T_F, where the thermal de Broglie wavelength becomes comparable to the mean interparticle spacing a_0 and quantum effects become important. Pairing superfluidity in Fermi fluids is therefore much more difficult to achieve, in contrast to the case for bosons where the onset of superfluidity occurs when the quantum fluid condition $\lambda_T \gtrsim a_0$ is satisfied. For the case of pairing superfluidity, two electrons near the Fermi surface can give up energy by forming a Cooper pair. The same energy, say 2Δ, must be supplied to break up a Cooper pair. The size of this energy gap 2Δ is a fundamental parameter in the theory of superconductivity. The energy gap parameter Δ approaches zero as we approach T_c, but Δ grows in size as a superconductor is cooled to absolute zero. Why should the energy gap be a function of temperature? We take the mean field or molecular field view which is so successful in explaining magnetism. For that case, the tendency toward further ordering increases with increasing ordering, which corresponds to a strengthening of the molecular field. Applying this to the case of pairing in a superconductor, we find that a larger number of pair states leads to a larger binding. The energy gap is a measure of the binding energy of a pair and will therefore increase as the number of pairs increases with decreasing temperature. This behavior is fully accounted for by the BCS theory.

Many years before the BCS theory, Fritz and Heinz London[19] had developed phenomenological equations for superconducting metals. Fritz London[20] showed how these equations could be discussed in terms of his idea of quantum mechanics on a macroscopic scale. In 1950 Ginzburg and Landau[14] proposed a complex order parameter ψ, consistent with London's discussion, representing the many electron state in a superconductor, where $|\psi|^2$ was equal to the local density of superconducting electrons. Here we have invoked the two fluid model familiar from our discussion of superfluid ^{4}He, where we now consider two interpenetrating electron fluids corresponding to normal and superconducting components. (We have already mentioned a Ginzburg Landau order parameter for superfluid ^{4}He.) Ginzburg and Landau derived a differential equation from an expansion of the free energy in powers of ψ given by

$$\frac{1}{2m^*}\left(\frac{\hbar}{i}\nabla - \frac{e^*}{c}\vec{A}\right)^2 \psi + \beta|\psi|^2\psi = -\alpha(T)\psi.$$

This equation resembles the Schrödinger equation but has an additional term in $|\psi|^2$. Although it is not the Schrödinger equation, the electric current obtained from this equation has exactly the same form as that for a wave function, namely

$$\vec{J} = \frac{e^*\hbar}{2m^*i}\left(\psi^*\nabla\psi - \psi\nabla\psi^*\right) - \frac{(e^*)^2}{m^*c}|\psi|^2\vec{A}.$$

Experiments later showed that $e^* = 2e$ and $m^* = 2m$, thus making contact with the BCS theory of superconductivity in metals and showing that the order parameter describes the correlated pairs.[21,22]

The order parameter is again given by the simple expression $\psi = \psi_0\, e^{i\phi}$ which possesses a phase ϕ and an amplitude ψ_0 which increases in magnitude with the gap parameter Δ. We have already interpreted $|\psi|^2 = \psi_0^2$ in terms of the two fluid model, so the amplitude ψ_0 is simply the square root of the density of superconducting electrons. As in the case for superfluid ^{4}He, the phase of the order parameter is of paramount importance for the superflow properties. Such phenomena as quantized flux[21,22] and the Josephson effect[23] (corresponding to the tunneling of pairs) require phase coherence throughout the superconductor.

Could the pairing theory be applied to other systems? Liquid ^{3}He was the most obvious candidate to be examined. It is composed of neutral atoms with a nuclear spin angular momentum of $\hbar/2$ and a nuclear magnetic moment. The ^{3}He atom has an odd number of elementary particles and so it obeys FD statistics and the Pauli exclusion principle. The atoms in the liquid are known to interact strongly so one cannot strictly apply the theory of an ideal Fermi gas to predict the properties of liquid ^{3}He in the normal non-superfluid Fermi liquid (NFL), meaning the liquid above any possible superfluid transition temperature. Lev Landau[24] formulated a theory of strongly interacting Fermi liquids which introduced the idea of quasiparticles corresponding to bare fermions "clothed" by their interactions with the others. The various properties of normal liquid ^{3}He qualitatively resembled the properties of ideal Fermi gases, but the numerical factors were entirely different. The Landau theory showed how these properties could be expressed in terms of a set of parameters called the Fermi liquid parameters. In a Fermi liquid at low temperatures, the thermally excited quasiparticles will occur in a narrow band near the Fermi surface with energy width of order kT. Only the quasiparticles in this narrow band of states can participate in scattering or in thermal excitations. As T is lowered the width of the band shrinks and fewer quasiparticles can participate in such events. As a result the specific heat and the entropy depend linearly on temperature ($c = \gamma T$) and the mean free path is proportional to T^{-2}. The thermal conductivity therefore has a $1/T$ dependence and the viscosity has a $1/T^2$ dependence. The numerical constants for an interacting Fermi fluid will differ from those for an ideal Fermi gas due to interactions described by Landau parameters, as discussed above. These and other experimental properties of normal liquid ^{3}He were studied in laboratories around the world, but the dominant group was led by John Wheatley,[25] first at the University of Illinois and then at the University of California at San Diego (La Jolla). A primary result of these investigations was the evaluation of the Landau Fermi liquid parameters. The Landau theory also made an important prediction; collisionless sound, called by Landau zero sound.[24] Ordinary sound in a gas for example involves the propagation of waves of compression and rarefaction in a state of local thermodynamic equilibrium brought about by collisions between the molecules. At the lowest temperatures in a Fermi liquid the collisions are substantially absent and ordinary sound dies away. According to Landau, a new mode of sound propagation arises at the lowest temperatures involving self-consistent rearrangements of the quasiparticles under the influence of Fermi liquid interactions. This prediction was dramatically confirmed in laboratory experiments by Keen, Matthews and Wilks,[26] and Abel, Anderson and Wheatley.[27]

Because of the strong interactions between ^{3}He atoms, it was soon realized that if Cooper pairs formed in liquid ^{3}He, they would be quite different in nature from the pairs associated with superconducting electrons. In ordinary superconductors, the Cooper pairs have zero orbital angular momentum ($l = 0$) so the members of a pair do not rotate around one another. The strong short range repulsion of the quasiparticles in liquid ^{3}He prevents this type of pairing from occurring, but higher angular momentum pairing is indeed possible, as was first proposed by Lev Pitaevskii.[28]

Over the years a number of higher orbital angular momentum pairing states were proposed for a hypothetical superfluid state of liquid ^{3}He. Both p wave ($l = 1$) and d wave ($l = 2$) states of relative orbital angular momentum were suggested. Among the early studies were those by Emery and Sessler[29] and Anderson and Morel.[30] The proposals for p wave pairing by Balian and Werthamer,[31] and Anderson and Morel[30] were later identified with the *actual* superfluid phases of liquid ^{3}He.

An important feature of odd l pairing is that it requires the total spin of the pair to be 1 (not zero as in even l pairing). Thus any order parameter representing odd l pairing will possess the internal degrees of freedom associated with non-zero spin and orbital angular momentum. This fact is of the utmost importance for understanding the properties of superfluid ^{3}He, and contrasts dramatically to ordinary superconductivity for which $S = L = 0$.

It is extremely difficult to calculate the transition temperature to a superfluid phase of liquid ^{3}He. Such an estimate depends very sensitively on the detailed nature of the interactions between the ^{3}He quasiparticles. Since there is no external crystal lattice to mediate these interactions, the pairing mechanism itself must be *intrinsic.* Layzer and Fay[32] considered the fact that the nuclear magnetic susceptibility of liquid ^{3}He was considerably higher than would be expected for an ideal FD gas of comparable density. This result indicated that there was at least some tendency for the liquid to be ferromagnetic. They considered a pairing mechanism based on spin fluctuations which goes something like this: As a ^{3}He quasiparticle passes through the liquid it tends to polarize spins of neighboring quasiparticles parallel to its own spin because of this ferromagnetic tendency. Another ^{3}He quasiparticle approaching this polarized cloud will be attracted to the cloud if its spin is parallel to that of the cloud and the original quasiparticle. Thus it becomes favorable to form Cooper pairs with non-zero spin which requires odd orbital angular momentum via the Pauli Exclusion Principle.

The most striking characteristics, then, of the hypothetical superfluid ^{3}He were that (1) it would need to have an intrinsic pairing mechanism not mediated by an ionic lattice for example and (2) the resulting Cooper pairs would probably have internal degrees of freedom. These two properties would distinguish superfluid ^{3}He from the other known superfluids, superfluid ^{4}He and superconducting electrons.

In spite of considerable progress on the theoretical front, before 1971 no evidence of a superfluid transition had been found by experimenters who were pushing the cooling technology to lower and lower temperatures. Experimentalists and theorists alike became dubious that the Holy Grail would be found in a reasonable temperature range. A mood of gloom and pessimism prevailed by about 1970.

EVENTS LEADING TO THE DISCOVERY OF SUPERFLUID ^{3}HE: A PERSONAL ACCOUNT

The rare isotope of helium, ^{3}He, first became available for research in low temperature physics after World War II as a byproduct of the nuclear weapons program. It was obtained from the radioactive decay of tritium which decays via beta decay with a twelve year half life to ^{3}He. Some of the earliest research on liquid ^{3}He was performed[33] at the national laboratories which were involved in nuclear weapons research. My Ph.D. thesis adviser Professor Henry A. Fairbank at Yale University was one of the pioneers in research involving ^{3}He. In his early work he specialized in studies of second sound in liquid ^{3}He–^{4}He mixtures. In the autumn of 1955, I had the good fortune to be selected as the first graduate student to investigate pure liquid ^{3}He. My first project was to study the thermal conductivity of liquid ^{3}He with the idea of searching for Fermi-Dirac degeneracy effects which would lead to a $\kappa \sim 1/T$ dependence at low temperatures. It was anticipated that the thermal conductivity of this liquid would be very small at temperatures below the liquefaction temperature of 3.2 K as is the case for liquid ^{4}He above the superfluid transition. The experimental set up was as simple as could be imagined. The liquid was contained in a thin-walled cupro-nickel tube which had a very small thermal conductivity, thus limiting the amount of heat transport through this tube. (Nevertheless, it was necessary to correct for this as well as the thermal boundary resistance between solid walls and the liquid.) Simple (semi-conducting) carbon resistors served as the thermometers. At the top of the tube was an electrical heating coil and at the bottom a heat sink consisting of a copper block attached to a paramagnetic salt which was used to cool the sample to 0.2 K or less by adiabatic demagnetization. The magnetic susceptibility of this salt, measured with a ballistic galvanometer, served as the primary thermometer and was used to calibrate the carbon resistance thermometers. The thermal conductivity was determined from the standard formula $\dot{Q} = \kappa\ A\ \Delta T/\Delta X$ with corrections for added heat flow through the cupro-nickel tube. Measurements of the resistance thermometers were accomplished with a home-made A.C. resistance bridge including a phase sensitive detector and a tuned amplifier to assure great sensitivity and low noise even when the voltages across the thermometers were small enough to prevent significant self heating.

The initial results of the experiment were not very interesting; the thermal conductivity simply continued to decrease as the temperature decreased in very much the same fashion as that of liquid ^{4}He above the superfluid transition, with no evidence of Fermi-Dirac statistics. Nevertheless, when the experiments were run at higher powers, a very intriguing thing happened. Below a certain temperature, T_m (~ 0.5 K), the heat conduction rapidly increased as the temperature was lowered further. This effect was attributed to convective heat flow, the onset of which corresponded to a maximum in the density. To verify this, an inverted cell was constructed with the heat flowing upward. For this new geometry, the convective heat flow appeared above T_m, as would be expected if a density maximum occurred at T_m. There was wonderful serendipity here. We were studying thermal conductivity but the most interesting result involved the density.[34]

The existence of a density maximum at T_m implied that for temperatures less than T_m, the thermal expansion coefficient was *negative* and via a Maxwell relation

$$\left(\frac{\partial V}{\partial T}\right)_P = -\left(\frac{\partial S}{\partial P}\right)_T$$

it was evident that the entropy should increase with pressure below T_m. The entropy is a very basic property which could be used to estimate the interactions between the ^{3}He particles and so we decided to measure the density directly by measuring the dielectric constant, which for helium is related to the density via the Clausius-Mossotti relation,

$$\frac{\varepsilon - 1}{\varepsilon + 2}\frac{M}{\rho} = \frac{4\pi}{3}A$$

where A is the atomic polarizability and M is the atomic mass.

The method chosen to measure ε was to employ a stable radiofrequency oscillator with a tank circuit whose capacitor contained the sample. The frequency varied as the temperature was changed and the dielectric constant of the liquid ^{3}He changed. A great deal of mechanical stability was required since the electronic circuitry involved vacuum tubes which were at room temperature, while the capacitor and inductor forming the tank circuit were at the bottom of the cryostat, a full meter away. With the apparatus a series of measurements was carried out over a range of temperatures and pressures which clearly showed the density maximum and enabled the entropy of compression to be obtained from the measured thermal expansion coefficients.[35]

At the time these experiments were being performed, there was a great deal of interest in determining the melting curve of ^{3}He. Because the nuclear moments are very small, it was expected that solid ^{3}He would undergo nuclear magnetic ordering only at very low temperatures (This nuclear magnetic ordering transition was later discovered[36] at about 1 mK by my Cornell colleague and fellow laureate Robert C. Richardson, his student William Halperin, and their associates). Therefore in the range of temperatures above 0.01 K, the nuclear spins of the ^{3}He atoms comprising the solid should be almost fully disordered. For spin 1/2 nuclei this required that the entropy S should be equal to $R\ln 2$ per mole.

On the other hand liquid ^{3}He obeys Fermi-Dirac statistics. The departure from classical behavior occurs roughly at the temperature where the thermal de Broglie wavelength is on the order of the mean interparticle spacing. This temperature is of order 1 K for liquid ^{3}He (depending on the density). Well below this temperature (called the Fermi degeneracy temperature T_F), the specific heat and the entropy will both be linear functions of the absolute temperature, i.e. $S = \gamma T$. Let us now consider the implications of the above discussion to the liquid-solid phase equilibrium which is determined by the famous Clausius-Clapeyron equation. According to this equation the slope of the melting curve is given by

$$\frac{dP}{dT} = \frac{S_{\text{liquid}} - S_{\text{solid}}}{V_{\text{liquid}} - V_{\text{solid}}} = \frac{\text{Latent Heat}}{T(V_{\text{liquid}} - V_{\text{solid}})}.$$

For ^{3}He, V_{liquid} is always greater than V_{solid}, so the denominator is always positive. On the other hand, the numerator will *change sign* as one cools into the Fermi degenerate region because $S_{liquid} = \gamma T$ will become less than the constant solid entropy $S_{solid} = R\ln 2$ corresponding to random spin orientation, so at the lowest temperatures the slope of the melting curve becomes *negative.* Furthermore, in this regime, the latent heat becomes negative, i.e. *it takes heat to freeze liquid ^{3}He.* At the higher temperatures, the entropy of the liquid will be greater than $R\ln 2$ per mole so that the melting curve will have a minimum. Idealized melting and entropy curves are shown schematically in Figure 1. Because the density of the solid is about 5 % higher than that of the liquid, it was possible to discern the presence of solid in our dielectric constant cell. Above the temperature of the minimum, solid could easily be formed in the cell as the pressure in the capillary tube connecting the ^{3}He pressurizing system at room temperature to the cell was increased. Below the temperature of the minimum, T_{min}, the capillary between the cell and room temperature became blocked as the pressure was increased and so no solid could be admitted into the cell. This difference in behavior above and below T_{min} allowed the temperature and the pressure of the minimum to be obtained. The best values of pressure and temperature as of 1996 for this minimum are 29.3 bar and 0.32 K respectively.

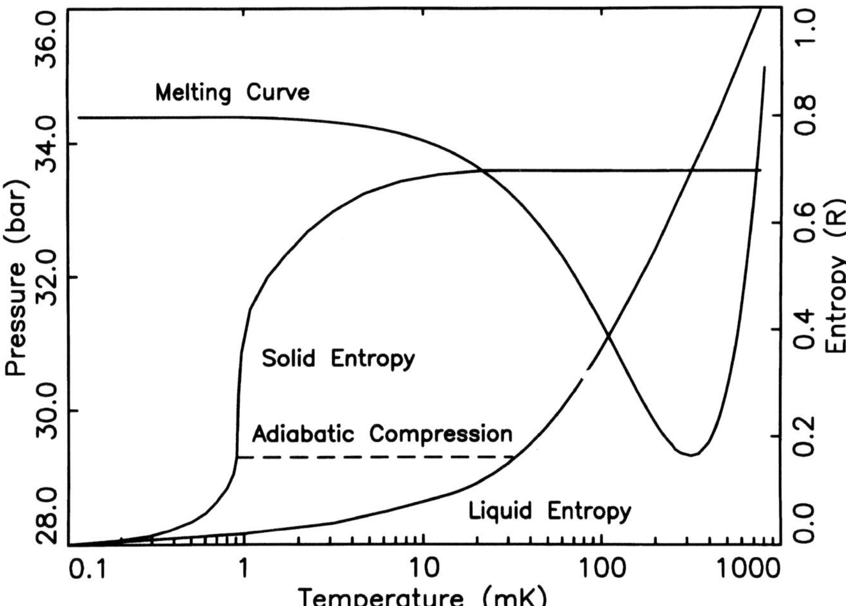

Figure 1. A semilog plot of the melting pressure of ^{3}He versus temperature showing the minimum at 0.32 K and 29.3 bar. An idealized semilog plot of the entropy of liquid ^{3}He and the entropy of solid ^{3}He at the melting curve is shown on the same graph. For the highest temperatures, the liquid entropy was calculated from the melting curve via the Clausius Clapeyron equation and the assumption that $S_{solid} = R\ln 2$. The two entropy curves cross at the melting curve minimum. The steep slope of the solid entropy curve near 1 mK corresponds to a magnetic phase transition in the solid. The dashed line is the path of an adiabatic compression from liquid to solid (Pomeranchuk cooling), to be discussed later.

The blocked capillary made extremely difficult any measurement near or along the melting curve which required a variation of the pressure or even a knowledge of the pressure for $T < 0.32$ K and $P > 29.3$ bar. This included the possibility of carrying out a suggestion by the Russian particle theorist, Isaac Pomeranchuk[37] that one could cool ^{3}He by pressurizing the liquid to form solid ^{3}He at temperatures below T_{min}. My Ph.D. thesis advisor, Henry Fairbank, first told me about Pomeranchuk's idea shortly before I left Yale for Cornell University as a newly minted Ph.D. in January 1959. He said that it would be a wonderful thing if I could invent a method for carrying out Pomeranchuk's suggestion in the laboratory. His enthusiasm was quite compelling, and I would often dream about Pomeranchuk cooling after moving to Cornell.

My first mission in 1959 as a young faculty member at Cornell was to convert an empty room into a low temperature laboratory capable of investigating liquid ^{3}He in the temperature region below 1 K. The most crucial task was to write a research proposal to the National Science Foundation to obtain funding to build up the laboratory and provide for graduate students to help in the research program. One of the major pieces of equipment required was an electromagnet for adiabatic demagnetization with sufficient field homogeneity for nuclear magnetic resonance (NMR) studies. Fortunately we were successful in obtaining a National Science Foundation grant which enabled us to begin taking measurements in 1960. The earliest experiments involved studies of nuclear magnetic resonance along the melting curve of ^{3}He. Later studies involved experiments on the melting and freezing properties of ^{3}He-^{4}He mixtures. These latter measurements made use of commercial strain gauges to determine the pressure of the sample in regions where the fill capillary was blocked. Development of a much more sensitive strain gauge was underway which utilized a stable tunnel diode oscillator to drive a tank circuit, one of whose capacitor plates was the flexible wall of the cell which thus formed the pressure sensor. Cell pressure changes would be registered as frequency changes of the oscillator. The standard modern design for pressure transducers as developed by Straty and Adams[38] is similar in principle. Nowadays, sensitive capacitance bridges are used in conjunction with these gauges. Unfortunately all of these experiments were interrupted when the entire laboratory began to collapse into the excavation for a new physics building (now Clark Hall). By good luck all of the equipment was saved before the laboratory was totally destroyed. A temporary laboratory was constructed in the metallurgy building where experiments were performed to finish mapping out the complete phase diagram of the melting and freezing properties of ^{3}He-^{4}He mixtures[39] and to make the first observation of transverse sound in solid ^{4}He.[40]

During this period, I began to think seriously about the possibility of performing Pomeranchuk cooling experiments after we finally moved into the new physics building. We have previously discussed the fact that pressurizing through the minimum in the melting curve resulted in a fill capillary blocked with solid. Therefore, to pressurize the sample along the melting curve at temperatures below the melting curve minimum, some external force had to be applied to the ^{3}He, independent of any external pressure communication through the fill capillary. We thought at the time that the best way to do this would be to construct a Pomeranchuk cell from a flexible thin-walled bellows. One could immerse the Pomeranchuk cell in liquid ^{4}He and raise the pressure of the liquid to 25 atmospheres pressure before solidification of the ^{4}He would occur. This was still less than the pressures of 29–34 atmospheres required to compress ^{3}He to pressures above the pressure of the melting curve minimum. An extra boost needed to be applied to attain the requisite pressure. The idea which we had at the time was to add an external spring to provide this extra force. A bellows and spring combination had been used previously by Grilly *et al.*[41] to study the properties of liquid ^{3}He near the melting curve minimum. (If properly placed, the spring would not contribute substantially to the heating of the ^{3}He sample). This provided the basis for our further thinking although as time went on there were many substantial improvements and modifications that went well beyond this early scheme.

The academic year 1966–67, my sabbatical year, was spent at the Brookhaven National Laboratory. There I had the time to interact strongly with Paul Craig, Thomas Kitchens, Myron Strongin, and Victor Emery. They and other staff members at Brookhaven made many extremely valuable suggestions and were helpful in many other ways. For example, one of the objections to Pomeranchuk cooling was the fact that the stretching of metal parts would lead to internal frictional heating, which would counter the cooling effect of compressing liquid ^{3}He into the solid phase. Discussions at Brookhaven convinced me that this problem could be overcome with careful design. Less work would be done and smaller energy losses could be achieved with a thinner and more flexible bellows. Basically, however, it was really an article of faith that Pomeranchuk cooling could be brought to fruition.

At a Solid State Sciences Panel meeting in the mid-1960's Philip W. Anderson and John C. Wheatley suggested that there was a great frontier opening up in ultra low temperature physics. This vision of the future by two such distinguished scientists greatly enhanced the prospects for our obtaining a higher level of research support. This allowed us to hire my colleague Robert C. Richardson as a research associate under a university-wide grant to the Cornell Materials Science Center by the Advanced Research Projects Agency. Bob had been a graduate student of Professor Horst Meyer at Duke University. Not only was he an expert on nuclear magnetic resonance (NMR) in solid ^{3}He and cryogenics but he also had exceptional vigor and scientific judgment. All of these skills and traits would be of the utmost importance to our low temperature program at Cornell. Within a short time, Bob became a member of our faculty. Shortly before Bob's arrival at Cornell, John Reppy had also joined our faculty. John's main area of expertise was experimental superfluid ^{4}He. We benefitted tremendously from his friendship, his wisdom, his sage advice and his extraordinary technical ingenuity.

The development of the ^{3}He–^{4}He dilution refrigerator in the mid-60's[42] and later improvements[43] had an enormous impact on ultra-low temperature physics. It had now become possible to continuously cool experimental samples to temperatures of order 10 mK. Previously, adiabatic demagnetization of paramagnetic salts was the only way to cool to this temperature range. Typically, a paramagnetic salt and the sample were cooled to the base temperature and then warmed slowly, a "one shot" experiment. Continuous adiabatic demagnetization refrigerators could be built but they were exceedingly cumbersome. The decision was therefore made to develop a dilution refrigerator system at Cornell. This effort was spearheaded by Bob Richardson. Our program was aided by very able new graduate students, James R. Sites, Linton Corruccini and Douglas D. Osheroff. The dilution refrigerator was to serve as a 10 mK low temperature platform from which to launch Pomeranchuk cooling.

The first Cornell Pomeranchuk cell was based on a brilliant but complex design involving two sets of thin, very flexible nested bellows suggested by John Reppy to minimize the effects of internal frictional heating in the metal bellows and to utilize a hydraulic press method for obtaining a mechanical advantage. It was first used successfully in the thesis experiment of Jim Sites who performed NMR susceptibility measurements on solid formed in the cell during compression. The purpose of the experiment was to study Curie-Weiss behavior as a precursor to the anticipated magnetic phase transition in solid ^{3}He. A Physical Review Letter on this experiment was published by Sites, Osheroff, Richardson and Lee.[44] Temperatures of about 2 mK were achieved. It was suspected that the lowest temperature that could be attained was limited by heating caused by solid being crushed in the bellows convolutions.

In the meantime, other laboratories were not standing still. Unbeknownst to us, Yuri Anufriyev at the Institute for Physical Problems (now the Kapitza Institute) in Moscow was the first to actually achieve Pomeranchuk cooling in 1965.[45] His cell was based on a stressed diaphragm technique in which the ^{3}He was forced into a cell with strong but flexible walls until the inlet capillary was blocked at high pressures. For this case, the flexible walls played the role of the spring mentioned in our earlier discussion. In the actual design, the flexible walled tube was place inside a larger rigid tube. The outer annular space contained the ^{3}He. Liquid ^{4}He in the inner tube was then pressurized to pressures approaching the melting pressure of ^{4}He. In spite of the stress applied to the walls of the ^{3}He cell to achieve the necessary volume change (5 %) for solidification, internal frictional heating did not seem to be a serious limiting factor. Later on John Wheatley and his associates modified the Anufriyev design somewhat and also were able to achieve Pomeranchuk cooling.[46] On the basis of these developments, several of us at Cornell realized that if the stressed diaphragm cells could achieve substantial Pomeranchuk cooling, it was no longer necessary to worry so much about internal frictional heating. Thus stronger bellows could be used in the design of any future Cornell Pomeranchuk cell.

An important consideration in the design of a Pomeranchuk experiment was the thermal isolation between the Pomeranchuk cell and the dilution refrigerator once the compressional cooling process was in operation. Fortunately the thermal boundary resistance between liquid ^{3}He and any metal heat link to the dilution refrigerator is quite large, which severely limits heat flow between the ^{3}He in the cell and the dilution refrigerator mixing chamber. Therefore the heat flow is very slow, or in the words of the late John Wheatley,[47] "Time is the thermal switch." In all of the early Cornell Pomeranchuk cells, even though the cells were bolted directly to the dilution refrigerator mixing chambers, many hours of experimentation below 3 mK were made available.

A BRIEF ACCOUNT OF THE DISCOVERY

It was clear from the outset that Douglas Osheroff was an extremely promising graduate student with tremendous potential. Once Jim Sites had obtained his Ph.D., Doug was next in line to take on the role as the lead graduate student in the Pomeranchuk cooling program. A great deal had been learned about Pomeranchuk cooling experiments as a result of work performed at Cornell and elsewhere. We knew that we had a powerful cooling technique which could cool a mixture of solid and liquid ^{3}He along the melting curve to temperatures of 2 mK or lower. Since we wanted to study ^{3}He, the cooling method had the advantage that the sample *was* the refrigerant, so that awkward heat transfer between sample and refrigerant could be avoided. The disadvantage was that the sample was confined to the melting curve so liquid and solid ^{3}He were simultaneously present in the cell.

With Doug Osheroff and Bob Richardson on board, it was time to start considering the next generation of Pomeranchuk cells to continue our program of cooling ^{3}He with the goal of searching for the long anticipated magnetic transition in solid ^{3}He. A number of ideas were considered including the rather whimsical suggestion of using a weight made of a heavy metal such as gold at the top of the flexible bellows which would supply the extra force needed to make up the difference between the melting pressure of ^{4}He and that of ^{3}He. There were no springs, stiff bellows or cell walls to be heated and the bellows could be very pliable. Certainly there would be no heating involving the gravitational field. Furthermore, the price of gold was rising rapidly at the time, so at the end of the experiment the gold weight could be sold to help support the research program!

We settled on a cell design which Doug Osheroff developed while he was recovering from a skiing-induced knee injury. This cell, shown in Figure 2 made good use of many of the lessons learned in previous work at Cornell and elsewhere. It employed two beryllium-copper bellows connected by a piston rod to transmit the force to pressurize the ^{3}He sample. The cross sectional area of the ^{4}He bellows was larger than that of the ^{3}He bellows, leading to a pressure amplification, as in a hydraulic press. When liquid ^{4}He in the upper bellows was externally pressurized, it forced the piston rod down, causing the lower bellows to distend downward into the ^{3}He cell. The opening up of the lower bellows prevented solid ^{3}He from being trapped and squeezed in the convolutions. Bob Richardson was very eager to have the most sensitive

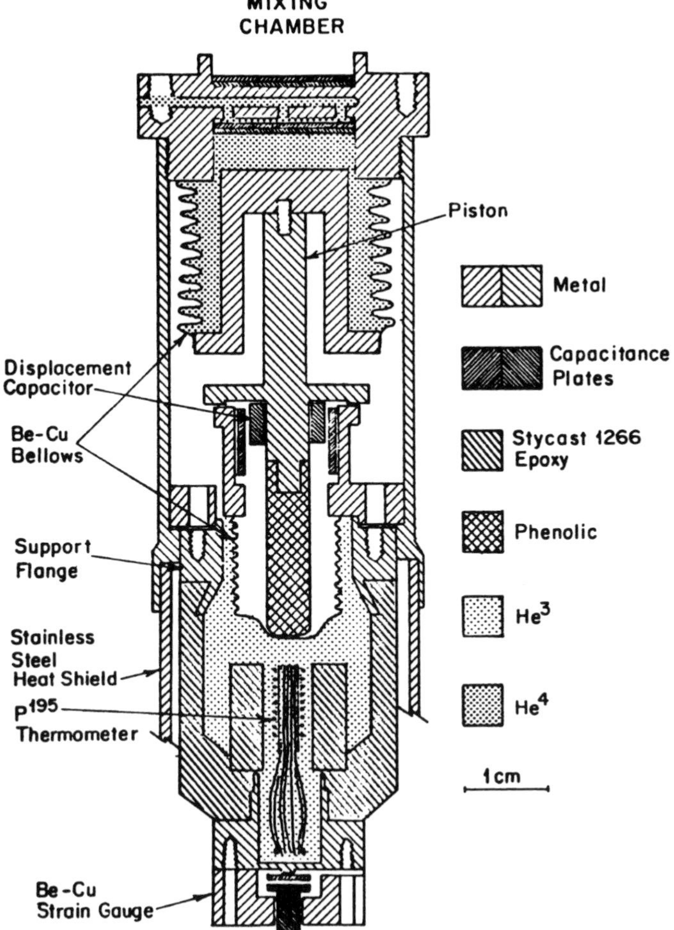

Figure 2. The Pomeranchuk cell used in the discovery experiments of Osheroff, Richardson and Lee. The pressure applied to the liquid ^{4}He in the upper bellows causes a piston rod to drive the lower bellows into the ^{3}He cell to increase the ^{3}He pressure. The ^{4}He bellows is larger than the ^{3}He bellows, thus providing a favorable compression ratio, in analogy to a hydraulic press.

melting pressure gauge possible. Therefore a sensitive capacitance strain gauge of the Straty-Adams[38] design was attached to the bottom of the cell containing the ^{3}He, allowing the ^{3}He pressure to be monitored during the compression process, thus providing a secondary melting curve thermometer. A platinum NMR thermometer which made use of a coil wrapped around a bundle of fine Pt wires (or copper wires in the first experiments) served as the primary thermometer down to 3 mK, below which it tended to lose thermal contact with the sample, possibly as a result of solid formation around the wires. In spite of the lack of a primary thermometer below 3 mK it was possible to obtain an estimate from melting pressure measurements of the temperature by extrapolating the melting curve. It was also possible to monitor the melting pressure as a function of time as the cell volume was changed at a constant rate. It was exactly this procedure which enabled Doug Osheroff to observe some peculiar but highly reproducible features on a chart recorder plot of the melting pressure vs. time in late November 1971.

I was heavily involved in preparing lectures for one of our large courses and did not find out about the observations immediately. When I did find out I was very excited. In fact all three of us were in a state of euphoria and knew we were on the brink of a major discovery. This was really the first defining moment in the experiment. A typical plot of the pressure vs. time observations for a complete cooling and warming cycle is shown in Figure 3. Anomalies labeled *A* and *B* in this figure were observed on cooling and corresponding features *A′* and *B′* were observed on warming. It certainly appeared that these effects were associated with new phase transitions, but *were they in the liquid or the solid?* The flattening in the pressure trace at *B′*, corresponding to a brief hesitation in the warming, was a rather subtle feature wich was not observed until several days after the initial observation of the signatures *A, A′ B.* I clearly recall coming in to the laboratory to lament to Doug that we had not as yet seen the signature for *B′* corresponding to the warming analog to *B.* At that very moment the chart recorder was beginning to display a small feature which was the first evidence for *B′*. The brief flattening was interpreted as a manifestation of latent heat associated with a first-order transition. Giving this idea further credence were the peculiar zig-zags seen at *B* which could be a characteristic of a supercooled transition where the latent heat was suddenly released, giving a brief rapid warming. Supporting the supercooling inter-

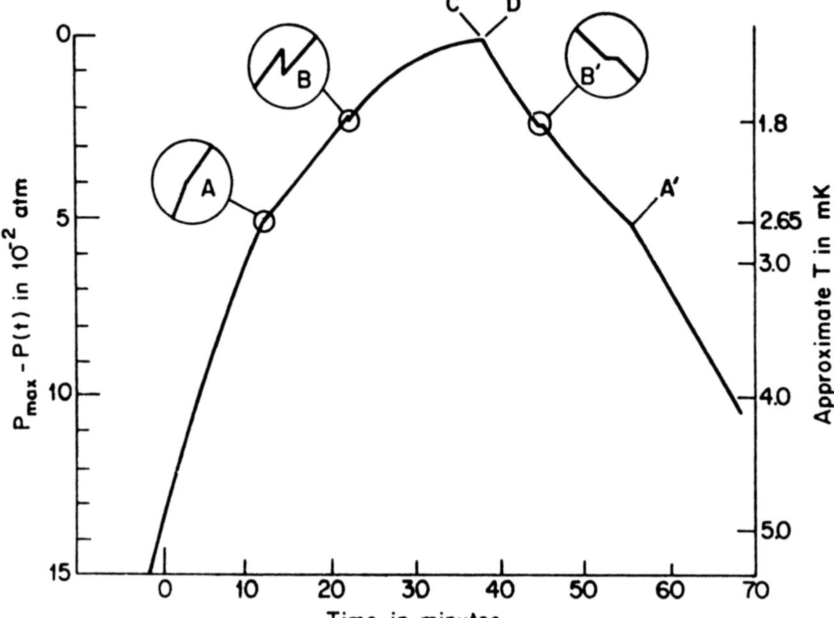

Figure 3. Features observed on a plot of cell pressure vs. time during compressional cooling to the minimum temperature followed by warming the cell during decompression. The A and A′ features were changes in slope. They always occurred at precisely the same pressure. Lower temperature features B and B′ were also observed. On cooling through B a sudden pressure drop appeared, and on warming through B′ a small plateau was observed. The pressure at B was always greater than or equal to the pressure at B′.

pretation was the fact that B' was reproducible whereas B occurred at a pressure that was not reproducible but was always greater than the pressure corresponding to B'. The features A and A', corresponding to slope changes on the pressure-time traces, were completely reproducible, showing no signs of supercooling. Although our thermometry was crude, we were able to estimate the temperature of the A transition, $T_{A'}$, to be about 2.7 mK and the temperature of the B' transition, $T_{B'}$ to be about 2.1 mK, based on the temperature scale in use in 1971.

Since both liquid and solid ^{3}He were present in the cell, we at first thought that the A transition corresponded to the long sought second order magnetic phase transition in solid ^{3}He. Possibly the variation of the specific heat of solid ^{3}He with temperature near such a phase transition could have given a signature similar to that observed in the melting pressure traces. The rapid response of the sample upon passing through the supercooled B transition was much more difficult to interpret in terms of a phase transition in solid ^{3}He. Nevertheless, our first paper, published in Physical Review Letters, was entitled "Evidence for a New Phase of Solid ^{3}He."[48] John Goodkind of the University of California San Diego in a private conversation with me at an American Physical Society meeting indicated that he was quite sure the action was taking place in the liquid, not the solid. We also received a letter from Victor Vvedenskii at the Institute for Physical Problems (now the Kapitza institute) in Moscow suggesting that the behavior at the A feature was associated with a step in the specific heat corresponding to a pairing transition in liquid ^{3}He in analogy with a similar specific heat step seen at the superconducting transition in metals.

The overall research plan from the beginning was to perform NMR studies in the cell. This became even more urgent because of the need to unambiguously establish the identity of the A and B features. The results of these NMR studies were to give, as is discussed below, a clear indication that both the A and B transitions were associated with *liquid* ^{3}He!

Since both liquid and solid were present in the cell, some means was required to distinguish between the liquid and the solid. Solid ^{3}He exhibits Curie-Weiss behavior at low temperatures and so its magnetic susceptibility is large, since all of the spins participate. On the other hand liquid ^{3}He obeys FD statistics, and so its magnetic susceptibility is governed by Pauli paramagnetism which requires that only those quasiparticles in the immediate vicinity of the Fermi surface are free to flip. Thus the liquid susceptibility must be very small, since only a small fraction of the spins are involved in a magnetic response.

If a field gradient is superposed on the homogeneous applied steady field, the Larmor frequency $\omega = \gamma H$ will vary across the cell. Thus different regions of the cell have different Larmor frequencies, so that the NMR response of each of the small regions will have to correspond to a different frequency. Sweeping the frequency or the magnetic field enabled us to monitor different regions of the cell. This was one of the first applications of magnetic resonance imaging (MRI) which is now an essential tool in medical diagnosis.[49]

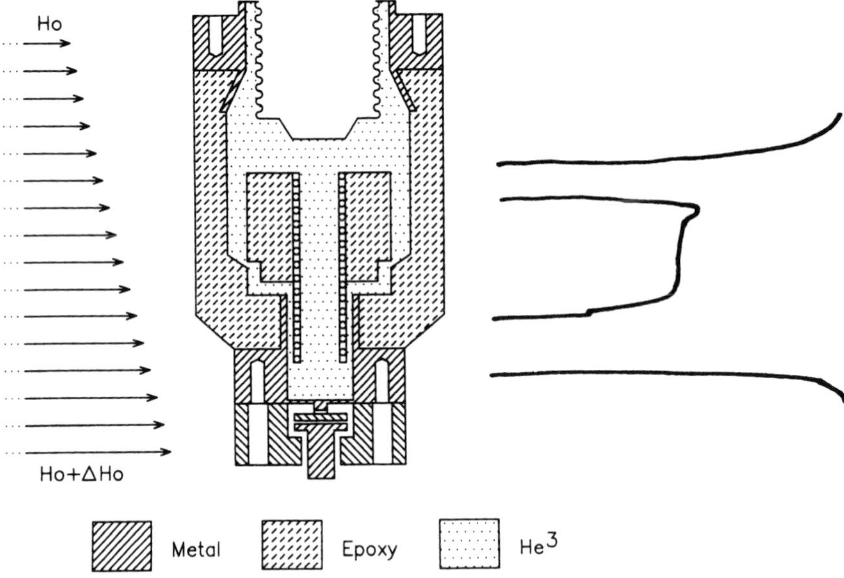

Figure 4. The lower portion of the Pomeranchuk cell of Figure 2 showing an NMR coil along the cell axis. A small field gradient was applied to the cell as shown by the arrows, providing for one dimensional magnetic resonance imaging. An idealized plot of the ^{3}He susceptibility vs. height is shown on the right. The solid, corresponding to the large susceptibility peaks, tends to clump at the ends, allowing the center of the coil to be relatively free of solid.

A radiofrequency coil was introduced into the ^{3}He cell as shown in Figure 4. The gradient in the applied dc magnetic field is indicated by the arrows. It was very fortunate in these experiments that the solid formed in localized regions of the cell. Thus it was possible to separately examine the behavior of the liquid corresponding to a small susceptibility and the solid corresponding to a large susceptibility as shown in Figure 5. As the sample cooled through *B*, the *liquid* susceptibility suddenly dropped by about a factor of two. Upon making this observation, Doug called me at home in the early hours of the morning. I was elated when I heard the news! It was clear that the *B* phase was a liquid phase. Furthermore, the susceptibility drop could perhaps be related to a BCS pairing transition. This was the second defining moment in the ex-

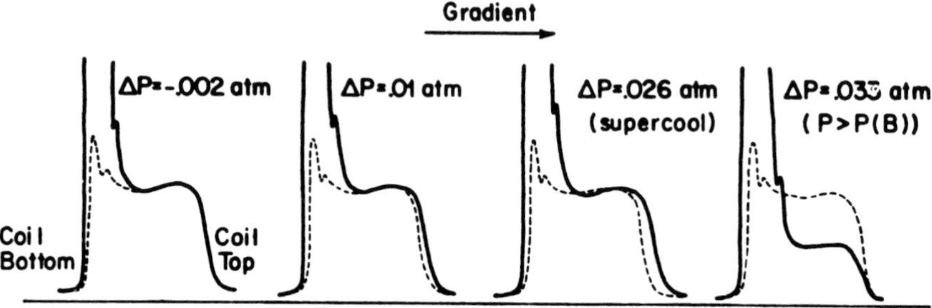

Figure 5. A sequence of NMR profiles taken with a field gradient in the cell of Figure 4, as the ^{3}He sample is cooled along the melting curve. We show a run for which solid happened to form at only one end of the coil. The susceptibility associated with the liquid drops abruptly as we cool through point B.

periment. At this point I vowed to myself that I would be present at any other such moment.

Since I believed that it was important to check for any possible frequency shift, I asked Doug to remove the field gradient in order to examine the response of the mixture of the liquid and solid in a homogeneous field. As the three of us watched, a truly dramatic thing happened when the sample was cooled into the *A* phase. At the *A* transition a satellite line emerged from the main mostly solid peak and steadily moved to higher frequencies as the sample cooled. At the *B* transition this satellite line abruptly disappeared back into the main peak, so that the *B* phase did not show a frequency shift away from the Larmor frequency. These effects are illustrated in Figure 6. This was the third defining moment in the experiment. The satellite line had the same amplitude and shape as the all liquid line in the normal Fermi liquid phase when no solid was present in the cell. Furthermore, the amplitude did not change with temperature. Thus the satellite line corresponded to the entire liquid line shifting in frequency as we traversed the *A* phase. At last we had something quantitative to deal with!

The experiment was performed at various magnetic fields corresponding to various different Larmor frequencies. At the suggestion of our Cornell colleague, Robert Silsbee, the results were plotted as the difference between squares of liquid and solid frequencies respectively vs.\ the increment in pressure above the pressure corresponding to point *A*. To our delight, all the points fell on a universal curve, shown in Figure 7 corresponding to the equation $\omega^2 - \omega_0^2 = \Omega_A^2(T)$, where T was obtained from the extrapolated melting pressure. In this relationship, ω is the observed satellite line frequency and ω_0 is the solid frequency corresponding to the Larmor frequency of ^{3}He. The right-hand side corresponded to an increasing function of $1 - T/T_A$. This was interpreted as the development of an order parameter as the temperature de-

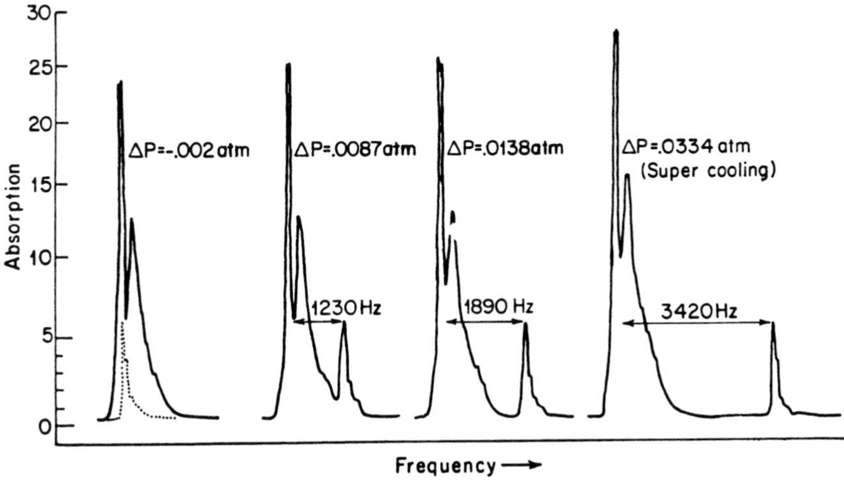

Figure 6. A sequence of NMR profiles taken in the cell of Figure 4 after the gradient was removed. The liquid signal in the A phase shifts away from the solid signal (corresponding to the Larmor frequency) as the sample is cooled. The shifted line abruptly disappears at point B.

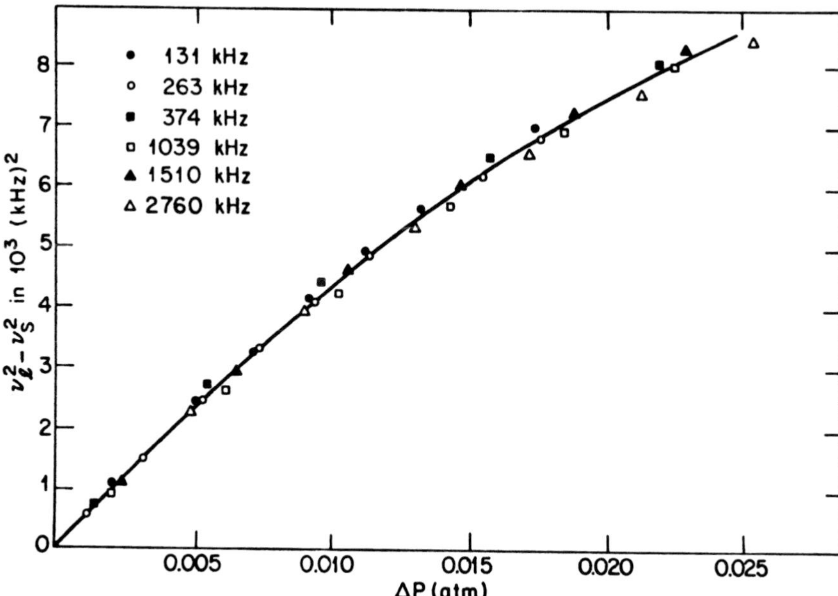

Figure 7. The differences between the squares of the liquid frequency and those of the solid (Larmor) frequency are plotted against the Pomeranchuk cell pressure as referred to the pressure at the A transition. The different symbols correspond to data taken at different magnet fields as indicated by the Larmor frequency associated with each symbol. All the data points fall on a *universal* curve.

creased. The Pythagorean relationship suggested that two magnetic fields were present in the problem, the external applied magnetic field and an internal field associated with an order parameter perpendicular to the applied magnetic field. As the temperature was lowered, the magnitude of the internal field varied from zero to approximately 30 gauss, which is much greater than the dipolar interaction field between two ^{3}He atoms separated by the mean distance between atoms in the liquid. A vector diagram demonstrating how the applied field and a perpendicular internal field combine to give the Pythagorean frequency shift formula is given in Figure 8.

By the early summer of 1972, it was completely clear to us that the strange phenomena seen during the previous six months could be clearly identified with liquid ^{3}He. We were therefore very anxious to correct the preliminary but erroneous interpretation given in our first publication. A second manuscript was prepared which described the results of our nuclear magnetic resonance experiments and which carefully argued that new phases of liquid ^{3}He had been discovered. This manuscript was submitted to Physical Review Letters, but unfortunately it was turned down by the referee. We were shocked by this development and spent a great deal of time trying to get the decision overturned. Ultimately, reason prevailed and the manuscript finally appeared.[50]

At the time there was still no theoretical understanding of the strange drop in susceptibility at the *B* transition nor of the even stranger frequency shift in the *A* phase. Could the low susceptibility of the *B* phase be associated with singlet pairing? As mentioned earlier, s wave pairing was excluded for ^{3}He,

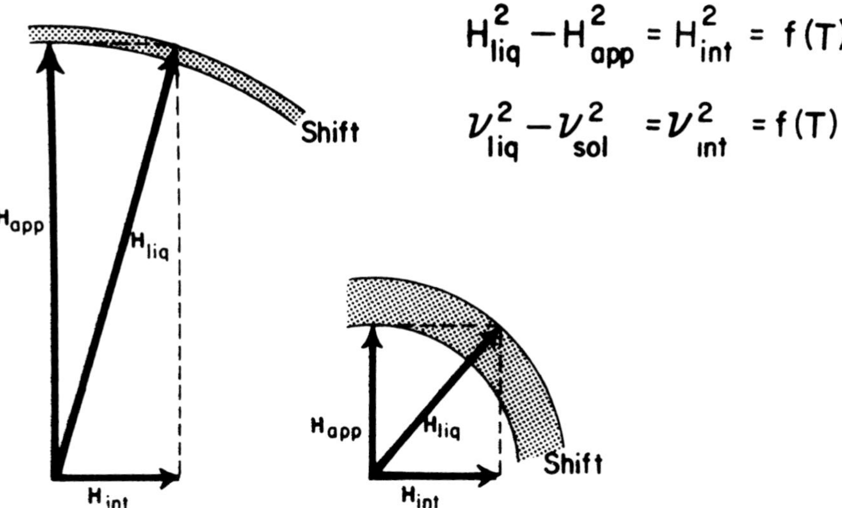

Figure 8. The effect of the internal field on the magnetic resonance frequency in ^3He A. For a given temperature, the width of the shaded region corresponds to the size of the frequency shift away from the Larmor frequency. Diagrams for a large applied field and for a small applied field are shown.

but d wave pairing was possible and had been discussed by Emery and Sessler.[29] As for explaining the frequency shift, there were only vague suggestions that ^3He A was exhibiting a new type of antiferromagnetic behavior. Of course we discussed these issues with members of our theory group. Neil Ashcroft suggested that we should consider the possibility of p wave pairing and Vinay Ambegaokar pointed out the paper by Balian and Werthamer[31] which actually treated p wave pairing. As it turned out, both of these suggestions were directly applicable to the B phase of liquid ^3He, at least.

A special p wave ($S = 1$) pairing state was hypothesized by Balian and Werthamer as mentioned in the introduction. This state corresponded to an order parameter with three spin components, $S_z = +1$, $S_z = 0$ and $S_z = -1$ i.e. ↑↑, $1/\sqrt{2}$ (↑↓ + ↓↑) and ↓↓ states. Since the $S_z = 0$ substate is "spinless", it does not contribute to the susceptibility. Thus the Balian Werthamer susceptibility should approach 2/3 that of the normal Fermi liquid as $T \rightarrow 0$. In fact the *B* phase susceptibility is even smaller as a result of Fermi liquid interactions. Another property of the *B* phase which fits in with the Balian Werthamer state is the suppression of this phase by a magnetic field which was observed in some of our early Pomeranchuk cooling experiments. By about 0.6 tesla, the *B* phase no longer existed. As the field increased, the *B* phase was pushed to lower temperatures in favor of the *A* phase. The interpretation is that the higher magnetic field tends to suppress the spin zero (↑↓ + ↓↑) pairs. The ↓↓ and ↑↑ pairs can easily respond to the field merely by reorienting themselves.

The puzzle of the *A* phase frequency shift still needed to be solved. Within an amazingly short time, Anthony Leggett[51,52] came forth with a brilliant solution to the problem. It was known that in conventional high temperature NMR experiments the main effect of the weak dipole-dipole interactions was to broaden the NMR line. How could these weak dipolar interactions all conspire to provide a frequency shift corresponding to an internal field of 30 gauss at the lowest temperature attained in ^{3}He A? Leggett introduced the hypothesis of spontaneously broken spin orbit symmetry (SBSOS) and, with the aid of sum rules, was able to obtain the proper order of magnitude for the frequency shift.[53] Let us consider how SBSOS can be responsible for a large frequency shift in the context of a superfluid with $l = 1$ pairing. The weak interaction between the tiny nuclear dipole moments is much less than one microkelvin, but somehow, in spite of much larger thermal fluctuations, the presence of Cooper pairs must lead to a coherent addition of all the dipole moments, giving rise to an effective internal field large enough to produce the observed frequency shifts. This comes about because all the Cooper pairs must be correlated, i.e. locked together so all of the nuclear moments must act together to provide the requisite effective internal magnetic field. By this **means, we self-consistently generate a macroscopic dipolar interaction.**

Let us consider two possible configurations for rotation of two nuclear dipoles about one another. One involves rotations such that the pair orbital angular momentum is parallel to the nuclear dipole moments and the other involves rotations where the pair orbital angular momentum is perpendicular to the dipole moments. Classically, this latter configuration has a lower energy and it turns out to be the favored state. Because of SBSOS, it will be highly favored, in fact. If we introduce an $\vec{l}$ vector corresponding to the pair orbital angular momentum and define a vector $\vec{d}$ corresponding to the direction of *zero* spin projection, then it would be energetically favorable for $\vec{d}$ and $\vec{l}$ to be parallel.

As a result of a set of fundamental equations of motion derived by Leggett[54] and discussed in the next section, we have the following picture for NMR in ^{3}He A: A nuclear magnetic resonance experiment corresponds to an oscillation in the direction of $\vec{d}$ about $\vec{l}$ where the direction of $\vec{l}$ is stabilized by the quasiparticles and the container boundaries. An important point here is that $\vec{d}$ and $\vec{l}$ are macroscopic vectors, with the restoring torque for oscillations of $\vec{d}$ with respect to $\vec{l}$ being provided by the macroscopic dipole interaction. This extra restoring torque is what gives rise to the frequency shift observed in ^{3}He A.

The magnetic susceptibility of the *A* phase remains constant at the normal Fermi liquid value throughout the temperature range where it occurs. The explanation is as follows: In contrast to ^{3}He B, the *A* phase belongs to a class of states known as equal spin pairing states containing only $\uparrow\uparrow$ and $\downarrow\downarrow$ pairs which can respond directly to any field change in a fashion similar to that of the normal Fermi liquid. The p wave pairing state put forward by Anderson and Morel[30] in 1961 is one possible member of this class. It is now believed that the early Anderson Morel state corresponds to the *A* phase of superfluid ^{3}He.

THE POST DISCOVERY PERIOD

The discovery experiments[55,56] and Leggett's explanation of the A phase frequency shift[51] were presented in the summer of 1972 at the 13th International Conference on Low Temperature Physics held in Boulder, Colorado. The discussions aroused great enthusiam for further investigations, both experimental and theoretical. The first anomalous flow properties associated with possible superfluidity in the new phases were seen in a vibrating wire experiment conducted in a Pomeranchuk cell by a group at the Helsinki University of Technology.[57] Actual superfluid behavior was later demonstrated in fourth sound experiments by Yanof and Reppy[58] and Kojima, Paulson and Wheatley.[59] Ultrasound experiments, performed at Cornell (also in a Pomeranchuk cell) showed a pronounced attenuation peak near the A transition.[60] The peak was associated with the breaking of Cooper pairs near T_c as well as with the collective modes (pair vibrations of the order parameter). Experiments below the melting pressure were performed at La Jolla by adiabatically demagnetizing a powdered cerium magnesium nitrate paramagnetic salt contained directly in the liquid ^{3}He sample. The small grain powder made it possible to overcome the large thermal boundary resistance between the ^{3}He and the salt by providing a large area of contact. Measurements of the specific heat at the transitions into the new phases of ^{3}He gave curves characteristic of a BCS type transition with behavior below the transition showing a rapid rise with temperature, associated with pair breaking and the greater availability of quasiparticles, followed by a sharp drop at the transition.[61] Above the transition, the typical linear temperature dependence of a normal Fermi liquid was found. These results are portrayed in Figure 9.

I have so far not mentioned the existence of a third phase, which could only exist in the presence of an applied magnetic field. Evidence for this was found in the discovery experiment where, in the presence of a magnetic field, instead of a single point corresponding to a change of slope in the pressure vs. time plot, there were two closely spaced points, each involving a change in slope in the melting pressure vs. time signature.[62] Thus the A transition splits into two transitions in a magnetic field. Before hearing of these results, Vinay Ambegaokar and David Mermin had actually predicted theoretically that the A phase *should* split linearly in a magnetic field.[63] They showed that this was required by l = 1 BCS pairing and named the newly discovered phase A_1. Particularly dramatic signatures of the splitting of the A transition into two transitions can be seen in the ultrasound data of Lawson *et al.*[64] (see Figure 10). The temperature width of the A_1 phase grows linearly with field at the rate of 60 μK per tesla all the way up to 10 tesla and beyond. The A_1 phase is believed to have only a single spin component, $|\uparrow\uparrow\rangle$.

Finally, the La Jolla group also investigated the phase diagram in a magnetic field[65] at pressures below melting pressure by studying the static magnetization of the liquid via superconducting quantum interference (SQUID) interferometry. The ^{3}He sample and the magnet supplying the applied magnetic field were contained in a separate tower surrounded by a superconduc-

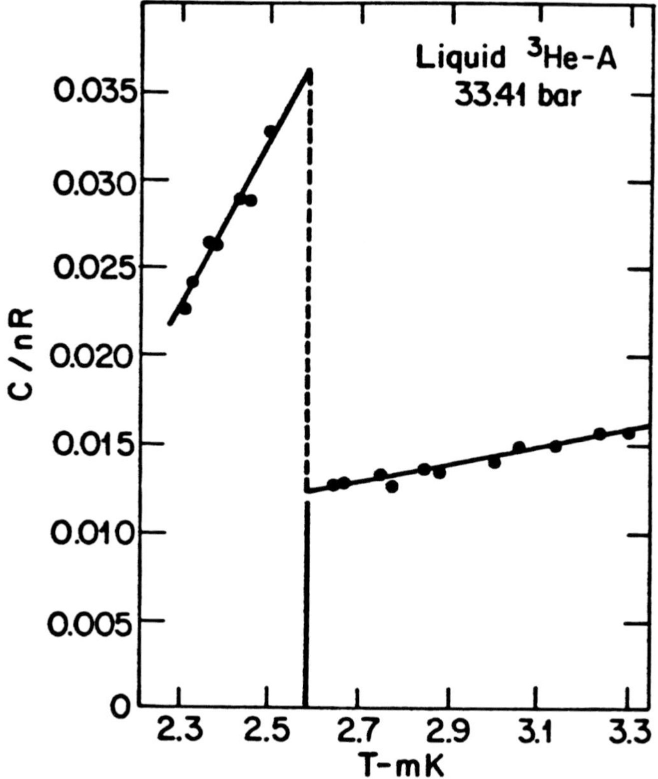

Figure 9. Early specific heat measurements of liquid ^{3}He near the superfluid transition.[61] The shape is characteristic of a BCS pairing transition.

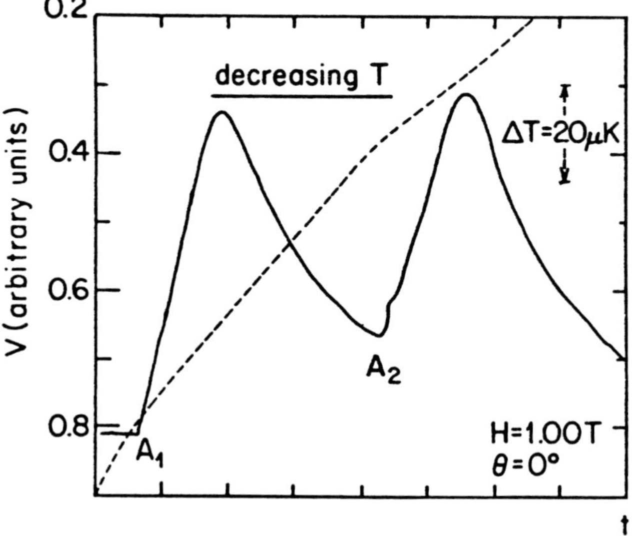

Figure 10. Sound attenuation data from a Pomeranchuk cooling run plotted against time in a 1 tesla magnetic field clearly showing the splitting of the A transition. A_1 and A_2 mark the two resulting transitions.[64] The attenuation peaks are associated with collective mode absorption and pair breaking near the transitions. The dashed line representing the melting pressure shows two kinks corresponding to the splitting of the A transition into the A_1 and A_2 transitions.

ting niobium magnetic shield. The liquid ^{3}He sample in the tower was maintained in good thermal contact with the cerium magnesium nitrate refrigerant in the main cell via a column of liquid ^{3}He. The most dramatic finding was the narrowing and finally the vanishing of the *A* phase in zero field at a point called the polycritical point as shown in Figure 11. All of the features discussed above are summarized by the schematic P-T-H phase diagram shown in Figure 12.

Soon after the discovery of the *A* and *B* phases at Cornell, low temperature laboratories all over the world began a broad effort to explore their properties. Condensed matter theorists became very actively involved in explaining the observed effects and predicting new phenomena. One of the main tasks to be undertaken was the proper identification of the respective order parameters corresponding to ^{3}He A and ^{3}He B. In our experimental group, we

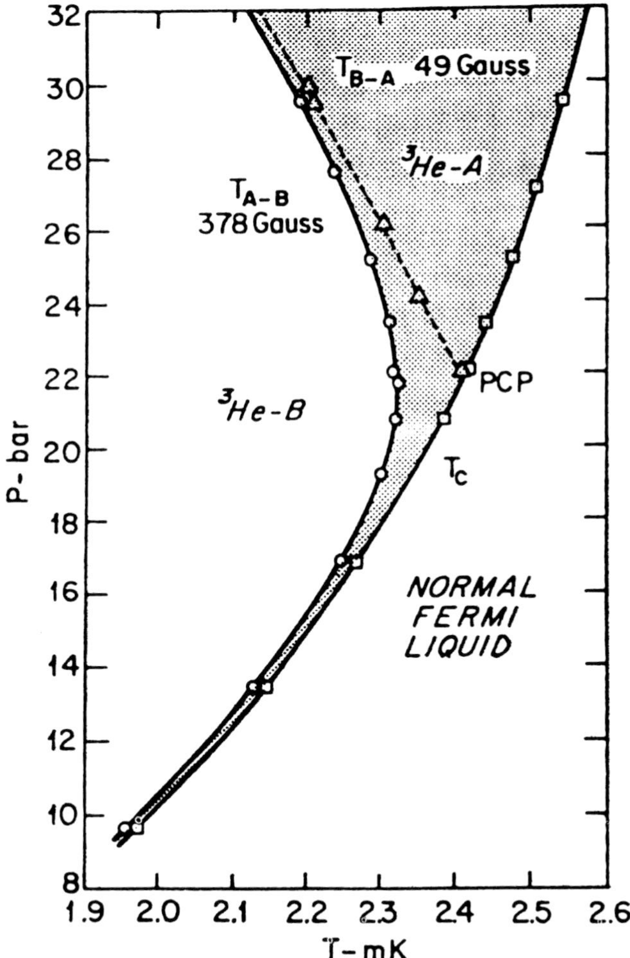

Figure 11. Experimental data of Paulson, Kojima and Wheatley.[65] At the lowest magnetic field, the A phase is not present below the polycritical point PCP at about 22 bar. In a larger magnetic field, the B phase is suppressed in favor of the A phase even at the lowest pressure and the polycritical point disappears.

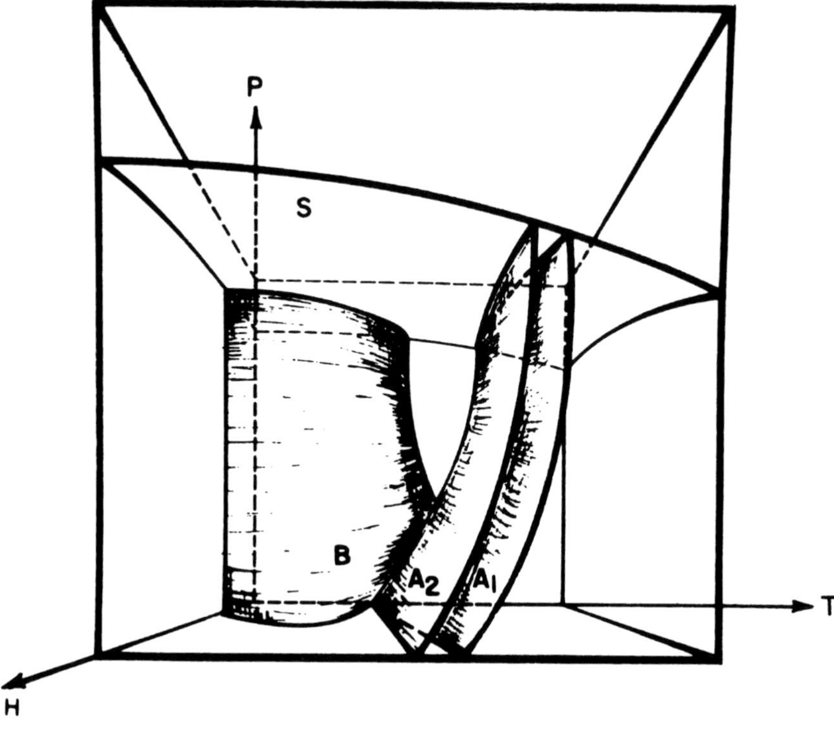

Figure 12. A schematic P-T-H diagram showing the general topology of the phase diagram of the superfluid phases, A, A_1 and B of liquid ^{3}He. The A_1 phase occurs between the surfaces labeled A_1 and A_2. The A phase occurs at temperatures below the boundary labeled A_2. The boundary between phases A and B is labelled B. The surface labelled S corresponds to the melting curve.

adopted the working hypothesis that ^{3}He A corresponded to the p wave equal spin pairing state first considered by Anderson and Morel[30] in 1961 and that ^{3}He B corresponded to the state suggested by Balian and Werthamer[32] in 1963. As mentioned in the previous section, these states were at least consistent with the Cornell discovery experiments.

Both the Anderson Morel and the Balian Werthamer states of p wave pairing are states with total $L = 1$ and total $S = 1$. The Anderson Morel state is an orbital $m = 1$ state along some direction $\hat{l}$ and a spin $m = 0$ state along some direction $\hat{d}$. Recall that we introduced $\hat{d}$ as the direction of zero spin projection earlier in our discussion. We express the Anderson Morel order parameter as the product between an orbital part in configuration or momentum space and a part in spin space, i.e.,

$$\psi_{AM} = (\text{orbital part}) \times (\text{spin part}).$$

If we consider only angular dependence, the Anderson Morel order parameter is defined as

$$\psi_{AM} \sim e^{i\varphi} \sin\theta \left[\frac{1}{\sqrt{2}} (\downarrow\uparrow + \uparrow\downarrow) \right],$$

where the spherical harmonic $Y_{11} \sim e^{i\varphi} sin\,\theta$ defines a polar axis $\hat{l}$ corresponding to the direction of the pair orbital angular momentum. In the above expression for the spin triplet pair wave function the spin part appears along the $\hat{d}$ axis, so that only the $(\downarrow\uparrow + \uparrow\downarrow)$ component occurs. For the case of the Anderson Morel state, we see that the spin part of the order parameter does *not* depend on any orbital variables but is a *constant* in orbital space; i.e., in k space, every point on the Fermi surface has the same $\hat{d}$. We discussed earlier how a classical argument involving the dipolar interaction combined with spontaneously broken spin orbit symmetry would favor the state for which $\hat{l} \,||\, \hat{d}$. Taking this into account we sketch the AM order parameter in k space in Figure 13a. The small arrows correspond to the $\hat{d}$ vector and the large arrow corresponds to $\hat{l}$. One of the striking features of this order parameter is the orbital anisotropy, with nodes at $\theta = 0$ and $\theta = \pi$. The behavior of the BCS energy gap follows that of the order parameter so that gap nodes also appear at $\theta = 0$ and π as shown in Figure 13b. The full three dimensional picture is obtained by a revolution about the $\hat{l}$ axis. The patterns in the orientation of $\hat{l}$ as a function of position in the liquid are highly analogous to patterns found in liquid crystals. These patterns have been named textures. Ambegaokar, de Gennes and Rainer[66] have shown that the $\hat{l}$ vector will be *perpendicular* to the walls of the containers. This boundary condition plays an important role in determining the texture pattern in liquid ^{3}He A. The direction of $\hat{l}$ is also sensitive to flow and to the applied magnetic field.

The spin state $\frac{1}{\sqrt{2}}(\uparrow\downarrow + \downarrow\uparrow)$ can be rotated in spin space to give the equal spin pairing version of the Anderson Morel order parameter,

$$\psi_{AM} \sim e^{i\varphi} \sin\theta\left[(|\uparrow\uparrow\rangle + e^{i\Phi}|\downarrow\downarrow\rangle)\right],$$

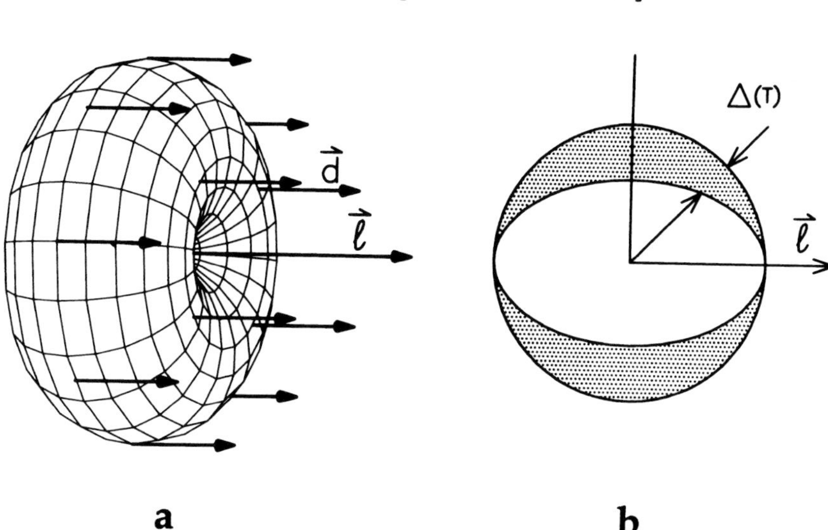

Figure 13. (a) A three dimensional representation of the Anderson-Morel order parameter. The vector $\hat{l}$ at the center defines the axis of the order parameter. Along this axis, the amplitude is zero corresponding to sin θ dependence where θ is the polar angle with respect to $\hat{l}$. The vector $\hat{d}$ has the *same* direction for all points on the Fermi surface.

(b) The anisotropic energy gap is indicated by the shaded region. The two nodes along $\hat{l}$ are clearly shown.

where Φ is a phase factor, which is helpful in discussing longitudinal NMR experiments. This representation shows that the Anderson Morel order parameter can be characterized by a spin configuration with only $|\uparrow\uparrow\rangle$ and $|\downarrow\downarrow\rangle$ states, as mentioned in our earlier discussions. The A_1 phase has the orbital properties including the gap nodes described by the Anderson Morel state but has only $|\uparrow\uparrow\rangle$ spin pairs.

We shall now discuss the Balian Werthamer state. The simplest possible Balian Werthamer state is the $^3P_0$ state, represented by the wave function

$$\psi_{BW} \sim Y_{1,-1} \mid \uparrow\uparrow\rangle + Y_{10} \mid \uparrow\downarrow + \downarrow\uparrow\rangle + Y_{11} \mid \downarrow\downarrow\rangle$$

so that all three spin species are included. Hence we do not have an equal spin pairing state. Since the $^3P_0$ state has total $J = 0$, it will be a spherically symmetric state. When this is taken into account, it is customary to specify this simple Balian Werthamer state in terms of the vector $\hat{d}$ by $\hat{d}\,(k)$ = constant $\times\ \hat{k}$ which has the necessary spherical symmetry. Notice that in contrast to the Anderson Morel state, $\hat{d}$ depends on $\hat{k}$.

The simple state discussed above does not perfectly represent the order parameter of superfluid ^{3}He B. As far as the most important interactions are concerned, the energy will not change when the spin and orbital coordinates are rotated with respect to one another. Thus we could rotate $\hat{d}$ about some axis $\hat{n}$ to get $\hat{d} = R\hat{k}$, where R is an arbitrary rotation about an arbitrary axis $\hat{n}$ for superfluid ^{3}He B. This degeneracy is broken when the small dipolar interaction is taken into account, which results in a rotation of the spin coordinates relative to the orbital coordinates by an angle of 104° as discussed below. This subtle anisotropy allows textures associated with liquid crystal like behavior to be observed in superfluid ^{3}He B. Nevertheless the overall orbital symmetry of the order parameter is still spherical, leading to an isotropic energy gap similar to that of s wave superconductors. Figure 14a shows the order parameter with $\hat{d}$ twisted about some axis $\hat{n}$ by 104°, and Figure 14b shows the isotropic energy gap.

I have now outlined the basic properties of the Anderson Morel and the Balian Werthamer states which were provisionally identified with ^{3}He A and ^{3}He B, respectively. An important question still remained to be addressed. The early studies of the possible order parameters of p wave pairing showed that the Balian Werthamer state would have a lower free energy and therefore should always be the preferred state. On the other hand the existence of an Anderson Morel type state in ^{3}He was firmly established by the experiments. The apparent discrepancy was resolved by Anderson and Brinkman[67] who introduced the idea of spin fluctuation feedback which led to a mechanism for a stable Anderson Morel phase. (Recall our previous discussion of the possible role of spin fluctuations by Layzer and Fay.) Since the pairing mechanism is intrinsic, thus involving the ^{3}He quasiparticles themselves, any modification in the status of the helium quasiparticles should affect the pairing mechanism, including the onset of pairing itself. Anderson and Brinkman showed that this feedback effect could indeed lead to a stable Anderson

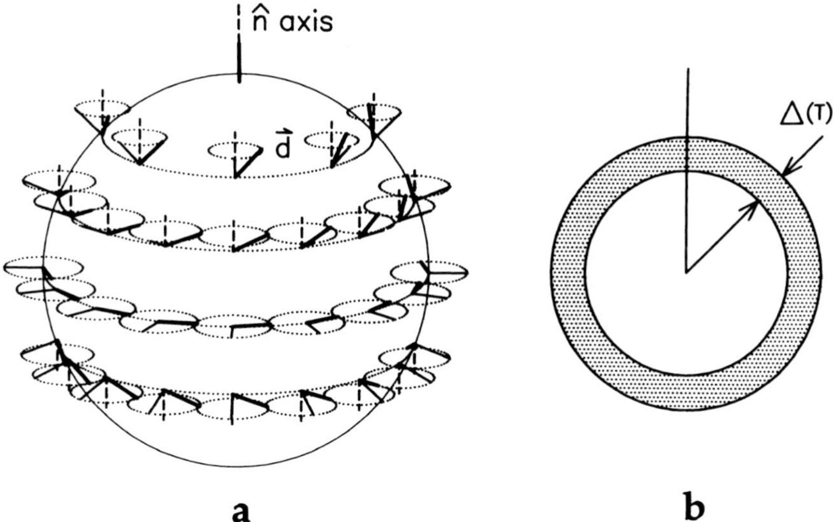

Figure 14. (a) The order parameter for superfluid ^{3}He B showing $\hat{d}$ vectors (represented by thick lines) rotated by 104° about a vector $\hat{n}$ from the radial directions (thin lines) for all points on the Fermi sphere. The rotation axis $\hat{n}$ points in the vertical direction.

(b) The isotropic energy gap of the Balian Werthamer order parameter is indicated by the shaded region. Ordinary s-wave superconductors also have isotropic energy gaps.

Morel phase in zero magnetic field, which was renamed the Anderson Brinkman Morel phase or ABM state. These studies led to the general acceptance that the Anderson Brinkman Morel state corresponded to ^{3}He A and the Balian Werthamer state corresponded to ^{3}He B. More recent comprehensive studies of a variety of pairing mechanisms conducted by Rainer and Serene have not changed this conclusion.[68]

No general discussion of superfluid ^{3}He would be complete without a treatment of the macroscopic nuclear dipole interaction and its role in the dramatic NMR effects observed experimentally. The general scheme for calculating the dipolar interaction is to take a quantum mechanical average of the dipolar Hamiltonian over the pair wave function (order parameter). It can then be shown that the dipolar free energies are given by

$$\Delta F_D = \begin{cases} -\frac{3}{5} g_D(T)[1 - (\vec{d} \cdot \vec{\ell})^2], & \text{A phase} \\ \frac{4}{5} g_D(T)\{\cos\theta + 2\cos^2\theta + \frac{3}{4}\}, & \text{B phase} \end{cases}$$

where

$$g_D \approx 10^{-3}\left(1 - \frac{T}{T_c}\right) \quad \text{ergs/cm}^3.$$

Therefore, to minimize the free energy, $\vec{l}$ and $\vec{d}$ must be parallel for the case of the ABM state (A phase) in agreement with our earlier qualitative discussion. For the BW state, a simple calculation shows that the dipole energy is minimized for $\theta = \cos^{-1}(-\frac{1}{4}) = 104°$ justifying our earlier statement.

Making use of the macroscopic dipolar interaction, Leggett[52] derived a set of coupled equations giving a complete description of the spin dynamics of superfluid ^{3}He. His equations of motion are

$$\dot{\vec{S}} = \gamma\vec{S} \times \vec{H} + R_D(T)$$

$$\dot{\vec{d}} = \vec{d} \times \gamma\vec{H}_{\text{eff.}} = \vec{d} \times \gamma\left(\vec{H} - \frac{\gamma\vec{S}}{\chi}\right)$$

The first term in the first Leggett equation corresponds to Larmor precession (ordinary NMR) whereas the second term is a restoring torque resulting from the dipolar interaction. The second equation describes the precession of $\hat{k}$ in an effective field. The motion of $\vec{d}$ and that of $\vec{S}$ are coupled. In the A phase $R_D(T)$ takes on a particularly simple form:

$$R_D(T) = \frac{6}{5}g_D(T)(\vec{d} \times \vec{\ell})\ (\vec{d} \cdot \vec{\ell})$$

It is this term which gives rise to the frequency shift found in superfluid ^{3}He A. The coupled motions of $\vec{S}$ and $\vec{d}$ for ^{3}He A as predicted by the Leggett equations are illustrated in Figure 15, which shows the free precession following a 10° tipping pulse. The spin vector precessing about the applied steady mag-

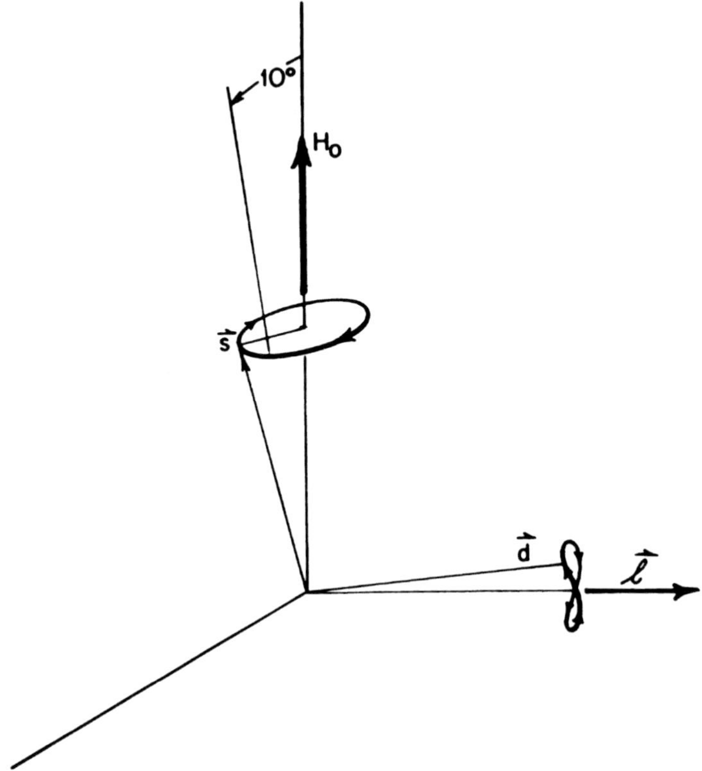

Figure 15. The free precession of the spin vector and the $\vec{d}$ vector after a small tipping pulse according to the Leggett equations.

netic field describes an ellipse, while the corresponding motion of the unit vector d describes a figure eight on the unit sphere.[69]

With this beautiful theory of the dynamics of the order parameter, Leggett calculated the frequency shift for the transverse (ordinary) NMR for both the ABM and the Balian Werthamer (BW) states. The calculation of the frequency shift for the ABM state gave the Pythagorean relationship $\omega^2 - \omega_0^2 = \Omega_A^2(T)$ found in the discovery experiments. The BW state showed no transverse frequency shift, again in agreement with experiment. The Leggett equations of motion also predicted the existence of longitudinal resonance in both the ABM and BW states. Longitudinal resonance experiments are performed by orienting the radiofrequency coil parallel to the applied steady magnetic field H_0 so that $H_{rf} \parallel H_0$. For the case of ordinary (transverse) NMR, $H_{rf} \perp H_0$. At the time of the discovery experiment we speculated on this possibility for the A phase based on the Pythagorean relationship which implied an internal field perpendicular to the applied field (see Figure 8). We had no idea that the B phase would also manifest a longitudinal magnetic resonance. Longitudinal resonance signals were observed by Osheroff and Brinkman[70] at the Bell Telephone Laboratories in both the A and B phases. Later we observed longitudinal magnetic resonance in the A phase at Cornell.[71] (Typical longitudinal resonance frequencies ranged up to about 100 kHz in ^{3}He A at melting pressure.) The longitudinal frequency in ^{3}He A, $\Omega_A(T)$, is related to the longitudinal frequency in ^{3}He B, $\Omega_B(T)$ by the ratio

$$\left[\frac{\Omega_B(T)}{\Omega_A(T)}\right]^2 = \frac{5}{2}\frac{\chi_B}{\chi_A}.$$

Longitudinal signals were also observed by Webb *et al.*[72] by applying a step in the steady field and observing radiofrequency longitudinal ringing with SQUID detectors.

An important feature which sets superfluid ^{3}He apart from the more conventional superconductors and superfluid ^{4}He is the presence of internal degrees of freedom of the order parameter. So far, we have discussed mainly the spin degrees of freedom and how they relate to the NMR experiments. Observations have also revealed a variety of interesting phenomena related to the orbital degrees of freedom. For example, the orbital anisotropy of the A phase leads to anisotropic flow properties. Mermin and Ho[73] showed that this completely altered the character of quantized circulation, leading to novel mechanisms for the decay of supercurrents. The anisotropy of the superfluid density was first discovered by Berthold, Giannetta, Smith and Reppy[74] at Cornell using torsional oscillator techniques. Figure 16 shows their results for two orientations of the A phase obtained by appropriately orienting the external magnetic field.

One area in which we have been particularly actively involved is the study of sound propagation in superfluid ^{3}He. As we mentioned earlier, the sound attenuation mechanisms in the superfluid are mainly associated with pair breaking when $\hbar\omega_{sound} \geq 2\Delta$ and with collective modes corresponding to in-

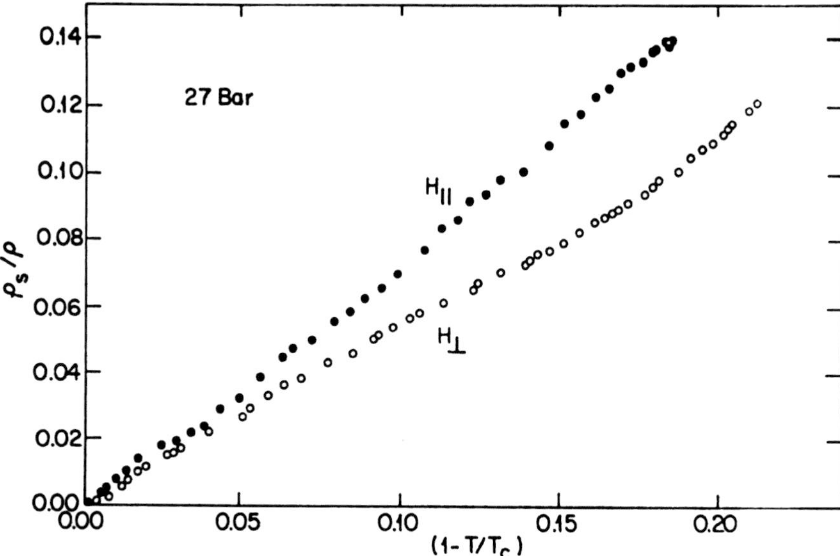

Figure 16. Anisotropy of the superfluid density as determined in torsional oscillator measurements by Berthold, Giannetta, Smith and Reppy.[74] Two different orientations of the order parameter are shown, where the orientation is controlled by the applied magnetic field via the dipolar interaction.

ternal vibrations of Cooper pairs. Because they are governed by a macroscopic order parameter, the Cooper pairs are coherently excited and thus vibrate in phase. The pair vibration modes which couple to density fluctuations can be excited by sound waves. The collective modes are analogous to excited states of atoms, but for the case of superfluid ^{3}He, the whole liquid sample participates in the collective motion, providing a spectacular example of London's quantum mechanics on a macroscopic scale, this time being associated with the orbital internal degrees of freedom.

The energy gap of ^{3}He A has two point nodes where the $\hat{l}$ axis intersects the Fermi surface. Thus pair breaking can take place via these nodes all the way to absolute zero. The pair vibration modes exhibit anisotropy in superfluid ^{3}He A. In fact, the first evidence for orbital anisotropy in ^{3}He A was seen in early ultrasound experiments by Lawson *et al.*[75] and Roach *et al.*[76] The B phase, on the other hand, has an isotropic energy gap and so the breaking of pairs takes place only above a finite threshold called the pair breaking edge. It is thus possible to study the collective modes of ^{3}He B without the complications of pair breaking as long as the energy of a sound quantum $\hbar\omega$ is less than 2Δ. At the lowest temperatures a vanishing number of excited quasiparticles is present so any quasiparticle collision broadening of the collective modes is almost completely negligible. Thus very narrow sound absorption lines associated with collective modes in superfluid ^{3}He B are expected. Therefore the analogy between pair vibrations and the excited states of atoms is best illustrated by the collective modes in superfluid ^{3}He B.

Our more recent ultrasound experiments utilized nuclear adiabatic demagnetization methods to cool the liquid ^{3}He down into the superfluid phase. Modern experiments on superfluid ^{3}He employ this cooling technique. The technique had been pioneered by Nicholas Kurti and his co-workers at Oxford University[77] but was first used to cool liquid ^{3}He by John Goodkind and his students at the University of California at San Diego.[78] Nuclear adiabatic demagnetization of a bundle of copper wires or plates to cool liquid ^{3}He well into the superfluid regime was developed into a standard technology by Olli Lounasmaa and his group at Helsinki University of Technology.[79] The technique had the advantage of making accessible pressures below melting pressure for the liquid ^{3}He sample just as in the case of adiabatic demagnetization of paramagnetic salts, but much lower temperatures (well below 1 mK) also became available. To overcome the large thermal boundary resistance between the ^{3}He sample and the copper bundle, a sintered silver heat exchanger with very high surface area was placed inside the ^{3}He cell and firmly anchored thermally to the wall of the cell.

The two collective modes we studied in superfluid B are called the imaginary squashing mode and the real squashing mode. They correspond to two different types of periodic distortions of the energy gap. These modes were first studied theoretically by Wölfle,[80] Serene,[81] and Maki.[82] Both of these modes are in the total angular momentum state $J = 2$ in contrast to the $J = 0$ ground state associated with the B phase order parameter. We are neglecting the effect of the 104° rotation of the spin coordinates of the ground state relative to the orbital coordinates which is not important for our ultrasound studies. The real squashing mode couples much more weakly to sound than the imaginary squashing mode. These modes are excited by ultrasound at frequencies given by the following expressions:

$$\hbar\omega_{\mathbf{sound}} = \begin{cases} \sqrt{\frac{12}{5}}\Delta(T), & \text{Imaginary Squashing Mode} \\ \sqrt{\frac{8}{5}}\Delta(T), & \text{Real Squashing Mode} \end{cases}$$

Fixed frequency quartz piezoelectric sound transducers driven at harmonic frequencies between 40 and 120 MHz were used in these experiments so that the profile of the sound absorption line had to be determined by sweeping $\Delta(T)$ via varying the temperature rather than by sweeping the frequency. Sound attenuation experiments were performed at Cornell by Giannetta *et al.*[83] A complete temperature sweep is shown in Figure 17 based on the Cornell data where the sound attenuation data is plotted against the temperature. Clearly displayed are three well-differentiated peaks. At the highest temperature is the broad peak associated with pair breaking. The intermediate peak is the imaginary squashing mode. The attenuation is so large for these two high temperature peaks that it could not be measured in our experiment. The small, very narrow low temperature peak is the real squashing mode. The small size of this peak is directly attributable to the weak coupling of the real squashing mode to sound. Observations of the real squashing mode were made by Mast *et al.*[84] at the same time as those of Giannetta *et al.*

The real squashing mode and the imaginary squashing mode display spectacular effects in the presence of applied magnetic fields. These effects were first revealed in the experiments of Avenel, Varoquaux and Ebisawa[85] for the

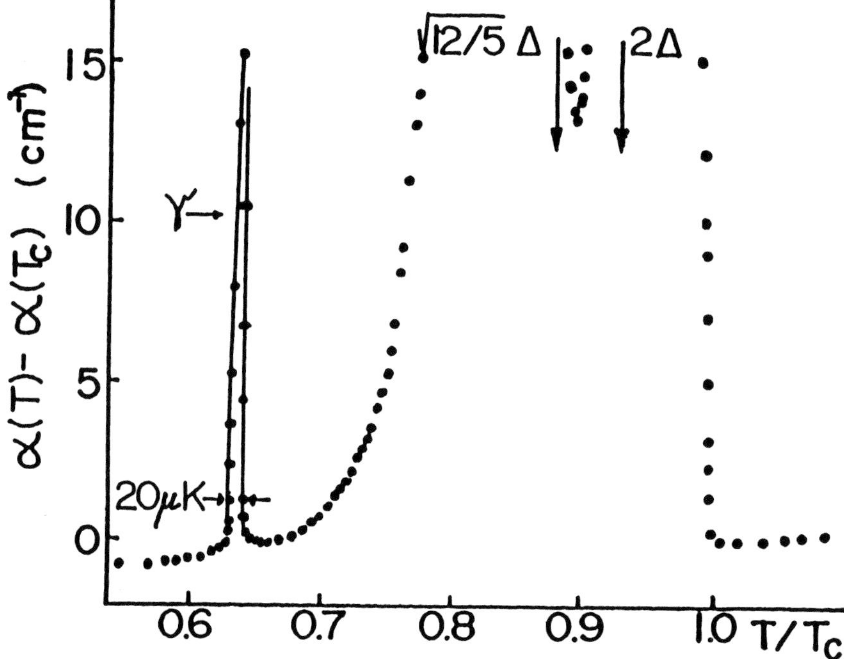

Figure 17. Attenuation of ultrasound in superfluid ^{3}He B showing the pair breaking peak, the imaginary squashing mode peak and the narrow real squashing mode peak.[83] The very strong attenuation associated with pair breaking and the imaginary squashing mode made it impossible to observe the peak maxima for these two peaks. The data were taken at a constant sound frequency during a temperature sweep.

real squashing mode, which exhibited a five-fold Zeeman splitting corresponding to a $J = 2$ excited state, linear in field for low fields as shown in Figure 18. A similar five-fold splitting for the imaginary squashing mode has been observed at Cornell by Movshovich *et al.*[86] using special techniques. The splittings of the real squashing mode and the imaginary squashing mode were first predicted by Tewordt and Schopohl.[87] A spectrum schematically showing these modes is provided in Figure 19. The left side of the figure shows the spectrum of a bulk sample of superfluid ^{3}He B for zero applied magnetic field and the right side shows the five-fold splittings of the $J = 2$ excited modes in a magnetic field. These spectra provide spectacular examples involving the roles of both magnetic and orbital degrees of freedom in the excited state pair wave functions. These phenomena are further dramatic manifestations of London's idea of quantum mechanics on a macroscopic scale, this time in excited modes involving spin and orbital degrees of freedom simultaneously.

SOME IMPORTANT DEVELOPMENTS

Superfluid ^{3}He is a complex and beautiful system where the internal degrees of freedom play a prominent role, leading to a large array of phenomena not seen in conventional s wave superconductivity or in superfluid ^{4}He. In con-

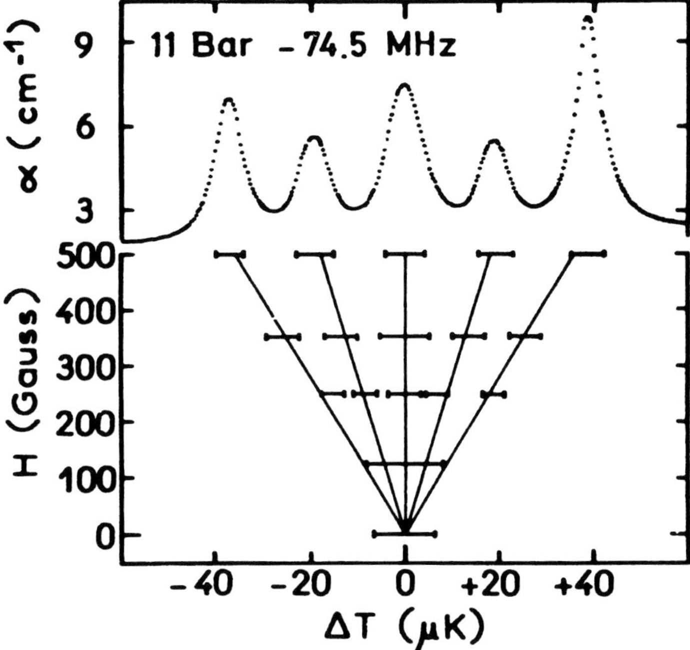

Figure 18. Splitting of the real squashing mode in a magnetic field as observed by Avenel, Varoquaux and Ebisawa.[85] Also shown are the attenuation peaks for the sound propagation direction perpendicular to the magnetic field.

trast to conventional superconductors, the pairing is intrinsic, not relying on the exchange of phonons via an ionic lattice. The complex phase diagram shows three different superfluid phases of liquid ^{3}He, each with its own unique set of properties. We have already discussed the way in which London's idea of quantum mechanics on a macroscopic scale applies to the nuclear magnetic resonance experiments and the studies of collective modes excited by sound. Interesting effects seen in other branches of condensed matter physics have also been observed in superfluid ^{3}He. We have only touched upon the anisotropic flow properties associated with the anisotropy of the energy gap and the gap nodes in the A and the A_1 phases. In addition, satellite NMR lines in ^{3}He A, first seen for longitudinal resonance by Avenel *et al.*[88] and for transverse resonance by Gould and Lee,[89] have been interpreted[90] as solitons resembling thin domain walls where the orientations of the $\hat{l}$ and $\hat{d}$ vectors abruptly change.

The phenomenon of second sound, seen previously in superfluid ^{4}He, has actually been observed in the A_1 phase of superfluid ^{3}He by Corruccini and Osheroff,[91] where it involves a wave-like propagation of heat and simultaneously the propagation of a spin density wave. This is related to the fact that the superfluid density in the two fluid model contains only spin up pairs. Since spin down pairs are not present in the order parameter of the A_1 phase, the spin down quasiparticle density comprising the normal fluid density remains large at the lowest temperatures. This condition is not obeyed for the A and B phases where second sound is highly damped due to the relative

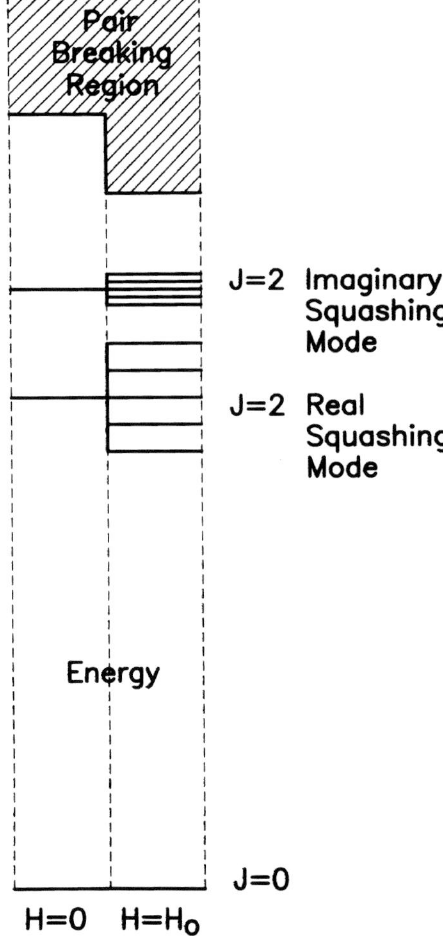

Figure 19. A schematic diagram showing the collective modes of superfluid ^{3}He B discussed in the text. The left hand side of the diagram shows the zero field spectrum. The five fold degenerate $J = 2$ real squashing and imaginary squashing modes are shown as well as the pair breaking edge which is crudely analogous to the ionization energy of an atom. The right hand side of the diagram shows the five fold Zeeman splitting of these collective modes. The g factor of the imaginary squashing mode is smaller than that of the real squashing mode. The pair breaking edge occurs at a lower energy in a magnetic field partially as a result of gap distortion by the field.

scarcity of normal fluid. According to the two fluid model, the heat is associated with the normal fluid density. A wave-like propagation of heat will involve a wave-like propagation of alternate regions of high normal fluid density and high superfluid density, leading to the propagation of a spin density wave.

In recent years, studies of superfluid ^{3}He have been conducted in rotating cryostats at Cornell, the Helsinki University of Technology, the University of California at Berkeley, and most recently the University of Manchester. The study of persistent currents in superfluid ^{3}He A and ^{3}He B by John Reppy and his coworkers at Cornell has furnished the final proof that both phases are indeed true superfluids.[92,93] The very complexity of superfluid ^{3}He leads to a rich variety of vortex behavior which has been the main effort at

Helsinki in recent years. A vast experimental and theoretical effort has been under way to classify and understand this fascinating vortex behavior.[94,95] Finally Richard Packard and his associates have studied vortices in ^{3}He B using the vibrating wire technique developed by Vinen[96] for superfluid ^{4}He and have shown directly the role of Cooper pairs in the vortices.[97] A major advance has also been made by physicists at the Kapitza Institute in Moscow. They have been able to stabilize a large region of superfluid ^{3}He with the spins uniformly precessing at a single frequency in spite of the presence of a significant magnetic field gradient. Spin supercurrents played a vital role in attaining these homogeneously precessing domains. One of the most important results to emerge from this work has been the observation of the Josephson effect involving spin currents.[98]

Superfluid ^{3}He research may have some bearing on studies of exotic superconductivity.[99] In particular, complex (non-s wave) order parameters have been proposed for heavy fermion and high T_c superconductors. In the latter case some very recent tunneling[100] and magnetometer[101] experiments provide compelling evidence that the high temperature superconducting cuprates can exhibit d wave pairing. Work at various laboratories around the world is under way to verify this result. With regard to the heavy fermion superconductors, UPt_3 has turned out to be a very promising candidate for $l > 0$ pairing.[99,102] It has a complex superconducting phase diagram with several superconducting phases as well as thermal properties which are consistent with gap nodes.

The cooling techniques and thermometry techniques developed by laboratories around the world for studies of superfluid ^{3}He have already been useful. For example, the phase transitions of superfluid ^{3}He along the melting curve can serve as thermometric fixed points. The nuclear adiabatic demagnetization techniques have been used at the Helsinki University of Technology to study nuclear magnetic ordering of solids in the sub-microkelvin temperature range.[103]

In the realm of astrophysics, theoretical studies of the neutron rich interior of neutron stars have indicated the possibility of a $^3P_2$ neutron superfluid ground state order parameter.[104] In addition, superfluid ^{3}He can serve as a model system for processes in the early universe. Gregory Volovik[105] has been particularly active in drawing analogies between the phases of liquid ^{3}He and phases of the vacuum in the early universe. As the universe cooled, it made transitions from a featureless hot universe to the universe we know today via phase transitions involving broken symmetry. Already some experimental work in superfluid ^{3}He has been performed to pursue this analogy.[106,107,108]

CONCLUSION

Superfluid ^{3}He has provided an enormous amount of fun to the participants in the field. The role of internal degrees of freedom of the Cooper pairs has given new meaning to London's proposal that quantum mechanics on a macroscopic scale can describe superfluidity.

In addition, studies of this exotic superfluid have stimulated investigators in a wide variety of other fields. We have discussed ways in which superfluid ^{3}He research can be linked to high T_c superconductivity, heavy fermion superconductivity and liquid crystals in condensed matter physics. Examples of soliton behavior have also been observed in superfluid ^{3}He. The NMR technique employed in our experiments involved one of the very earliest applications of magnetic resonance imaging, now a widely accepted diagnostic technique in medicine and biology.

Phenomena in superfluid ^{3}He can also be related to astrophysics. The states of p wave pairing in superfluid ^{3}He may have their analogies to possible pairing states in neutron stars. The superfluid phase transitions in ^{3}He may also serve as a model for transitions which occurred in the early universe.

One may take the point of view that the importance of a field of research is related to its impact on other areas of science. In spite of the fact that superfluid ^{3}He exists in a difficult temperature range attainable only by highly specialized techniques, both the techniques and the results of superfluid ^{3}He investigations have indeed had important influences on diverse fields of research.

We gratefully acknowledge the National Science Foundation for providing support to our research program for more than 36 years. We have also benefitted from support by the Advanced Research Projects Agency for a shorter time, during an important period spanning some of the most exciting developments discussed herein.

REFERENCES

1. Kamerlingh Onnes, H., Proc. Acad. Sci. Amsterdam **11**, 168 (1908).
2. Kamerlingh Onnes, H., Leiden Comm. **1226**, 124c (1911) Suppl. **35** (1913).
3. Simon, F., Nature, London **133**, 529 (1934).
4. Atkins, K. R., Liquid Helium, Cambridge University Press, Cambridge, 1959, p. 2.
5. Keesom, W. H. and K. Clusius, Proc. Royal Acad. Amsterdam **35**, 307 (1932). \ Keesom, W. H., and A. P. Keesom, Proc. Royal Acad. Amsterdam **35**, 736 (1932).
6. Daunt, J. G. and K. Mendelssohn, Proc. Royal Society A **170**, 423, 439 (1939).
7. Kapitza, P. L., Nature, London **141**, 74 (1938), and Kapitza, P. L., J. Phys. Moscow **4**, 181 (1941).
8. Allen, J. F. and H. Jones, Nature, London **141**, 75 (1938) and Proc. Camb. Phil. Soc. **34**, 299 (1938).
9. Landau, L. D., J. Phys., Moscow **5**, 71 (1941).
10. Tisza, L., J. Phys. Radium I, 165 and 350 (1940).
11. Andronikashvili, E. L., J. Phys. Moscow **10**, 201 (1946).
12. Peshkov, V., J. Phys. (USSR) **8**, 381 (1944).
13. London, Fritz, Superfluids, Vol. II, John Wiley and Sons, Inc., New York, (1954).
14. Ginzburg, V. L. and L. D. Landau, Zh. Eksperim i Teor. Fiz. **20**, 1064 (1950).
15. Maxwell, E., Phys. Rev. **78**, 477 (1950).
16. Reynolds, C. A., B. Serin, W. H. Wright and L. B. Nesbitt, Phys. Rev. **78**, 487 (1950).
17. Bardeen, J., L. N. Cooper and J. R. Schrieffer, Phys. Rev. **108**, 1175 (1957).
18. Cooper, L. N., Phys. Rev. **104**, 1189 (1956).
19. London, F. and H. London, Proc. Roy. Soc. (London) **A149**, 71 (1935).
20. London, F., Superfluids, Vol. I, John Wiley and Sons Inc., New York (1950).
21. Deaver, B. S. and W. M. Fairbank, Phys. Rev. Lett. **7**, 43 (1961).

22. Doll, R. and M. Näbauer, Phys. Rev. Lett. **7**, 51 (1961).
23. Josephson, B. D., Phys. Lett. **1**, 251 (1962).
24. Landau, L. D., Zh. Eksp. i Teor. Fiz. **30**, 1058 (1956); Soviet Phys. JETP **3**, 920 (1957), and Landau, L. D., Zh. Eksp, i Teor. Fiz. **32**, 59 (1957); Soviet Phys. JETP **5**, 101 (1957).
25. Wheatley, J. C. in Progress in Low Temp. Physics, Vol. VI, ed. C. J. Gorter, North-Holland Publishing, Amsterdam (1966), p. 183.
26. Keen, B. E., P. W. Mathews and J. Wilks, Phys. Lett. **5**, 5 (1963).
27. Abel, W. R., A. C. Anderson and J. C. Wheatley, Phys. Rev. Lett. **17**, 74 (1966).
28. Pitaevskii, L. P., Zh. Eksp. i Teor. Fiz. **37**, 1794 (1959); Soviet Phys. JETP **10**, 1267 (1960).
29. Emery, V. J. and A. M. Sessler, Phys. Rev. **119**, 43 (1960).
30. Anderson, P. W. and P. Morel, Phys. Rev. **123**, 1911 (1961).
31. Balian, R. and N. R. Werthamer, Phys. Rev. **131**, 1553 (1963).
32. Layzer, A. and D. Fay, Int. J. Mag. **1**, 135 (1971).
33. Grilly, E. R., E. F. Hammel and S. G. Sydoriak, Phys. Rev. **75**, 1103 (1949).
34. Lee, D. M. and H. A. Fairbank, Phys. Rev. **115**, 1359 (1959).
35. Lee, D. M., H. A. Fairbank and E. J. Walker, Phys. Rev. **121**, 1258 (1961).
36. Halperin, W. P., C. N. Archie, F. B. Rasmussen, R. A. Buhrman and R. C. Richardson, Phys. Rev. Lett. **32**, 927 (1974).
37. Pomeranchuk, I., Zh. Eksp. i Teor. Fiz. **20**, 919 (1950).
38. Straty, G. E., and E. D. Adams, Rev. Sci. Instr. **40**, 1393 (1969).
39. Tedrow, P. M. and D. M. Lee, Phys. Lett. **9**, 193 (1964).
40. Lipschultz, F. P. and D. M. Lee, Phys. Rev. Lett. **14**, 1017 (1965).
41. Grilly, E. R., S. G. Sydoriak and R. L. Mills, in Helium Three, ed. J. G. Daunt, Ohio State University Press, Columbus, Ohio (1960), p. 120.
42. Hall, H. E., P. J. Ford and K. Thompson, Cryogenics **6**, 80 (1966).
43. Wheatley, J. C., R. E. Rapp and R. T. Johnson, J. Low Temp. Phys. **4**, 1 (1971).
44. Sites, J. R., D. D. Osheroff, R. C. Richardson and D. M. Lee, Phys. Rev. Lett. **23**, 836 (1969).
45. Anufriyev, Yuri D., Soviet Physics JETP Letters **1**, 155 (1965).
46. Johnson, R. T., R. Rosenbaum, O. G. Symko and J. C. Wheatley, Phys. Rev. Lett. **22**, 449 (1969).
47. Wheatley, J. C., private communication.
48. Osheroff, D. D., R. C. Richardson and D. M. Lee, Phys. Rev. Lett. **28**, 885 (1972).
49. Lauterbur, P. C., Nature **242**, 190 (1973).
50. Osheroff, D. D., W. J. Gully, R. C. Richardson and D. M. Lee, Phys. Rev. Lett. **29**, 920 (1972).
51. Leggett, A. J., Phys. Rev. Lett. **29**, 1227 (1972).
52. Leggett, A. J., Rev. Mod. Phys. **47**, 331 (1975).
53. Leggett, A. J., J. Phys. C **6**, 3187 (1973).
54. Leggett, A. J., Ann. Physics (New York) **85**, 11 (1974).
55. Lee, D. M., Proceedings of the 13th International Conference on Low Temperature Physics, Boulder, Colorado, edited by K. D. Timmerhaus, W. J. O'Sullivan and E. F. Hammel (Plenum Press) 1972, Vol. 2, p. 25.
56. Osheroff, D. D., W. J. Gully, R. C. Richardson and D. M. Lee, Proceedings of the 13th International Conference on Low Temperature Physics, Boulder, Colorado, edited by K. D. Timmerhaus, W. J. O'Sullivan and E. F. Hammel (Plenum Press) 1972, Vol. 2, p. 134.
57. Alvesalo, T. A., Yu. D. Anufriyev, H. K. Collan, O. V. Lounasmaa and P. Wennerström, Phys. Rev. Lett. **30**, 962 (1973).
58. Yanof, A. and J. D. Reppy, Phys. Rev. Lett. **33**, 631 and 1030 erratum (1974).
59. Kojima, H., D. N. Paulson and J. C. Wheatley, Phys. Rev. Lett. **32**, 141 (1974).
60. Lawson, D. T., W. J. Gully, S. Goldstein, R. C. Richardson and D. M. Lee, Phys. Rev. Lett. **30**, 541 (1973).

61. Webb, R. A., T. J. Greytak, R. J. Johnson and J. C. Wheatley, Phys. Rev. Lett. **30**, 210 (1973).
62. Osheroff, D. D., Cornell University Ph.D thesis, (1973). See also Gully, W. J., D. D. Osheroff, D. J. Lawson, R. C. Richardson and D. M. Lee, Phys. Rev. A **8**, 1633 (1973).
63. Ambegaokar, V. and N. D. Mermin, Phys. Rev. Lett. **30**, 81 (1973).
64. Lawson, D. T., H. M. Bozler and D. M. Lee in Quantum Statistics and the Many Body Problem, ed. S. B. Trickey, W. P. Kirk and J. W. Duffy, Plenum Press, New York, 1975, p. 19.
65. Paulson, D. N., H. Kojima and J. C. Wheatley, Phys. Rev. Lett. **32**, 1098 (1974).
66. Ambegaokar, V., P. G. de Gennes and D. Rainer, Phys. Rev. A **9**, 2676 (1974); erratum, *ibid*, **A12**, 345 (1975).
67. Anderson, P. W. and W. F. Brinkman, Phys. Rev. Lett. **30**, 1108 (1973).
68. Rainer, D. and J. W. Serene, Phys. Rev. B **13**, 4745 (1976).
69. Lee, D. M. and R. C. Richardson in The Physics of Liquid and Solid Helium, ed. K. H. Bennemann and J. B. Ketterson, John Wiley and Sons, New York 1978, pp. 287–496.
70. Osheroff, D. D. and W. F. Brinkman, Phys. Rev. Lett. **32**, 584 (1974).
71. Bozler, H. M., M. E. R. Bernier, W. J. Gully, R. C. Richardson and D. M. Lee, Phys. Rev. Lett. **32**, 875 (1974).
72. Webb, R. A., R. L. Kleinberg and J. C. Wheatley, Phys. Rev. Lett. **33**, 145 (1974).
73. Mermin, N. D. and T. L. Ho, Phys. Rev. Lett. 36, 594 (1976); erratum, *ibid*, **36**, 832 (1976).
74. Berthold, J. E., R. W. Giannetta, E. N. Smith and J. D. Reppy, Phys. Rev. Lett. **37**, 1138 (1976).
75. Lawson, D. T., H. M. Bozler and D. M. Lee, Phys. Rev. Lett. **34**, 121 (1975).
76. Roach, P. R., B. M. Abraham, P. D. Roach and J. B. Ketterson, Phys. Rev. Lett. **34**, 715 (1975).
77. Kurti, N., F. N. Robinson, F. Simon and D. A. Spohr, Nature, London, **178**, 450 (1956).
78. Dundon, J. M., D. L. Stolfa and John Goodkind, Phys. Rev. Lett. **30**, 843 (1973).
79. Ahonen, A. I., M. T. Haikala, M. Krusius and O. V. Lounasmaa, Phys. Rev. Lett. **33**, 628 (1974) and *ibid* **33**, 1595 (1974).
80. Wölfle, P., Phys. Rev. Lett. **30**, 1169 (1973) and *ibid* **31**, 1437 (1973).
81. Serene, J. W., Cornell University Ph.D thesis (1974).
82. Maki, K., J. Low Temp. Phys. **16**, 465 (1974).
83. Giannetta, R. W., A. Ahonen, E. Polturak, J. Saunders, E. K. Zeise, R. C. Richardson and D. M. Lee, Phys. Rev. Lett. **45**, 262 (1980).
84. Mast, D. B., B. K. Sarma, J. R. Owers-Bradley, I. D. Calder, J. B. Ketterson and W. P. Halperin, Phys. Rev. Lett. **45**, 266 (1980).
85. Avenel, O., E. Varoquaux and H. Ebisawa, Phys. Rev. Lett. **45**, 1952 (1980).
86. Movshovich, R., E. Varoquaux, N. Kim and D. M. Lee, Phys. Rev. Lett. **61**, 1732 (1988).
87. Tewordt, L. and N. Schopohl, J. Low Temp. Phys. **37**, 421 (1979), and N. Schopohl and L. Tewordt, *ibid* **45**, 67 (1981).
88. Avenel, O., M. E. Bernier, E. J. Varoquaux and C. Vibet in Proceedings of the 14th International Conference on Low Temp. Physics, ed. M. Krusius and M. Vuorio, North Holland, Amsterdam, 1975, p. 429.
89. Gould, C. M. and D. M. Lee, Phys. Rev. Lett. **37**, 1223 (1976).
90. Maki, K. and P. Kumar, Phys. Rev. B **16**, 182 (1977).
91. Corruccini, L. R. and D. D. Osheroff, Phys. Rev. Lett. **45**, 2029 (1980).
92. Gammel, P. L., H. E. Hall and John D. Reppy, Phys. Rev. Lett. **52**, 121 (1984).
93. Gammel, P. L., T. L. Ho and John D. Reppy, Phys. Rev. Lett. **55**, 2708 (1985).
94. Salomaa, M. M. and G. E. Volovik, Reviews of Modern Phys. **59**, 533 (1987).
95. Hakonen, P. J., M. Krusius, M. M. Salomaa, J. T. Simola, Yu. M. Bunkov, V. P. Mineev and G. E. Volovik, Phys. Rev. Lett. **51**, 1362 (1983), and Hakonen, P. J., M. Krusius and H. K. Seppälä, J. Low Temp Phys. **60**, 187 (1985).
96. Vinen, W. F., Proc. Roy. Soc. A **260**, 218 (1961).
97. Davis, J. C., J. D. Close, R. Zieve and R. E. Packard, Phys. Rev. Lett. **66**, 329 (1991).

98. Borovik-Romanov, A. S., Yu. Bunkov, V. V. Dmitriev, Yu. M. Mukharsky and D. A. Sergatskov, Phys. Rev. Lett. **62**, 1631 (1989), and Borovik-Romanov, A. S., Yu. M. Bunkov, A. de Vaard, V. V. Dmitriev, V. Makrotsieva, Yu. M. Mukharskii, and D. A. Sergatskov, Soviet Phys. JETP Letters **47**, 478, 1988.
99. Sigrist, M. and K. Ueda, Reviews of Modern Phys. **63**, 239 (1991).
100. Wollman, D. A., D. J. Van Harlingen, W. C. Lee, D. M. Ginsberg and A. J. Leggett, Phys. Rev. Lett. **71**, 2134 (1993).
101. Tsuei, C. C., J. R. Kirtley, C. C. Chi, L. S. Yu-Jahnes, A. Gupta, T. Shaw, T. Z. Sun and M. B. Ketchen, Phys. Rev. Lett. **73**, 593 (1994).
102. Lin, S.-W., C. Jin, H. Zhang, J. B. Ketterson, D. M. Lee, D. J. Hinks, M. Levy and B. K. Sarma, Phys. Rev. B **49**, 10001 (1994).
103. Oja, A. S. and O. V. Lounasmaa, Rev. Mod. Phys., **69**, 1 (1997).
104. Sauls, J. A., D. Stein and J. W. Serene, Phys. Rev. D **25**, 967 (1982).
105. Volovik, G., Exotic Properties of Superfluid ^{3}He (World Scientific, Singapore 1992).
106. Ruutu, V. M. H., V. B. Eltsov, A. J. Gill, T. W. B. Kibble, M. Krusius, Yu. G. Makhlin, B. Placais, G. E. Volovik and W. Xu, Nature **382**, 334 (1996).
107. Bäuerle, G., Yu. M. Bunkov, S. N. Fisher, H. Godfrin and G. R. Pickett, Nature **382**, 332 (1996).
108. Bevan, T. D. C., A. J. Manninen, J. B. Cook, J. R. Hook, H. E. Hall, T. Vachaspati and G. E. Volovik, preprint, submitted to Nature.

DOUGLAS D. OSHEROFF

Ethnically, I come from a mixed family. My father was the son of Jewish immigrants who left Russia shortly after the turn of the century, and my mother was the daughter of a Lutheran minister whose parents were from what is now Slovakia. Mostly, however, I grew up in a medical family. My father's father and all his children either became physicians or married them. My parents had met in New York where my father was a medical intern and my mother was a nurse. At the end of World War II, my parents settled in Aberdeen, a small logging town on the west coast of Washington State, where medical doctors were in short supply. Surrounded by natural beauty, it was a perfect place to raise a family, and I was the second of five children.

To this day I grow pale at the sight of blood, and never for a moment considered a career in medicine. Despite this, my father, who was usually engrossed in his medical career, inspired in me passions for both photography and gardening, which were his hobbies when time permitted, as they are mine. Natural science interested me intensely from a very early age. When I was six I began tearing my toys apart to play with the electric motors. From then on, my free hours were occupied by a myriad of mechanical, chemical and electrical projects, culminating in the construction of a 100 keV X-ray machine during my senior year in high school.

My projects often involved an element of danger, but my parents never seemed too concerned, nor did they inhibit me. Once a muzzle loading rifle I had built went off in the house, putting a hole through two walls. On another occasion a make-shift acetylene 'miners' lamp blew up on my chemistry bench in the basement, embedding shards of glass in the side of my face, narrowly missing my right eye. With blood running down my face, I came up the stairs cupping my hands to keep the blood off the carpet. My mother was by then at the top of the stairs. Knowing my propensity for practical jokes, she exclaimed loudly "If you're kidding I'll kill you!" As usual, my father lectured me about safety as he sewed the larger wounds closed, and there was always an unspoken understanding that that particular phase of my experimentation was over.

In high school I was a good student, but only really excelled in physics and chemistry classes. While I liked physics much more than chemistry, the chemistry teacher, William Hock, had spent quite a bit of time telling us what physical research was all about (as opposed to my experimentation), and that effort made a deep impression on my young mind. My interest in experimentation helped me to develop excellent technical skills, but I did not feel motivated to do independent reading in those areas of physics or chemistry associated with my projects. I was intellectually rather lazy, and in high school I would always take one free class period so that I could get my homework out of the way, freeing the evenings for my many projects.

My parents were generous, and the home for me was filled with scientific toys and gadgets. In addition, their children were allowed to attend any university to which they could get admitted. I chose Caltech over Stanford to avoid a continuing comparison of my academic record with that of my older brother, then a Stanford undergraduate.

It was a good time to be at Caltech, as Feynman was teaching his famous undergraduate course. This two-year sequence was an extremely important part of my education. Although I cannot say that I understood it all, I think it contributed most to the development of my physical intuition. The Feynman problem sets were very challenging, but I had the good fortune to know Ernest Ma, who was an undergraduate one year ahead of me. Ernest would never tell me how to solve problems, but would give obscure hints when I got stuck, at least they seemed obscure to me at the time.

It was a shock to suddenly have to work so hard in my studies. I had the most trouble in math, and only through considerable trauma did I gradually improve my performance from a grade of C+ to A+ over a three-year period. Years later, when Caltech was offering me a faculty position, I confided that I did not have a very illustrious career as an undergraduate. To this remark the division chair replied "That's OK Doug, we are not hiring you to be an undergraduate."

The pressure at Caltech was extreme, and I am not sure I would have survived had I not joined a group of undergraduates working with Gerry Neugebauer on his famous infra-red star survey during my junior year. This experience made me recognize how satisfying research could be, and how different it was from doing endless problem sets. In my senior year, in order to get out of a third term of senior physics lab, I also began working in David Goodstein's low temperature lab (David was in Italy). Two professors, Don McCullum from U.C. Riverside and Walter Ogier from Pamona College, were spending their sabbatical leaves there trying to reach a temperature of 0.5K by pumping on a helium bath in which the superfluid film had been carefully controlled. They filled my mind with the wonders of the low temperature world, and I decided I would go into solid state physics.

I chose to attend Cornell for graduate school largely because it was so far away from the Pasadena smog. In the end, it was a good choice, and a good time to be at Cornell. Soon after my arrival I met two people who were to become very important in my life. While still looking for housing, I met Phyllis Liu, a pretty young woman from Taiwan, who had also just arrived in Ithaca. We dated a bit, but then she found herself too busy with her studies for such diversions. We met again three years later, and were married in August, 1970, two weeks after she obtained her Ph.D. The other person was David Lee, the head of the low temperature laboratory at Cornell and the professor under whom I was to work as a teaching assistant my first year. Dave seemed to think that I was bright, and encouraged me to join the low temperature group.

Low temperature physics seemed even more exciting at Cornell than it had been at Caltech. New technologies and interesting physics made the field easy to choose, and I found myself thoroughly enjoying every minute of my work. In the spring of my fourth year Dave Lee asked me to talk to the Bell Labs recruiter, who came to campus in the fall and spring of each year. I was not ready to graduate, but we talked a bit, especially about making tiny electrical plugs to be used throughout the Bell Telephone system. It seemed interesting to me, although not really physics. In the fall, Dave suggested I start interviewing in earnest. I first talked with General Electric, who seemed to have no jobs whatsoever. I then talked to Bell Labs again, but this time to a new recruiter, Venky Narayanamurti, who had recently received his Ph.D. in physics at Cornell. Venky was enthusiastic about what I was doing, and felt that I might be able to get a postdoc doing Raman spectroscopy. I didn't confess that I knew nothing about the subject.

We discovered our mysterious phase transitions in my Pomeranchuk cell in November 1971, and almost by magic, Venky called me up in early December with good news. The hiring freeze which had been in place for almost two years at Bell had been lifted. How soon could I be ready to come down for a job interview? I told Venky that we had stumbled on to something that was pretty exciting, and we fixed the date: January 6, 1972.

At Bell Labs, a job interview began with a thesis defense, and it could at times turn nasty. I was lucky that no one questioned my association of the A and B features with the solid. In particular, Dick Werthamer was in the audience, and he had done early work on the p-wave BCS state soon to be associated with the B phase. I think my enthusiasm carried the day, and ultimately Bell Labs offered me not a postdoc position in Raman spectroscopy, but a permanent position which would allow me to continue my studies on ^{3}He.

Phyllis and I moved to New Jersey in September, 1972; Phyllis to a postdoc position at Princeton University, and I to Bell Laboratories at Murray Hill. This was the golden era at Bell Labs. The importance of the transistor, invented in the research area there, made management extremely supportive of basic research. The only requirement was that work done should be 'good physics' in that it changed the way we thought about nature in some important way. I joined the Department of Solid State and Low Temperature Research under the direction of C. C. Grimes, and began purchasing the equipment I would need to continue what I by then knew were studies of superfluidity in ^{3}He. Some instrumentation was even purchased before I arrived in New Jersey. Yet I knew it would take at least a year to set up my laboratory, and I feared that most of the important pioneering work would be done before my own lab became operational.

I was surprised to find that by the time my laboratory did become operational, few of the studies that interested me had been done. Indeed, there seemed to be some question as to whether or not these new phases were all p-wave BCS states. In addition, theorists Phil Anderson and Bill Brinkman at Bell Labs had become interested in the theory of superfluid ^{3}He. This set the stage for what was to be an extremely productive period in my career. Over a five year period, beginning in 1973, we measured many of the important characteristics of the superfluid phases which helped identify the microscopic states involved. We found the superfluid phases to be almost unbelievably complex, and at the same time extremely well described by the BCS theory and extensions to that theory developed during that period.

In about 1977 I began to feel pressure from Bell Laboratories management to go on to study other physical systems. I decided to study solid ^{3}He, my original thesis topic, and at the same time Gerry Dolan and I began a modest program to test some of the ideas that David Thouless had discussed on electron localization in disordered one dimensional systems. This latter study had to fit within the extremely slow time scale of the solid ^{3}He work. By late 1979, both of these efforts had succeeded beyond my wildest expectations. We discovered antiferromagnet resonance in nuclear spin ordered solid ^{3}He samples which we grew from the superfluid phase directly into the spin-ordered solid phase. At the same time, the low temperature group at the University of Florida also discovered these resonances, but because we cooled our samples by adiabatic nuclear demagnetization of copper rather than Pomeranchuk cooling, only we were able to form and study single crystals, and could thus identify the allowed magnetic domain orientations. In the end, Mike Cross, Daniel Fisher and I were able to determine the symmetry of the magnetic sublattice structure, and correctly guessed the precise ordered structure, later confirmed by polarized neutron scattering. The frequency shifts resulting from this antiferromagnetic resonance have made solid ^{3}He an extremely useful model magnetic system, and to understand them theoretically, we had borrowed some of the same formalism which Leggett used to understand the frequency shifts in superfluid ^{3}He.

At almost the same time that Cross, Fisher and I made our breakthrough in our solid ^{3}He studies, Dolan and I discovered the log(T) temperature dependence to the electrical resistivity in disordered 2D conductors which Phil Anderson and his 'gang of four' had just predicted would exist, as a result of what they termed 'weak localization'. I did not continue the work on weak localization, as I only had one cryostat, and to do so would have meant that I could not continue my studies on nuclear spin ordering in solid ^{3}He, since the two sets of experiments would have vastly different time scales. Somewhat ironically, I got a second cryostat two years later.

In 1987, after fifteen years, I left Bell Laboratories to accept a position at Stanford University. I had received informal offers of university positions periodically while at Bell Labs, but always found Bell to be the ideal place to do research. The combination of in-house support for basic science and first rate collaborators made Bell Labs unbeatable as an environment for doing research. However, my wife recognized in me a teacher waiting to be born. In addition, she was not happy with her job in New Jersey, and we agreed that she would apply for positions elsewhere. When she received offers from two biotech companies in California, Amgen and Genentech, I suggested that she accept the Genentech offer and that I would start talking to Stanford and U.C. Berkeley. Stanford, which has a small physics department, had just begun a search for a low temperature physicist. Ultimately, I received offers from both institutions, and chose Stanford because we liked the atmosphere better, and it was a better commute for Phyllis.

At Stanford my students and I have continued work on superfluid and solid ^{3}He, studying how the B superfluid phase is nucleated from the higher temperature A phase and diverse properties of magnetically ordered solid ^{3}He in two and three dimensions. In addition, we have developed a program to study the low temperature properties of amorphous solids. Our work has shown that interactions between active defects in these systems create a hole in the density of states vs. local field, just as is seen in spin-glasses. In amorphous materials, it may be possible to measure the size of coupled clusters of such defects, something which has been difficult in spin-glasses.

I have thoroughly enjoyed all aspects of university life, except for having to apply for research support. In particular, I have been fortunate to have had excellent graduate students, and to be able to teach bright undergraduates. Of course, with undergraduates one always has a few students who do not appreciate the professor's efforts. In 1988, after teaching my first large lecture course, one student wrote in his course evaluation: "Osheroff is a typical example of some lunkhead from industry who Stanford University hires for his expertise in some random field." Despite this minority opinion, in 1991 Stanford presented me their Gores Award for excellence in teaching. From 1993–1996 I served as Physics Department chair, and stepped down in September 1996, hoping to spend more time with my graduate students. The day I learned I was to receive the Nobel Prize, after just two and a half hours sleep the night before, I taught my class on the physics of photography, although the lecture was not on photographic lenses, but the discovery of superfluidity in ^{3}He.

SUPERFLUIDITY IN ^{3}HE: DISCOVERY AND UNDERSTANDING

Nobel Lecture, December 7, 1996

by

DOUGLAS D. OSHEROFF

Stanford University, Department of Physics, Stanford, California 94305-4060, USA

1. DISCOVERY

In starting to compose this lecture, I am reminded of the general excitement which permeated the field of low temperature physics in 1971. There were new cooling technologies being developed, and everyone felt that interesting and important new physics was just waiting to be discovered in the ultra-low temperature world to be made available for study by these techniques.

I had come to Cornell to do graduate work in what was then called solid state physics, but was quickly attracted to the promising field of low temperature physics by a talk which Bob Richardson gave in the fall of 1967, in which he described how dilution refrigerators work. At the time, I was working as a teaching assistant for Dave Lee. Dave seemed to think I was fairly bright, and invited me to join the group. By the end of my first year at Cornell, Jim Sites, a senior graduate student, and I had built the dilution refrigerator which would later be used in the discovery of superfluidity in ^{3}He.

A dilution refrigerator is a device which utilizes the non-vanishing (~6%) solubility of liquid ^{3}He in superfluid ^{4}He at low temperatures to effectively 'evaporate' liquid ^{3}He down to arbitrarily low temperatures. Because ^{3}He at mK temperatures is a degenerate Fermi fluid whose entropy is proportional to the temperature, such a device has a cooling capacity which decreases at least as fast as the square of the temperature. In those days, such refrigerators could only reach about 15 mK.

Sites intended to measure the magnetic susceptibility of solid ^{3}He down to as close to the nuclear spin ordering temperature as possible, expected to be at about 2 mK. The effective spin interactions in this system result from atom-atom exchange at rates as high as 40 MHz. By contrast, one calculates an atom-atom exchange rate for silicon which is less than one over the age of the universe! The dilution refrigerator was only the first step in Jim's new cooling process. The second refrigeration stage was to be through the adiabatic solidification of liquid ^{3}He, or Pomeranchuk refrigeration, named for the Russian theorist that had suggested the process[1)] in 1950, before anyone had even created liquid ^{3}He! Pomeranchuk reasoned that liquid ^{3}He would become a degenerate Fermi fluid at low temperatures owing to its half-integral spin. Thus at sufficiently low temperatures its entropy would depend linearly on temperature. The solid would possess an entropy dominated at low temperatures by the disordered nuclear spins, S_{solid}~Rln2, and hence at sufficiently low temperatures the liquid entropy would drop below that of the solid. In this rather unique situation, the latent heat of melting would be negative, and hence if one compressed to solidify part of a liquid sample at constant entropy, the liquid would have to cool. The actual cooling rate is about 1 mK for every percent of the liquid sample converted to solid. Pomeranchuk had argued that this process would continue down to the temperature at which the solid spin system ordered, making it an ideal vehicle for studying nuclear ordering in the solid.[2)]

The main problem with Pomeranchuk's proposal was that the work done on the system to solidify the liquid, $P_{melt}\bullet(V_{solid}-V_{liquid})$, exceeded the latent heat by two to three orders of magnitude near the solid ordering temperature. How reversible would the compression process be? Anufriyev in Russia had showed that the process worked[3)] in 1964, but he only reached a temperature of about 20 mK in his experiments. No one had demonstrated that this process could reach temperatures significantly below those available with a dilution refrigerator, much less down to the solid ordering temperature. Both John Wheatley's group at La Jolla (U.C. San Diego) and we at Cornell were betting on positive results. Despite these and other complications, Sites' experiment worked[4)] during my second year of graduate study, and Jim graduated in the summer of 1969, leaving his cryostat to me. Jim was the only member of the low temperature group to obtain a Ph.D. in just four years.

In my third year, I set about to improve the dilution refrigerator Sites and I had built, and also built a new Pomeranchuk cell based on a design I had developed the previous year, while recovering from knee surgery following a skiing accident. Sites had used a fairly complicated arrangement of three helium filled chambers and two metal bellows to decrease the volume of the ^{3}He region, necessary to compress and solidify the liquid. In his cell the convolutions of the bellows containing the ^{3}He closed as compression occurred. If solid formed in the convolutions, plastic deformation of that solid, and hence irreversible heating, would occur. My cell was simpler, relying on only two helium chambers with bellows in the form of a hydraulic press as shown in Figure 1. Superfluid ^{4}He was introduced under pressure to the large bellows assembly at the top, forcing the central piston downward, and extending the lower bellows into the ^{3}He filled region. A photograph of the upper bellows attached to the dilution refrigerator is shown in Figure 2.

Beginning my fourth year of graduate study, the new double refrigeration stage was operational, and I had also improved our NMR thermometry scheme. To utilize the system to study physics, something I was not yet very comfortable doing on my own, I teamed up with Linton Corruccini, a student in the group who was one year ahead of me. Linton wanted to test an unusual prediction by Leggett and Rice[5)] resulting from the effects of molecular fields in the collisionless regime arising from Fermi liquid behavior of ^{3}He, in which the spin diffusion coefficient in the liquid would vary with the angle by

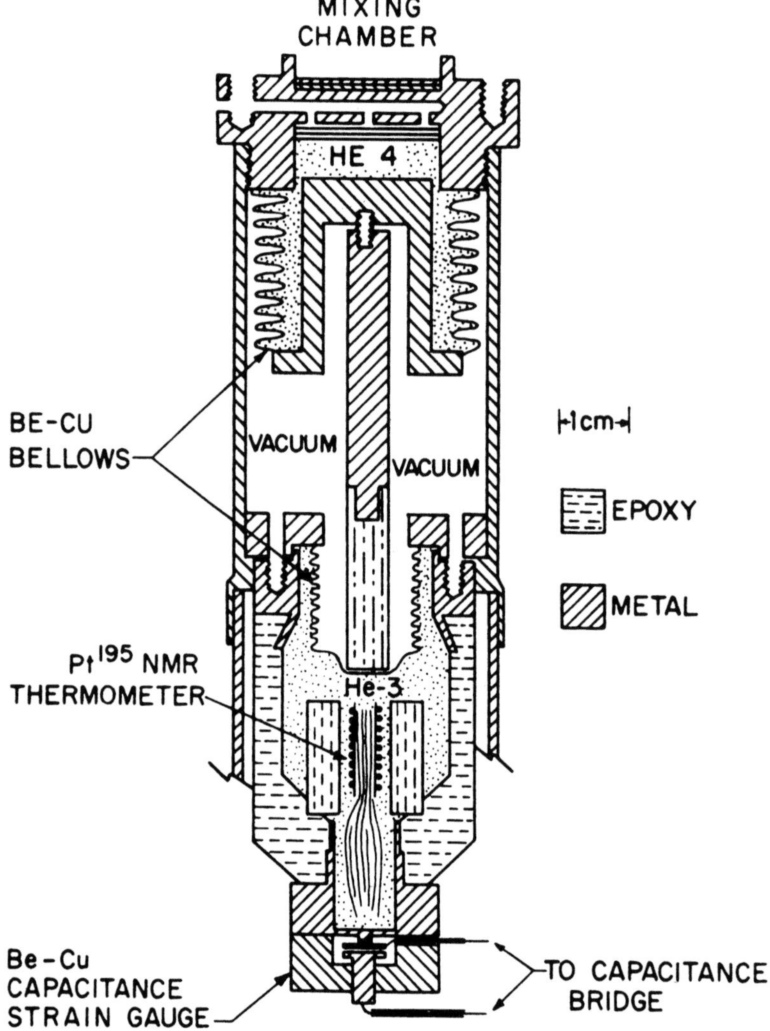

Figure 1: Schematic diagram of the Pomeranchuk cell used in the discovery of superfluidity in ^{3}He. Superfluid ^{4}He is added to the upper bellows assembly under pressure, forcing the lower bellows into the ^{3}He-filled region. The capacitive pressure transducer is at the bottom. (Reproduced with permission from reference 11).

which the spins where tipped in a pulsed NMR experiment. We used the Pomeranchuk cell to indirectly cool a separate sample of liquid ^{3}He down to about 6 mK. The experiment worked[6)], and we were able to estimate the Landau Fermi liquid parameter $F_1{}^a$ for the first time using the Leggett-Rice theory.

It was soon to be my fifth and presumably final year of graduate study. I had married Phyllis S. K. Liu in August of 1970, just two weeks after she successfully defended her Ph.D. thesis, and she was working as a postdoc in biochemistry waiting for me to complete my thesis. Dave Lee had wanted me to detect nuclear spin ordering in solid ^{3}He using my apparatus. I wasn't exactly sure how to do this, but before long Dave showed me a preprint of an article

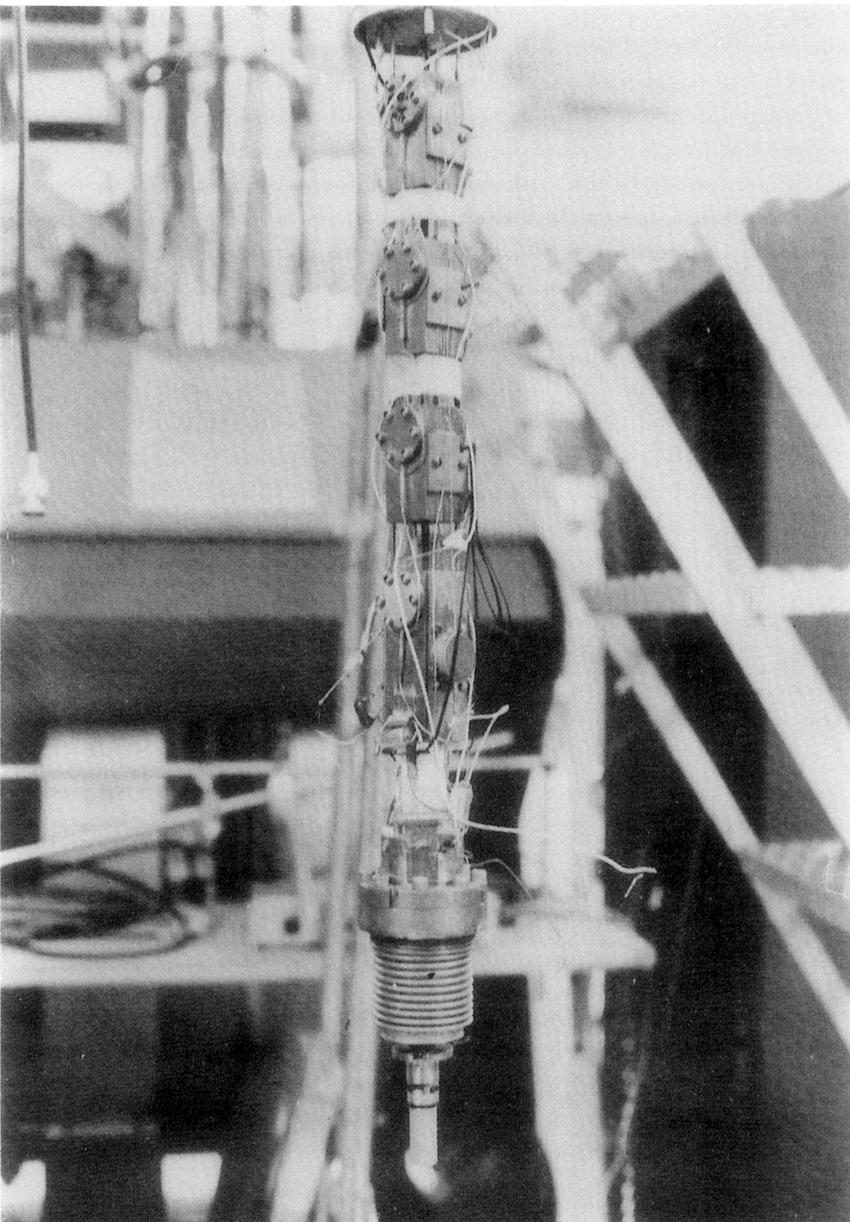

Figure 2: Photograph of the lower portion of the dilution refrigerator used in the discovery, showing the heat exchange column and, at the bottom, the mixing chamber with the ^{4}He bellows chamber and piston for the Pomeranchuk cell attached.

by John Wheatley's group[7)] at U.C. San Diego in which the depression of the ^{3}He melting pressure was measured as a function of temperature at several magnetic fields. This depression should be a direct measure of the difference in magnetization between the Fermi liquid and the solid. The group's results suggested an anomalously large magnetization in the solid at low magnetic fields. We decided we would check this unexpected result with our apparatus, which used a very different thermometry scheme than that of the La Jolla group.

Unfortunately, it turned out that the effect reported by the La Jolla group had actually been an artifact of that group's thermometry, and the degree of suppression which I found was small and difficult to measure[8)]. Despite the fact that this did not look like a good thesis experiment, I persisted doggedly. Finally Bill Tomlinson, a postdoc in the group, and Jim Kelly, another graduate student, argued that I had monopolized the lab's only NMR magnet long enough, and that it was their turn to use it. I reluctantly agreed to give up the magnet, but kept my apparatus cold in case their apparatus leaked, as often happened in those days.

While I was waiting for the verdict on their experiment, I decided to see just how low a temperature I could reach with my Pomeranchuk cell. We knew our copper wire NMR thermometer lost thermal contact with the liquid ^{3}He in the cell below about 2.7 mK, but I felt we could extrapolate our thermometry to lower temperatures using the expected slope of the ^{3}He melting curve, which had already been measured at La Jolla[9)] and by myself to below 3 mK. My experiment consisted of forming solid ^{3}He at a very steady rate, and plotting the melting pressure vs. time on a strip-chart recorder. It is important to note here that I was using a capacitive pressure transducer of the sort first developed by Straty and Adams[10)] at the University of Florida. In such a device the hydrostatic pressure flexes a thin metal diaphragm to which one plate of a parallel plate capacitor is attached, thus changing the capacitor gap, which was measured with an AC capacitance bridge. It had far better resolution than anything which had been available before. The parts of my pressure transducer are shown in Figure 3.

Figure 3: Photograph of the capacitive pressure transducer for the ^{3}He cell during assembly. The moving capacitor plate attached to the metal diaphragm is seen at the left, while the stationary plate is on the right.

The first such experiment was carried out on November 24, 1971, the day before the American Thanksgiving holiday. As I watched, the pressure rose steadily as the cell cooled. Suddenly, at a temperature I estimated to be about 2.6 mK, the rate of cooling abruptly dropped by about a factor of two. I guessed that this decrease in the cooling rate signaled the onset of heating due to the plastic deformation of solid ^{3}He by the moving bellows, and soon decided to terminate the compression. A portion of the resulting pressurization curve first showing this 'kink' is seen in Figure 4. The handwritten numbers in the figure were added four days later. After melting the solid in the cell by decompression, I decided to let the cell pre-cool to as low a temperature as could be reached with my dilution refrigerator over the entire four-day holiday, and then to try the experiment again on Monday. If I started at 15 mK rather than 20 mK, there would be 30% less solid in my cell at 2.6 mK than there had been in the compression on Nov. 24.

On that fateful Monday I got into the lab at about noon, ate a quick lunch as was my habit, and started the compression at about 12:35 pm. By 5:50 pm I neared the pressure at which the sudden decrease in cooling rate had been

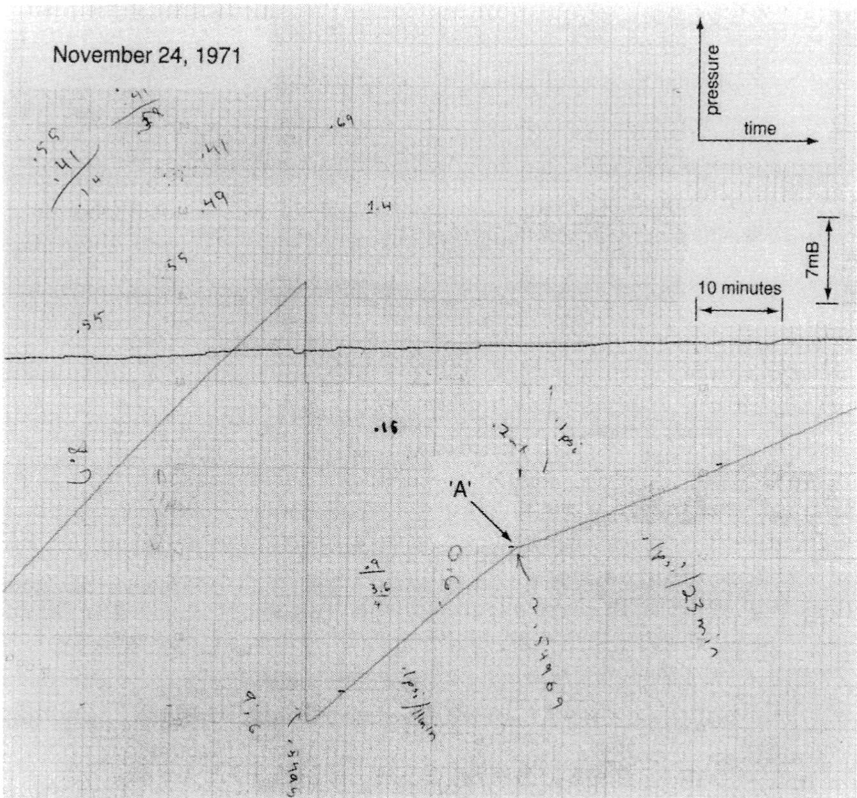

Figure 4: Pressurization curve taken Nov. 24, 1971, showing the first observation of the 'A' transition. Pressure increases vertically while time increases to the right. The abrupt jumps in the pressurization curve occurred when the capacitance bridge was re-balanced. The ragged line is the temperature of the dilution refrigerator.

seen in the previous run. I did not expect the kink to occur at the same pressure, if at all. Nonetheless, I soon saw another kink in the pressurization curve, and could tell that it was close to the same pressure at which it had occurred before. My heart sank. I then made a careful determination of the pressures at which these 'glitches' had occurred, and found the two pressures were the same to within about one part in 50,000!

At this moment adrenaline began to flow through my veins, as I immediately recognized that the probability that plastic deformation would just begin in my cell at exactly the same pressure on successive compressions with very different starting conditions was vanishingly small. A more logical explanation for this coincidence was that this glitch signaled some highly reproducible phase transition in my cell. Had I managed to reach the temperature of the nuclear magnetic phase transition in solid ^{3}He? The temperature seemed too high. I then repeatedly compressed and de-compressed through the region of the glitch to insure that it was indeed repeatable, and to measure its pressure more accurately. The initial pressurization curve through the 'glitch' that day is shown in Figure 5. I then found Bob Richardson, and we discussed the possible nature of the new transition I had discovered. We agreed that if there was a first order transition in the solid in which the spin system lost per-

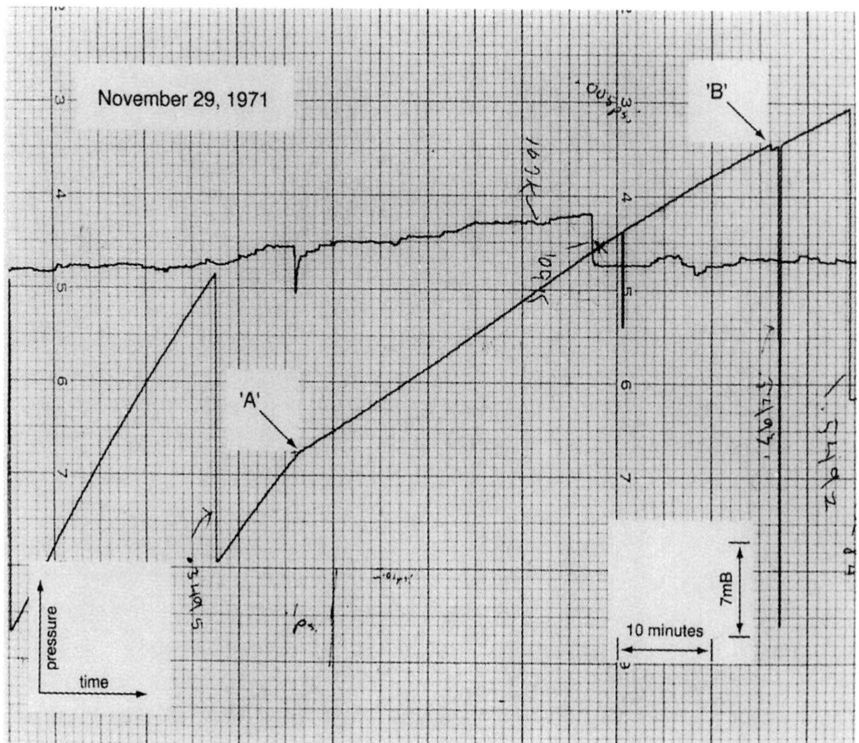

Figure 5: Pressurization curve taken Nov. 29, 1971, showing the second observation of the 'A' transition. Note that the 'A' features in this and Fig. 4 occur at the same pressure, and note the tiny but abrupt drop in pressure labeled 'B'. This was the first time the 'B' transition was ever observed.

haps 30% of its entropy, we could understand the change in slope of the pressurization curve. This discussion resulted in a possible magnetic phase diagram for the solid spin system being recorded in my lab book, which is shown in Figure 6. There was no discussion what-so-ever at that time of a possible phase transition in the liquid!

Note that at the far right in Figure 5 one can see a small abrupt drop in the pressure vs. time curve. I soon realized that this feature, which was always seen, but not at the same pressure, was also a transition. But this transition exhibited a substantial degree of supercooling. There was always associated with this tiny back-step a narrow plateau in the de-pressurization curve upon warming, and this plateau did occur at a highly reproducible pressure. This apparent transi-

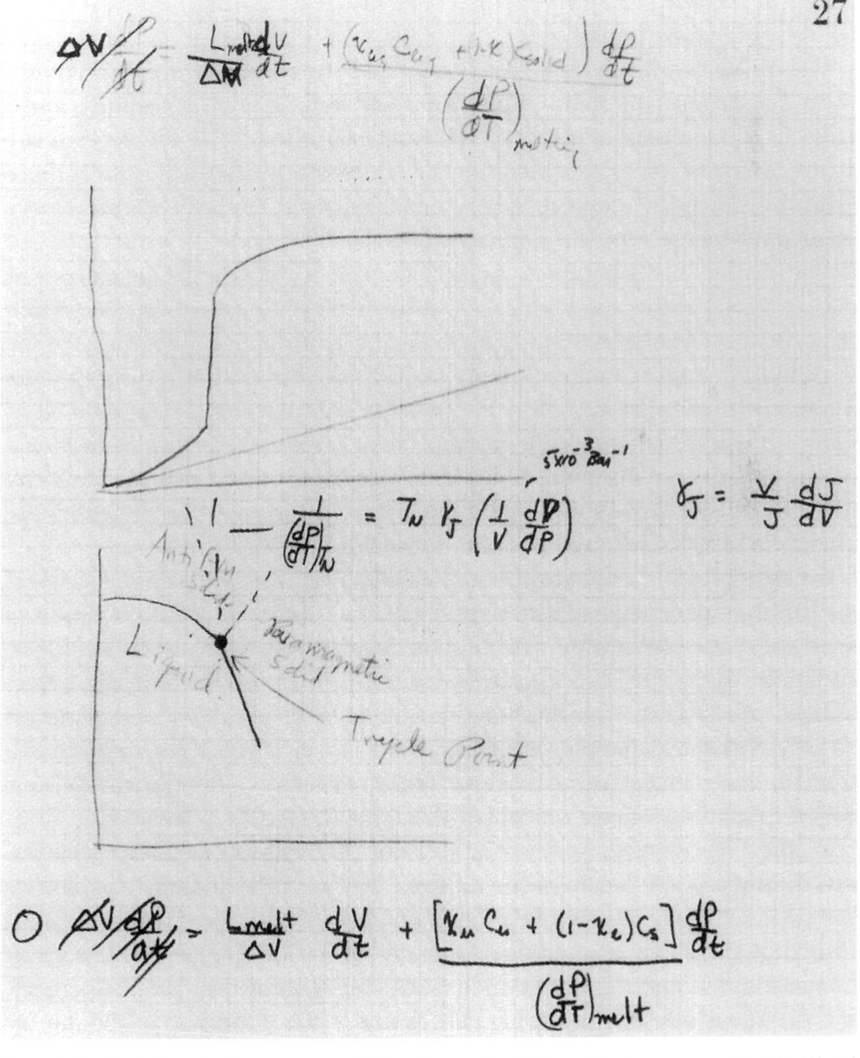

Figure 6: Result of discussion between me and Richardson on Nov. 29, in which it was assumed that the 'A' and 'B' features represented phase transitions in solid ^{3}He.

tion seemed far more difficult to explain by our model of the solid spin system. While we had originally called the higher temperature transition a 'glitch' and this lower temperature transition the 'glitch prime', we soon renamed the 'glitch' the A transition, and the 'glitch prime' the B transition. Those designations remain to this day, and as we shall see, they were actually quite prophetic of the microscopic identities of the two superfluid phases.

We soon found that if we continued compressing after the A and B transitions, the pressure would continue to rise to a maximum value, though slowly, and that the total change in pressure from the pressure of the A transition exceeded the change possible if the A transition was the 'expected' solid spin ordering transition, or even if it was only the 'expected' transition at 2 mK. The solid seemed to hold onto its entropy to too low a temperature. To kill two birds with one stone, we assumed that at the A transition something happened in the solid that prevented further nuclear ordering. Perhaps a crystallographic phase transition.

The above revelations completely changed the course of my Ph.D. thesis. Soon Tomlinson and Kelly completed their experiment successfully, and I again took possession of the lab NMR magnet on Dec. 2. We set about to study the effects of a magnetic field on the A and B transitions, for surely there would be effects if these were magnetic transitions in the solid. Perhaps more importantly at the time, we hoped that the field dependence would rule out the possibility that the transitions we observed could actually be artifacts of the capacitive pressure transducer. This was indeed the case, as we found that the pressures of both transitions decreased by an amount proportional to the square of the applied magnetic field. In addition, however, we found that the A transition split ever so slightly into two transitions upon the application of a large magnetic field, with the splitting being linear in field and about 60 microkelvin wide in a field of 1 T. We also were able to show that the temperature in our cell, as indicated by our NMR thermometer, also reflected the change in pressurization seen at the A transition.

On Dec. 17, 1971, I warmed up to 1 K and moved the ^{3}He pressure, now off the melting curve, through the region where we had seen the A transition to see if the A transition could still, despite the magnetic field results, be due to an artifact of the pressure measurement. The results were reassuring. I then warmed the cryostat to room temperature in order to add a new platinum NMR thermometer, which we hoped would stay in thermal contact with the liquid to lower temperatures and give better resolution. This began a difficult period in the experiment which was to last for almost three months.

I cooled the cryostat back down on Dec. 21. By Dec. 24 I had gotten the new platinum NMR thermometer working and spent the next few days testing for thermal contact to the ^{3}He bath and for rf eddy current heating. By Christmas Eve I had reached the A transition, but left the lab early, at about 10 pm. I had hoped to use the same NMR coil to study the ^{3}He NMR signal, but the signal was very small. I finally found a ^{3}He NMR line a 1:55 am on New Years night. It seemed pretty useless. I left the lab at about 3:00 am, and when I came in the next afternoon someone had written "Happy New Year,

Doug" in the lab book. Two days later we compressed and found that very little solid ^{3}He formed between the platinum wires of the NMR thermometer. We could not use a single NMR coil to measure temperature and simultaneously study the ^{3}He NMR signal!

On January 5 I flew to New Jersey to interview for a job at Bell Laboratories, fully believing that the A and B transitions were in the solid nuclear spin system. Luckily, no one argued those identifications, and within two months I was offered a permanent job as a regular member of their technical staff. At this point we decided to write up our results, without the NMR confirmation of our interpretation, and submitted to Physical Review Letters a paper entitled "Evidence for a New Phase of Solid ^{3}He". This paper sailed through the peer review process and was quickly published[11)].

On Jan. 21 I warmed up again to replace the ^{3}He cell with one which had a separate ^{3}He NMR coil. We had tried a new epoxy resin for this cell, and it cracked as we cooled down. I then made another cell with the more traditional 'Stycast 1266' resin, and replaced our beryllium copper pressure transducer with one made of 304 stainless steel, to further ensure that the effects we had seen were not simply an artifact of our strain gauge.

I cooled down again on Feb. 10, but started having problems with the dilution refrigerator. Later it became clear that corrosion due to the solder flux we used had nearly occluded the fine capillary tubes which connected the heat exchangers together. I also had troubles with the NMR electronics. Finally, on Feb. 18 (my wife's birthday) at 11:58 pm I wrote in the lab book: "I've got the ^{3}He NMR line–it's a big mother!!" (I didn't think back in those days that these volumes might have historical significance!) This euphoria was short lived, however, as two days later, Cornell University lost electrical power, allegedly due to a squirrel shorting out a high voltage power line, and the tube which pumped on our ^{4}He pot became blocked with solid air. I had to warm up to 77 K to eliminate the block, but when I again transferred liquid helium into the dewar, the cryostat had sprung a leak in the main vacuum flange which admitted the pumping lines to the experimental vacuum space. The leak defied my every effort to eliminate it, and in late February I began to machine the parts necessary to replace the main vacuum flange, a very difficult and risky procedure.

At this point Willie Gully, a second year student whom Dave Lee had assigned to work with me, asked if he could try his hand at fixing the leak. Up to this point I had made every effort to insure that Willie's 'uneducated' hands stayed off my precious cryostat. (This unfortunate attitude persists amongst my own senior graduate students to this day.) Since I had already written off the vacuum flange, however, I told Willie to go ahead. I had already tried the old Wheatley trick of applying a warm mixture of glycerin and soap (Ivory Flakes) to the offending joint, but Willie used a different mixture, and poured a rather large beaker of warm glycerin and soap over the entire vacuum flange, to a depth of about 1 cm as I recall. When we cooled back down on Feb. 29, the leak had vanished, never to return. From that day forward, Willie was a full collaborator in the project.

I took NMR data throughout the first two weeks in March, recording the NMR peak height as a function of time as we cooled and warmed through the A transition. In the ^{3}He cell the liquid contribution to the NMR signal was small, and expected to be independent of temperature, while the solid signal rose as one cooled almost as $1/T$. The stainless steel pressure transducer produced a fairly severe magnet field gradient, but we were clearly able to show that the growth rate of the ^{3}He NMR peak, presumably due to the increasing solid signal, increased slightly just as we cooled through the A transition; every time. The change was not large, but consistently present and highly correlated with the kink in the pressurization curve. Much later we were to realize that this increase in growth rate occurred because the liquid NMR frequency shifted as we cooled through the A transition, and superimposed itself on top of the solid NMR signal.

By this time we were becoming increasingly aware of the possibility that the A transition might be in the liquid, and I considered the possibility that a change in the thermal conductivity of the liquid at the A transition could result in a change in the rate of solid growth in the region seen by the NMR coil, thus producing the small change in peak growth rate discussed above. To rule out this possibility, we decided we needed to have spatial resolution, and I began working on a fourth cell design. In this new cell, the ^{3}He NMR coil consisted of five separate solenoidal coils, each oriented vertically, over a total length of about 2.5 cm. The idea was to look at solid formed in each of the five coils. To differentiate the signals from the various coils, I put soft iron shims between the pole faces of the NMR magnet, so that the (horizontal) NMR field would be larger at the bottom of the cell than at the top. Then the NMR signal from each coil would occur at a different frequency. We cooled this cell down on March 27, and I saw the expected five NMR peaks on April 3, but warmed up once more, to install a much better, final modification to the ^{3}He cell. In reading the lab books 25 years later, I am amazed at how often I was willing to thermally cycle such a fragile piece of apparatus!

I had by this time spoken with Michael Fisher about how one would expect solid ^{3}He to grow in a Pomeranchuk cell. Almost everyone imagined solid nucleating spontaneously at the warmest point in the cell (since there the pressure would be above the melting pressure), resulting in ^{3}He snow. Michael gently told me about the surface energy which would exist at the solid-liquid interface, and how it would act to prevent spontaneous solid nucleation. He expected the solid to nucleate only on surfaces, and at only a few places in the cell, with all subsequent solid growing from these seeds. To take advantage of this growth characteristic, I once again changed the ^{3}He NMR coil, this time to a single vertical solenoid roughly 0.5 cm in diameter and 2.5 cm long. In this geometry, the field gradient would allow an NMR resonance to occur only in a thin horizontal layer, whose position could be moved continuously either up or down by sweeping the NMR frequency, allowing us to obtain a true 1D profile of the magnetization in the NMR region.

On April 7, I once again started putting the cryostat back together, however at about midnight, as I tightened the bolts which held the liquid helium dewar in place, the dewar suddenly exploded, sending a shower of glass into the pit below. The only spare dewar we had was too short for my long inner vacuum can, and I was lucky to notice this before I broke it as well! After machining a spacer to lower the position of the helium dewar, I again put up the dewars, but then had problems with the pressure transducer.

I began to really feel the pressure at this point. Our Physical Review Letter had just come out, and people were beginning to criticize our interpretation that the transition in the solid was first order. In addition, John Goodkind and Victor Vvedensky had both suggested that the pressure signature of the A transition was consistent with a BCS transition in the liquid. While this interpretation could not explain the high melting pressures we had measured (which suggested more entropy in the solid at low temperatures than one would expect based on a phase transition at 2 mK), I felt we needed to make a definitive NMR test of our model, and soon!

I finally succeeded in cooling the cryostat back down on April 10, and on April 14 started a compression to see if Michael Fisher's ideas regarding solid growth were correct. I was nervous as I began to form solid. With great relief, we soon found that Michael was exactly right! We almost always saw only two or three solid growth sites within the NMR region, with an all-liquid signal between the solid peaks.

I studied the solid growth characteristics throughout the middle of April. Much to my relief, the increase in growth rate of the solid peak heights as we cooled through the A transition was equally evident in all of our now spatially differentiated solid peaks. In addition, however, there was a curious, though small, drop in the solid peak heights which correlated with the backstep in pressure seen at the B transition, typically 2–3% of the total peak height. In these runs, we would decrease the gain on our chart recorder to keep the ever-growing solid peak heights on scale, but on the night of April 20 my eyes were attracted to the tiny liquid signal between the solid peaks in the data of April 17, reproduced in Figure 7.

What I saw in the liquid at the B transition hit me like a bolt of lightning: While the solid NMR signals seemed to respond to the B transition with a very small fractional change in peak height, the liquid signal almost completely disappeared at this point! Figure 8 shows the entry in my lab book that night: "2:40 am: Have discovered the BCS transition in liquid ^{3}He tonight. The pressure phenomena associated with B and B′ are accompanied by changes in the ^{3}He susceptibility both on and off the peaks approximately equal to the entire liquid susceptibility." I checked all the other data I had taken, and then I looked around for someone with whom to share my good news. No one was anywhere to be found in the entire building. At 4:00 am, I decided to call Dave Lee and Bob Richardson, perhaps a risky move for any graduate student. Both agreed that the identification was a strong one, and at 6:00 am Dave called back for more details.

It was now almost the time of the April American Physical Society Meeting in Washington D.C., and Dave and Bob arranged for me to give a post-deadline invited talk on our work at that meeting. We still believed that the A tran-

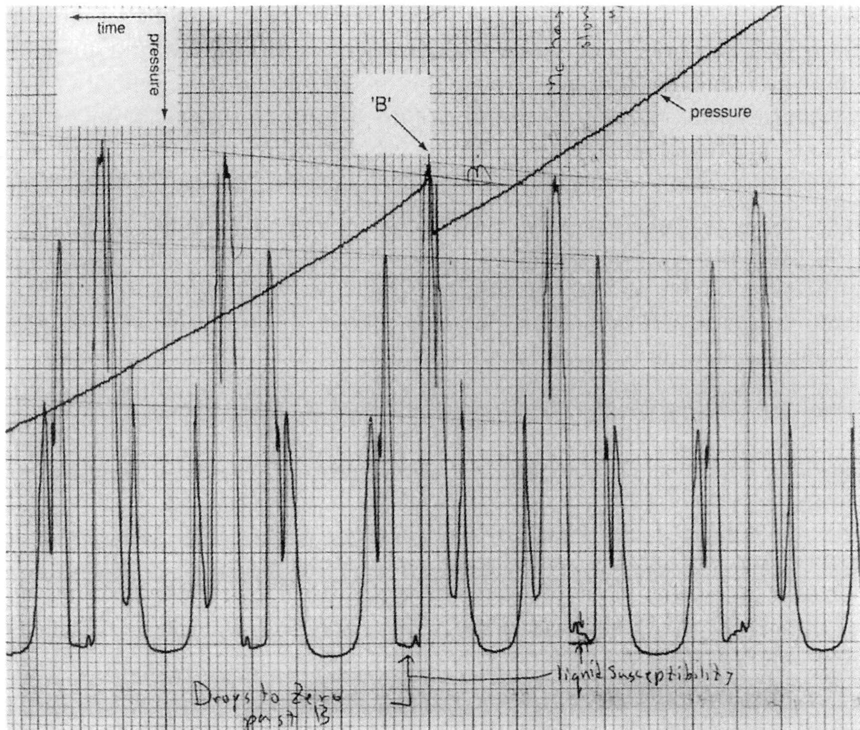

Figure 7. Continuous NMR sweeps of the cell showing behavior of the solid susceptibility (peaks) and liquid (region between two highest peaks) through the B transition. Note the subtle change in the liquid signal before and after the B transition.

sition was in the solid, so I reported that we had discovered transitions in both solid and liquid ^{3}He. When I returned from Washington, I began to study with renewed interest the supercooling properties of the B transition, and once again handed off the NMR magnet to Jim Kelly. In the meantime, I developed a multiplexer which would allow me to look at the NMR signal from two different places in the cell almost simultaneously.

When I got the magnet back in early May, I found using the multiplexer that the B transition occurred at the bottom of the cell first, and that the A-B interface moved upward at a velocity of many cm/sec. We also found a very strange magnetic signature as the A-B interface moved through the NMR resonance region: The NMR absorption signal would first rise *above* the level of the normal liquid signal, and then drop to well below that level. This effect seemed extremely difficult to understand, and we eventually tried to get the theorists down from the 5th floor in hopes of getting an explanation by offering beer and popcorn. They came, ate the popcorn, drank the beer; and then shook their heads and departed.

Beginning to feel the potential importance of these experiments, I decided to make motion pictures of the data we were obtaining in real time. Those reels of film still lie, largely unseen, in a chest in my laundry room at home. I

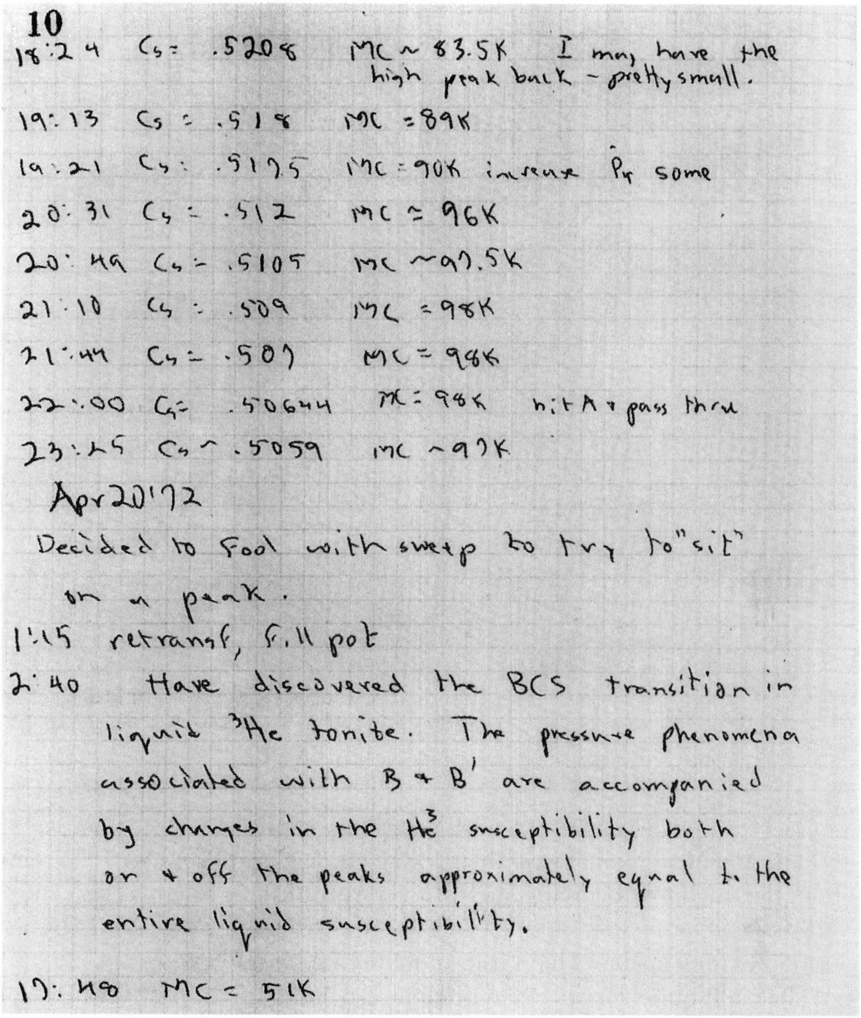

10

18:24 Cs = .5208 MC ~ 83.5K I may have the high peak back – pretty small.

19:13 Cs = .518 MC = 89K

19:21 Cs : .5175 MC = 90K increase Pr some

20:31 Cs = .512 MC ≃ 96K

20:49 Cs = .5105 MC ~97.5K

21:10 Cs = .509 MC = 98K

21:44 Cs = .507 MC = 98K

22:00 Cs = .50644 MC = 98K hit A + pass thru

23:45 Cs ~ .5059 MC ~97K

Apr 20 '72

Decided to fool with sweep to try to "sit" on a peak.

1:15 retransf, fill pot

2:40 Have discovered the BCS transition in liquid ^{3}He tonite. The pressure phenomena associated with B + B′ are accompanied by changes in the He^3 susceptibility both on + off the peaks approximately equal to the entire liquid susceptibility.

17:40 MC = 51K

Figure 8: Photograph from my lab book showing entry the night of April 20, 1972, when I realized the B transition was in the liquid.

also took a portrait of myself next to the cryostat. I reproduce that photograph in Figure 9. The haggard expression on my face was quite genuine.

The final revelation in our odyssey came sometime near the beginning of June. Curiously, nothing is mentioned in the lab book. Dave Lee encouraged me to remove the iron shims from between the NMR magnet pole faces to eliminate the gradient in the magnetic field. He wanted to test for an NMR frequency shift such as one could get in a magnetically ordered system. Indeed, we had already seen 'distortions' of the liquid ^{3}He NMR profile with the gradient applied. Both Dave and Bob were there as I cooled through the A transition. What we saw was almost too much for words: As we cooled below the A transition, a small satellite line shifted gradually to higher and higher frequencies above the larger solid peak. It resembled the all-liquid signal in both

Figure 9: Self-photograph of myself taken some time in April, 1972 with my left hand on the NMR magnet used in our work. The cryostat, suspended from above, is inside the glass dewar seen entering the magnet field region.

its shape and area. Then, just as the pressurization curve indicated the B transition, the satellite line disappeared! In Figure 10 I show roughly every third NMR trace as a function of time from the compression of June 13. The conclusion was inescapable: The A transition was also in the liquid.

Willie Gully and I spent most of June investigating this unprecedented frequency shift in the liquid. At a suggestion by Bob Silsbee, we found that it obeyed what Dave Lee termed a 'Pythagorean' relationship: $(\nu_{liquid})^2 - (\nu_{larmor})^2$ was found to be a function only of temperature. Here ν_{larmor} is the precession frequency of the spins in the normal state. This difference rose from zero at the A transition to about 10^{10} Hz^2 at the lowest temperatures attainable. Vinay Ambegaokar assured us that one could not get such a shift from any conventional BCS transition. He was right. On July 14 we terminated the run and I began writing my thesis rather feverishly as Willie Gully began to modify the Pomeranchuk cell to include a vibrating wire viscometer.

Aware of how important our new understanding of the A and B features were, particularly in light of our previous published erroneous interpretations, we quickly wrote up a new manuscript and submitted it to Physical Review Letters in early July. Having learned our lesson, and certainly not able to understand the frequency shifts we had discovered, we simply focused on presenting the data, and avoided ever suggesting that these might be superfluid transitions. Ironically, although the earlier manuscript had sailed

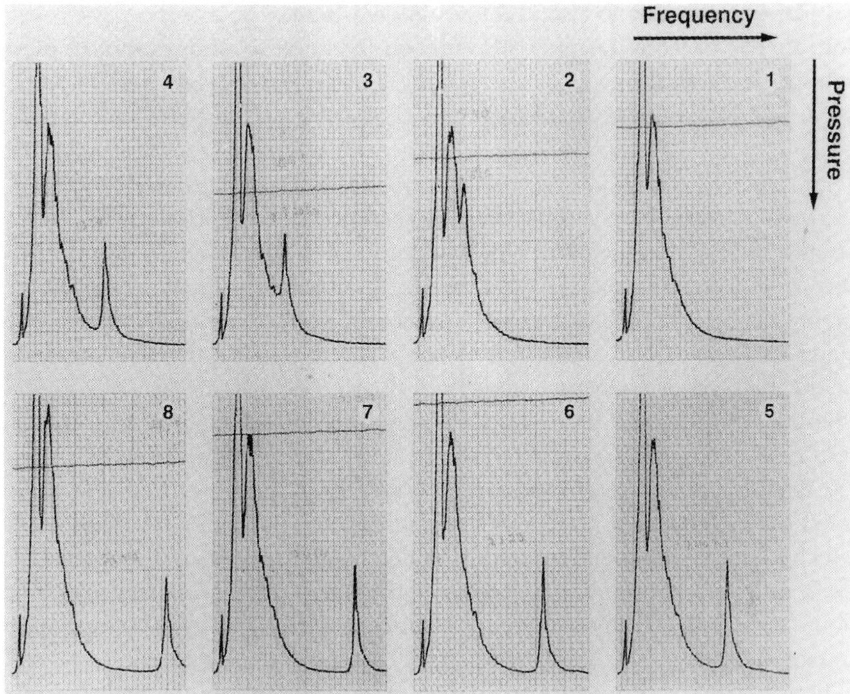

Figure 10: Sequence of NMR traces without applied magnetic field gradient as the temperature is slowly decreased below the A transition. As the liquid cools, a satellite NMR line can be seen to shift to higher frequencies. The nearly horizontal line in the traces is the cell pressure, increasing slowly from trace 1 to 8. Between traces 3 and 4 the capacitance bridge was rebalanced.

through the review process, this manuscript, in which we had much more to present, and had been very careful in doing so, was rejected. As the referee stated: "I read very carefully the previous Letter (PRL 28, 885 (1972)) and compared it with the contents of the present letter. Although the letter is clearly written, and, I presume, gives proper credit to others, I think the difference in the results is not large enough to warrant fast publication, in particular if one takes into account your rule against serial publication." Eventually, after both editors upheld the referee's conclusion, Jim Krumhansl, an associate editor of Physical Review Letters, interceded on our behalf, and the paper was published[12].

In August 1972 the 13th International Conference on Low Temperature Physics was held in Boulder, Colorado. I attended the conference before reporting for work at Bell Laboratories. David Lee presented our results in a plenary invited talk. John Wheatley, who was as fast to check on our results as we had been to check on his, also spoke with supporting evidence. But the most spectacular talk of the conference, for me at least, was one by Tony Leggett, read by his colleague Mike Richards. Tony showed how our NMR frequency shifts could be produced by a p-wave BCS state in the liquid. My own talk came on the last afternoon of the conference, and even I had to change my plane reservation to attend the session in what was largely an empty room! But, thanks to Tony Leggett, we were on the road to understanding these strange new fluids.

2. UNDERSTANDING

In the next three years almost every low temperature laboratory with the capability to reach the necessary low temperatures studied aspects of superfluidity in ^{3}He, but for much of this time the theorists were ahead of the experimentalists. Several questions had been raised: Were these really p-wave BCS states as Leggett had suggested? What was the pairing mechanism? How could there be two separate superfluid phases? What were the microscopic identities of the A and B phases? Did these BCS states actually support persistent mass currents? How well did the heat capacity agree with the BCS theory? What happened to ultra-sound propagation in the superfluid phases? In addition, Leggett soon predicted[13)] an entirely new 'longitudinal' NMR mode which would be independent of the applied magnetic field, and Ambegaokar, de Gennes and Rainer[14)] soon predicted that the A phase was likely to exhibit liquid crystal-like textures. Were these and other predictions to be supported by experiment?

Progress in providing answers to these questions was rapid, largely because so much theory and experiment existed from which to draw. In the 1960s John Wheatley and others had studied the normal state properties of liquid ^{3}He extensively[15)]. In 1961 Anderson and Morel[16)] had discussed a manifold of possible p-wave BCS states which had a curious property that the angular momenta of all the Cooper pairs pointed in a single direction in space, and formed only with parallel spin orientations. This state was ultimately identified with superfluid ^{3}He A. In 1963, Balian and Werthamer[17)] had shown that within the entire manifold of p-wave states, the state with the lowest free energy within the weak coupling BCS limit was a state in which the Cooper pairs formed with orbital angular momentum $L = 1$, spin angular momentum $S = 1$, but total angular momentum $J = 0$. A variation of this state was ultimately identified with superfluid ^{3}He B, and the formalism which Balian and Werthamer developed was adopted by everyone entering the field. In 1965 Leggett[18)] had studied the expected magnetic susceptibility of the nuclei in this phase, including Fermi liquid corrections. In 1971, Layzer and Fay[19)] had shown how ferromagnetic spin fluctuations in liquid ^{3}He could lead to an attractive interaction between quasiparticles for odd-l pairing, which they predicted would lead to superfluidity in a p-wave state. This paper provided the correct pairing mechanism for the p-wave superfluids as we understand it.

The question of how there could be two separate superfluid phases was answered very elegantly by Anderson and Brinkman[20)], who assumed that ferromagnetic spin fluctuations in the liquid produced the attractive interactions leading to Cooper pair formation, but noted that in ^{3}He, *but not in conventional superconductors*, the formation of the condensate wave function would modify this pairing interaction. The two concluded that the A phase should be an Anderson-Morel state which has been termed the Anderson-Brinkman-Morel, or ABM state; and that the B phase should be the Balian-Werthamer, or BW state.

At Bell Laboratories I was given an empty lab and enough money to fill it with whatever equipment I needed to continue my studies of the new phases of ^{3}He. In addition, I was allowed to hire Wolfgang Sprenger, an excellent technician who had been working for Robert Pohl at Cornell. However, the lab space I had been given was occupied by one of Bernd Matthias's people, who did not want to move out until his new lab was completed, and seemed in no rush to complete the design work. Matthias had so much influence at Bell Labs, even after leaving for U.C. San Diego, that there was nothing I could do but wait. Even so, by July 1973 my dilution refrigerator was installed and operational, thanks largely to Sprenger's efforts. By September I had reached the A transition in my new Pomeranchuk cell.

My main interest was in establishing the microscopic identities of the A and B phases. Anderson and Brinkman had suggested possible state identifications, but these identifications needed experimental verification. I also wanted to investigate the narrow region formed when the A transition split in the presence of an intense magnetic field, but needed a high-homogeneity superconducting magnet to do so. It would take one and a half years for this magnet to be delivered. There seemed only two ways for me to proceed: The first was to measure accurately the NMR susceptibility of the B phase and compare it with the 1965 Leggett prediction. The second was to search for the longitudinal NMR resonance which Leggett[13)] had predicted and compare it to the NMR shifts we had seen at Cornell in the A phase.

To understand the spin dynamics of p-wave superfluids, one must realize that the formation of Cooper pairs prevents the net nuclear dipole-dipole energy from averaging to zero, due to what Leggett termed a 'spontaneously broken spin-orbit symmetry'. To understand this, consider a fictitious Cooper pair in which the atoms actually orbit about one another as shown in Figure 11a. The dipole-dipole energy will be minimized over one period of the orbit when the projection of the total spin along the direction of the orbital angular momentum is zero. This spin direction is usually referred to as **d**, and in equilibrium **d** should point parallel to the orbital angular momentum of the Cooper pair. In the ABM state, the angular momenta of all the Cooper pairs are oriented in the same direction locally, a direction which we call **l**. Thus in the A phase one might expect that in equilibrium **d** would point parallel to **l**. Since the spin projection along **d** is zero, the Zeeman energy will be minimized when **d** is perpendicular to the static magnetic field, **B**. This forces both **d** and **l** to orient perpendicular to **B** in bulk A liquid[21)].

In an NMR experiment, one applies a magnetic field oscillating at the precession frequency of the nuclear spins in the static magnetic field, **B**, which will cause the magnetization to tip away from **B** and to precess about it. This disequilibrium perturbs the spin system, causing **d** to oscillate as shown in Figure 11b for weak oscillating fields. For conventional or 'transverse' NMR where the oscillating field lies in the plane normal to **B**, **d** oscillates in the

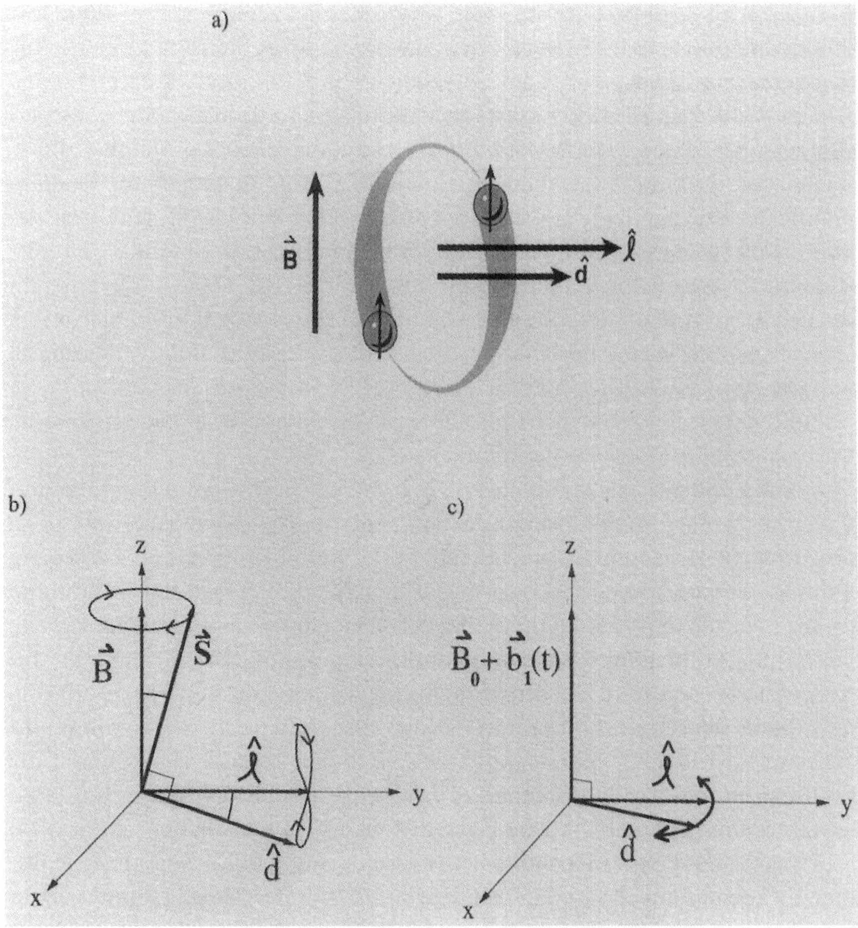

Figure 11: Superfluid A phase spin dynamics: a) A fictitious Cooper pair. The dipole energy of the two orbiting nuclei is minimized when the component of total spin, $m_s = 0$ along the orbital angular momentum, **l**. b) In transverse resonance, the magnetization, **S**, precesses about the magnetic field **B**, causing **d** to trace out a shallow figure '8' nearly in the **B**-**l** plane. c) When the oscillating field $\mathbf{b_1}(t)$ is parallel to the static field, the disequilibrium in the magnetization oscillates along the z-axis, causing **d** to oscillate back and forth in the x-y plane.

plane containing both **l** and **B**. However, if one applies an oscillating field along **B**, then **d** tends to rotate in the plane normal to **B** as shown in Figure 11c. This is what should happen in longitudinal resonance. When **d** is driven away from its equilibrium orientation, it exerts a torque on the precessing magnetization for transverse resonance, whose net effect is to produce the transverse resonance shift. For longitudinal resonance, the oscillating **d** causes the magnetization along **B** to oscillate in magnitude.

The longitudinal resonance was expected to be a difficult thing to find, for its frequency would be independent of any applied external magnetic field. Thus one could not search for it by sweeping the field. There was some speculation that the resonance might be extremely narrow. If the A phase were the ABM state, then $(\nu_{\text{longitudinal}})^2 = (\nu_{\text{liquid}})^2 - (\nu_{\text{larmor}})^2$, where ν_{liquid} is the conventional, or transverse, resonance we had measured in the

A phase. This suggested a value which varied from zero to about 100 kHz at the lowest temperatures. I searched hard and long for such a mode, but found nothing. In the end, I gave up the idea that the mode would be sharp, and assumed instead that it would be very broad. In this case, one could not detect it by sweeping the NMR frequency, since one would then sweep over the broad resonance of the NMR tank circuit. I decided to hold the frequency fixed and sweep the temperature, thereby sweeping the longitudinal resonance through my probe frequency. This strategy worked quite well, and the resonance was very broad. These results[22)] seemed to confirm that the A phase was indeed the ABM state, and soon the Cornell group[23)] joined in the study of this new resonance mode.

It turned out to be more difficult to identify the B phase with the Balian-Werthamer (BW) state. The B phase was a quasi-isotropic state consisting of $J=0$ Cooper pairs, except that as Leggett[21)] had showed, to minimize the nuclear dipole-dipole energy for this state, one had to rotate the spin coordinate system in which the pairs had been formed with respect to the orbital coordinate system about an arbitrary axis, **n**, by an angle of $\cos^{-1}(-1/4) \simeq 104°$. In this state, one third of the Cooper pairs should be in a $m_s = 0$ state with respect to the applied magnetic field, and ignoring interactions in the Fermi liquid, one would thus expect a susceptibility, χ_B, at $T = 0$ of 2/3 that of the normal state liquid. The Fermi liquid corrections which Leggett[18)] had made in 1965 reduced this value to only about 1/3 that of the normal state.

I went back and looked at the NMR absorption values I could glean from my old data, and concluded that χ_B was at least close to that which one would expect for the BW state. However, very soon Wheatley's group at U.C. San Diego (La Jolla) measured the B phase static magnetization using an RF SQUID magnetometer, and found that the drop in magnetization was considerably larger than expected for the BW state[24)]. Later, Lounasmaa's group in Helsinki reported[25)] an NMR measurement of χ_B which also showed a larger drop that expected. There were suggestions that perhaps the B phase might be a mixed p and d-wave state.

When I began to look at the B phase NMR susceptibility, I found a very strange thing: If I measured it in a low magnetic field, I would find a very low value relative to the normal phase, while if I measured it in a high magnetic field, I would find a value much closer to the Leggett prediction for the BW state. In addition, I found that there was always a high-frequency tail on the B phase NMR line. This is precisely the effect which plagued the early Helsinki measurements.

At this point I enlisted the help of Bill Brinkman in order to understand my strange observations. Within the Leggett theory, the NMR frequency in the BW state should depend upon how the rotation axis, **n**, is oriented with respect to **B**. There should be a shift proportional to $\sin^2(\theta)$, where θ is the angle between **B** and **n**. In the bulk, **n** should orient parallel to **B** to re-minimize the dipole energy as a result of depairing of the spin component for which $m_s = 0$ along the magnetic field[21)]. Thus, in the bulk, the BW phase should show no NMR frequency shift. However, if surfaces favored a different orientation, there could be a variation of orientation across the sample, which would result in a broad high frequency tail on the NMR line, with a resulting decrease in the spectral weight at the Larmor frequency. This might help explain my observations of NMR in the B phase. We were able to determine a lower bound for the B phase NMR susceptibility by going to high fields (about 0.1 T), however, and this limit was much closer to the predicted values for the BW state than to the La Jolla value based on static magnetization measurements[22)].

I and others were to make successively better and better measurements of the B phase NMR susceptibility[26, 27)], and Wheatley and others[28)] were to make better and better measurements of the static magnetization. They never came into agreement. While it is fairly universally agreed that the B phase is the BW state, and that the NMR susceptibility measurements support this, there seems to be a change in static magnetization at the A-B transition which exceeds the nuclear spin contribution, and which has never been understood.

By 1974 Pierre de Gennes[29)] had already described how one should carry the idea of Ginzburg-Landau bending energies over from liquid crystals to the ABM state to describe liquid-crystal-like textures in the A phase. Brinkman, Smith, and Blount[30)] produced a similar description for the BW state to try to understand my early NMR experiments in the B phase. They were able to show that within the cylindrical geometry used in my experiments, their theory fit the data, provided only that one assume that the equilibrium orientation of **n** near the cylindrical walls was not parallel to the magnetic field (which was oriented along the cylinder axis). This was a triumph for us, because it would be quite some time before these ideas could be tested in the A phase, owing to much stronger orienting effects in that phase.

Later experiments at Bell Labs and in Helsinki were to demonstrate the specific orienting effects by surfaces on the B phase textures[31,32)], and at Bell Labs we would observe spin-wave modes trapped in a B phase texture which existed between parallel surfaces[33)], while the Helsinki group would detect a spin mode associated with spin-waves trapped on quantized vortices in the A phase[34)]. The frequency shifts of the B liquid between closely spaced plates resulting from the surface orientation of **n** suggested B phase longitudinal resonance frequencies which agreed well with direct measurements of the B phase longitudinal resonant frequency[35)], which in turn agreed well with values calculated from the A phase longitudinal resonant frequencies using relationships due to Leggett.

During most of this period, the Helsinki group was the only group to be using copper adiabatic nuclear demagnetization for refrigeration. They were thus the only group which could observe superfluidity in ^{3}He at zero pressure, and were quick to do so, extending the phase diagram determined by the La Jolla group to zero pressure. The superfluid transition temperature diminished from about 2.5 mK at melting pressure to 0.9 mK at zero pressure, while the equilibrium B′ transition temperature rose nearly linearly with decreasing pressure, until below about 21 Bars only the B phase was stable in low magnetic fields. The La Jolla and Helsinki groups both made measurements of the heat capacity of the liquid through the superfluid transition, and found that the superfluid looked very much like a weak coupling BCS state at low pressures, but strong coupling effects at high pressures caused the gap in the A phase to open up faster than at lower pressures. In addition, the Helsinki group extended many of my NMR measurements along the melting curve to lower pressures[36)].

In late 1974 my high-homogeneity NMR magnet finally arrived, and I decided to look closely at the tiny region called the A_1 phase that existed between the split A transition in a magnetic field. It was presumed that in this region only one of the two spin species had undergone pairing, but it was not known which. When I had planned the development of my lab, I had been concerned that by the time I got running, the *only* phase left to study would be the A_1 phase. I wanted to measure the slope of the frequency shift vs. temperature in this phase, $d(\nu_{liquid}^2\text{-}\nu_{larmor}^2)/dT$, and compare it to the equivalent slope in the A phase at low fields. This ended up being a formidable task, however, because the maximum shift in the A_1 region turned out to be only about 3 Hz, independent of magnetic field. As usual, I was lucky, and found that owing to copper foils I had used to shield the oscillating NMR field from solid ^{3}He in my cell, a portion of my NMR line was extremely narrow, allowing frequency shifts to be measured to 0.1 Hz, even at an NMR frequency of 24 MHz. In the end, I found the ratio of the slopes to be 0.188, but I had no idea what this number meant.

I enlisted the help of Phil Anderson to interpret the A_1 phase data. He worked out the theory in a Sunday afternoon. His theory not only showed that the ratio stated above was consistent with the Anderson-Brinkman identifications of the A and B phases as the ABM and BW states, but correctly predicted the shape of the curves I had measured[37)]. Unfortunately, this combination of theory and measurement still did not indicate which spin species had undergone pairing in the A_1 phase, and while there is some evidence today that it is the spins parallel to the magnetic field, the evidence is not entirely convincing[38, 39)]. Ironically, the most important contribution from this early work on the A_1 phase was probably not in the detailed measurements, but in the simple fact that we actually saw a transverse shift at all. In 1975 David Mermin[40)] was to show that the most likely L = 3 (f-wave pairing) candidate for the A_1 phase would not have a transverse shift. This is probably still the best evidence that the A and B phases could not be f-wave states.

By mid-1974, there were at least four groups studying the new phases. Groups at Cornell, La Jolla and Argonne had found that they could excite normal modes of the superfluid order parameters with ultrasonic sound waves, and this technique provided another incisive probe of the superfluids. The Cornell group later found very interesting non-linear response to sound waves, while the La Jolla group had studied linear and non-linear 'ringing' of the magnetization in the superfluid phases when a small magnetic field was abruptly turned off.

In the summer of 1974 Linton Corruccini, with whom I had worked at Cornell as a graduate student, came to Bell Labs and we had a wonderfully productive summer. Henrik Smith was also there from Copenhagen, and he and Brinkman were working together on the theory of superfluid ^{3}He. Corruccini and I decided to try some pulsed NMR experiments, which had never been used to study the superfluid phases. It was necessary to completely shield the solid ^{3}He which formed as we compressed from the oscillating magnetic field, and we created a very clever design to accomplish this.

Corruccini and I first measured the B phase magnetic susceptibility with pulsed NMR[27)], and got extremely good agreement with my previous continuous wave NMR results. We then set out to study spin dynamics in which the magnetization was tipped far from equilibrium. Brinkman and Smith predicted how the A phase frequency shift should depend upon tipping angle, and we found their prediction to describe the behavior we observed remarkably well[41)]. However, when we asked them to predict what we would see in the B phase, they weren't even close. As we increased the angle by which we tipped the magnetization away from the static field, no frequency shift was observed at all, until we reached an angle of about $(104\pm1.5)^\circ$, at which point a rapidly increasing frequency shift was observed[42)]. This angle was remarkably close to the angle by which one had to rotate the spin coordinate system with respect to the orbital coordinate system to minimize the dipole energy in the BW state, and Bob Richardson was elated to see a direct measurement of this angle. Brinkman and Smith[43)] were able to determine what the spin system was doing in our experiments, and it became clear that we had indeed made a direct measurement of the rotation angle. But the two could never really understand why the spin system behaved as it did. This understanding was finally supplied by a Russian theorist Igor Fomin[44)]. Corruccini and I also made the first studies of longitudinal spin relaxation in both superfluid phases[45)]. In all, work that summer resulted in seven publications, including three Physical Review Letters.

In the course of our pulsed NMR experiments, Corruccini and I observed in the B phase that, for sufficiently large tipping angles, the tipped magnetization would precess coherently for times much longer than should be possible, based on the magnetic field gradients which existed across our sample. We called this behavior a 'zero k spin-wave' but never really understood it, and its existence was buried in conference proceedings[42,46)]. Years later, Bunkov and co-workers in Russia were to re-discover this phenomenon[47)], and Fomin explained the behavior[48)] as being due to spin supercurrents in the B liquid driven by gradients in the order parameter as the magnetization in the regions of field gradient wound up in a helix. These spin supercurrents effectively increased the tipping angle in regions of lower magnetic field. Once the tipping angle exceeded 104°, the NMR frequency in that portion of the sample would begin to increase. Once the frequency there became equal to that in regions of higher magnetic field, the spin supercurrents would cease, and the entire sample would precess in-phase. They named this beautifully orchestrated dance the 'homogeneously precessing domain', and used it to study many unusual phenomena in the B superfluid.

There is not enough space nor time to complete even this abbreviated account of our growing understanding of these remarkable fluids. I encourage the reader to learn more in a series of excellent review articles written during this period of discovery by some of its most important contributors.[49)] I will mention briefly in closing, however, what happened to the search for anti-ferromagnetism in solid ^{3}He; the topic which Dave Lee had assigned for my Ph.D. thesis.

In 1974 Bill Halperin[50)], one of Bob Richardson's graduate students, did use Pomeranchuk cooling to observe the drop in spin entropy which signaled nuclear ordering in solid ^{3}He, but not at 2 mK as had been expected based on measurements above T_C, but at about 1 mK. In addition, the transition was strongly first order, rather than second order as had been expected. Halperin developed a clever self-consistent technique for extending the melting curve measurements to below the solid ordering temperature, which did not rely on any secondary thermometer. Later, Dwight Adams' group at the University of Florida[51)] was to apply this technique in high magnetic fields, and in the process discovered that above about 0.4 T there was a second magnetically ordered solid phase with a much higher magnetization than the low field phase. The nature of the magnetic order in this high field phase was first guessed by theorists[52)], however the nature of the low field phase remained a mystery.

Finally, in 1979, I again began to think about the solid, as did Dwight Adams at the University of Florida. Borrowing and modifying a strategy which I had first heard mentioned by Bill Halperin, and using copper nuclear demagnetization for refrigeration, I learned to grow single crystals of solid ^{3}He directly from the superfluid into the magnetically ordered solid phase. Magnetic resonance experiments at both Florida[53)] and Bell Labs[54)] then showed a rich NMR spectrum in the low field ordered phase, immediately showing that the sublattice structure had a symmetry lower than cubic. Mike Cross, Daniel Fisher and I were able to determine the symmetry of the ordered state from our NMR spectra, and to ultimately guess the exact sublattice structure. We found that the low field phase consisted of ferromagnetic planes of spins normal to any one of the principal lattice directions, with the spin orientations in these planes alternating two planes up, then two planes down, etc.[55)] This work was done at about the time that the movie 'Star Wars' was being shown, and in honor of the robot R2D2 in that movie, we named the ordered phase 'U2D2' meaning 'up 2, down 2'.

In solid ^{3}He the atom-atom exchange energy, which produces the spin order, is nearly four orders of magnitude larger than the direct nuclear magnetic dipole-dipole energy, and four orders of magnitude smaller than the characteristic lattice energy, the Debye energy. This has made these magnetic systems particularly convenient model systems for studying cooperative magnetic behavior[56)]. Without question we have been very fortunate that, within three thousandths of a degree of absolute zero, there exist a total of five beautifully ordered phases of solid and liquid ^{3}He, whose behavior continues to challenge our understanding and to provide a proving ground for new ideas.

REFERENCES

1. I. Pomeranchuk, Zh. Eksperim. i Theor. Fiz. **20**, 919 (1950).
2. See R.C. Richardson, *Les Prix Nobel*, this volume.
3. Y. D. Anufriyev, JETP Lett. **1**, 155 (1965).
4. J.R. Sites, D.D. Osheroff, R.C. Richardson and D.M. Lee, Phys. Rev. Lett. **23**, 836 (1969).
5. A.J. Leggett and M.J. Rice, Phys. Rev. Lett. **20**, 586 and **21**, 506 (1968) also A.J. Leggett, J. Phys. C **3**, 448 (1970).
6. L.R. Corruccini, D.D. Osheroff, D.M. Lee and R.C. Richardson, Phys. Rev. Lett. **27**, 650 (1971).
7. R.T. Johnson, R.E. Rapp, and J.C. Wheatley, J. Low Temp. Phys. **6**, 445 (1971).
8. D.D. Osheroff (unpublished) Ph.D. thesis, Cornell University, Ithaca, NY (1973).
9. R.T. Johnson, O.V. Lounasmaa, R. Rosenbaum, O.G. Symko and J.C. Wheatley, J. Low Temp. Phys. **2**, 403 (1970).
10. G.C. Straty and E.D. Adams, Rev. Sci. Instrum. **40**, 1393 (1969).
11. D.D. Osheroff, R.C. Richardson and D.M. Lee, Phys. Rev. Lett. **28**, 885 (1972).
12. D.D. Osheroff, W.J. Gully, R.C. Richardson and D.M. Lee, Phys. Rev. Lett. **29**, 920 (1972). My thanks to Gene Wells, the present editor of Physical Review Letters, for providing the text of the referee's report.
13. A.J. Leggett, Phys. Rev. Lett. **31**, 352 (1973).
14. V. Ambegaokar, P.G. de Gennes, and D. Rainer, Phys. Rev. A **9**, 2676 (1974).
15. See, for example, the London Prize lecture by J.C. Wheatley, Proceedings of the 14th International Conference on Low Temperature Physics, edited by Matti Krusius and Matti Vuorio (North Holland, Amsterdam), **5**, 6 (1975), and references therein.
16. P.W. Anderson and P. Morel, Phys. Rev. **123**, 1911 (1961).
17. R. Balian and N.R. Werthamer, Phys. Rev. **131**, 1553 (1963). There was an earlier determination of this state by Yu. Vdovin, however the formalism he used was not broadly applicable, and his results were not well known in the West: Yu. A. Vdovin in *Applications of Methods of Quantum Field Theory to Problems of Many Particles*, ed. A.I. Alekseyeva (Moscow: GOS ATOM ISDAT), 94 (1963). (In Russian)
18. A.J. Leggett, Phys. Rev. Lett, **14**, 536 (1965).
19. A. Layzer and D. Fay, Int. J. Mag. **1**, 135 (1971).
20. P.W. Anderson and W.F. Brinkman, Phys. Rev. Lett., **30**, 1108 (1973) and later: W.F. Brinkman, J. Serene and P.W. Anderson, Phys. Rev. A **10**, 2386 (1974).
21. A.J. Leggett, Annals of Phys. **85**, 11 (1974).
22. D.D. Osheroff and W.F. Brinkman, Phys. Rev. Lett. **32**, 584 (1974).
23. W. J. Gully, C. Gould, R.C. Richardson and D.M. Lee, J. Low Temp. Phys. **24**, 563 (1976).
24. D.N. Paulson, R.T. Johnson and J.C. Wheatley, Phys. Rev. Lett. **31**, 746 (1973).
25. A. I. Ahonen, T. A. Alvesalo, M.T. Haikala, M. Krusius, and M.A. Paalanen, Phys. Lett., **51A**, 279(1975)
26. A.I.Ahonen, M. Krusius and M.A. Paalanen, J. Low Temp. Phys., **25**, 421 (1976).
27. L.R. Corruccini and D.D. Osheroff, Phys. Rev. Lett., 34, 695 (1975).
28. Inseob Hahn, S.T.P. Boyd, H.M. Bozler, and C.M. Gould, J. Low Temp. Phys. **101**, 781 (1995) and references therein.
29. P.G. de Gennes, Phys. Lett., **44A**, 271 (1973).
30. W.F. Brinkman, H. Smith, D.D. Osheroff and E.I. Blount, Phys. Rev. Lett. **33**, 624 (1974).
31. D.D. Osheroff, S. Engelsberg, W.F. Brinkman and L.R. Corruccini, Phys. Rev. Lett. **34**, 190 (1975).
32. A.I. Ahonen, T.A. Alvesalo, M.T. Haikala, M. Krusius, and M.A. Paalanen, in J. Phys. C: Solid State Phys. **8**, L269 (1975)
33. D.D. Osheroff, W.van Roosbroeck, H. Smith and W. F. Brinkman, Phys. Rev. Lett. **38**, 134 (1977).

34. P.J. Hakonen, O.T. Ikkala and S.T. Islander, Phys. Rev. Lett. **48**, 1838 (1982).
35. D.D. Osheroff, Phys. Rev. Lett. **33**, 1009 (1974).
36. For a 1975 perspective see: O.V. Lounasmaa, Contemp. Phys. **15**, 353 (1974); J.C. Wheatley, Rev. Mod. Phys. **47**, 415 (1975); A.J. Leggett, Rev. Mod. Phys. **47**, 331 (1975). For a more current view of the field, see D. Vollhardt and P. Wölfle, *The Superfluid Phases of Helium 3* (Taylor and Francis, London, 1990).
37. D.D. Osheroff and P.W. Anderson, Phys. Rev. Lett. **33**, 686 (1974).
38. L.R. Corruccini and D.D. Osheroff, Phys. Rev. Lett. **45**, 2029 (1980).
39. R. Ruel and H. Kojima, Phys. Rev. BRC **28**, 6582 (1983) and Q. Jiang and H. Kojima, J. Low Temp. Phys. **88**, 317(1992).
40. N.D. Mermin, Phys. Rev. Lett. **34**, 1651 (1975), and N.D. Mermin, Phys. Rev. B **13**, 112(1976).
41. D.D. Osheroff and L.R. Corruccini, Phys. Lett., **A51**, 447 (1975).
42. D.D. Osheroff and L.R. Corruccini, Proceedings of the 14th International Conference on Low Temperature Physics, edited by Matti Krusius and Matti Vuorio (North Holland, Amsterdam), Vol.1, p.100 (1975).
43. W.F. Brinkman and H. Smith, Phys. Lett. **53A**, 43 (1975).
44. I.A. Fomin, J. Low Temp. Phys. 31, 509 (1978).
45. L.R. Corruccini and D.D. Osheroff, Phys. Rev. Lett. **34**, 564 (1975).
46. D.D. Osheroff, *Quantum Fluids and Solids*, edited by S.B. Trickey, E.D. Adams and J.W. Duffy, (Plenum Press, New York) p.161 (1977).
47. A.S. Borovik-Romanov, Yu. M. Bun'kov, V.V. Dmitriev, and Yu. M. Mikharskii, JETP Lett. **40**,1033 (1985).
48. I.A. Fomin, JETP Lett.**40**, 1037 (1985).
49. A.J. Leggett, Rev. Mod. Phys. **47**, 331 (1975); John Wheatley, Rev. Mod. Phys. **47**, 415 (1975); P.W. Anderson and W.F. Brinkman, *Physics of Liquid and Solid Helium Part II*, edited by K.H. Bennemann and J.B. Ketterson (John Wiley & Sons, New York)177(1978); David M. Lee and Robert C. Richardson, *Physics of Liquid and Solid Helium Part II*, edited by K.H. Bennemann and J.B. Ketterson (John Wiley & Sons, New York),287 (1978).
50. W.P. Halperin, C.N. Archie, F.B. Rasmussen, R.A. Buhrman, and R.C. Richardson, Phys. Rev. Lett. **32**, 927 (1974).
51. R.B. Kummer, E.D. Adams, W.P. Kirk, A.S. Greenberg, R.M. Mueller, C.V. Britton and D.M. Lee, Phys. Rev. Lett. **34**, 527 (1975).
52. J.M. Delrieu, M. Roger, and J.H. Hetherington, J. Phys. (Paris) **41**, C7-231 (1978).
53. E.D. Adams, E.A. Schubert, G.E. Haas, and D.M. Bakalyar, Phys. Rev. Lett. **44**, 789 (1980).
54. D.D. Osheroff, M.C. Cross and D.S. Fisher, Phys. Rev. Lett. **44**, 792 (1980).
55. For a reasonably up-to-date review article on the magnetic properties of solid ^{3}He see: D.D. Osheroff, "Magnetic Properties of Solid Helium Three", *J. Low Temp. Phys.*, **87**, 297 (1992).
56. Y.P. Feng, P. Schiffer, and D.D. Osheroff, Phys. Rev. B, **49**, 8790 (1994).

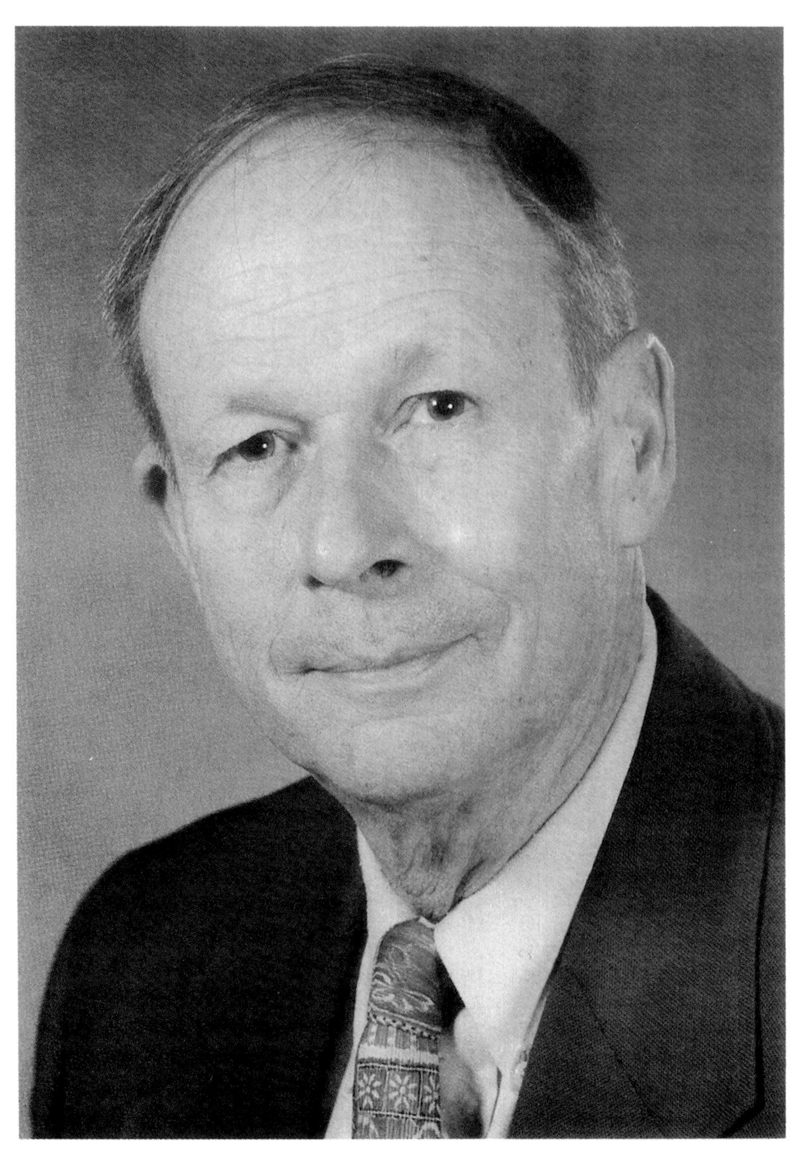

Robert C. Richardson

ROBERT C. RICHARDSON

I was born on June 26, 1937 in Georgetown University Hospital in Washington, DC. My parents, Lois Price Richardson and Robert Franklin Richardson, lived in Arlington, VA. My sister and only sibling, Addie Ann Richardson, was born on May 6, 1939, also in Georgetown University Hospital.

My earliest memories are of the apartment building in Arlington where my mother, sister, and I lived during the years of World War II while my father was away in the US Army. He was an officer in the Signal Corps. We lived across the street from the fire department and became accustomed to the blast of the siren at all hours of the day and night. It is fortunate that we lived so close to the fire department because one morning while my mother was visiting neighbors my sister set the apartment on fire while playing with the gas stove. Little damage was done, though I am certain that my mother was thoroughly embarrassed.

My father was a native Virginian. Branches of his family could be traced back to the early colonial times. His father, Robert Coleman Richardson, after whom I was named, owned a general store in a small rural village, Penola, VA. My father attended Roanoke College for two years during the Great Depression. When his mother became seriously ill, he left college because of the increased family expenses. He became interested in electricity and began work as a 'line man' for the Chesapeake and Potomac Telephone Company in Richmond, VA.

My mother's family was from North Carolina. She was an orphan, practically from birth, and was shuttled among relatives in North Carolina. As was a common practice in the rural South, she was taught at home by various aunts. She attended only one year of public school before going off to college. The one year of high school was in Reidsville, NC in 1918. She attended various colleges–Gulf Park College, the University of Alabama, the University of Mississippi, and the University of Virginia. She was one of the first women to attend the latter and obtained a Master's Degree in History there. During her college career she was brought in to the large and warm family of Ernest H. Mathewson in Richmond, and thus gained three brothers and two sisters. The Mathewsons were known by my sister and me as our other grandparents during our youth.

My parents met in Richmond and were married there in 1935. Shortly thereafter, my father was transferred by the telephone company to their branch in Washington, DC. As an army reservist my father was called to active duty during World War II and again during the Korean War. During his service for the latter he was assigned to the Pentagon so that it did not become necessary for him to leave home. During his second tour of duty with the army he took advantage of the educational benefits associated with the 'G.I. Bill of Rights' to finish college. He graduated from the University of Maryland in 1955.

I do not remember having any special scientific interests during childhood but I did love school. In 1946, when I was in the fourth grade, my family moved from the apartment building we had lived in during the war years. My father bought a new house in one of the housing tract developments so common to the post-war suburbs of American cities. We still lived in Arlington, VA. My new elementary school, Walter Reed, was overcrowded. The fourth and fifth grades met in the same room with the same teacher. I paid as much attention to the fifth grade instruction as the fourth. I especially loved the history lessons because Mrs. Walton, our teacher, was a remarkable storyteller. During the summer between fourth and fifth grade, I went to summer school just to have something to do. The teacher of the summer session was confused about my grade status and inadvertently promoted me to the sixth grade. The Arlington County School system accepted her decision. So I skipped a grade. Had I remained in the same grade, one of my classmates in Walter Reed School would have been Warren Beatty (of film star fame), whose family had just moved to our neighborhood in Arlington.

With my parent's encouragement, I became very active in the Boy Scouts. Scouting did not exist in rural Virginia, where my father grew up. In his youth, he had always envied boys from larger cities who could be in scouting. My involvement gave him, vicariously, the scouting experience he had missed. With his help, I became an Eagle Scout in the minimum amount of time permitted by the rules. I especially enjoyed the outdoor activities of scouting–hiking, camping, and even bird watching.

I spent the enjoyable summers of my high school years working as a counselor in Camp Letts, a Boy Scout Camp on the western shore of the Chesapeake Bay in Maryland. I was a nature counselor. I spent my days leading tours on nature trails through the camp. My ankles were covered with a minor poison ivy rash from June through August. In the evenings I led groups in 'star gazing;' and one morning each week I led a ten mile canoe trip through the Maryland marshland to look at birds. I liked the canoe trips best. We would arrive at the entrance of the marsh just at sunrise when the maximum number of birds would be out feeding. The marshes had large water birds like egrets and herons, three kinds of wrens, more than twenty different warblers, vireos, plus large birds of prey like hawks and eagles. It was possible in a single morning for a scout to spot enough birds on a single trip to qualify for the bird-watching merit badge. I learned where all of the birds hung out and how to tell them by their songs. Although I am color blind, I memorized their descriptions in the bird manual. I would describe subtle pastel features of warblers and vireos flitting about in the tree tops 60 feet above the ground to the amazement of even the adult scout leaders. There is a famous painting by James Audobon of a bald eagle diving toward an osprey just after the osprey has caught a fish. Each summer I was fortunate enough to see that scene re-enacted at least once. It made a special impression on the groups I led because I showed them a copy of the painting before we left on the trips.

My high school class at Washington-Lee High School had 925 students in it. I graduated, as I recall, in a six-way tie for 19th place. There was nothing exceptional about the math and science training at Washington-Lee. The idea of 'advanced placement' had not yet been invented. I did not take a calculus course until my second year of college. The biology and physics courses were very old fashioned. The idea of a 'photon' was said to be controversial. This in 1953! I was taught that absolute zero is the temperature at which all motion stops. It is most fortunate that the statement was wrong. Otherwise ^{3}He could not become a superfluid.

I entered Virginia Polytechnic Institute, also called Virginia Tech, in the Fall of 1954. In those days, the Reserve Officers Training Corps program was compulsory for all physically fit entering students at VPI. Moreover, all ROTC students lived in a cadet corps with fairly rigorous military discipline. I surprised myself by really enjoying life in the VPI Corps of Cadets. I learned an easy and democratic camaraderie. As we were assigned to live in cadet companies in alphabetical order, my closest friends were those in the bottom third of the alphabet.

In class, I started out as an electrical engineer but soon became bored and impatient with the mechanical drawing course and the rote application of a single principle, Kirchoff's Laws, in a five-hour course. I tried to become a chemistry major but ran into great difficulty in a course called quantitative analysis because of my color-blindness. I could not tell when the color of the indicator solution turned from pink to blue unless I made a very strong overconcentration of acid or base. When I complained to the professor he told me that I was very fortunate to discover my disability early in my college career because I certainly was not suited to be a chemist.

Finally, I turned to physics as a major. I was not an especially diligent student but nevertheless obtained a reasonable education in physics. I graduated with a B average and fourth in a group of about 9 physics majors. My education through the Cadet Corps was at least as valuable as that in formal class training. I was a leader in several campus organizations. The rigorous honor code at VPI in those days was almost exhilarating. We were all very proud of it. I never saw anyone cheat on a test in my years there.

In summers, while in college, I had a very interesting job with the National Bureau of Standards. I worked in the Electricity Division calibrating electrical resistance standards which power companies sent to NBS once each year. The NBS program for summer students was quite wonderful. First, we were well paid. Next, we actually did useful research. Finally, we attended a weekly seminar series which was given at our level of understanding. In my spare time at NBS, I read the scientific literature on electrical instrumentation and even met some of the authors of some of the classic articles. The experience at NBS gave me some notion of what a scientific research career could be.

After graduating from college, I had a vague idea of going to a graduate program in business–with hopes of becoming an executive in a large corporation. First, though, I felt that I had not quite given physics and research a chance so I decided to remain at VPI for one more year to obtain a Master's Degree before going off to military service as an Army Officer. The project I worked on was the measurement of the lifetime of photo-excited carriers in germanium. In the process I had to build a great deal of equipment because Tom Gilmer, my advisor, had just come to VPI to a practically empty lab. Tom was a good mentor, but he was very busy as department chairman and VPI professors had quite a large teaching load. I learned a great deal about how do things with my own hands–operate a lathe, solder, make simple electronic circuits, etc. I knew about keeping a lab book from my summer jobs at NBS. In that year I became a good deal more confident that I could learn physics at advanced levels, but still was not in any way special. I think I was still fourth in the group of graduate students. With the feeling that I would probably be a mediocre physicist, at best, I left VPI with the intent of attending a Masters in Business Administration, MBA, program after finishing military service.

A great piece of good fortune fell for me during my year of graduate work at VPI. The Army ran short of money. Thus, rather than having to spend two years on active duty, I was only assigned for six months of active duty in the US Army Ordnance Corps between November 1959 and May 1960. This was a time well after the Korean War and well before the Viet Nam War. There was no likelihood of actually having to see any combat. At Aberdeen Proving Ground, the Ordnance Corps training base, I took courses in how to manage a platoon which would do things like repairing jeeps and tanks. I hated the course and the being in the Army. Wearing a uniform and the military discipline did not bother me; I had become used to both while in the VPI Cadet Corps. But I did not enjoy the training in how to run a small business–for that's what a repair platoon in the Ordnance Corps was. Therefore, I decided to return to graduate school to obtain a Ph. D. in physics.

I had no opportunity while in the Army to take tests like the Graduate Record Exam to qualify me for admission to one of the top graduate schools –like MIT, Harvard, or Cornell. Besides, I probably would not have been admitted even if I had taken the tests. Therefore, I looked for smaller research universities with strong specialties. In my graduate research project, I had made a simple liquid nitrogen dewar, and found the area of low temperature physics to be interesting. I had read some articles about the work going on at Duke University so decided to apply there. I received a warm letter from Horst Meyer, a new Assistant Professor at Duke, encouraging me to come to work for him. The letter was very flattering–the first strong encouragement I had ever received about my potential as a physicist. Therefore, I entered Duke in the Fall of 1960 as a full-time graduate student.

I had a glorious time at Duke. I made strong friendships which have been maintained through the rest of my life. I met my wife, Betty McCarthy, there. One of only two physics majors in her class at Wellesley College, Betty was also a graduate student in Physics. We were married in 1962 and our daughters Jennifer and Pamela were born in Durham, NC in 1965 and 1966.

Horst was a very conscientious mentor. He taught me a great deal of the craft of low temperature technology he had learned as a research associate at the Clarendon Laboratory in Oxford. In all of the subsequent years he has been a valued friend. We had the best of two worlds in our low temperature group at Duke in those days. Bill Fairbank had been there but left before I arrived. Much of the old equipment and the residue of the experimental technology from Bill Fairbank remained. Horst brought a different set of techniques with him and we had our choice of which way to do things – for example the use of wood's metal to attach vacuum cans along with Epiezon J-oil for thermal contact were the Oxford technique. Indium O-rings and vacuum grease were the Fairbank method. Both had advantages.

Horst put me on a good problem–the NMR study of the exchange interaction in solid ^{3}He. Earle Hunt came to Duke as a research associate with Horst and taught me about the new methods for pulsed NMR–spin echos and all of that. The combination of training with Horst and Earle put me in business for practically the rest of my research career.

I finished my thesis in 1965 and remained at Duke for another year as a research associate in order to clean up some of the loose ends of the research and to look for a good job. In the latter, I was fortunate indeed. Cornell University, with its special funding as an Interdisciplinary Laboratory (IDL) had decided to expand its effort in low temperature physics. In the Spring of 1966 the Laboratory of Atomic and Solid State Physics invited me to join them to work with Dave Lee and John Reppy on very low temperature helium research. As far as I was concerned, there could be no better career opportunity.

I moved my family to Ithaca in October 1966 and have remained there ever since. I received sound career advice from Dave and John from the day I arrived. The research environment at Cornell has been superb with an unbroken string of talented graduate students, close colleagues in both theory and experiment, and a team of technical support specialists who helped make everything work. During my thirty years at Cornell I even learned how to teach undergraduate physics courses, an activity which my wife and I enjoy a great deal. After our daughters entered Junior High School, Betty turned to teaching physics at Cornell also. She is now a Senior Lecturer.

My children grew to adulthood in Ithaca. It is a wholesome college town with few of the problems of large cities. Jennifer went to college back at Duke and later attended a Master of Fine Arts in Creative Writing program at Columbia University. Jenny married James Merlis in June 1994. We had a beautiful wedding reception among my large rhododendron bushes in our back garden. In addition to her writing and other activities, she now plays violin in an all female rock band called Splendora.

Pamela went to college at Cornell. After graduation, she went to the New York School of Interior Design for a year and then decided to become a nurse. She returned home to take the science courses she had skipped at Cornell. She spent a year at our local community college taking chemistry, biology, anatomy, etc., displaying a surprising scientific talent. After the year at home she went to Vanderbilt University where she entered a Masters of

Nursing program. In November of 1994–after one year in the Vanderbilt nursing program–she died tragically, of heart failure. Though she had been born with a heart defect, her death came without warning.

In an effort to drag ourselves out of our grief and despondency over losing Pam, we have taken on a major family project in the past year: the production of an introductory college physics text book. Betty is the co-author of the book, with Alan Giambattista of Cornell; and I have been working on a companion CD ROM. When completed, the work will be published by McGraw Hill.

THE POMERANCHUK EFFECT

Nobel Lecture, December 7, 1996

by

ROBERT C. RICHARDSON

Cornell University, Department of Physics, Ithaca, New York 14853, USA

INTRODUCTION

A central part of the story of the discovery of superfluid ^{3}He is the cooling technique used for the experiments, the Pomeranchuk Effect. Although it is not an especially useful technique for obtaining low temperatures today, it contains my favorite example of the use of the Clausius-Clapeyron equation. The cooling technique is fun to describe in undergraduate physics classes on thermodynamics.

In 1950, I. Pomeranchuk, a well known particle theorist, suggested that melting ^{3}He could be cooled by squeezing it[1]. At the time of his suggestion ^{3}He was quite rare and had not yet even been liquefied. He observed that at low enough temperatures the thermal phenomena in condensed ^{3}He would be dominated by spin properties instead of phonon properties. The liquid of ^{3}He would obey Fermi statistics with an entropy proportional to the temperature, much like the free electrons in a good metal. On the other hand, the entropy of solid ^{3}He would be that of the disordered collection of weakly interacting spin 1/2 nuclei. At temperatures greater than the [then] expected nuclear magnetic ordering temperatures less than 1 μK, the entropy per mole of solid ^{3}He would be $S = R \log_e 2$, independent of temperature until the high temperature phonon modes of the solid become important. (The Debye temperature of solid ^{3}He is approximately 30 K.)

The idea of the method is represented in Figure 1. The entropy of solid ^{3}He exceeds that of liquid ^{3}He at temperatures less than 0.3 K. If the mixture is compressed without heat input it will cool as liquid is converted into solid.

DISCUSSION OF POMERANCHUK'S PROPOSAL

Fifteen years passed before anyone took up the suggested cooling technique[2]. There were several reasons. The most important was the availability of ^{3}He. It comes from tritium decay. Tritium was being produced for the most deadly part of the weapons industry. By 1965 copious quantities had been made. Low temperature physicists took advantage of the waste product of the arms race, the ^{3}He extracted from the gases prepared for hydrogen bombs.

The second reason for the late date of attempts at the method was the skep-

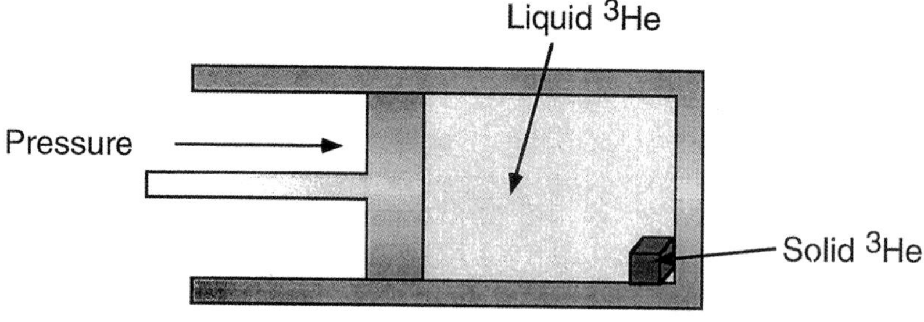

$$\frac{\Delta V}{V} = \frac{1.3}{24} = 5\% \quad \text{(Fraction volume change for complete conversion of liquid to solid)}$$

Figure 1. Pomeranchuk's suggestion for cooling a melting mixture of ^{3}He. The solid phase has a higher entropy than the liquid at low temperatures. As the liquid-solid mixture is compressed, heat is removed from the liquid phase as solid crystallites form. The fractional change of volume required to completely convert liquid into solid is approximately 5%. Unlike melting water, the solid phase forms at the *hottest* part of the container.

ticism of experimentalists about practical considerations. The entropies of the liquid and solid phases are illustrated in Figure 2. Liquid ^{3}He is rather accurately described by Landau's Fermi Liquid Theory[3,4]. At low temperatures[5,6] the entropy per mole of the liquid at melting pressure is approximately given by $S \approx 3RT$. The entropies of the liquid and solid phases are equal at 0.32 K. At lower temperatures Pomeranchuk's suggestion for cooling will work.

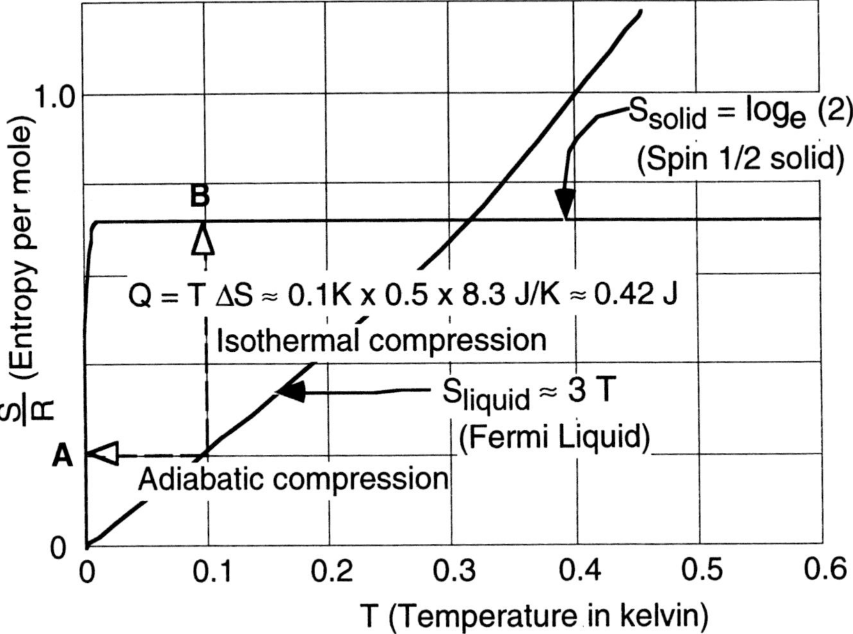

Figure 2. The entropies of liquid and solid ^{3}He. At $T < 0.32$ K, liquid ^{3}He has a lower entropy than the solid phase. The figure shows an example at $T = 0.1$ K. The latent heat associated with converting 1 mole of ^{3}He liquid into solid is 0.42 Joules, a substantial amount of heat removal at these low temperatures.

The adiabatic cooling path is indicated with the arrow **A** marked on the vertical axis. In the example, liquid compressed at an initial temperature of 0.1 K, with an entropy of 0.2R, will form a liquid-solid mixture which eventually cools to very low temperatures. The maximum amount of heat which can be removed is the latent heat of conversion of the liquid to solid and is indicated by the isothermal path, labelled **B**. The latent heat per mole at T = 0.1 K is 0.42 Joules.

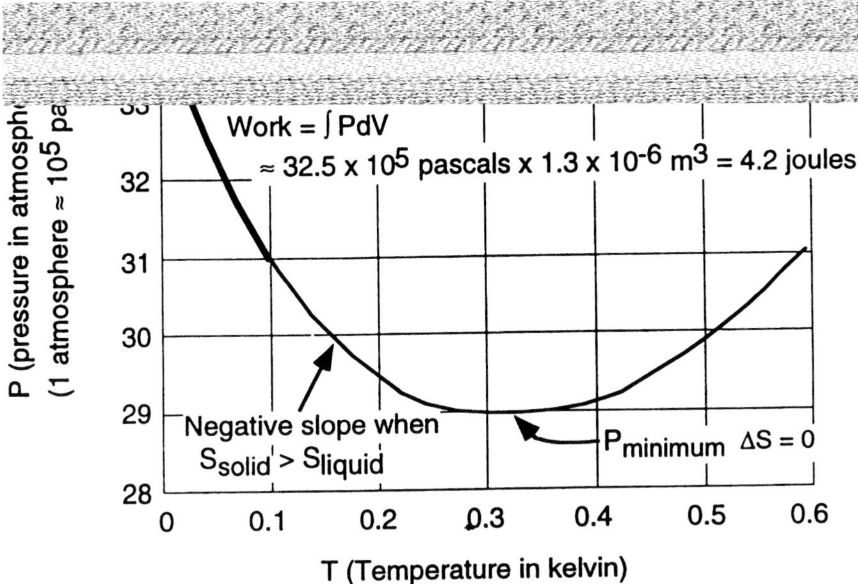

Figure 3. The melting pressure of ^{3}He. The figure can be constructed from Figure 2 through the Clausius-Clapeyron equation [see text]. The molar volume of liquid ^{3}He exceeds that of the solid by 1.3 cm^3 per mole. Thus, the slope of the melting curve is negative at temperatures less than 0.32 K. The work of compression in forming the solid is approximately 4.2 Joules, an order of magnitude larger than the heat which might be extracted.

The cooling effectiveness of the method must be compared with the possible heat losses in the process. At this point the natural skepticism of experimentalists arises. The amount of work involved during the compression is large. The melting pressure versus temperature of ^{3}He is illustrated in Figure 3. The melting curve may be calculated from the liquid and solid entropies using the Clausius-Clapeyron equation,

$$\left.\frac{dP}{dT}\right)_{melting} = \frac{S_{liquid} - S_{solid}}{V_{liquid} - V_{solid}}, \qquad [1]$$

where, $S(T)_{liquid}$ and $S(T)_{solid}$ are the molar entropies at melting. V_{liquid} and V_{solid} are the molar volumes of the two phases at melting. The difference, $V_{liquid}-V_{solid}$ is nearly independent of temperature and has the value 1.3 cm^3 per mole. We will return to Equation [1] later when I describe our experiments designed to measure the entropy of solid ^{3}He.

The work performed in converting the liquid to solid, starting at 0.1 K can be obtained from the integral, $\int PdV$ along the melting curve. It's value is approximately 4.2 Joules. An order of magnitude more work must be done than will be extracted during the cooling process. The ratio, W/Q, of work to heat extracted is near a minimum at the temperature illustrated. When the process is performed at lower starting temperatures, as was the usual practice, W/Q becomes larger than 100. The challenge of the experimental design, thus apparently became the avoidance of frictional heat losses during the compression process.

COMPRESSIONAL COOLING IN PRACTICE

In the Spring of 1966, David Lee invited me to join him at Cornell to begin experiments on the cooling of solid ^{3}He using the compressional cooling technique. The goal was to reach the temperature of the nuclear magnetic ordering transition in solid ^{3}He. My Ph. D. thesis at Duke University with Horst Meyer[7] had been concerned with NMR measurements of the size of the exchange interaction in solid ^{3}He. We knew from these measurements that the magnetic phase transition in solid ^{3}He at the melting pressure should occur at temperatures closer to 1 mK than 1 μK. Despite my certainty that Pomeranchuk's technique for cooling was probably doomed to failure, I was anxious to join Dave in searching for the transition. As a back-up we would attempt to cool solid ^{3}He with magnetic cooling schemes.[2] The latter had not yet been used successfully to obtain such low temperatures in liquid or solid helium but we began a parallel effort to use nuclear demagnetization[8].

By the time I arrived at Cornell in October 1966, the method had been successfully demonstrated by Anufriev in Moscow.[9] A cross section of Anufriev's apparatus is represented in Figure 4. The volume available for liquid ^{3}He is exaggerated in the figure. The ^{3}He space contained 30 cm^2 of metal foil to

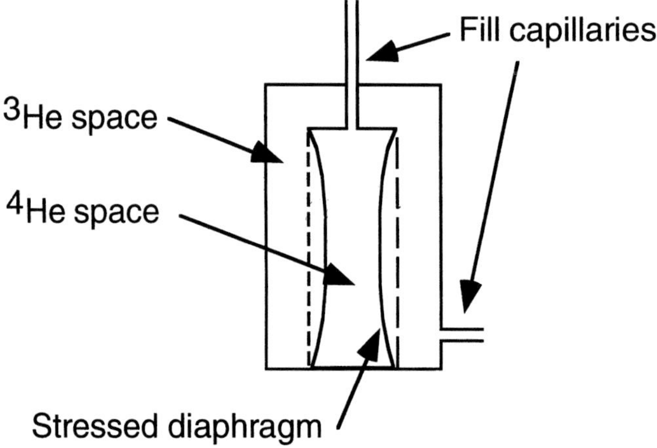

Figure 4. Cross-sectional representation of the compressional cooling cell used by Anufriev. Both inner and outer chambers have a rectangular shape. Thin stainless diaphragms of the inner cell were displaced inward with the application of ^{3}He at the melting pressure. The walls of the stressed diaphragm were then forced outward by filling the inner chamber with pressurized ^{4}He.

act as heat exchangers to cool the outside wall of the chamber. The outer chamber was first filled with liquid ^{3}He at the melting pressure at temperatures greater than 0.32 K.. As the cell was cooled using the demagnetization of a paramagnetic salt, a block formed in the fill capillary trapping a fixed quantity of ^{3}He in the cell. At low temperatures carefully cooled ^{4}He was admitted within the inner chamber to act as a hydraulic fluid. Pressurized liquid ^{4}He forced the diaphragm walls outward to decrease the volume available for the melting ^{3}He. The cell cooled to temperatures less than 15 mK, the minimum temperature which could be measured with the thermometers in contact with the exterior of the cell.

It is interesting to notice the use of liquid ^{4}He as the hydraulic fluid for forcing changes in the ^{3}He cell dimensions. At temperatures less than 0.3 K, the normal fluid fraction of ^{4}He liquid is very small. The heat capacity and thermal conductivity of liquid ^{4}He are negligible. Anufriev's pioneering experiment tested the method, showed that it would work, and demonstrated the use of liquid ^{4}He as a hydraulic fluid at low temperatures. The fears of excessive frictional heating associated with the movement of a metal diaphragm were unfounded. We now know that practically all metals have a very high quality factor at temperatures less than 4 K. In the subsequent years, *every* apparatus built upon the principle of forcing a metal diaphragm or bellows to move produced successful cooling. There is an important caveat in using liquid ^{4}He. The melting pressure of ^{4}He is 4 atmospheres less than that of ^{3}He. Spring tension or pressure amplifiers must be used in the experimental design so that ^{4}He does not solidify during the compression process.

POMERANCHUK CELLS AT CORNELL

The Cooling Cell of Jim Sites

Our first venture with Pomeranchuk cooling came with the thesis project of Jim Sites. The goal of his experiment was to measure the magnetic susceptibility of melting solid ^{3}He at temperatures near or below the nuclear magnetic phase transition. We had long discussions about the design of an apparatus. I have often regretted that we did not follow one of Dave Lee's original suggestions, the use of a weight to compress a bellows filled with melting ^{3}He. A heavy mass would be suspended by a wire and slowly lowered on the bellows in a controlled manner. The mass would have to be a metal of high density with negligible magnetic properties. Gold seemed to be the only really suitable metal. Imagine the profit if we had purchased 5 kilograms of gold at the 1967 price, $30 per troy ounce!

Before we completed our first experiments at Cornell the Wheatley group in La Jolla reported highly successful compressional cooling of ^{3}He down to temperatures less than 2 mK. [10, 11] Their design employed a tube of ^{3}He with an ellipsoidal cross section. The tube was surrounded by pressurized liquid ^{4}He. The entire cooling assembly was placed within the mixing chamber of a dilution refrigerator[2] and pre-cooled, over a matter of days, to 24 mK.

The design first used at Cornell was one suggested by our colleague John

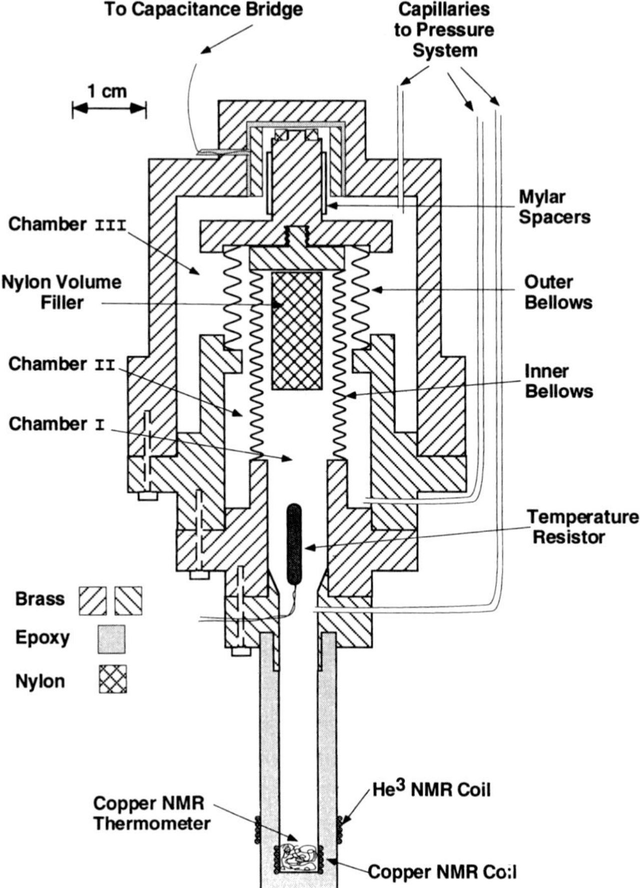

Figure 5. The Pomeranchuk cooling cell of Jim Sites.

Reppy. It is illustrated in Figure 5. A simplified schematic of the apparatus is shown in Figure 6. The cell contained two concentric beryllium-copper bellows and three helium chambers.[12, 13] The ^{3}He sample was contained in the innermost chamber, Chamber I, and ^{4}He in the outer two chambers. Initially all three chambers were pressurized to approximate the melting curves of their contents, and the sample cell was precooled with a dilution refrigerator to 25 mK. Compression of the ^{3}He was achieved by releasing the pressure on the ^{4}He in Chamber II. We had thought that there might be some advantage in removing fluid from the cell. The position of the bellows assembly was monitored by measuring the change in capacitance of a short rod attached to the 'Top Plate' of the assembly. The minimum temperature recorded by the copper NMR thermometer was 7 mK. Extrapolating from the size of the magnetic susceptibility of the solid ^{3}He, the average temperature of the solid was sometimes as low as 2 mK. Like the Wheatley group[10], our attention was focused solely on the solid ^{3}He. The liquid component was viewed merely as the cooling agent!

Our first Pomeranchuk cell had several significant disadvantages. The most important was that it tended to warm up quite rapidly after only several hours and achieved far less volume change than the maximum 5% required for conversion of all of the liquid into solid. Apparently, solid helium was being trapped within the convolutions of the bellows to become crushed as the bellows contracted. With regard to our desire to measure the temperature dependence of the magnetic susceptibility of the solid the cell had two further design flaws. The place where solid grew in the cell was unpredictable and the time constant for the copper NMR thermometer was very long. Fortunately, some of the solid ^{3}He nucleated on the "rat's nest" of copper wires at the bottom of the NMR tail section. The thermal equilibrium time for nuclear magnetization in metals, T_1, is inversely proportional to temperature. In copper the product $T_1 \circ T$ is approximately 1 sec-K. With only 10 minutes available at the bottom temperatures near 2 mK, the copper thermometer never caught up with the temperature changes.

A final conceptual mistake in our first compressional cooling cell is that there was no provision for the direct measurement of the ^{3}He pressure. We could monitor the volume change by keeping track of the bellows displacement. But the valuable thermodynamic information available with a knowledge of the melting pressure was not available in this set of experiments.

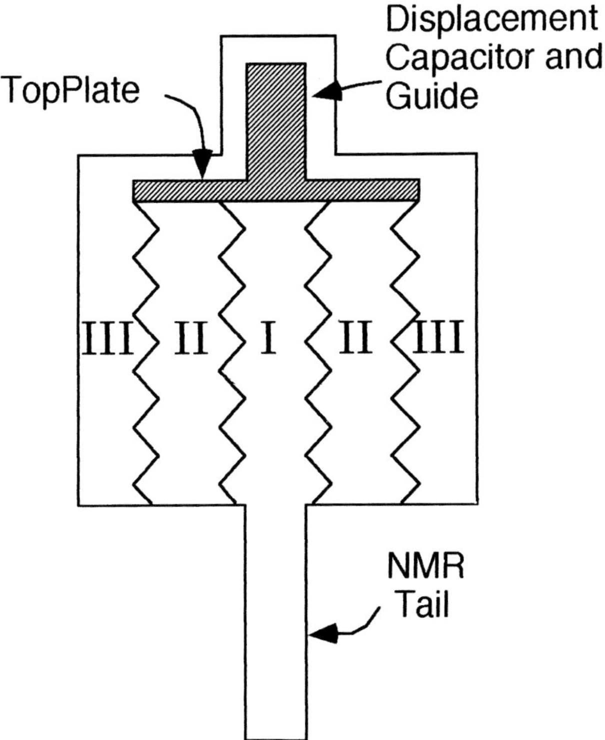

Figure 6. Schematic representation of the cell of Jim Sites.

THE CORRUCCINI-OSHEROFF COOLING CELL

Our next cooling attempts were made with a cell which was designed to investigate some unusual spin diffusion phenomena predicted by Leggett and Rice.[14] The cell is illustrated in Figure 7. We used the cold liquid ^{3}He in a Pomeranchuk cell to cool a separate chamber of either liquid ^{3}He or dilute liquid mixtures of ^{3}He in ^{4}He.[15,16] Doug worked on the development of the cooling technique while Linton Corruccini worked on the design of the chamber for NMR measurements. Following our experience with Jim Site's cell, Osheroff decided to try compressing the ^{3}He with a bellows which expands. The idea was to avoid crushing solid ^{3}He within contracting bellows. The ^{3}He compression cell was filled with liquid at the melting pressure and ^{4}He pressure was applied to the upper chamber. The differences in the melting pressures of the two isotopes were taken into account by having the diameter of the ^{4}He bellows larger than that for the ^{3}He by a ratio of 3.5:1. The ^{3}He could be completely solidified without raising the ^{4}He pressure above 10 bar.

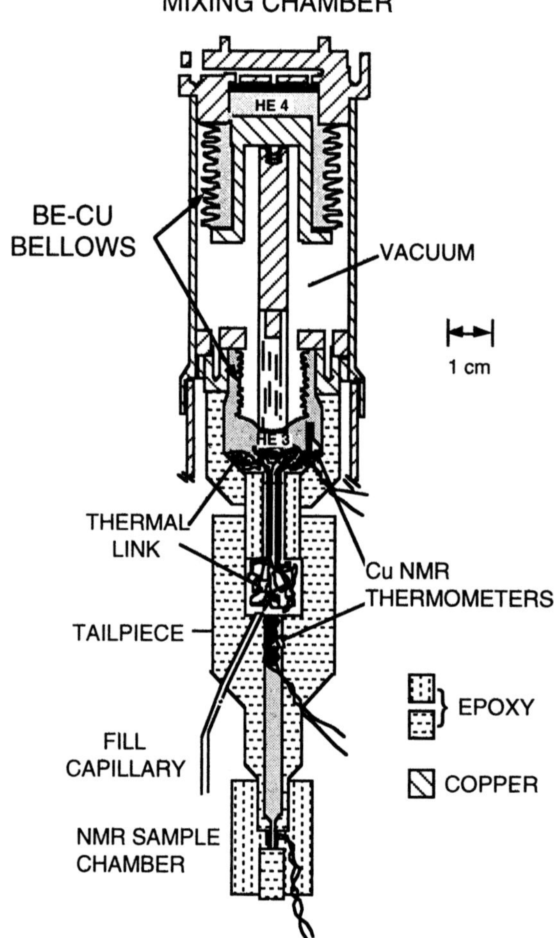

Figure 7. The Corruccini-Osheroff cooling cell.

A bundle of small copper wires made a thermal connection between the ^{3}He compression region and the NMR sample region. The temperatures of the compression region and sample region were measured with copper NMR signals.

The apparatus worked very well indeed, as did the Leggett-Rice theory. It was an easy experiment. Before compression, the cell was first cooled to 25 mK with a dilution refrigerator. We succeeded in cooling the sample region to temperatures as cold as 4 mK. We typically spent 8 to 12 hours in compressing the liquid ^{3}He in order to maintain thermal equilibrium and maximize the cooling efficiency between the cooling cell and sample cell. The chamber would remain at the lowest temperature for up to 4 hours, then slowly warm up to 10 mK over a period of 5 hours. Further warming was achieved by partial decompression of the cooling cell. The minimum temperature recorded by the copper NMR thermometer in the cooling cell was 3 mK. The cell was a precursor for the one used in the discovery of superfluid ^{3}He. Since it was not intended for studies of the ^{3}He there was minimal instrumentation in the compression cell.

OSHEROFF'S COMPRESSION CELL

The compressional cooling cell used by Osheroff[17] during the course of the experiments on melting ^{3}He was a variation of the one we used in Corruccini's spin diffusion measurements. The epoxy bottom of the cell could be readily replaced and in the course of six months no less than five different epoxy tail sections were used. The cell illustrated in Figure 8 is the version published with the results purporting to measure the phase transition in solid ^{3}He. It had two very important design changes from the previous cells. The first was that we changed the metal NMR thermometer from copper to platinum. The thermal equilibrium time for ^{195}Pt nuclei is a factor of 30 shorter than that of copper. It is far less susceptible to small eddy current heating effects than copper. It eventually read temperatures well under 2 mK.

The second change was probably more important. It was the inclusion of a gauge to measure the pressure of the melting ^{3}He. A thin metal diaphragm on the bottom of the cell deflected as the pressure changed. The amount of the deflection was measured capacitively. One plate was attached to the center of the diaphragm while the other was fixed to a mounting arrangement on the epoxy tail section. The design is one which had been invented by Straty and Adams[18] and became widely used at Cornell and elsewhere. The melting pressure is a unique function of temperature. At higher temperatures the vapor pressure of ^{3}He and ^{4}He gases in equilibrium with liquid helium are routinely used to calibrate other thermometers. Adams had previously suggested that the ^{3}He melting pressure be used as a temperature standard.[19] During the interval in late November 1971 in which Doug Osheroff was 'practicing' the use of the apparatus the pressure measurements gave information about changes in temperature of the apparatus.

Dave Lee and Doug Osheroff have described many of the details about those early measurements in their Nobel Prize lectures. The well known 'pressure versus time' curve is reproduced in Figure 9. The experiment was conducted with a constant rate of compression of the ^{3}He bellows, that is with a constant cooling rate. The pressure scale can be interpreted as a measurement of temperature change. The temperature scale on the right was our best guess at the thermodynamic temperature and was based upon measurements of the ^{195}Pt magnetic susceptibility. The pressure measurements are relative to the maximum melting pressure of ^{3}He.

The points labelled A and A′ are transitions of liquid ^{3}He from the normal liquid phase to the superfluid A phase and then back again. The cooling or warming rate changes at these points because of the change in heat capacity of the liquid ^{3}He. dT = (1/C)•dQ, where T is the temperature, C is the heat capacity, and Q is the heat input. For a constant rate of heating dQ/dt, the rate of temperature change becomes (dT/dt) = (1/C)•(dQ/dt). A sudden in-

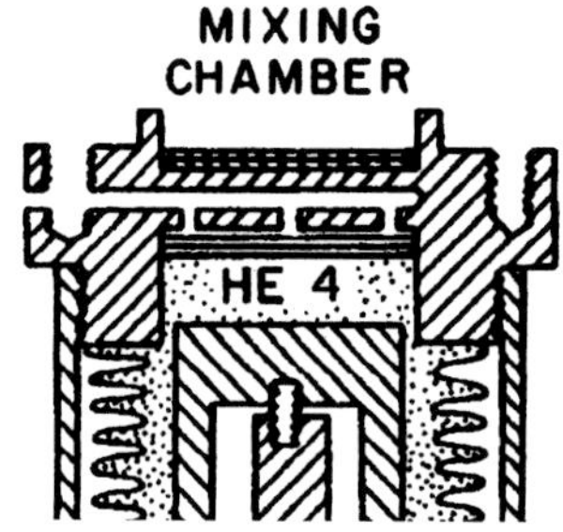

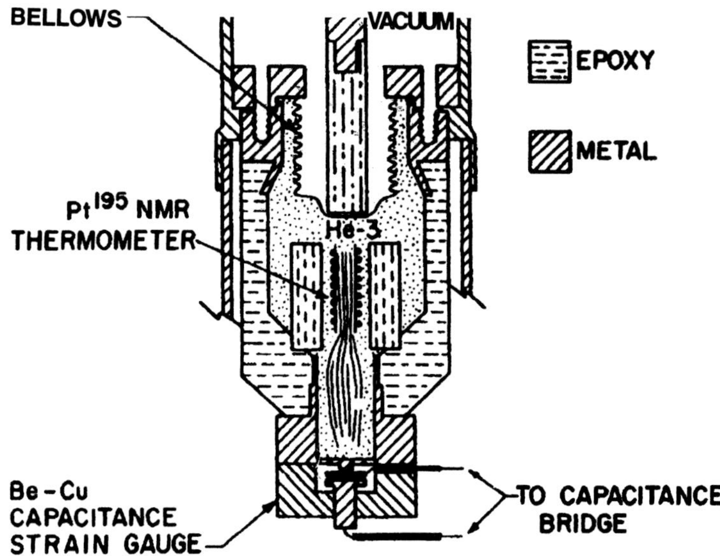

Figure 8. Doug Osheroff's Pomeranchuk Cell.

crease in heat capacity will cause the rate of cooling at A to decrease. At the time, we mistakenly identified the heat capacity change with the long sought nuclear magnetic transition in the solid phase. Points B and B′ are related to another thermal event. At point B there must be an evolution of latent heat because there is a sudden but small decrease in the temperature in the cell. The pressure at which the B type event took place varied, generally depending upon the cooling rate. We attributed this, correctly, to a super cooling. Point B′ is the equilibrium transition, we now know, from the superfluid B phase to the superfluid A phase. The temperature change pauses briefly as the B phase absorbs extra heat to pass through a first order phase change (like the melting of ice, or the freezing of liquid ^{3}He). Points C and D in Figure 9 correspond to the maximum melting pressure achieved and the time at which a slow decompression was begun.

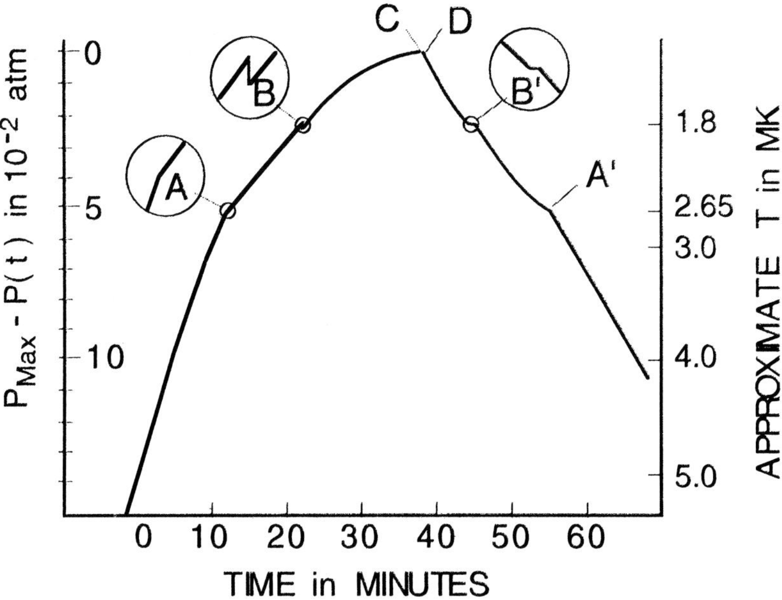

Figure 9. Measurement of the change in ^{3}He melting pressure with time with a constant cooling rate in an Osheroff's Pomeranchuk cell.

This measurement in Figure 9 contained an embarrassing amount of contradictory detail. Since nothing was previously known about the nuclear ordering process we supposed that points B and B′ marked the transition to a second magnetic phase. Still, the total pressure change between point A and the maximum melting pressure was surprisingly large. In SI units the pressure difference is 0.00527 MPa. In this connection we made the following observation, "In order to obtain sufficient pressure change from 2.7 mK to 0 mK through integration of the Clausius-Clapeyron equation,

$$\Delta P = \int \frac{(S_{solid} - S_{liquid})dT}{(V_{solid} - V_{liquid})} \approx \int \frac{S_{solid}}{\Delta V} dT, \qquad [2]$$

to agree with the value presented above, one is forced to hold the solid entropy nearly constant over a broad temperature region below the 2.7 mK transition temperature. This possible behavior of the solid entropy is, in fact, also suggested by the nearly constant slope of P(t) between A and B in Figure 9. We know of no physical system which furnishes a precedent for the entropy behavior we postulate here."[20] Using our 'approximate' temperature scale would require the solid entropy to remain at the value $R \bullet \log_e 2$ over the temperature interval between 2.7 mK and 1.5 mK!

Our misgivings about the interpretation of the data were well founded. In our subsequent paper about the NMR properties of the ^{3}He in the compression cell we finally got it right.[21] The change in heat capacity signaled at point A corresponded with a change in the properties of liquid ^{3}He and point B (or B′) marked the phase boundary to a liquid phase with even different behavior.

Less than a year after the report of the A and B transitions in ^{3}He, the group in Helsinki[22] used Pomeranchuk cooling to study the pressurized liquid phase in the cell to show that the viscosity of liquid ^{3}He decreases by a factor of 1000 in the new phases. The viscosity measurements were made with a vibrating wire immersed within the liquid.

FINALLY, THE REAL PHASE TRANSITION IN SOLID ^{3}HE

After Anufriev, Jim Sites, and the Wheatley group had shown us that the compressional cooling method was an effective way to cool melting helium, we decided to begin a second set of measurements using a cell designed primarily for the optimization of studies of solid ^{3}He.[23] The compression cell designed by Bill Halperin is shown in Figure 10. The design thought was that by having a lens-shaped compression region there would be a minimum of heating related to crushing solid ^{3}He during compression. The cell contained both a pressure gauge for the ^{3}He and a method for measuring the absolute volume changes, a second set of capacitor plates attached to the moving diaphragm. We also had a provision for measuring the changes in the magnetization of the ^{3}He. The basic operation of the cell was the same as the others I have described. ^{3}He at the melting pressure was trapped in the lower region while the upper was filled with liquid ^{4}He. The assembly was precooled to 25 mK and compression was achieved by forcing liquid ^{4}He in the upper region.

In the sequence of measurements with this cell, we measured the entropy of solid ^{3}He down to temperatures below the phase transition. In addition we were able to measure the heat capacity of liquid ^{3}He through the superfluid transition, the latent heat of the transition between the A and B phases, and to determine a 'first principles' temperature scale.[23–25] Through

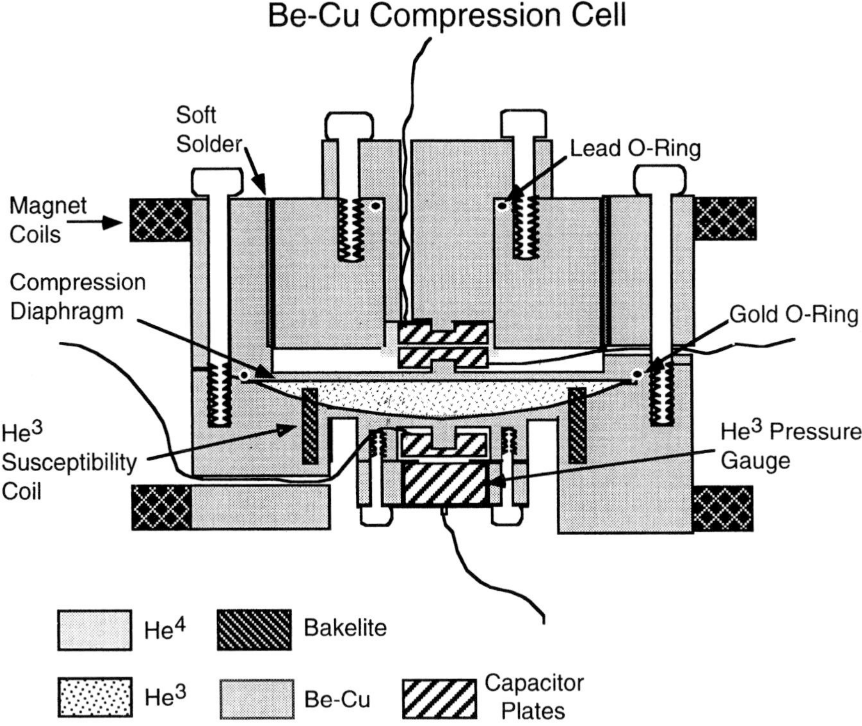

Figure 10. The Pomeranchuk Cell of Bill Halperin.

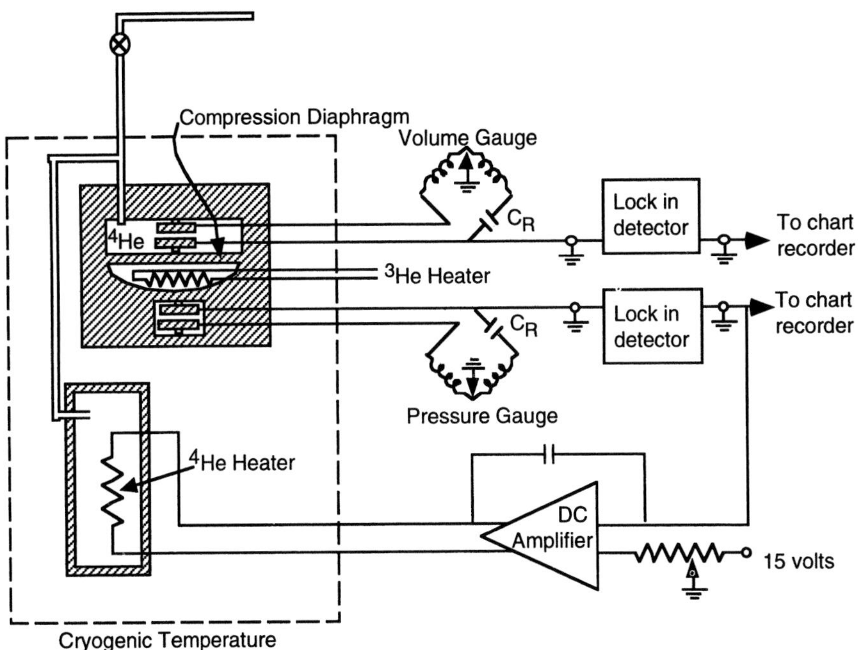

Figure 11. Servo Loop for Controlling Compression Rate.

experience, we grew to understand the different time constants for thermal equilibrium of the liquid and solid phases. A separate container for liquid ^{4}He was located in the cryostat. Heat applied to the ^{4}He in that vessel would rapidly change the ^{3}He volume in the compression cell.

Heat pulses and 'cool pulses' could be applied by means of short bursts of decompression or compression of the diaphragm. Heating was also accomplished by passing a calibrated current through a heater wire in the ^{3}He compression cell. The melting pressure (and hence temperature) could be maintained at a constant value by passing the error signal from the pressure gauge back through the DC amplifier to the ^{4}He heater. An example of such a measurement is shown in Figure 12. A signal from the pressure bridge has been sent to the ^{4}He heater to maintain a constant ^{3}He pressure. At $t = 0$ a short heat pulse of 32.26 ergs was applied to the ^{3}He heater. In response, the servo control system briefly accelerated the rate of ^{3}He compression. The volume change associated with the heat pulse was 2.56×10^{-4} cm^3 as additional liquid was suddenly converted into solid ^{3}He. The output of the pressure bridge is also shown, the pressure has been converted to temperature units. During the measurement the maximum temperature excursion of the cell was less than 5 μK.

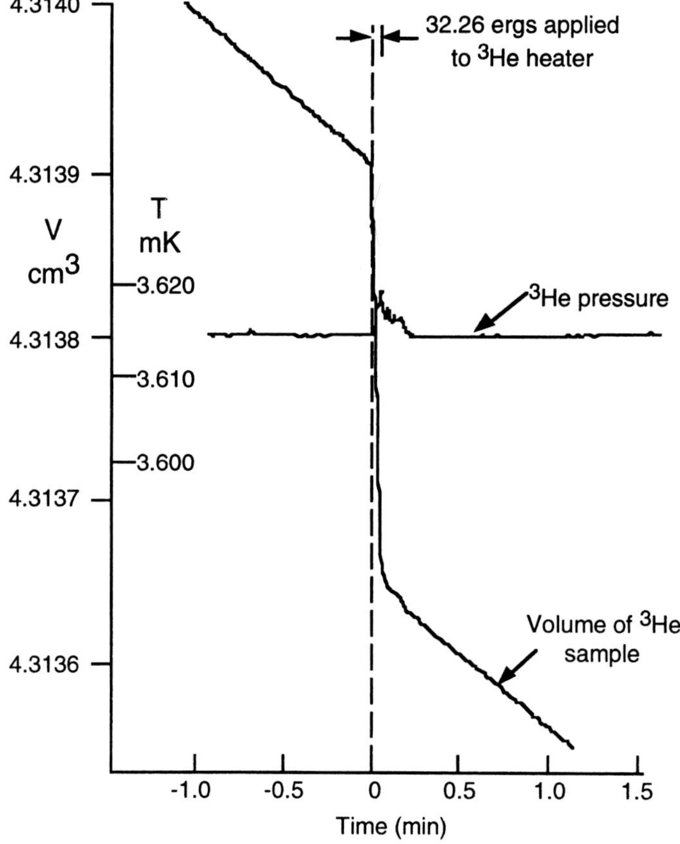

Figure 12. Measurement of volume change following a heat pulse with the compression cell constrained at constant pressure.

The Clausius-Clapeyron equation can be invoked, once again, with the data obtained in Figure 12. If we multiply both sides of [1] by the temperature, T, we obtain, $T \cdot \left.\frac{dP}{dT}\right)_{melting} = \frac{T \cdot \Delta S}{\Delta V} = \frac{\Delta Q}{\Delta V}$. The quantity on the right is the ratio of the heat input to the volume change. The measurement was repeated from temperatures near 25 mK down to the temperature of the maximum melting pressure to generate a table of values of P versus T(dP/dT).

With regard to the elusive phase transition of solid ^3He, measurements like that illustrated in Figure 12 and other non-equilibrium measurements with pulsed volume changes were used to generate the data shown in Figure 13.[24] At a pressure near the maximum melting pressure, the value of T(dP/dT) decreases rapidly, corresponding to an entropy decrease of more than half of the spin entropy. The experiment was the first quantitative identification of the point of magnetic order.

The entropy versus temperature curve shown in the inset curve in Figure 13 was obtained through integration of the P versus T(dP/dT) data. A relative temperature scale is given by $\frac{T}{T_{solid}} = \exp \int_{P_{solid}}^{P} \left[T \frac{dP'}{dT} \right]^{-1} dP'$. A single fixed point in temperature is sufficient to generate the complete low temperature melting curve from the data. The fixed point we used was that of the

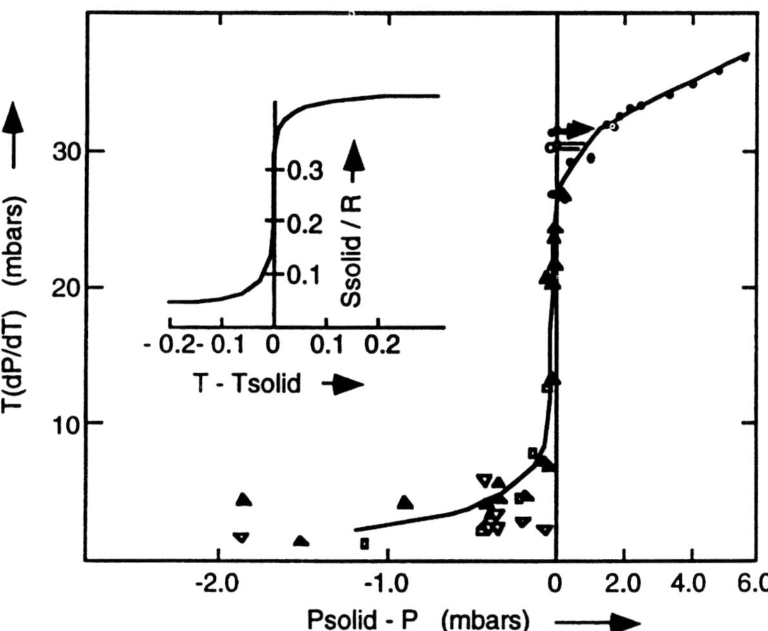

Figure 13. The entropy decrease at the nuclear magnetic phase transition of solid ^3He. The entropy versus temperature data were calculated from the measurements of T(dT/dP) versus pressure.

high temperature solid entropy, $S_{solid} \to R \cdot \log_e 2$. [The Debye temperature for solid helium is of order 30 K so that the phonon contribution is negligible at 30 mK.].

THE POMERANCHUK EFFECT AND MELTING CURVE IN 1996

The Pomeranchuk effect is no longer the preferred cooling technique for studying either liquid or solid ^{3}He. It is too restrictive because it permits measurements of ^{3}He only at the melting pressure. Moreover, an even more important reason is that the method of nuclear demagnetization is far more efficient and permits cooling of ^{3}He to temperatures in the range of 10 μK.[26]

The ^{3}He melting plays the same role in thermometry as it did in the days of our discovery of the superfluid transitions. There are four easily measured and reproduced fixed points on the melting curve: the minimum in the melting pressure; the superfluid A transition; the superfluid B transition; and the magnetic ordering transition of solid ^{3}He. Melting curve thermometers[27] have become the temperature standard for very low temperature work. In the days since Halperin's integration of T(dP/dT) along the melting curve there have been many independent measurements of the melting curve fixed points.[6,28] The most recent[28] give: $T_{minimum}$ = 0.31517 mK; T_A = 2.41 mK; T_B = 1.87 mK; and T_{solid} = 0.88 mK. The associated pressures are also know with great precision so that the strain gauge can also be calibrated with the fixed points. It is amusing to realize that many modern low temperature physicists routinely reproduce the data of Figure 9 as a standard temperature calibration of their apparatus.

REFERENCES

1. I. Pomeranchuk, *Zh. Eksp. i Teor. Fiz.* (USSR) **20**, 919 (1950).
2. There are several elegant and systematic discussions of the compressional cooling technique. Some especially useful discussions about compressional cooling and cryogenic methods in general are contained in the textbooks by Betts and Lounasmaa. Betts, D. S., *Refrigeration and Thermometry Below 1K*, Sussex University Press, London (1974), and O. V. Lounasma, *Experimental Principles and Methods Below 1K*, Academic Press, NY (1974).
3. L. D. Landau, *Zh. Eksp. i Teor. Fiz.* (USSR) **30**, 1058 (1956).
4. My first encounter with a discussion of the ^{3}He melting curve came through **Exercise 9** of Pippard's excellent undergraduate thermodynamics book, *The Elements of Classical Thermodynamics*, A. B. Pippard, Cambridge University Press, London, 161 (1957). There are only 14 problems in the 'Exercise' section for the entire book!
5. A good source of data about liquid ^{3}He remains the compilation of Wheatley in the review: J. C. Wheatley, "Experimental Properties of Superfluid ^{3}He", *Review of Modern Physics*, **47**,415 (1975).
6. D. S. Greywall, *Phys. Rev.* **B33**, 7520 (1986).
7. R. C. Richardson, E. R. Hunt, and H. Meyer, *Physical Review* **138**, A1326 (1965).
8. N. Kurti, *Cryogenics* **1**, 2 (1960).
9. Yu. D. Anufriev, *Sov. Phys. J. E. T. P. Letters* **1**, 155 (1965).

10. R. T. Johnson, R. Rosenbaum, O. G. Symko, and J. C. Wheatley, *Phys. Rev. Lett.* **22,** 449 (1969).
11. R. T. Johnson, O. V. Lounasmaa, R. Rosenbaum, O. G. Symko, and J. C. Wheatley, *J. Low Temp. Phys.* **2**, 403 (1970).
12. James R. Sites, "Magnetic Susceptibility of Solid Helium Three Cooled by Adiabatic Compression," Ph. D. Thesis, Unpublished, Cornell University (1969).
13. J. R. Sites, D. D. Osheroff, R. C. Richardson, and D. M. Lee, *Phys. Rev. Lett.* **23**, 836 (1969).
14. A. J. Leggett and M. J. Rice, *Phys. Rev. Lett.* **20,** 586; **21**, 506 (1968).
15. L. R. Corruccinni, "Spin Wave Phenomena in Liquid ^{3}He Systems", Ph. D. Thesis, Unpublished, Cornell University (1972).
16. L. R. Corruccini, D. D. Osheroff, D. M. Lee, and R. C. Richardson, *J. Low Temp. Phys.* **8**, 119 (1972).
17. D. D. Osheroff, "Compressional Cooling and Ultralow Temperature Properties of ^{3}He", Ph. D. Thesis, Unpublished, Cornell University (1972).
18. G. C. Straty and E. D. Adams, *Rev. Sci. Instrum.* **40**, 1393 (1969).
19. R. A. Scribner, M. Panzick, and E. D. Adams, *Phys. Rev. Lett.* **21**, 427 (1968).
20. D. D. Osheroff, R. C. Richardson, and D. M. Lee, *Phys. Rev. Lett.* **28,** 885 (1972).
21. D. D. Osheroff, W. J. Gully, R. C. Richardson, and D. M. Lee, *Phys. Rev. Lett.* **29**, 920 (1972).
22. Yu. D. Anufriev, T. A. Alvesalo, H. K. Collan, N. T. Opheim, P. Wennerström, *Phys. Lett.* **43A**, 175 (1973); T. A. Alvesalo, Yu. D. Anufriev, H. K. Collan, O.V. Lounasmaa, P. Wennerström, *Phys. Rev. Lett.* **30**, 962 (1973).
23. W. P. Halperin, *"Melting Properties of 3He: Specific Heat, Entropy, Latent Heat, and Temperature "*, Ph. D. Thesis, Unpublished, Cornell University (1975).
24. W. P. Halperin, C. N. Archie, F. B. Rasmussen, R. A. Buhrman, and R. C. Richardson, *Phys. Rev. Lett.* **32**, 927 (1974).
25. W. P. Halperin, C. N. Archie, F. B. Rasmussen, and R. C. Richardson, *Phys. Rev. Lett.* **34**, 718 (1975).
26. An excellent review of nuclear magnetic cooling is given in the text by F. Pobell, Matter and Methods at Low Temperatures, Springer-Verlag, New York (1992).
27. J. S. Souris and T. T. Tommila, *Experimental Techniques in Condensed Matter Physics at Low Temperatures,* [R. C. Richardson and E. N. Smith, editors; Addison-Wesley, New York], 245 (1988).
28. G. Schuster, A. Hoffmann, and D. Hechtfischer, *Czech. J. Phys.* **46-S1**, 481 (1996).

Physics 1997

STEVEN CHU, CLAUDE N. COHEN-TANNOUDJI and WILLIAM D. PHILLIPS

"for development of methods to cool and trap atoms with laser light"

THE NOBEL PRIZE IN PHYSICS

Speech by Professor Bengt Nagel of the Royal Swedish Academy of Sciences. Translation of the Swedish text.

Your Majesties, Your Royal Highnesses, Ladies and Gentlemen,

The Royal Swedish Academy of Sciences has decided to award this year's Nobel Prize for Physics jointly to three physicists for "development of methods to cool and trap atoms with laser light".

The air around us consists of molecules moving back and forth with breathtaking velocities around a mean of about 500 m/s, distributed around this velocity roughly according to a bell curve, so dear to statisticians. A molecule collides with some other molecule about one billion times per second, and there are around twenty-five trillion (= million million million) molecules in each cubic centimeter. As result of this disordered rapid motion we get properties such as air pressure, the ability to transmit sound, a (fairly bad) capacity for conducting heat, etc.

If we want to study the individual or small numbers of molecules or atoms we must slow down their motion, both in order to get time to observe them, and also because the properties of our communication with them, which occurs with the help of light, are influenced by their velocity. The frequency of the light that a moving atom sends out or can absorb changes because of the Doppler effect, with which we are familiar from acoustics. We can make an analogy with an absurd opera performance–maybe something for this year's prize winner in literature–where the opera star sings her aria while moving at great velocity around the stage, colliding with and rebounding from her fellow actors. Visually this might be a spectacular performance, but acoustically it could be a disaster: the audience would think that the star didn't sing in tune. (If somebody would like to try this out, I can mention that if the singer moves with the speed of a good sprinter, the maximum variation in pitch would be half a tone interval.)

Last year's Nobel Prize in physics dealt with low temperatures, and with conventional cooling methods one might cool the atoms down to some millionths of a degree above absolute zero, which formally would correspond to velocities of some centimeters per second. The problem is that long before this temperature is reached, the atoms or molecules would have condensed into a liquid and finally a solid body. Atoms in a chorus sing different tunes from solo, and it is the solo performance we want. We must reduce the motion of the atoms while keeping them apart .

The idea of using laser light and the Doppler effect to achieve this was discussed as early as the 1970s and is based on the fact that when an atom absorbs a light particle, a photon, it receives an impulse from the absorbed photon, and if it has a velocity opposite to that of the photon, its motion is slowed down. We adjust the frequencey of the laser beam so that the photon resonates with, and hence can be absorbed only by, atoms moving in the opposite direction from that of the laser beam. – This basic idea has demanded many years of extensions and experimental developments to result in an effective cooling and trapping of atoms. An important device is the magneto-optical trap, where atoms are slowed down in an "optical molasses" consisting of three pairs of mutually opposite laser beams, and are kept trapped by a magnetic field. In the latest development, subrecoil cooling, equivalent temperatures of fractions of a millionth of a degree have been obtained.

Besides the direct importance for increasing our knowledge of the interaction between atoms and radiation which is a result of increased control of the atomic motion, one could of course ask about possible practical uses of these new advances.

The classical answer : "One day, sir, you may tax it" was given in the 1850s by Michael Faraday in reply to a question by William Gladstone, then British minister of finance–Chancellor of the Exchequer–if electricity had any practical value. Faraday's studies of electricity and magnetism laid the foundations of electrotechnology.

I cannot yet see any directly taxable object resulting from the contributions awarded this year's physics prize, but I believe it will come.

Because we can make atoms practically stand still, we can design much more precise atomic clocks; this enables us, for example, to make a more accurate position determination on earth via satellites. The awarded cooling techniques supplemented by other cooling techniques, have led to achievement of what could be called the "dream mile" of atomic physics, Bose-Einstein condensation of gases, a phenomenon predicted by Einstein more than 70 years ago. This in turn can lead to the construction of "atomic lasers", coherent intense atom beams, which among other things could be used for the construction of very small electronic components.

Professors Steven Chu and Claude Cohen-Tannoudji, Dr. William Phillips,

You have been awarded this year's Nobel Prize in Physics as leaders and representatives of the most successful groups and collaborations in the extensive work of developing methods for cooling and trapping of atoms, which has opened up a new area of control and study of atoms and atomic gases.

It is my pleasure and my privilege to convey to you the felicitations of the Royal Swedish Academy of Sciences and to ask you to step forward and receive the Prize from the hands of His Majesty the King.

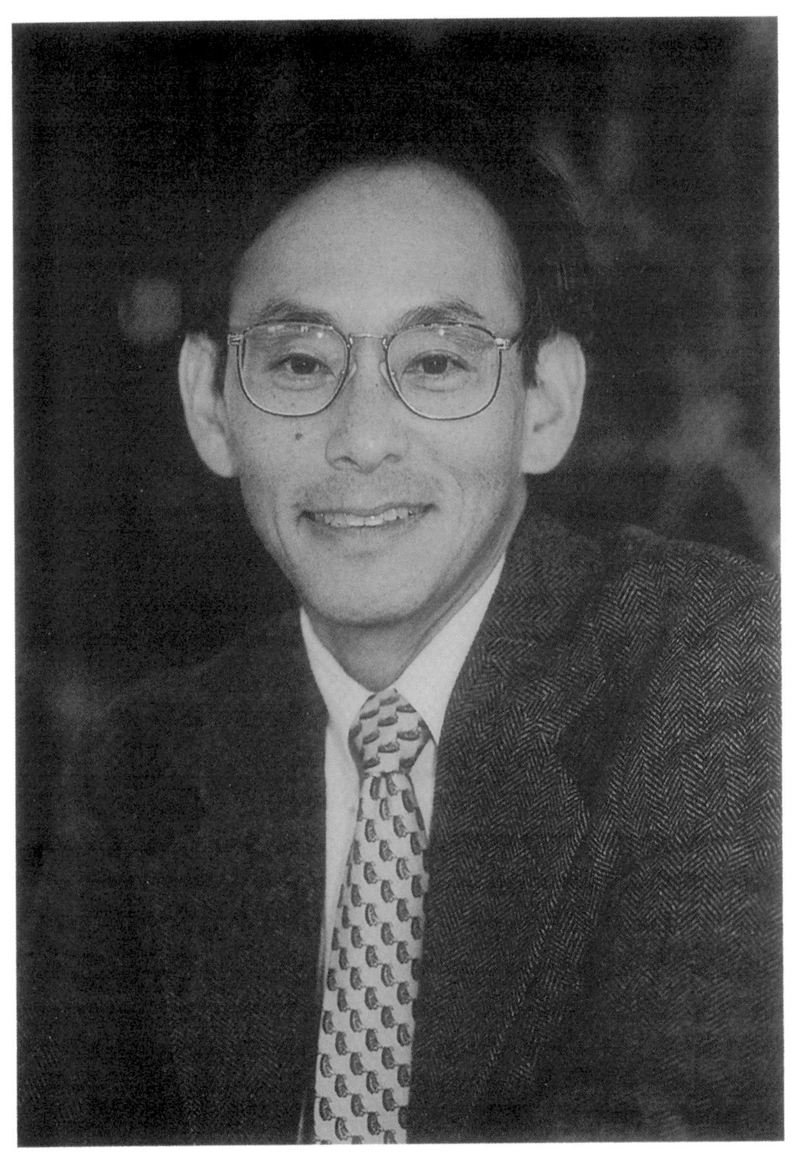

STEVEN CHU

My father, Ju Chin Chu, came to the United States in 1943 to continue his education at the Massachusetts Institute of Technology in chemical engineering, and two years later, my mother, Ching Chen Li, joined him to study economics. A generation earlier, my mother's grandfather earned his advanced degrees in civil engineering at Cornell while his brother studied physics under Perrin at the Sorbonne before they returned to China. However, when my parents married in 1945, China was in turmoil and the possibility of returning grew increasingly remote, and they decided to begin their family in the United States. My brothers and I were born as part of a typical nomadic academic career: my older brother was born in 1946 while my father was finishing at MIT, I was born in St. Louis in 1948 while my father taught at Washington University, and my younger brother completed the family in Queens shortly after my father took a position as a professor at the Brooklyn Polytechnic Institute.

In 1950, we settled in Garden City, New York, a bedroom community within commuting distance of Brooklyn Polytechnic. There were only two other Chinese families in this town of 25,000, but to our parents, the determining factor was the quality of the public school system. Education in my family was not merely emphasized, it was our raison d'être. Virtually all of our aunts and uncles had Ph.D.'s in science or engineering, and it was taken for granted that the next generation of Chu's were to follow the family tradition. When the dust had settled, my two brothers and four cousins collected three MDs, four Ph.D.s and a law degree. I could manage only a single advanced degree.

In this family of accomplished scholars, I was to become the academic black sheep. I performed adequately at school, but in comparison to my older brother, who set the record for the highest cumulative average for our high school, my performance was decidedly mediocre. I studied, but not in a particularly efficient manner. Occasionally, I would focus on a particular school project and become obsessed with, what seemed to my mother, to be trivial details instead of apportioning the time I spent on school work in a more efficient way.

I approached the bulk of my schoolwork as a chore rather than an intellectual adventure. The tedium was relieved by a few courses that seem to be qualitatively different. Geometry was the first exciting course I remember. Instead of memorizing facts, we were asked to think in clear, logical steps. Beginning from a few intuitive postulates, far reaching consequences could be derived, and I took immediately to the sport of proving theorems. I also fondly remember several of my English courses where the assigned reading often led to binges where I read many books by the same author.

Despite the importance of education in our family, my life was not completely centered around school work or recreational reading. In the summer after kindergarten, a friend introduced me to the joys of building plastic model airplanes and warships. By the fourth grade, I graduated to an erector set and spent many happy hours constructing devices of unknown purpose where the main design criterion was to maximize the number of moving parts and overall size. The living room rug was frequently littered with hundreds of metal "girders" and tiny nuts and bolts surrounding half-finished structures. An understanding mother allowed me to keep the projects going for days on end. As I grew older, my interests expanded to playing with chemistry: a friend and I experimented with homemade rockets, in part funded by money my parents gave me for lunch at school. One summer, we turned our hobby into a business as we tested our neighbors' soil for acidity and missing nutrients.

I also developed an interest in sports, and played in informal games at a nearby school yard where the neighborhood children met to play touch football, baseball, basketball and occasionally, ice hockey. In the eighth grade, I taught myself tennis by reading a book, and in the following year, I joined the school team as a "second string" substitute, a position I held for the next three years. I also taught myself how to pole vault using bamboo poles obtained from the local carpet store. I was soon able to clear 8 feet, but was not good enough to make the track team.

In my senior year, I took advanced placement physics and calculus. These two courses were taught with the same spirit as my earlier geometry course. Instead of a long list of formulas to memorize, we were presented with a few basic ideas or a set of very natural assumptions. I was also blessed by two talented and dedicated teachers.

My physics teacher, Thomas Miner was particularly gifted. To this day, I remember how he introduced the subject of physics. He told us we were going to learn how to deal with very simple questions such as how a body falls due to the acceleration of gravity. Through a combination of conjecture and observations, ideas could be cast into a theory that can be tested by experiments. The small set of questions that physics could address might seem trivial compared to humanistic concerns. Despite the modest goals of physics, knowledge gained in this way would become collected wisdom through the ultimate arbitrator – experiment.

In addition to an incredibly clear and precise introduction to the subject, Mr. Miner also encouraged ambitious laboratory projects. For the better part of my last semester at Garden City High, I constructed a physical pendulum and used it to make a "precision" measurement of gravity. The years of experience building things taught me skills that were directly applicable to the construction of the pendulum. Ironically, twenty five years later, I was to develop a refined version of this measurement using laser cooled atoms in an atomic fountain interferometer.

I applied to a number of colleges in the fall of my senior year, but because of my relatively lackluster A- average in high school, I was rejected by the Ivy

League schools, but was accepted at Rochester. By comparison, my older brother was attending Princeton, two cousins were in Harvard and a third was at Bryn Mawr. My younger brother seemed to have escaped the family pressure to excel in school by going to college without earning a high school diploma and by avoiding a career in science. (He nevertheless got a Ph.D. at the age of 21 followed by a law degree from Harvard and is now a managing partner of a major law firm.) As I prepared to go to college, I consoled myself that I would be an anonymous student, out of the shadow of my illustrious family.

The Rochester and Berkeley Years

At Rochester, I came with the same emotions as many of the entering freshman: everything was new, exciting and a bit overwhelming, but at least nobody had heard of my brothers and cousins. I enrolled in a two-year, introductory physics sequence that used *The Feynman Lectures in Physics* as the textbook. The *Lectures* were mesmerizing and inspirational. Feynman made physics seem so beautiful and his love of the subject is shown through each page. Learning to do the problem sets was another matter, and it was only years later that I began to appreciate what a magician he was at getting answers.

In my sophomore year, I became increasingly interested in mathematics and declared a major in both mathematics and physics. My math professors were particularly good, especially relative to the physics instructor I had that year. If it were not for the Feynman Lectures, I would have almost assuredly left physics. The pull towards mathematics was partly social: as a lowly undergraduate student, several math professors adopted me and I was invited to several faculty parties.

The obvious compromise between mathematics and physics was to become a theoretical physicist. My heroes were Newton, Maxwell, Einstein, up to the contemporary giants such as Feynman, Gell-Man, Yang and Lee. My courses did not stress the importance of the experimental contributions, and I was led to believe that the "smartest" students became theorists while the remainder were relegated to experimental grunts. Sadly, I had forgotten Mr. Miner's first important lesson in physics.

Hoping to become a theoretical physicist, I applied to Berkeley, Stanford, Stony Brook (Yang was there!) and Princeton. I chose to go to Berkeley and entered in the fall of 1970. At that time, the number of available jobs in physics was shrinking and prospects were especially difficult for budding young theorists. I recall the faculty admonishing us about the perils of theoretical physics: unless we were going to be as good as Feynman, we would be better off in experimental physics. To the best of my knowledge, this warning had no effect on either me or my fellow students.

After I passed the qualifying exam, I was recruited by Eugene Commins. I admired his breadth of knowledge and his teaching ability but did not yet learn of his uncanny ability to bring out the best in all of his students. He was ending a series of beta decay experiments and was casting around for a new direction of research. He was getting interested in astrophysics at the time and asked me to think about proto-star formation of a closely coupled binary pair. I had spent the summer between Rochester and Berkeley at the National Radio Astronomy Observatory trying to determine the deceleration of the universe with high red-shift radio source galaxies and was drawn to astrophysics. However, in the next two months, I avoided working on the theoretical problem he gave me and instead played in the lab.

One of my "play-experiments" was motivated by my interest in classical music. I noticed that one could hear out-of-tune notes played in a very fast run by a violinist. A simple estimate suggested that the frequency accuracy, $\Delta\nu$ times the duration of the note, Δt did not satisfy the uncertainty relationship $\Delta\nu\Delta t \geq 1$. In order to test the frequency sensitivity of the ear, I connected an audio oscillator to a linear gate so that a tone burst of varying duration could be produced. I then asked my fellow graduate students to match the frequency of an arbitrarily chosen tone by adjusting the knob of another audio oscillator until the notes sounded the same. Students with the best musical ears could identify the center frequency of a tone burst that eventually sounded like a "click" with an accuracy of $\Delta\nu\Delta t \sim 0.1$.

By this time it was becoming obvious (even to me) that I would be much happier as an experimentalist and I told my advisor. He agreed and started me on a beta-decay experiment looking for "second-class currents", but after a year of building, we abandoned it to measure the Lamb shift in high-Z hydrogen-like ions. In 1974, Claude and Marie Bouchiat published their proposal to look for parity non-conserving effects in atomic transitions. The unified theory of weak and electromagnetic interactions suggested by Weinberg, Salam and Glashow postulated a neutral mediator of the weak force in addition to the known charged forces. Such an interaction would manifest itself as a very slight asymmetry in the absorption of left and right circularly polarized light in a magnetic dipole transition. Gene was always drawn to work that probed the most fundamental aspects of physics, and we were excited by the prospect that a table-top experiment could say something decisive about high energy physics. The experiment needed a state-of-the-art laser and my advisor knew nothing about lasers. I brashly told him not to worry; I would build it and we would be up and running in no time.

This work was tremendously exciting and the world was definitely watching us. Steven Weinberg would call my advisor every few months, hoping to hear news of a parity violating effect. Dave Jackson, a high energy theorist, and I would sometimes meet at the university swimming pool. During several of these encounters, he squinted at me and tersely asked, "Got a number yet?" The unspoken message was, "How dare you swim when there is important work to be done!"

Midway into the experiment, I told my advisor that I had suffered enough as a graduate student so he elevated me to post-doc status. Two years later, we and three graduate students published our first results. Unfortunately, we were scooped: a few months earlier, a beautiful high energy experiment at the Stanford Linear Collider had seen convincing evidence of neutral weak interactions between electrons and quarks. Nevertheless, I was offered a job as assistant professor at Berkeley in the spring of 1978.

I had spent all of my graduate and postdoctoral days at Berkeley and the faculty was concerned about inbreeding. As a solution, they hired me but also would permit me to take an immediate leave of absence before starting my own group at Berkeley. I loved Berkeley, but realized that I had a narrow view of science and saw this as a wonderful opportunity to broaden myself.

A Random Walk in Science at Bell Labs

I joined Bell Laboratories in the fall of 1978. I was one of roughly two dozen brash, young scientists that were hired within a two year period. We felt like the "Chosen Ones", with no obligation to do anything except the research we loved best. The joy and excitement of doing science permeated the halls. The cramped labs and office cubicles forced us to interact with each other and follow each others' progress. The animated discussions were common during and after seminars and at lunch and continued on the tennis courts and at parties. The atmosphere was too electric to abandon, and I never returned to Berkeley. To this day I feel guilty about it, but I think that the faculty understood my decision and have forgiven me.

Bell Labs management supplied us with funding, shielded us from extraneous bureaucracy, and urged us not to be satisfied with doing merely "good science." My department head, Peter Eisenberger, told me to spend my first six months in the library and talk to people before deciding what to do. A year later during a performance review, he chided me not to be content with anything less than "starting a new field". I responded that I would be more than happy to do that, but needed a hint as to *what* new field he had in mind.

I spent the first year at Bell writing a paper reviewing the current status of x-ray microscopy and started an experiment on energy transfer in ruby with Hyatt Gibbs and Sam McCall. I also began planning the experiment on the optical spectroscopy of positronium. Positronium, an atom made up of an electron and its anti-particle, was considered the most basic of all atoms, and a precise measurement of its energy levels was a long standing goal ever since the atom was discovered in 1950. The problem was that the atoms would annihilate into gamma rays after only 140×10^{-9} seconds, and it was impossible to produce enough of them at any given time. When I started the experiment, there were 12 published attempts to observe the optical fluorescence of the atom. People only publish failures if they have spent enough time and money so their funding agencies demand something in return.

My management thought I was ruining my career by trying an impossible experiment. After two years of no results, they strongly suggested that I abandon my quest. But I was stubborn and I had a secret weapon: his name is Allen Mills. Our strengths complemented each other beautifully, but in the end, he helped me solve the laser and metrology problems while I helped him with his positrons. We finally managed to observe a signal working with only ~4 atoms per laser pulse! Two years later and with 20 atoms per pulse, we refined our methods and obtained one of the most accurate measurements of quantum electrodynamic corrections to an atomic system.

In the fall of 1983, I became head of the Quantum Electronics Research Department and moved to another branch of Bell Labs at Holmdel, New Jersey. By then my research interests had broadened, and I was using picosecond laser techniques to look at excitons as a potential system for observing metal-insulator transitions and Anderson localization. With this apparatus, I accidentally discovered a counter-intuitive pulse-propagation effect. I was also planning to enter surface science by constructing a novel electron spectrometer based on threshold ionization of atoms that could potentially increase the energy resolution by more than an order of magnitude.

While designing the electron spectrometer, I began talking informally with Art Ashkin, a colleague at Holmdel. Art had a dream to trap atoms with light, but the management stopped the work four years ago. An important experiment had demonstrated the dipole force, but the experimenters had reached an impasse. Over the next few months, I began to realize the way to hold onto atoms with light was to first get them very cold. Laser cooling was going to make possible all of Art Ashkin's dreams plus a lot more. I promptly dropped most of my other experiments and with Leo Holberg, my new post-doc, and my technician, Alex Cable, began our laser cooling experiment. This brings me to the beginning of our work in laser cooling and trapping of atoms and the subject of my Nobel Lecture.

Stanford and the future

Life at Bell Labs, like Mary Poppins, was "practically perfect in every way". However, in 1987, I decided to leave my cozy ivory tower. Ted Hänsch had left Stanford to become co-director of the Max Planck Institute for Quantum Optics and I was recruited to replace him. Within a few months, I also received offers from Berkeley and Harvard, and I thought the offers were as good as they were ever going to be. My management at Bell Labs was successful in keeping me at Bell Labs for 9 years, but I wanted to be like my mentor, Gene Commins, and the urge to spawn scientific progeny was growing stronger.

Ted Geballe, a distinguished colleague of mine at Stanford who also went from Berkeley to Bell to Stanford years earlier, described our motives: "The best part of working at a university is the students. They come in fresh, enthusiastic, open to ideas, unscarred by the battles of life. They don't realize it, but they're the recipients of the best our society can offer. If a mind is ever free to be creative, that's the time. They come in believing textbooks are authoritative but eventually they figure out that textbooks and professors don't know everything, and then they start to think on their own. Then, I begin learning from them."

My students at Stanford have been extraordinary, and I have learned much from them. Much of my most important work such as fleshing out the details of polarization gradient cooling, the demonstration of the atomic fountain clock, and the development of atom interferometers and a new method of laser cooling based on Raman pulses was done at Stanford with my students as collaborators.

While still continuing in laser cooling and trapping of atoms, I have recently ventured into polymer physics and biology. In 1986, Ashkin showed that the first optical atom trap demonstrated at Bell Labs also worked on tiny glass spheres embedded in water. A year after I came to Stanford, I set about to manipulate individual DNA molecules with the so-called "optical tweezers" by attaching micron-sized polystyrene spheres to the ends of the molecule. My idea was to use two optical tweezers introduced into an optical microscope to grab the plastic handles glued to the ends of the molecule. Steve Kron, an M.D./Ph.D. student in the medical school, introduced me to molecular biology in the evenings. By 1990, we could see an image of a single, fluorescently labeled DNA molecule in real time as we stretched it out in water. My students improved upon our first attempts after they discovered our initial protocol demanded luck as a major ingredient. Using our new ability to simultaneously visualize and manipulate individual molecules of DNA, my group began to answer polymer dynamics questions that have persisted for decades. Even more thrilling, we discovered something new in the last year: identical molecules in the same initial state will choose several distinct pathways to a new equilibrium state. This "molecular individualism" was never anticipated in previous polymer dynamics theories or simulations.

I have been at Stanford for ten and a half years. The constant demands of my department and university and the ever increasing work needed to obtain funding have stolen much of my precious thinking time, and I sometimes yearn for the halcyon days of Bell Labs. Then, I think of the work my students and post-docs have done with me at Stanford and how we have grown together during this time.

THE MANIPULATION OF NEUTRAL PARTICLES

Nobel Lecture, December 8, 1997

by

STEVEN CHU

Stanford University, Departments of Physics and Applied Physics, Stanford, CA 94305-4060, USA

The written version of my lecture is a personal account of the development of laser cooling and trapping. Rather than give a balanced history of the field, I chose to present a personal glimpse of how my colleagues and I created our path of research.

I joined Bell Laboratories in the fall of 1978 after working with Eugene Commins as a graduate student and post-doc at Berkeley on a parity non-conservation experiment in atomic physics.[1] Bell Labs was a researcher's paradise. Our management supplied us with funding, shielded us from bureaucracy, and urged us to do the best science possible. The cramped labs and office cubicles forced us to rub shoulders with each other. Animated discussions frequently interrupted seminars and casual conversations in the cafeteria would sometimes mark the beginning of a new collaboration.

In my first years at Bell Labs, I wrote an internal memo on the prospects for x-ray microscopy and worked on an experiment investigating energy transfer in ruby with Hyatt Gibbs and Sam McCall as a means of studying Anderson Localization.[2, 3] This work led us to consider the possibility of Mott or Anderson transitions in other exciton systems such as GaP:N with picosecond laser techniques.[4] During this work, I accidentally discovered that picosecond pulses propagate with the group velocity, even when the velocity exceeds the speed of light or becomes negative.[5]

While I was learning about excitons and how to build picosecond lasers, I began to work with Allan Mills, the world's expert on positrons and positronium. We began to discuss the possibility of working together while I was still at Berkeley, but did not actually begin the experiment until 1979. After three long and sometimes frustrating years, a long time by Bell Labs standards, we finally succeeded in exciting and measuring the 1S–2S energy interval in positronium.[6]

MOVING TO HOLMDEL AND WARMING UP TO LASER COOLING

My entry into the field of laser cooling and trapping was stimulated by my move from Murray Hill, New Jersey, to head the Quantum Electronics Research Department at the Holmdel branch in the fall of 1983. During conversations with Art Ashkin, an office neighbor at Holmdel, I began to learn about his dream to trap atoms with light. He found an increasingly attentive listener and began to feed me copies of his reprints. That fall I was also joined by my new post-doc, Leo Hollberg. When I hired him, I had planned to construct an electron energy-loss spectrometer based on threshold ionization of a beam of atoms with a picosecond laser. We hoped to improve the energy resolution of existing spectrometers by at least an order of magnitude and then use our spectrometer to study molecular adsorbates on surfaces with optical resolution and electron sensitivity. However, Leo was trained as an atomic physicist and was also developing an interest in the possibility of manipulating atoms with light.

Leo and I spontaneously decided to drive to Massachusetts to attend a workshop on the trapping of ions and atoms organized by David Pritchard at MIT. I was ignorant of the subject and lacked the primitive intuition that is essential to add something new to a field. As an example of my profound lack of understanding, I found myself wondering about the dispersive nature of the "dipole force". The force is attractive when the frequency of light is tuned below the resonance, repulsive when tuned above the resonance, and vanishes when tuned directly on the atomic resonance. Looking back on these early fumblings, I am embarrassed by how long it took me to recognize that the effect can be explained by freshman physics. On the other hand, I was not alone in my lack of intuition. When I asked a Bell Labs colleague about this effect, he answered, "Only Jim Gordon really understands the dipole force!"

By 1980, the forces that light could exert on matter were well understood.[7] Maxwell's calculation of the momentum flux density of light,[8] and the laboratory observation of light pressure on macroscopic objects by Lebedev[9] and by Nichols and Hull[10] provided the first quantitative understanding of how light could exert forces on material objects. Einstein[11] pointed out the quantum nature of this force: an atom that absorbs a photon of energy $h\nu$ will receive a momentum impulse $h\nu/c$ along the direction of the incoming photon $\mathbf{p}_{in}$. If the atom emits a photon with momentum $\mathbf{p}_{out}$, the atom will recoil in the opposite direction. Thus the atom experiences a net momentum change $\Delta\mathbf{p}_{atom} = \mathbf{p}_{in}\text{-}\mathbf{p}_{out}$ due to this incoherent scattering process. In 1930, Frisch[12] observed the deflection of an atomic beam with light from a sodium resonance lamp where the average change in momentum was due to the scattering of one photon.

Since the scattered photon has no preferred direction, the net effect is due to the absorbed photons, resulting in scattering force, $\mathbf{F}_{scatt} = N\mathbf{p}_{in}$, where N is the number of photons scattered per second. Typical scattering rates for atoms excited by a laser tuned to a strong resonance line are on the order of 10^7 to 10^8/sec. As an example, the velocity of a sodium atom changes by 3 cm/sec per absorbed photon. The scattering force can be 10^5 times the gravitational acceleration on Earth, feeble compared to electromagnetic forces on charged particles, but stronger than any other long-range force that affect neutral particles.

There is another type of force based on the lensing (i.e. coherent scattering) of photons. A lens alters the distribution of momentum of a light field, and by Newton's third law, the lens must experience a reaction force equal and opposite to the rate of momentum change of the light field. For example, a positive lens will be drawn towards regions of high light intensity as shown in Fig. 1.[13] In the case of an atom the amount of lensing is calculated by adding the amplitude of the incident light field with the dipole field generated by the atomic electrons driven by the incident field.

Figure 1. A photograph of a 10 μm glass sphere trapped in water with green light from an argon laser coming from above. The picture is a fluorescence image taken using a green blocking, red transmitting filter. The exiting (refracted) rays show a notable decrease in beam angles relative to the incident rays. The increased forward momentum of the light results in an upward force on the glass bead needed to balance the downward scattering force. The stria in the forward-scattered light is a common Mie-scattering ring pattern. (Courtesy A. Ashkin).

This reaction force is also called the "dipole force". The oscillating electric field **E** of the light induces a dipole moment **p** on the particle. If the induced dipole moment is in phase with **E**, the interaction energy -**p**·**E** is lower in high field regions. If the induced dipole moment is out of phase with the driving field, the particle's energy is increased in the electric field and the particle will feel a force ejecting it out of the field. If we model the atom or particle as a damped harmonic oscillator, the sign change of the dipole force is easy to understand. An oscillator driven below its natural resonant frequency responds in phase with the driving field, while an oscillator driven above its natural frequency oscillates out of phase with the driving force. Exactly on resonance, the oscillator is 90 degrees out of phase and **p**·**E** = 0.

The dipole force was first discussed by Askar'yan[14] in connection with plasmas as well as neutral atoms. The possibility of trapping atoms with this force was considered by Letokhov,[15] who suggested that atoms might be confined along one dimension in the nodes or antinodes of a standing wave of light tuned far from an atomic transition. In 1970, Arthur Ashkin had succeeded in trapping micron-sized particles with a pair of opposing, focused beams of laser light, as shown in Fig. 2. Confinement along the axial direction was due to the scattering force: a displacement towards either of the focal points of the light would result in an imbalance of scattered light that would push the particle back to the center of the trap. Along the radial direction, the outwardly directed scattering force could be overcome by the attractive dipole force. In the following years, other stable particle trapping geometries were demonstrated by Ashkin,[16] and in 1978, he proposed the first three-dimensional traps for atoms.[17] In the same year, with John Bjorkholm and Richard Freeman, he demonstrated the dipole force by focusing an atomic beam using a focused laser beam.[18]

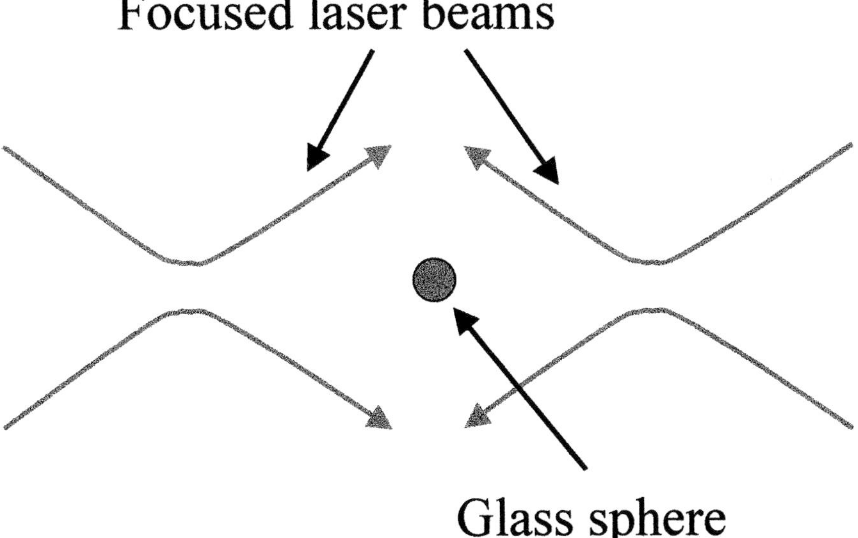

Figure 2. A schematic diagram of the first particle trap used by Ashkin. Confinement in the axial direction is due to an imbalance of the scattering forces between the left and right propagating beams. Confinement in the radial direction results from the induced dipole force which must overcome the outwardly directed scattering force.

Despite this progress, experimental work at Bell labs stopped a year later because of two major obstacles. First, the trapping forces generated by intense focused laser beams are feeble. Atoms at room temperature would have an average energy $\frac{3}{2}k_BT\sim\frac{1}{2}mv^2$, orders of magnitude greater than could be confined by the proposed traps. A cold source of atoms with sufficiently high flux did not exist and a trap with large volume was needed to maximize the number of atoms that could be trapped. Second, the relatively large-volume optical trap made from opposing light beams was found to have serious heating problems. An atom could absorb a photon from one beam and be stimulated back to the initial state by a photon from the opposing beam. In this process, it would receive two photon impulses in the same direction. However, the same atom could have been excited and stimulated by the two beams in the reverse order, resulting in a net impulse in the other direction. Since the order of absorption and stimulated emission is random, this process would increase the random velocity of the atom and they would quickly heat and boil out of the trap. This heating effect was rigorously calculated by Jim Gordon with Ashkin for a two-level atom.[19]

TAKING THE PLUNGE INTO THE COLD

My first idea to solve the trap loading problem was modest at best, but it got me to think seriously about trapping atoms. I proposed to make a cold source of atoms by depositing sodium atoms into a rare-gas matrix of neon.[20] By heating the cryogenic surface supporting this matrix of atoms with a pulsed laser, I thought it should be possible to "puff" the neon and sodium atoms into a vapor with a temperature of a few tens of Kelvin. Once a vapor, a reasonable fraction of the sodium would become isolated atoms and the puffed source would contain the full Maxwell-Boltzmann distribution of atoms, including the very slowest atoms. In a conventional atomic beam, the slowest atoms are knocked out of the way by faster moving atoms overtaking them. In a puffed source, the surface could be quickly heated and cooled so that there would be no fast atoms coming from behind. An added advantage was that the source would turn off quickly and completely so that the detection of even a few trapped atoms would be possible.

Soon after my passage from interested bystander to participant, I realized that the route to trapping was through laser cooling with counter-propagating beams of light. If the laser beams were tuned below the atomic resonance, a moving atom would Doppler-shift the beam opposing its motion closer to resonance and shift the beam co-propagating with the motion away from resonance. Thus, after averaging over many impulses of momentum from both beams, the atom would experience a net force opposing its motion. In the limit where the atoms were moving slowly enough so that the difference in the absorption due to the Doppler effect was linearly proportional to the velocity, this force would result in viscous damping, $\mathbf{F} = -\alpha\mathbf{v}$. This elegant idea was proposed by Hänsch and Schawlow in 1975.[21] A related cooling scheme was proposed by Wineland and Itano in the same year.[22]

An estimate of the equilibrium temperature is obtained by equating the cooling rate in the absence of heating with the heating rate in the absence of cooling,

$$dW_{heating}/dt = dW_{cooling}/dt = -\mathbf{F}\cdot\mathbf{v}.$$

The heating rate is due to the random kicks an atom receives by randomly scattering photons from counter-propagating beams that surround the atoms.[23, 19] The momentum grows as a random walk in momentum space so the average random momentum p would increase as

$$\frac{dW_{heating}}{dt} = \frac{d}{dt}\left(\frac{p^2}{2M}\right) = \frac{N(p_r)^2}{2M}$$

where p_r is the momentum recoil due to each photon and N is the number of photon kicks per second. By equating the heating rate to the cooling rate, one can calculate an equilibrium temperature as a function of the laser intensity, the linewidth of the transition, and the detuning of the laser from resonance. The minimum equilibrium temperature $k_B T_{min} = \hbar\Gamma/2$, where Γ is the linewidth of the transition, was predicted to occur at low intensities and a detuning $\Delta\nu = \Gamma/2$ where the Doppler shift asymmetry was a maximum. In the limit of low intensity, all of the laser beams would act independently and the heating complications that would result from stimulated transitions between opposing laser beams could be ignored.

Not only would the light cool the atoms, it would also confine them. The laser cooling scheme was analogous to the Brownian motion of a dust particle immersed in water. The particle experiences a viscous drag force and the confinement time in a region of space could be estimated based on another result in elementary physics: the mean square displacement $\langle x^2 \rangle$ after a time Δt described by a random walk, $\langle x^2 \rangle = 2Dt$, where the diffusion constant is given by the Einstein relation $D = k_B T/\alpha$. For atoms moving with velocities $\mathbf{v}$ such that $\mathbf{k}\cdot\mathbf{v}<\Gamma$, the force would act as a viscous damping force $\mathbf{F} = -\alpha\mathbf{v}$. By surrounding the atoms with six beams propagating along the ± x, y and z directions, we could construct a sea of photons that would act like an exceptionally viscous fluid: an "optical molasses".[24] If the light intensity was kept low, the atoms would quickly cool to temperatures approaching T_{min}. Once cooled, they would remain confined in a centimeter region of space for times as long as a fraction of a second.

At this point, Leo and I shelved our plans to build the electron spectrometer and devoted our energies to making the optical molasses work. We rapidly constructed the puffing source of sodium needed to load the optical molasses. To simplify matters, we began with a pellet of sodium heated at room temperature. Rather than deal with the complications of a rare gas matrix, Leo and I decided to increase the number of cold atoms by slowing atoms from the puff source before attempting the optical molasses experiment. There were several early experiments that slowed atomic beams with laser light,[25] but sodium atoms had to be slowed to velocities on the order of 200–300 cm/sec (essentially stopped!) before an atom trap could be loaded. Two groups achieved this milestone in late 1984: a group at the National Bureau of Standards in Gaithersburg, Maryland, led by Bill Phillips using a tapered magnetic field[26] and another NBS group in Boulder, Colorado, led by Jan Hall.[27] We decided to copy the technique of Ertmer, *et al.*,[27] and use an electro-optic generator to produce a frequency-shifted sideband. The frequency-shifted light is directed against the atoms coming off the sodium surface, and as the atoms slow down, the frequency is changed in order to keep the light in resonance with the Doppler shifting atoms.

Leo was better at electronics than I and assumed the responsibility of the radio-frequency part of the project while I set out to build a wideband, transmission line electro-optic modulator. One of the advantages of working at Bell Laboratories was that one could often find a needed expert consultant within the Labs. Much of the electro-optic modulator development was pioneered at the Labs in Holmdel in the 1960s and we were still the leaders of the field in 1983. I learned about making electro-optic modulators by reading *the* book written by a colleague, Ivan Kaminow.[28] I enlisted Larry Buhl to cut and polish the $LiTaO_3$ crystal for the modulator. Rod Alferness taught me about microwave impedance matching and provided the SMA "launchers" needed to match Leo's electronics with my parallel-plate transmission line modulator. One month after we decided to precool the atoms with a frequency-swept laser beam, we had a functioning, wideband gigahertz electro-optic modulator and driver and could begin to precool the atoms from our puffing source.

In the early spring of 1984, Leo and I started with a completely bare optical table, no vacuum chamber, and no modulator. Later that spring, John Bjorkholm, who had previously demonstrated the dipole force by focusing an atomic beam, joined our experiment. In the early summer, I recruited Alex Cable, a fresh graduate from Rutgers. Officially he was hired as my "technician": unofficially, he became a super-graduate student. In less than one year, we submitted our optical molasses paper.[29, 30] The two papers reporting the stopping of atomic beams [26, 27] were published one month earlier.

The apparatus we built to demonstrate optical molasses is shown in Figs. 3a and 3b. We had an ultra-high-vacuum chamber, but did not want to be hampered by long bake-out times to achieve good vacuum. Instead, we built a cryo-shield painted with Aquadag, a graphite-based substance. When cooled

Figure 3 a. A photograph of the apparatus used to demonstrate optical molasses and the first optical trap for atoms. The photograph is a double exposure made by photographing the apparatus under normal lighting conditions and then photographing the laser beams by moving a white card along the beam path in a darkened room. The 10 Hz pulsed laser used to evaporate the sodium pellet (doubled YAG at 532nm) appears as dots of light.

Figure 3 b. Art Ashkin and the author in front of the apparatus in 1986, shortly after the first optical trapping experiment was completed.

to liquid-nitrogen temperatures, the shield became a very effective sorption pump: we could open the vacuum chamber one day and be running by the next day. Fast turnaround time has always been important to me. Mistakes are unavoidable, so I wanted an apparatus that would allow mistakes to be corrected as rapidly as possible.

The first signals of atoms confined in optical molasses showed confinement times of a few tens of milliseconds, but shortly afterwards we improved the storage time by over an order of magnitude. Surprisingly, it took us a week after achieving molasses to look inside the vacuum can with our eyes instead of with a photomultiplier tube. When we finally did, we were rewarded with the sight shown in Fig. 4.

In this early work, the laser beams were aligned to be as closely counter-propagating as we could manage. A year later, we stumbled onto a misalignment configuration that produced another order of magnitude increase in the storage time. This so-called "super-molasses" alignment of our beams also created a compression of the atoms into a region of space on the order of 2 mm diameter from an initial spread of 1 cm. We were never able to understand this phenomena and after a number of attempts, published a brief summary of these results in conference proceedings.[31]

In our first molasses work, we realized that the traditional method of measuring the temperature by measuring the Doppler broadening of an atomic resonance line would not work for the low temperatures we hoped to achieve. Instead we introduced a time-of-flight technique to directly measure the vel-

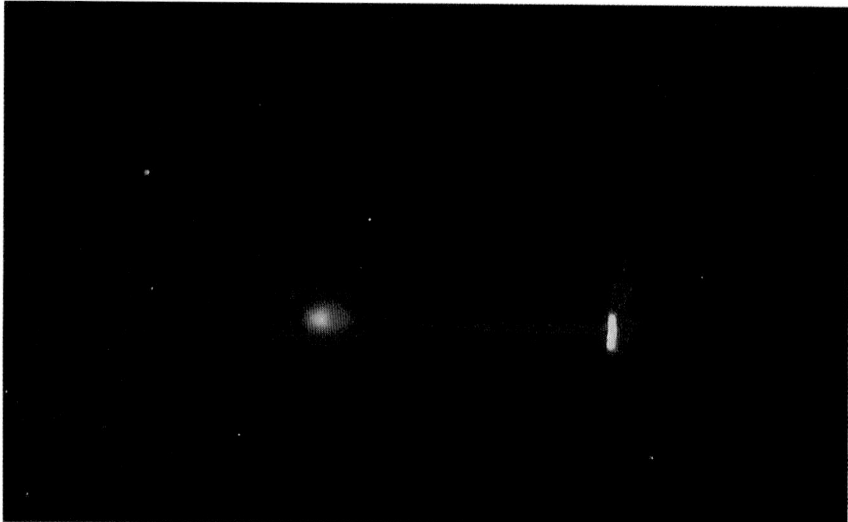

Figure 4. A photograph of sodium atoms confined in optical molasses experiment. The atoms were pre-cooled by a counterpropagating laser before entering the region of six crossed laser beams.

ocity distribution of the atoms. After allowing the atoms in the molasses to come to equilibrium, we turned off the light for a variable amount of time. The fast atoms escaped ballistically during this time while the slower atoms were recaptured by the molasses. Our first measurements showed a temperature of 185μK, slightly lower than the minimum temperature allowed by the theory of Doppler cooling. We then made the cardinal mistake of experimental physics: instead of listening to Nature, we were overly influenced by theoretical expectations. By including a fudge factor to account for the way atoms filled the molasses region, we were able to bring our measurement into accord with our expectations.

ON TO OPTICAL TRAPPING

Once we demonstrated optical molasses, we began to explore ways to achieve our original goal of optically trapping atoms. As a point of reference, Bill Phillips and his collaborators had reported the magnetic trapping of sodium atoms[32] two weeks before our optical molasses paper came out. Although the $1/e$ storage of the molasses confinement in our first experiment was a respectable $\tau \sim 0.36$ sec, optical molasses does not provide a restoring force that would push the atoms to the center of the trap.

Despite the fact that we were in possession of a great source of cold atoms, the path to trapping was not yet clear to us for a number of reasons: (*i*) Optical traps based strictly on the scattering force seemed to be ruled out because of a no-trapping theorem referred to as the "Optical Earnshaw Theorem". This theorem was published in response to earlier proposals to make atom traps based on the scattering force.[33, 34] (*ii*) We believed a trap based on an opposing-beam geometry was not viable because of the severe stimulated heating effects. (*iii*) Finally, we ruled out a single focused laser beam because of the tiny trapping volume. We were wrong on all counts.

Immediately after the molasses experiment, we tried to implement a large-volume ac light trap suggested by Ashkin.[35] Our attempt failed, and after a few months, we began to cast around for other alternatives. One possibility was another type of ac trap we proposed at a conference talk in December of 1984,[36] but we wanted something simpler. Sometime in the winter of 1986, during one of our brainstorming sessions on what to do next, John Bjorkholm tried to resurrect the single focused beam trap first proposed in Ashkin's 1978 paper.[17] I promptly rejected the idea because of the small trapping volume. A ~1-watt laser focused to produce a ~5-mK-deep trap would have a trapping volume of ~10^{-7}cm^3. Since the density of atoms in our optical molasses was 10^6 atoms/cm^3, we would capture fewer than one atom in a trap surrounded by 10^6 atoms in molasses. A day or two after convincing the group that a trap based on a focused laser beam would not work, I realized that many more atoms would be captured by the trap than my original estimate. An atom close to the trap might not be immediately captured, but it would have repeated opportunities to fall into the trap during its random walk in optical molasses.

The trap worked. We could actually see the random walk loading with our own eyes. A tiny dot of light grew in brightness as more atoms fell into the trap. During the first days of trapping success, I ran up and down the halls, pulling people into our lab to share in the excitement. My director, Chuck Shank, showed polite enthusiasm, but I was not sure he actually picked out the signal from the reflections in the vacuum can windows and the surrounding fluorescence. Art Ashkin came down with the flu shortly after our initial success. He confessed to me later that he began to have doubts: as he lay in bed with a fever; he wasn't sure whether the fever caused him to imagine we had a working trap.

We tried to image the tiny speck of light onto an apertured photomultiplier tube, but the slightest misalignment would include too much light from the surrounding molasses. It was a frustrating experience not to be able to produce a repeatable signal on a photomultiplier tube if we could actually see the atoms with our eyes. Then it dawned on me: if we could see the signal with our eyes, we could record it with a sensitive video camera and then analyze the video tape! A local RCA representative, tickled by the experiment, loaned us a silicon-intensified video camera. Our trapping paper included a photo of our trapped atoms, the first color picture published in Physical Review Letters.[37]

As we began the atom trapping, Art decided to trap micron-sized particles of glass in a single focused beam as a "proof of principle" for the atom trap. Instead of an atom in optical molasses, he substituted a silica (glass) sphere embedded in water. A micron-sized sphere is far more polarizable than an atom and Ashkin felt that it could be trapped at room temperature if the intensity gradient in the axial direction that would draw the glass bead into the focus of the light could overcome the scattering force pushing the particle out of the trap. This more macroscopic version of the optical tweezers trap was demonstrated quickly and gave us more confidence that the atom trap might work.[38] At that time, none of us realized how this simple "toy experiment" was going to flower.

Shortly after we demonstrated the optical trap, I hired Mara Prentiss as a new permanent staff member in my department. She began to work on the super-molasses riddle with us when I got a phone call from Dave Pritchard at MIT. He told me he and his student, Eric Raab, had been working on a scattering-force trap that would circumvent the Optical Earnshaw Theorem.[33] This theorem states that a scattering-force trap is impossible provided the scattering force $\mathbf{F}_{scatt}$ is proportional to the laser intensity **I**. The proof was straight forward: $\nabla \cdot \mathbf{F}_{scatt} = 0$ since any region in empty space must have the net intensity flux inward equal to the flux outward. Thus there cannot be a region in space where all force lines $\mathbf{F}_{scatt}$ point inward to a stable trapping point. Pritchard, Carl Wieman, and their colleagues had noted that the assumption $\mathbf{F}_{scatt} \propto \mathbf{I}$ need not be true.[39] They went on to suggest possible combinations of external magnetic or electric fields that could be used to create a stable optical trap.

Raab had had difficulties in getting a scattering-force trap to work at MIT and, as a last attempt before giving up, they asked if we were interested in collaborating with them on this work. The basic idea is illustrated in Fig. 5 for the case of an atom with F = 1 in the ground state and F = 2 in the excited state, where F is the total angular momentum quantum number. A weak spherical quadrupole trap magnetic field would split the Zeeman sublevels of a multilevel atom illuminated by counterpropagating circularly polarized laser beams. Due to the slight Zeeman-shift, an atom to the right of the trap center optically pump predominantly into the $m_F = -1$ state. Once in this state, the large difference in the scattering rates for σ^- light and σ^+ light causes the atom to experience a net scattering force towards the trap center. Atoms to the left of center would scatter more photons from the σ^+ beam. Since the laser beams remain tuned below all of the Zeeman split resonance lines, optical molasses cooling would still be occurring. The generalization to three dimensions is straightforward.

All we needed to do, was insert a pair of modest magnetic-field coils into our apparatus to test this idea. I wound some refrigeration tubing for the magnetic-field coils, but had to tear myself away to honor a previous commitment to help set up a muonium spectroscopy experiment with Allan Mills, Ken Nagamine and collaborators.[40] A few days later, the molasses was running again, and I received a call at the muon facility in Japan from Alex, his voice trembling with excitement. The trap worked spectacularly well and the atom cloud was blindingly bright compared to our dipole trap. Instead of the measly 1000 atoms we had in our first trap, they were getting 10^7 to 10^8 atoms.[41]

The basic idea for the trap was due to Jean Dalibard, a protegé of Claude Cohen-Tannoudji. His idea was stimulated by a talk given by Dave Pritchard on how the Earnshaw theorem could be circumvented. I called Jean in Paris

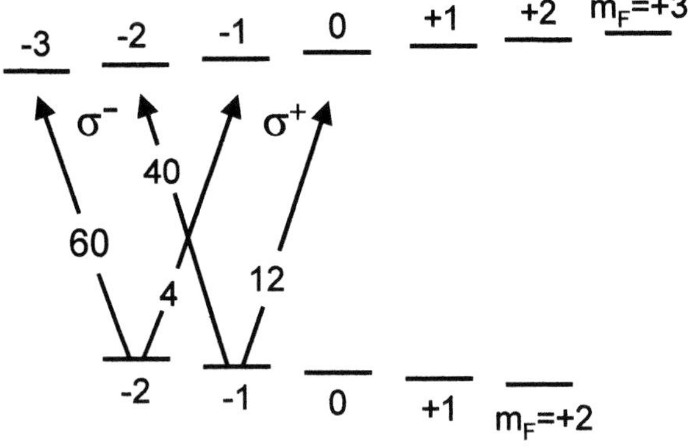

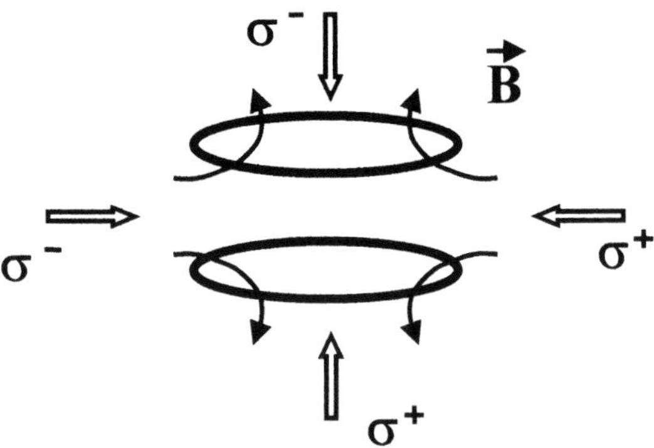

Figure 5. The magneto-optic trap for atoms with an F = 2 ground state and a F = 3 excited state. The slight energy level shifts of the Zeeman sub-levels cause symmetry to be broken and the atoms to optically pump predominantly into either the $m_F = +2$ or $m_F = -2$ state for **B**<0 or **B**>0. Once in the optically pumped states, the atoms are pushed towards the **B** = 0 region due to the large difference in relative strengths of the transition rates. The relative transition rates for σ^+ and σ^- light for the $m_F = -2$ and -1 states are shown.

to convince him that his name should go on the paper we were writing. Jean is both brilliant and modest, and felt it would be inappropriate to be a co-author since he did not do any of the work.[42]

The magneto-optic trap (commonly referred to as the MOT) immediately seized the attention of the growing community of coolers and trappers. Carl Wieman's group showed that atoms could be directly loaded from a tenuous vapor without the intermediate step of slowing an atomic beam.[43] By increasing the size of the laser beams used in the trap, Kurt Gibble and I showed that as many as $\sim 4 \times 10^{10}$ atoms could be trapped.[44] Wolfgang Ketterle, Pritchard, *et al.*[45] showed that the density of atoms in the MOT could be increased significantly by causing atoms to scatter less light in the central portion of the trap by blocking the repumping beam in that region of space. Stimulated by their shadowing idea, my collaborators and I at Stanford showed that by simply turning off the repumping light at the end stage of molasses dramatically increases the density of low-temperature atoms in the MOT.[46]

Figure 5 b. A photograph of atoms confined in a magneto-optic trap. The line of fluorescence below the ball of trapped atoms is due to the atomic beam used to load the trap.

The invention and development of the MOT exemplifies how the field of laser cooling and trapping grew out of the combined ideas and cooperation of an international set of scientists. For this reason, I find it especially fitting that the magneto-optic trap is the starting point of most experiments using laser-cooled atoms.

OPTICAL MOLASSES REVISITED

In the winter of 1987, I decided to leave the "ivory tower" of Bell Labs and accept an offer to become a professor at Stanford. When I left Bell Labs, we had just demonstrated the magneto-optic trap and it was clear the trap would provide an ideal starting point for a number of experiments. I arrived at Stanford in the fall of 1987, not knowing how long it would take to build a new research team.[47] Bill Phillips and Claude Cohen-Tannoudji were assembling powerful teams of scientists that could not be duplicated at Stanford. Dave Pritchard had been cultivating a powerful group at MIT. Other "home run hitters" in atomic physics such as Carl Wieman and Alain Aspect had just entered the field. Meanwhile, I had to start over again, writing proposals and meeting with prospective graduate students. If I had thought carefully about starting a new lab in the face of this competition, I might not have moved.

As with many aspects of my career, I may not have made the smart move, but I made a lucky move. From 1988 to 1993, I entered into the most productive time in my scientific career to date. My first three graduate students were Mark Kasevich, Dave Weiss, and Mike Fee. I also had two post-docs, Yaakov Shevy and Erling Riis who joined my group my first year at Stanford.

By January of 1988, Dave and Yaakov had a magneto-optic trap going in the original chamber we used to demonstrate optical molasses and the dipole trap. The plan was to improve the optical trapping techniques and then use the new laser cooling and trapping technology to explore new physics that could be accessed with cold atoms.

Another chamber was being assembled by Mark and Erling with the intent of studying the "quantum reflection" of atoms from cold surfaces. While I was at Bell Labs, Allan Mills and Phil Platzman began to get me interested in studying quantum reflection with ultracold atoms. The problem can be simply posed as follows: consider an atom with a long de Broglie wavelength λ incident on an idealized, short-range, *attractive* potential. In general there is a transmitted wave and reflected wave, but in the limit where λ is much greater than the length scale of the potential, one gets the counterintuitive result that the probability of reflection goes to unity. A real surface potential has a power-law attraction of the form $1/z^n$ and has no length scale. An atom near a surface will experience an attractive van der Waals force with a $1/z^3$ potential and, further away, the attractive potential would become $1/z^4$ due to "retarded potential" effects first discussed by Casimir. There are also subtleties to the problem when inelastic scattering channels are included. This problem had attracted the attention of a considerable number of theorists and experimentalists.

My plan of research was soon tossed out the window by a discovery that sent shock waves through the laser cooling community. By 1987, other groups began to produce optical molasses in their labs and measured atom temperatures near the expected limit,[48, 49] but in the spring of 1988, Bill Phillips and co-workers reported that sodium atoms in optical molasses could be cooled to temperatures far below the limit predicted by theory.[50] The NIST group reported the temperature of sodium atoms cooled in optical molasses to be 43 ± 20 μK and that the temperature did not follow the frequency dependence predicted by theory. The result was so surprising, they performed three different time-of-flight methods to confirm their result. Within a few months, three separate groups led by Wieman, Cohen-Tannoudji, and myself verified that sodium and cesium atoms in optical molasses could be cooled well below the Doppler limit.

As with many big "surprises", there were earlier hints that something was amiss. My group had been discussing the "super-molasses" problem at conferences since 1986. At a laser spectroscopy conference in Åre, Sweden in 1987, the NIST group reported molasses lifetimes with a very different frequency dependence from the one predicted by the simple formula $\langle x^2 \rangle = 2D\tau/\alpha^2$ that we published in our first molasses paper.[51] This group also found that the trap was more stable to beam imbalance than had been expected. In our collective euphoria over cooling and trapping atoms, the research community had not performed the basic tests to measure the properties of optical molasses, and I was the most guilty.

At the end of June, 1988, Claude and I attended a conference on spin-polarized quantum systems in Torino, Italy and gave a summary of the new surprises in laser cooling known at that time.[52] After our talks, Claude and I had lunch and compared the findings in our labs. The theory that predicted the minimum temperature for two-level atoms was beyond reproach. We felt the lower temperatures must be due to the fact that the atoms we were playing with were *real* atoms with Zeeman sub-levels and hyperfine splittings. Our hunch was that the cooling mechanism probably had something to do with the Zeeman sub-level structure and not the hyperfine structure, since cesium ($\Delta\nu_{hfs}$ = 9.19 GHz) and sodium ($\Delta\nu_{hfs}$ = 1.77 GHz) were both cooled to temperatures corresponding to an rms velocity on the order 4 to 5 times the recoil velocity, $v_{recoil} = \hbar k/M$ and $k = 2\pi/\lambda$. By then we also knew that the magnetic field had to be reduced to below 0.05 Gauss to achieve the best cooling.

After the conference, Claude returned to Paris while I was scheduled to give several more talks in Europe. My next stop was Munich, where I told Ted Hänsch that I thought it had to be an optical pumping effect. My knowledge of optical pumping was rudimentary, so I spent half a day in the local physics library reading about the subject. I was getting increasingly discouraged when I came across an article that referred to "Cohen-Tannoudji states". It was beginning to dawn on me that Claude and Jean were better positioned to figure out this puzzle.

After Munich, I went to Pisa where I gave a talk about our positronium and muonium spectroscopy work.[53, 54, 55] There, I finally realized how molasses was cooling the atoms. The idea was stimulated by a intuitive remark made by one of the speakers during his talk, "...the atomic polarization responds in the direction of the driving light field..." The comment reminded me of a ball-and-stick model of an atom as an electron (cloud) tethered by a weakly damped harmonic force to a heavy nucleus. I realized the cooling was due to a combination of optical pumping, light shifts, and the fact that the polarization in optical molasses changed at different points in space. A linearly polarized laser field drives the atomic cloud up and down, while a circularly polarized field drives the cloud in a circle. In optical molasses, the x, y, and z-directed beams all have mutually perpendicular linear polarizations and the polarization of the light field varies from place to place. As a simple example, consider a one-dimensional case where two opposing light beams have mutually perpendicular linear polarizations as shown in Fig. 6. The electron cloud wants to rotate with an elliptical helicity that is dependent on the atom's position in space.

Another effect to consider is the ac Stark (light) shift. In the presence of light, the energy levels of an atom are shifted, and the amount of the shift is proportional to the coupling strength of the light. Suppose an atom with angular momentum F = 2 is in σ^+ light tuned below resonance. It will optically pump into the $m_F = +2$ state and its internal energy in the field will be lowered as shown in Fig. 6 c. If it then moves into a region in space where the light is σ^-, the transition probability is very weak and consequently the ac Stark shift is small. For sodium the $m_F = +2$ energy is lowered 15 times more in σ^+ light than in σ^- light. Hence, the atom gains internal energy by moving into the σ^- region. This increase in internal energy must come at the expen-

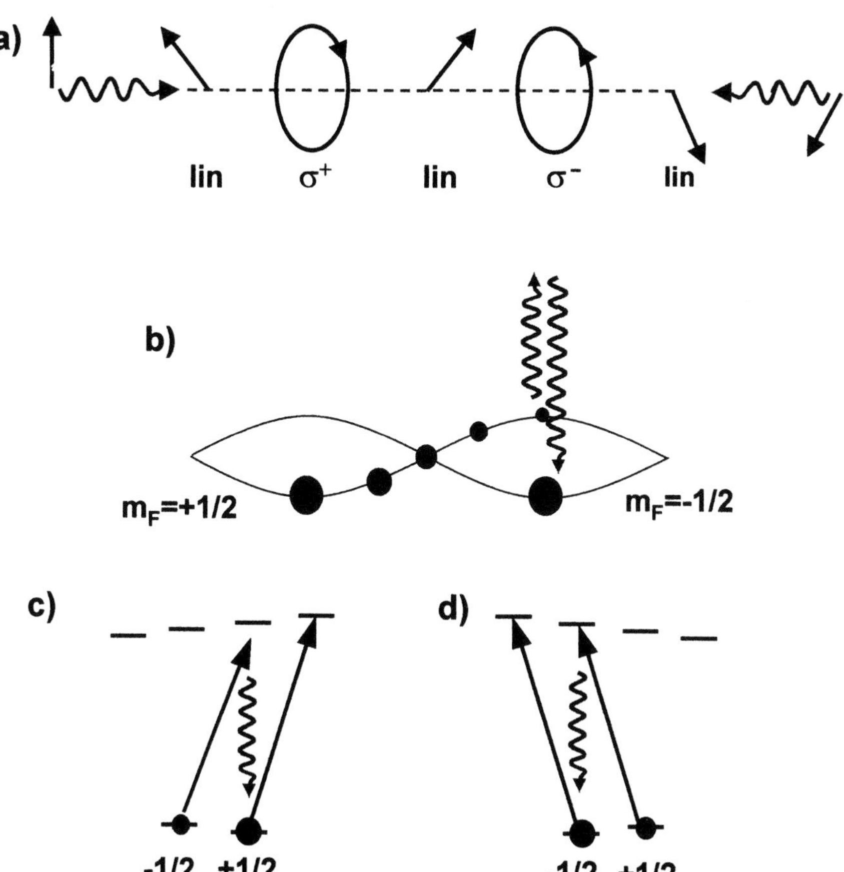

Figure 6. Polarization gradient cooling for an atom with an F = 1/2 ground state and an F = 3/2 excited state. a) The interference of two linearly polarized beams of light with orthogonal polarizations creates a field of varying elliptical polarization as shown. Under weak excitation, the atom spends most of its time in the ground states. b) The energy of the $m_F = \pm 1/2$ ground states as a function of position in the laser field is shown. c) An atom in a σ^+-field will optically pump mostly into the $m_F = +1/2$ ground state, the lower internal energy state. d) If the atom moves into a region of space where the light is σ^-, its internal energy increases due to the decreased light shift (AC Stark shift) of the $m_F = +1/2$ state in that field. The atom slows down as it goes up a potential hill created by the energy level shift. As the atom nears the top of the hill, it begins to optically pump to the $m_F = -1/2$ state, putting the atom into the lowest energy state again. Laser cooling by repeated climbing of potential hills has been dubbed "Sisyphus cooling" after the character in the Greek myth condemned forever to roll a boulder up a hill.

se of its kinetic energy. The final point is that the atom in the new region in space will optically pump into the $m_F = -2$ state (Fig. 6 d). Thus the atom will find itself again in a low-energy state due to the optical pumping process. The ensemble of atoms loses energy, since the spontaneously emitted photons are slightly blue-shifted with respect to the incident photons.

Cooling in polarization gradients is related to a cooling mechanism that occurs for two-level atoms in the presence of two counter-propagating laser beams. In the low-intensity limit, the force is described by our intuitive notion of scattering from two independent beams of light, as first discussed by

Hänsch and Schawlow.[21] However, at high intensities, the force reverses sign so that one obtains a cooling force for positive detuning. This cooling force has been treated by Gordon and Ashkin for all levels of intensity in the low velocity limit,[19] and by Minogin and Serimaa in the high-intensity limit for all velocities.[56] A physical interpretation of the cooling force in the high-intensity limit based on the dressed-atom description was given by Dalibard and Cohen-Tannoudji.[57] In their treatment, the atom gains internal energy at the expense of kinetic energy as it moves in the standing-wave light field. The gain in internal energy is dissipated by spontaneous emission, which is more likely to occur when the atom is at the maximum internal energy. When the atom makes a transition, it will find itself most often at the bottom of the dressed-state potential hills. Following Albert Camus, Jean and Claude again revived *Le Mythe de Sisyphe* by naming this form of cooling after the character in Greek mythology, Sisyphus, who was condemned eternally roll a boulder up a hill.

The name "optical molasses" takes on a more profound meaning with this new form of cooling. Originally, I conceived of the name thinking of a viscous fluid associated with cold temperatures: "slow as molasses in January". With this new understanding, we now know that cooling in optical molasses has two parts: at high speeds, the atom feels a viscous drag force, but at lower speeds where the Doppler shift becomes negligible, the optical pumping effect takes over. An atom sees itself walking in a swamp of molasses, with each planted foot sinking down into a lower energy state. The next step requires energy to lift the other foot up and out of the swamp, and with each sinking step, energy is drained from the atom.

The Pisa conference ended on Friday, and on Sunday, I went to Paris to attend the International Conference on Atomic Physics. That Sunday afternoon, Jean and I met and compared notes. It was immediately obvious that the cooling models that Jean and Claude and I concocted were the same. Jean, already scheduled to give a talk at the conference, gave a summary of their model.[58] I was generously given a "post-post-deadline" slot in order to give my account of the new cooling mechanism.[59]

Detailed accounts of laser cooling in light fields with polarization gradients followed a year later in a special issue of JOSA B dedicated to cooling and trapping. Dalibard and Cohen-Tannoudji provided an elegant quantum mechanical treatment of simple model systems.[60] They discussed two different types of cooling, depending on whether the counter-propagating light beams were comprised of mutually perpendicular linear polarizations or opposing σ^+ -σ^- beams. Their approach allowed them to derive the cooling force and diffusion of momentum (heating) as a function of experimental parameters such as detuning, atomic linewidth, optical pumping time, etc., that could be experimentally tested.

My graduate students and I presented our version of the Sisyphus cooling mechanism in the same issue.[61] In order to obtain quantitative calculations that we could compare to our experimental results, we chose to calculate the cooling forces using the optical Bloch equations generalized for the sodium F

= 2 ground state → F = 3 excited state transition. We derived the steady-state cooling forces as a function of atomic velocity for the same two simple polarization configurations, but for sodium atoms instead of a model system. However, we also showed considered by Dalibard and Cohen-Tannoudji that steady-state forces *cannot* be used to estimate the velocity distribution and that the transient response of the atom in molasses with polarization gradients was significant.[62] A weak point of our paper was that we made *ad hoc* assumptions in our treatment of the diffusion of momentum, and the predicted Monte Carlo calculations of the velocity distributions were sensitive to the details of these assumptions. Since that time, more rigorous quantum Monte Carlo methods have been developed.

In a companion experimental paper,[63] we measured non-thermal velocity distributions of atoms cooled in the laser fields treated in our theory paper. We also measured the velocity distributions of atoms cooled in σ^+–σ^+ light.[64] Under these conditions, sodium atoms will optically pump into an effective two-level system consisting of the $3S_{1/2,}$ F = 2, m_F = +2 and $3P_{3/2,}$ F = 3, m_F = +3 states. This arrangement allowed us finally to verify the predicted frequency dependence of the temperature for a two-level atom, three years after the first demonstration of Doppler cooling. In the course of those experiments, Dave Weiss discovered a magnetic-field-induced cooling[63] that could be explained in terms of Sisyphus-like effects and optical pumping.[61] This cooling mechanism was explored in further detail by Hal Metcalf and collaborators.[65]

The NIST discovery of sub-Doppler temperatures showed that the limiting temperature based on the Doppler effect was not actually a limit. What is the fundamental limit to laser cooling? One might think that the limit would be the recoil limit $k_BT \sim (p_r)^2/2M$, since the last photon spontaneously emitted from an atom results in a random velocity of this magnitude. However, even this barrier can be circumvented. For example, an ion tightly held in a trap can use the mass of the trap to absorb the recoil momentum. The so-called sideband cooling scheme proposed by Dehmelt and Wineland[66] and demonstrated by Wineland and collaborators[67] can in principle cool an ion so that the fractional occupancy of two states separated by an energy ΔE can have an effective temperature T_{eff} less than the recoil temperature, where T_{eff} is defined by $e^{-\Delta E/k}$ T_{eff}, i.e. the fraction of time the ion spends in the ground state.

For free atoms, it is still possible to cool an ensemble of atoms so that their velocity spread is less than photon recoil velocity by using velocity-selection techniques. The Ecolé Normale group devised a clever velocity-selection scheme based on a process they named "velocity selective coherent population trapping".[68] In their first work, metastable helium atoms were cooled along one dimension of an atomic beam to a transverse (one-dimensional) temperature of 2 μK, a factor of 2 below the single photon recoil temperature. The effective temperature of the velocity-selected atoms decreases roughly as the square root of the time that the velocity-selection light is on, so much colder temperatures may be achieved for longer cooling times. In subsequent experiments, they used atoms precooled in optical molasses and achieved much colder temperatures in two and three dimensions.[69] An important point to emphasize is that this method has no strict cooling limit: the longer the cooling time, the smaller the spread in velocity. However, there is a trade-off between the final temperature and the number of atoms cooled to this temperature since the atom finds it harder to randomly walk into a progressively smaller section of velocity space. Eventually, the velocity-selective cooling becomes velocity "selection" in the sense that the number of atoms in the velocity-selected state begins to decrease.

APPLICATIONS OF LASER COOLING AND TRAPPING

During the time we were studying polarization gradient cooling, my group of 3 students and one post-doc at Stanford began to apply the newly developed cooling and trapping techniques, but even those plans were soon abandoned.

After the completion of the studies of polarization gradient optical molasses cooling, Erling Riis, Dave Weiss and Kam Moler constructed a two-dimensional version of the magneto-optic trap where sodium atoms from a slowed atomic beam were collected, cooled in all three dimensions and compressed radially before being allowed to exit the trap in the axial direction.[70] This "optical funnel" increased the phase space density of an atomic beam by five orders of magnitude. Another five orders of magnitude are possible with a cesium beam and proper launching of the atoms in a field with moving polarization gradients. Ertmer and colleagues developed a two-dimensional compression and cooling scheme with the magneto-optic trap.[71] These two experiments demonstrated the ease with which laser cooling can be used to "focus" an atomic beam without the limitations imposed by the "brightness theorem" in optics.

In our other vacuum chamber, Mark Kasevich and Erling Riis were given the task of producing an atomic fountain as a first step in the quantum reflection experiment. That was to be Mark's thesis. The idea was to launch the atoms upwards in an atomic fountain with a slight horizontal velocity. When the atoms reached their zenith, they would strike a vertically oriented surface. As they were setting up this experiment, I asked them to do the first of a number of "quickie experiments". "Quickies" were fast diversions I promised would only take a few weeks, and the first detour was to use the atomic fountain to do some precision spectroscopy.

In the early 1950s, Zacharias attempted to make an "atomic fountain" by directing a beam of atoms upwards. Although most of the atoms would crash into the top of the vacuum chamber, the very slowest atoms in the Maxwell distribution were expected to follow a ballistic trajectory and return to the launching position due to gravity. The goal of Zacharias' experiment was to excite the atoms in the fountain with Ramsey's separated oscillatory field technique, the method used in the cesium atomic clock.[72] Atoms initially in state $|1\rangle$ would enter a microwave cavity on the way up and become excited into a superposition of two quantum states $|1\rangle$ and $|2\rangle$. While in that superposition state, the relative phases of the two states would precess with a frequency $\hbar\omega = E_1 - E_2$. When the atoms passed through the microwave cavity on the way down, they would again be irradiated by the microwave field. If the microwave generator were tuned exactly to the atomic frequency ω, the second pulse would excite the atoms completely into the state $|2\rangle$. If the microwave source were π radians (half a cycle) out of phase with the atoms, they would be returned to state $|1\rangle$ by the second pulse. For a time Δt separating the two excitation pulses, the oscillation "linewidth" $\Delta\omega_{rf}$ of this transition satisfies $\Delta\omega_{rf}\,\Delta t = \pi$.

This behavior is a manifestation of the Heisenberg uncertainty principle: the uncertainty ΔE in the measurement of an energy interval times the quantum measurement time Δt must be greater than Planck's constant $\Delta E \Delta t \geq \hbar$. An atomic fountain would increase the measurement time by more than two orders of magnitude as compared to conventional atomic clocks with horizontally moving thermal beams. Zacharias hoped to measure the gravitational redshift predicted by Einstein: identical clocks placed at different heights in a gravitational field will be frequency shifted with respect to each other. The atom fountain clock, during its trajectory, would record less time than the stationary microwave source driving the microwave cavity.

Unfortunately, Zacharias' experiment failed. The slowest atoms in the Maxwell-Boltzmann distribution were scattered by faster atoms overtaking them from behind and never returned to the microwave cavity. The failure was notable in several respects. The graduate student and post-doc working on the project still got good jobs, and the idea remained in the consciousness of the physics community.[73] With our source of laser cooled atoms, it was a simple matter for us to construct an atomic fountain.[74] Atoms were first collected in a magneto-optic trap and then launched upwards by pushing from below with another laser beam. At the top of the ballistic trajectory, we irradiated the atoms with two microwave pulses separated by 0.25 seconds, yielding a linewidth of 2 Hz. Ralph DeVoe at the IBM Almaden Research center joined our experiment and provided needed assistance in microwave technology. With our demonstration atomic fountain, we measured the sodium ground-state hyperfine splitting to an accuracy of one part in 10^9.

After the theory of polarization-gradient cooling was developed, we realized that there was a much better way to launch the atoms. By pushing with a single laser beam from below, we would heat up the atoms due to the random recoil kicks from the scattered photons. However, by changing the frequency of the molasses beams so that the polarization gradients would be in a frame of reference moving relative to the laboratory frame, the atoms would cool to polarization-gradient temperatures in the moving frame. The atoms could be launched with precise velocities and with no increase in temperature.[75]

André Clairon and collaborators constructed the first cesium atomic fountain.[76] Kurt Gibble and I analyzed the potential accuracy of an atomic fountain frequency standard and suggested that the phase shifts due to collisions might be a limiting factor to the ultimate accuracy of such a clock.[77] We then constructed an atomic fountain frequency source that surpassed the short term stability of the primary Cs references maintained by standards laboratories.[78] In that work, we also measured the frequency shift due to ultracold collisions in the fountain, a systematic effect which may be the limiting factor of a cesium clock. The group led by Clairon has recently improved upon our short-term stability. More important, they achieved an accuracy estimated to be $\Delta\nu/\nu \leq 2\times10^{-15}$, limited by the stability of their hydrogen maser reference.[79] Such a clock started at the birth of the universe would be off by less than four minutes today, ~15 billion years later.

The next "quickie" to follow the atomic fountain was the demonstration of normal incidence reflection of atoms from an evanescent wave. Balykin, *et al.*[80] deflected an atomic beam by a small angle with an evanescent sheet of light extending out from a glass prism. If the light is tuned above the atomic resonance, the induced dipole **p** will be out of phase with the driving field. The atom with energy $-\mathbf{p}\cdot\mathbf{E}$ is then repelled from the light by the dipole force. The demonstration of normal incidence reflection with laser-cooled atoms was a necessary first step towards the search for quantum reflection. With our slow atoms, we wanted to demonstrate an "atomic trampoline" trap by bouncing atoms from a curved surface of light created by internally reflecting a laser beam from a plano-concave lens. Unfortunately, the lens we used produced a considerable amount of scattered light, and the haze of light "levitated" the atoms and prevented us from seeing bouncing atoms. Mark ordered a good quality lens and we settled for bouncing atoms from the dove-prism surface[81] with the intent of completing the work when the lens arrived. We never used the lens he ordered because of another exciting detour in our research. A few years later, a trampoline trap was demonstrated by Cohen-Tannoudji's group.[82] The evolution of gravito-optic atom traps is summarized in Fig. 7.

While waiting for the delivery of our lens, we began to think about the next stage of the quantum reflection experiment. The velocity spread of the atoms in the horizontal direction would be determined by a collimating slit, but I was unhappy with this plan. The quantum reflection experiment would require exquisitely cold atoms with a velocity spread corresponding to an effective temperature of a small fraction of a micro-kelvin. Given the finite size of our atoms confined in the MOT, very narrow collimating slits would reduce the flux of atoms to distressingly low levels. Ultimately, collimating slits would cause the atoms to diffract.

While flying home from a talk, the solution to the velocity-selection problem came to me. Instead of using collimating slits, we could perform the velocity selection with the Doppler effect. Usually, the Doppler sensitivity is limited by the linewidth of the optical transition. However, if we induced a two-photon transition between two ground states with lasers beams at frequencies ν_1 and ν_2, there is no linewidth associated with an excited state. If the frequency of ν_2 is generated by an electro-optic modulator so that $\nu_2 = \nu_1 + \nu_{rf}$, the frequency jitter of the excitation laser would not enter since the transition would depend on the frequency difference $\nu_2-\nu_1 = \nu_{rf}$. The linewidth $\Delta\nu$ would be limited by the transition time Δt it took to induce the two-photon transition, and our atomic fountain would give us lots of time. Despite the fact that the resonance would depend on the frequency *difference*, the

Balykin, Letokhov, Ovchinnikov and Sidorov (1987)

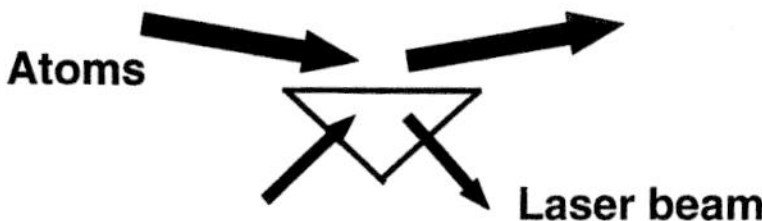

Kasevich, Weiss and Chu (1989)

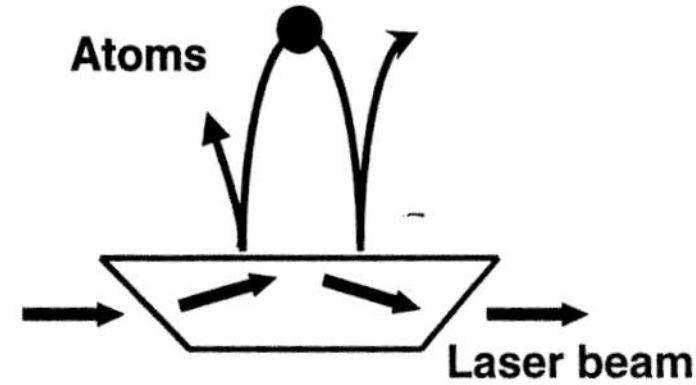

Aminoff, *et al.*, (1993)

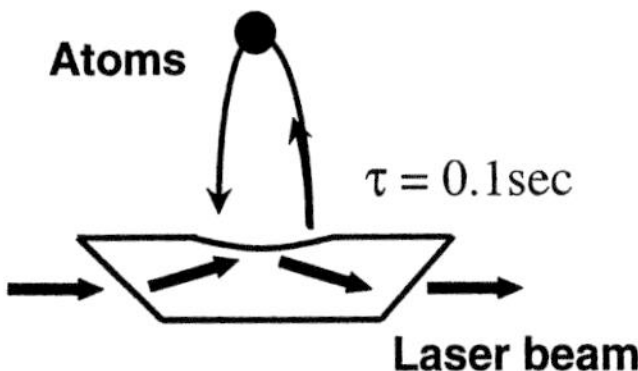

H.J. Lee, *et al.*, (1996)

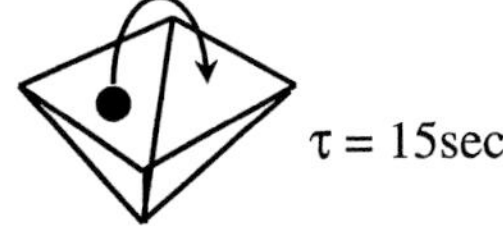

Inverted pyramid trap from 4 sheets of light.

Figure 7. Some of the highlights in the development of the reflection of atoms from optical sheets of light tuned to the blue side of the atomic resonance. Since the atoms spend a considerable amount of time in free fall, blue de-tuned optical traps allow fairly efficient cooling with stimulated Raman pulses. Over 3×10^6 sodium atoms have been Raman cooled to the recoil temperature at densities of 2×10^{11} atoms/cm^3 in an inverted pyramid trap.

Doppler sensitivity would depend on the frequency *sum* if the two laser beams were counter-propagating. This idea would allow us to achieve Doppler sensitivity equivalent to an ultraviolet transition, but with the frequency control of the microwave domain, and it would be easy. In a proof of principle experiment, we created an ensemble of atoms with a velocity spread of 270 μm/sec corresponding to an effective 1-dimensional "temperature" of 24 picokelvin and a de Broglie wavelength of 51 μm.[83] We also used the Doppler sensitivity to measure velocity distributions with sub-nanokelvin resolution.

By 1990, we were aware of several groups trying to construct atom interferometers based on the diffraction of atoms by mechanical slits or diffraction gratings, and their efforts stimulated us to think about different approaches to atom interferometry. We knew there is a one-to-one correspondence between the Doppler sensitivity and the recoil an atom experiences when it makes an optical transition. With a two-photon Raman transition with counter-propagating beams of light, the recoil is $\Delta p = \hbar k_{eff}$, where $k_{eff} = (k_1+k_2)$, and it is this recoil effect that allowed us to design a new type of atom interferometer.

If an atom with momentum p and in state $|1\rangle$, described by the combined quantum state $|1,p\rangle$, is excited by a so-called "$\pi/2$" pulse of coherent light, the atom is driven into an equal superposition of two states $|1,p\rangle$ and $|2,p+\hbar k_{eff}\rangle$. After a time Δt, the two wave packets will have separated by a distance $(\hbar k_{eff}/M)\Delta t$. Excitation by a π pulse induces the part of the atom in state $|1,p\rangle$ to make the transition $|1,p\rangle \rightarrow |2,p+\hbar k_{eff}\rangle$ and the part of the atom in $|2,p+\hbar k_{eff}\rangle$ to make the transition $|2,p+\hbar k_{eff}\rangle \rightarrow |1,p\rangle$. After another interval Δt, the two parts of the atom come back together and a second $\pi/2$ pulse with the appropriate phase shift with respect to the atomic phase can put the atom into either of the states $|1,p\rangle$ or $|2,p+\hbar k_{eff}\rangle$. This type of atom interferometer is the atomic analog of an optical Mach-Zender interferometer and is closely related to an atom interferometer first discussed by Bordé.[84] In collaboration with the PTB group led by Helmcke, Bordé used this atom interferometer to detect rotations.[85]

By January of 1991, shortly after we began seeing interference fringes, we heard that the Konstanz group led by Jürgen Mlynek[86] had demonstrated a Young's double slit version atom interferometer and that the MIT group, led by Dave Pritchard[87] had succeeded in making a grating interferometer. Instead of using atoms in a thermal beam, we based our interferometer on an atomic fountain of laser-cooled atoms. We knew we had a potentially exquisite measuring device because of the long measurement time and wanted to use our atom interferometer to measure something before we submitted a paper.

As we began to think of what we could easily measure with our interferometer, Mark made a fortuitous discovery: the atom interferometer showed a phase shift that scaled as Δt^2, the delay time between the $\pi/2$ and π pulses, and correctly identified that this phase shift was due to the acceleration of the atoms due to gravity. An atom accelerating will experience a Doppler shift with respect to the lasers in a laboratory frame of reference propagating in the direction of **g**. Even though the laser beams were propagating in the nominally horizontal direction, the few milliradian "misalignment" created enough of phase shift to be easily observable.

Our analysis of this phase shift based on Feynman's path-integral approach to quantum mechanics was outlined in the first demonstration of our atom interferometer[88] and expanded in our subsequent publications.[89, 90, 91, 92] Storey and Cohen-Tannoudji have published an excellent tutorial paper on this approach as well.[93] Consider a laser beam propagating parallel to the direction of gravity, as shown in Fig 8. The phase shift of the atom has two parts: (*i*) a free-evolution term $e^{iS_{Cl}/\hbar}$, where

$$S_{Cl} = \int_r Ldt$$

is the action evaluated along the classical trajectory *r*, and (*ii*), a phase term due to the atom interacting with the light. The evaluation of the integrals for both paths shows that the free-evolution part contributes no net phase shift between the two arms of the interferometer. The part of the phase shift due to the light/atom interaction is calculated by using the fact that an atom that makes a transition $|1,p\rangle \rightarrow |2,p+\hbar k\rangle$ acquires a phase factor $e^{-i(k_L z-\omega t)}$, where z is the vertical position of the atom and $k_L = k_1 + k_2$ is the effective *k* vector of the light. A transition $|2,p+\hbar k\rangle \rightarrow |1,p\rangle$ adds a phase factor $e^{+i(k_L z-\omega t)}$. If the atom does not make a transition, the phase factor due to the light is unity. If k_L is parallel to g, the part of the atom in the upper path will have a total phase $\phi_{upper} = k_L \cdot (z_A - z_B)$ read into the atom by the light. The part of the atom in the lower path will have a given phase angle $\phi_{lower} = k_L \cdot (z_{B'} - z_C)$. In the absence of gravity, $z_A - z_B = z_{B'} - z_C$, and there is no net phase shift between the two paths. However, with gravity, $z_B - z_A = g\Delta t^2/2$, while $z_{B'} - z_C = 3g\Delta t^2/2$. Thus the net phase shift is $\Delta\phi = k_L g \Delta t^2$. Notice that the acceleration is measured in the

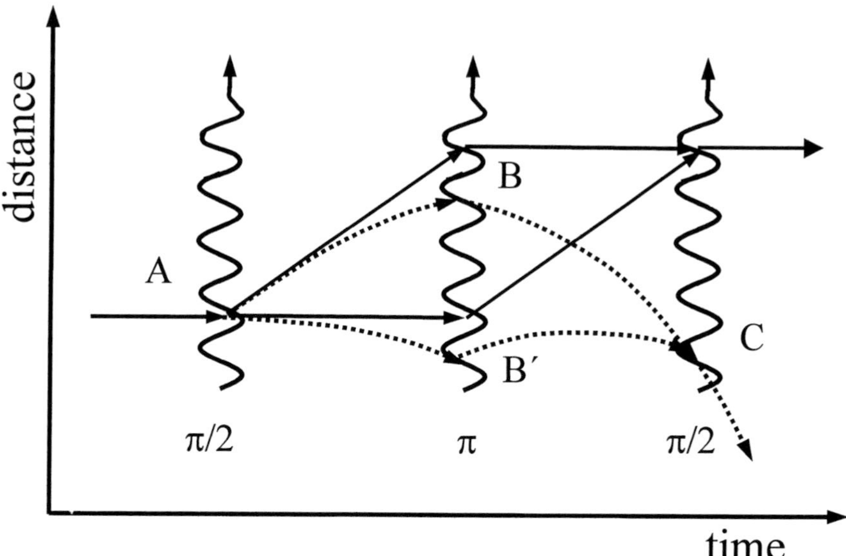

Figure 8. An atom interferometer based on optical pulses of light. The phase of the optical field is read into the atom during a transition from one state to another. In the absence of gravity (solid lines), the part of the atom moving along the upper path from the first π/2 pulse to the π pulse will experience 3 cycles of phase less than the part of the atom that moves along the lower path. The lower path experiences an identical loss of phase during the time between the π pulse and the second π/2 pulse. If the excitation frequency is exactly on resonance, the atom will be returned back to the initial state. With gravity (dotted lines), the phase loss of the upper path is more than the phase loss of the lower path since $z_C - z_{B'}$ is greater than $z_B - z_A$. Thus, by measuring displacements during a time Δt in terms of a phase difference, we can measure velocity changes due to g or photon recoil effects (Fig. 9) in the time domain.

time domain: we record the change in phase $\Delta\phi = k_L\Delta z$ that occurs after a time Δt.

The phase shift of the interferometer is measured by considering the relative phase between the atomic phase of the two parts of the atom and the phase of the light at position C. If the light is in phase with the atom, the second $\pi/2$ pulse will cause the atom to return to state $|1,p\rangle$. If it is π out of phase, the atom will be put into state $|2,p+\hbar k\rangle$. Thus the phase shift is measured in terms of the relative populations of these two states.

For the long interferometer times Δt that are obtainable in an atomic fountain, the phase shift can be enormous. For $\Delta t = 0.2$ seconds, over 4×10^6 cycles of phase difference accumulate between the two paths of the interferometer. In our first atom interferometer paper, we demonstrated a resolution in g of $\Delta g/g = 10^{-6}$, and with improved vibration isolation, achieved a resolution of $\Delta g/g<3\times10^{-8}$.[89] With a number of refinements, including the use of an actively stabilized vibration isolation system,[94] we have been able maintain the full **fringe contrast for times up to Δt = 0.2 seconds and have improved the fractional resolution to $\Delta g/g \sim 10^{-10}$.[92]**

Soon after the completion of our first atom interferometer measurements, Mark Kasevich thought of a way to cool atoms with stimulated Raman transitions.[95] Since the linewidth of a Raman transition is governed by the time to make the transition, we had a method of addressing a very narrow velocity slice of atoms within an ensemble already cooled by polarization-gradient cooling. Atoms are initially optically pumped into a particular hyperfine state $|1\rangle$. A Raman transition $|1\rangle \rightarrow |2\rangle$ is used to push a small subset of them towards v = 0. By changing the frequency difference $\nu_1-\nu_2$ for each successive pulse, different groups of atoms in the velocity distribution can be pushed towards $v = 0$, analogous to the "frequency chirp" cooling methods used to slow atomic beams. A critical difference is that Raman pulses permit much higher-resolution Doppler selectivity. After each Raman pulse, a pulse is used to optically pump the atom back into $|1\rangle$. During this process, the atom will spontaneously emit one or more photons and can remain near $v = 0$. The tuning of the optical pulses are adjusted so that an atom that scatters into a velocity state near $v = 0$ will have a low probability of further excitation. This method of cooling is analogous to coherent population trapping except that the walk in velocity space is directed towards $v = 0$. In our first demonstration of this cooling process, sodium atoms were cooled to less than $0.1T_{recoil}$ in one dimension, with an 8-fold increase in the number of atoms near $v = 0$. In later work, we extended this cooling technique to two and three dimensions.[96]

This cooling technique has also been shown to work in an optical dipole trap. We were stimulated to return to dipole traps by Phillips' group[97] and by Dan Heinzen's group,[98] who demonstrated dipole traps tuned very far from resonance. In this type of trap, the heating due to the scattering of trapping light is greatly reduced. A nondissipative dipole trap turns out to be useful in a number of applications. In traps formed by sheets of blue detuned light (a successor to the trampoline traps), we showed that atomic coherences can be preserved for 4 seconds despite hundreds of bounces.[99] We also demonstrated evaporative cooling in a red detuned dipole trap made from two crossed beams of light.[100, 101] Atoms can be Raman cooled in both red and blue detuned dipole traps.[102, 46] In our most recent work, over 10^6 atoms have been Raman cooled in a blue-detuned dipole trap to less than T_{recoil}. This is a factor of ~300 below what is needed for Bose condensation, but a factor of 400 improvement over the "dark-spot" magneto-optic trap phase space densities. Unfortunately, a heating process prevented us from evaporatively cooling to achieve Bose condensation with an optical trap. Recently, Wolfgang Ketterle and collaborators have loaded an optical dipole trap with a Bose condensate created in a magnetic trap.[103] With this trap they have been able to find the Feshbach resonances[104] calculated for sodium[105] in which the s-wave scattering length changes sign. Since this resonance is only one gauss wide, a non-magnetic trap makes its detection much easier. Optical traps could also be used to hold Bose condensates in magnetic-field-insensitive states for precision atom interferometry.

Our ability to measure small velocity changes with stimulated Raman transitions suggested another application of atom interferometry. If an atom absorbs a photon of momentum $p_\gamma = h\nu/c$, it will receive an impulse $\Delta p = M\Delta v$. Thus $h/M = c\Delta v/\nu$, and since Δv can be measured as a frequency shift, the possibility of making a precision measurement of h/M dropped into our lap. After realizing this opportunity, I called Barry Taylor at NIST and asked if this measurement with an independent measurement of Planck's constant could put the world on an atomic mass standard. He replied that the first application of a precise h/M measurement would be a better determination of the fine-structure constant α, since α can be expressed as

$$\alpha^2 = (2R_\infty/c)\,(m_p/m_e)\,(M_{atom}/m_p)\,(h/M_{atom}).$$

All of the quantities[106] in the above relation can be measured precisely in terms of frequencies or frequency shifts.

Dave Weiss' thesis project was changed to measure h/M with our newly acquired Doppler sensitivity. The interferometer geometry he chose was previously demonstrated as an extension of the Ramsey technique into the optical domain.[84] If two sets of $\pi/2$ pulses are used, two interferometers are created with displaced endpoints as shown in Fig. 9. The displacement is measured in terms of a phase difference in the relative populations of the two interferometers, analogous to the way we measured the acceleration of gravity. This displacement was increased by sandwiching a number of π pulses in between the two pairs of $\pi/2$ pulses. With Brent Young, Dave Weiss obtained a resolution of roughly a part in 10^7 in h/M_{Cs}.[107, 108] In that work, systematic effects were observed at the 10^6 level, but rather than spending significant time to understand those effects, we decided to develop a new atom interferometer method with a vertical geometry to measure h/M.

Instead of using impulses of momentum arising from off-resonant Raman pulses, I wanted to use an adiabatic transfer method demonstrated by Klaus Bergmann and collaborators.[109] The beauty of an adiabatic transfer method

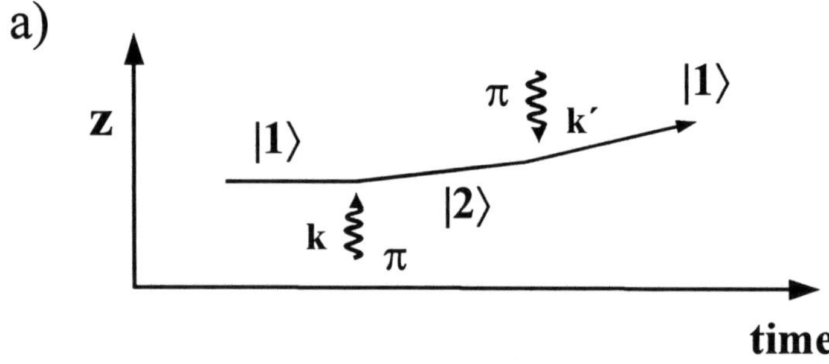

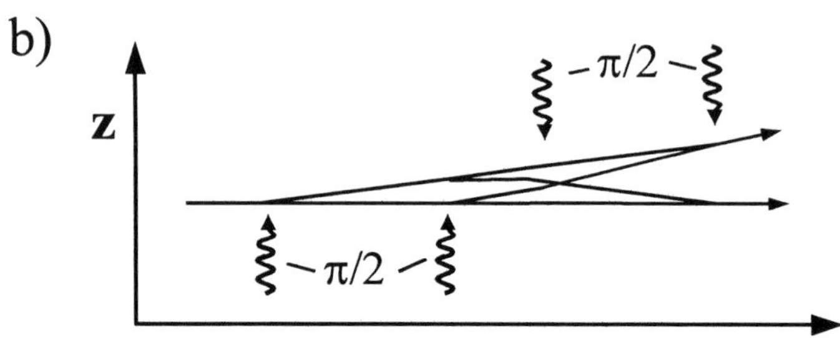

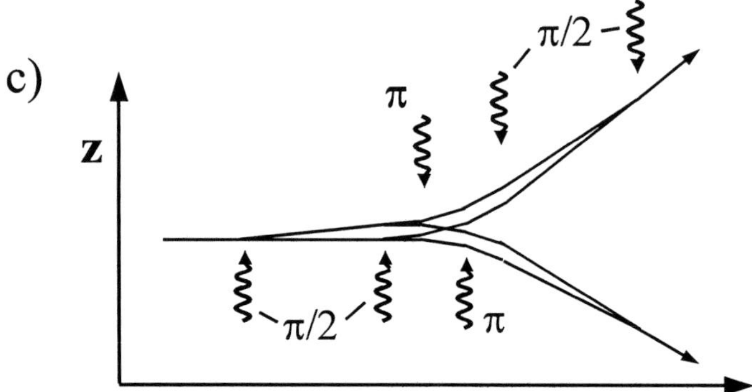

Figure 9 a. The basic method for measuring the recoil velocity requires two counterpropagating pulses of light. Because of energy and momentum conservation, the excitation light must satisfy $\hbar\omega - \hbar\omega_{12} = \mathbf{k}\cdot\mathbf{v} \pm (\hbar k)^2/2m$, where v is the velocity of the atom relative to the k-vector of the light and the sign depends on whether the initial state is higher or lower in energy than the final state. An atom at rest in ground state $|1\rangle$ is excited by a π pulse at frequency $\omega = \omega_{12} + \hbar k^2/2m$. The atom, recoiling with velocity $v = \hbar k/m$ in state $|2\rangle$ is returned back to $|1\rangle$ with a counterpropagating photon $\omega' = \omega_{12} - \hbar kk'/m + \hbar k^2/2m$. The two resonances are shifted relative to each other by $\Delta\omega = \omega' - \omega = \hbar(k+k')^2/2m$.

Figure 9 b. In order to increase the resolution without sacrificing counting rate, two sets of counterpropagating π/2 pulses are used instead of two π pulses. Thus, we are naturally led to use two atom interferometers whose end points are separated in space due to the photon recoil effect. Since the measurement is based on the relative separation of two similar atom interferometers, there are a number of "common mode" subtractions that add to the inherent accuracy of the experiment.

Figure 9 c. To further increase the resolution of the measurement, we sandwich many π pulses, each pulse coming from alternate directions. Only 2 π pulses are shown in the figure, but up to 60 π pulses are used in the actual experiment, where each π pulse separates the two interferometers by a velocity of $4\hbar k/m$.

is that it is insensitive to the small changes in experimental parameters such as intensity and frequency that adversely affect off-resonant π pulses. In addition, we showed that the ac Stark shift, a potentially troublesome systematic effect when using off-resonant Raman transitions, is absent when using adiabatic transfer in a strictly three level system.[110]

Consider an atom with two ground states and one excited state as shown in Fig 10 a. Bergmann *et al.* showed that the rediagonalized atom/light system will always have a "dark" eigenstate, not connected to the excited state. Suppose, for simplicity, that the amplitudes $A_1 = \langle e|H_{EM}|g_1\rangle$ and $A_2 = \langle e|H_{EM}|g_2\rangle$ are equal, where H_{EM} is the Hamiltonian describing the light/atom interaction. An atom initially in state $|g_1\rangle$ is in the dark state provided only ω_2 light is on. If we then slowly increase the intensity of ω_1 light until the beams have equal intensities, the dark state will adiabatically evolve into $\frac{1}{\sqrt{2}}[|g_1\rangle - |g_2\rangle]$. If we then turn down the intensity of ω_2 while leaving ω_1 on, the atom will evolve into state $|g_2\rangle$. Thus we can move the atom from state $|g_1\rangle$ to state $|g_2\rangle$ without ever going through the excited state.

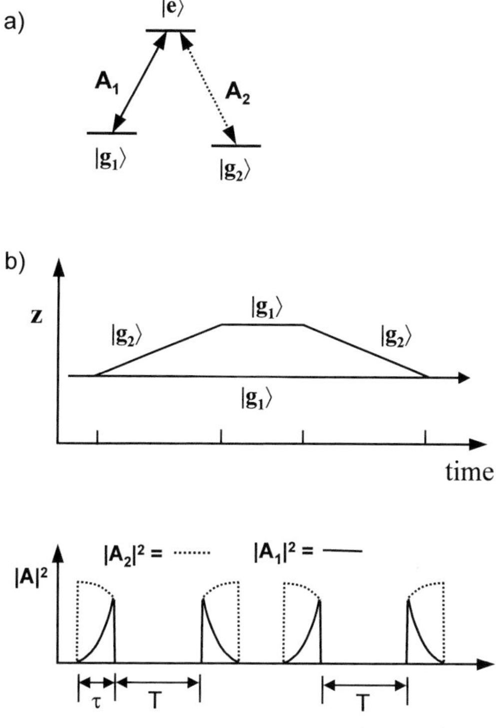

Figure. 10 a. An atomic system consisting of ground states $|g_1\rangle$ and $|g_2\rangle$ and an excited state $|e\rangle$ connected by amplitudes A_1 and A_2.

Fig. 10 b. In this space-time diagram of our adiabatic interferometer, the first interaction transfers the atom from $|g_1\rangle$ to the superposition state $(1/\sqrt{2})(|g_1\rangle + |g_2\rangle)$. In the second region, both frequencies are turned on simultaneously, projecting the atomic state $|\Psi(T+\tau)\rangle$ onto the dark superposition state $|\Psi(T+\tau)\rangle_D$ evaluated at the start of the pulse. This state is then adiabatically evolved into $|g_1\rangle$ by the light profiles shown in the second interaction region. To complete the interferometer, the sequence is repeated for adiabatic transfers with k_{eff} in the opposite direction as in Fig 9. Atom paths that do not contribute to the interference signal are not shown.

The work of Bergmann prompted Marte, Zoller and Hall[111] to suggest that this transfer process could also be used to induce a momentum change on an atom. Groups led by Mara Prentiss[112] and Bill Phillips[113] soon demonstrated the mechanical effect of this transition. In these experiments, the time-delayed light pulses were generated by having the atoms intersect spatially displaced laser beams. With atoms moving slowly in an atomic fountain, we could independently vary the intensity of each beam with acousto-optic modulators. This freedom allowed us to construct an atom interferometer using the adiabatic transfer method.[114] At different interaction points, differently shaped pulses are required, as shown in Fig. 10. For example, the first "adiabatic" beamsplitters require that ω_2 turn on first, but both ω_1 and ω_2 turn off in unison, while the second interaction point has both ω_1 and ω_2 turn on together and then ω_1 turn off first. With an adiabatic transfer interferometer, we are able to separate the two atom interferometers by up to ~$250\hbar k$ units of momentum without significant degradation of signal.

Currently, our atom interferometric method of measuring h/M_{Cs} has fractional resolution of ~$2\ 10^{-9}$ (1 ppb in α), corresponding to a velocity resolution less than 1/30 of an atom diameter/second. In terms of Doppler spectroscopy, this precision corresponds to a resolution in a Doppler shift of less than 100 μHz out of ~10^{15} Hz. We have been looking for systematic effects for the past five months and there are a few remaining tests to perform before publishing a value for h/M. The other measurements needed to determine α, such as the mass ratios m_e/m_p and m_p/M_{Cs}, and the frequency of the Cs D_1 **line will be measured with a fractional resolution better than ~10^{-9} in the near future. Curiously, some of the most accurate methods of determining α are** direct applications of three Nobel Prizes: the Josephson effect (56 ppb), the quantized Hall effect (24 ppb) and the equating of the ion-trap measurement of the electron magnetic moment with the QED calculation (4.2 ppb).[115]

OTHER APPLICATIONS IN ATOMIC PHYSICS

The topics I have discussed above are a small, personally skewed sampling of the many applications of the new technology of laser cooling and atom trapping. These techniques have already been used in nonlinear optics and quantum optics experiments. Laser-cooled atoms have spawned a cottage industry in the study of ultracold collisions. Atom traps offer the hope that radioactive species can be used for more precise studies of parity nonconservation effects due to the weak interactions and more sensitive searches for a breakdown of time-reversal invariance.

A particularly spectacular use of cold-atom technology has been the demonstration of Bose condensation in a dilute gas by Eric Cornell, Carl Weiman and collaborators,[116] and later by teams led by Wolfgang Ketterle[117] and Randy Hulet.[118] The production of this new state of matter opens exciting opportunities to study collective effects in a quantum gas with powerful diagnostic methods in laser spectroscopy. The increase in phase-space density of Bose condensed atoms will also generate new applications, just as the phase-space density increase due to laser cooling and trapping started a number of new areas of research.

APPLICATIONS IN BIOLOGY AND POLYMER SCIENCE

In 1986, the world was excited about atom trapping. During this time, Art Ashkin began to use optical tweezers to trap micron-sized particles. While experimenting with colloidal tobacco mosaic viruses,[119] he noticed tiny, translucent objects in his sample. Rushing into my lab, he excitedly proclaimed that he had "discovered *Life*". I went into his lab, half thinking that the excitement of the last few years had finally gotten the better of him. In his lab was a microscope objective focusing an argon laser beam into a petri dish of water. Off to the side was an old Edmund Scientific microscope. Squinting into the microscope, I saw my eye lashes. Squinting harder, I occasionally saw some translucent objects. Many of these objects were "floaters", debris in my vitreous humor that could be moved by blinking my eyes. Art assured me that there were *other* objects there that would not move when I blinked my eyes. Sure enough, there were objects in the water that could be trapped and would swim away if the light were turned off. Art had discovered bugs in his apparatus, but these were *real* bugs, bacteria that had eventually grown in his sample beads and water.

His discovery was quickly followed by the demonstration that infrared light focused to megawatts/cm^2 could be used to trap live e-coli bacteria and yeast for hours without damage.[120] Other work included the internal cell manipulation of plant cells, protozoa, and stretching of viscoelastic cytoplasm.[121] Steve Block and Howard Berg soon adapted the optical tweezers technique to study the mechanical properties of the flagella motor,[122] and Michael Burns and collaborators used the tweezers to manipulate live sperm.[123] Objects on the molecular scale could also be manipulated with optical tweezers. Block *et al.* sprinkled a low-density coverage of kinesin motor molecules onto a sphere and placed the sphere on a microtubule. When the kinesin was activated with ATP, the force and displacement generated by a single kinesin molecule could be measured.[124] Related experiments on the molecular motor actin/myosin associated with skeletal muscles have also been performed by Jeff Finer, Bob Simmons and Jim Spudich[125] using an active feedback optical tweezers developed in my lab.[126] Steve Kron and I developed a method to hold and simultaneously view a single molecule of DNA by attaching polystyrene handles to the ends of the molecule.[127, 128] These early experiments introduced an important tool for biologists, at both the cellular and the molecular level. The applications of this tool in biology have exploded and may eventually overtake the activity in atomic physics.[129]

My original goal in developing methods to manipulate DNA was to study, in real time, the motion of enzymes moving on the molecule. However, once we began to play with the molecules, we noticed that a stretched molecule of DNA would spring back like a rubber band when the extensional force was turned off, as shown in Fig 11. The "springiness" of the molecule is due to

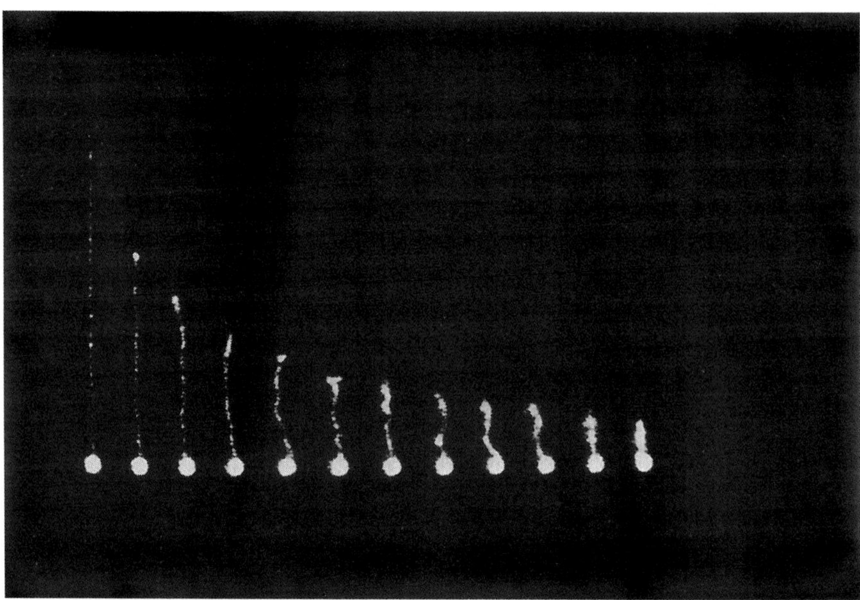

Fig. 11a: A series of video images showing the relaxation of a single molecule "rubber band" of DNA initially stretched by flowing fluid past the molecule. The DNA is stained with approximately one dye molecule for every 5 base-pairs and visualized in an optical microscope.[132]

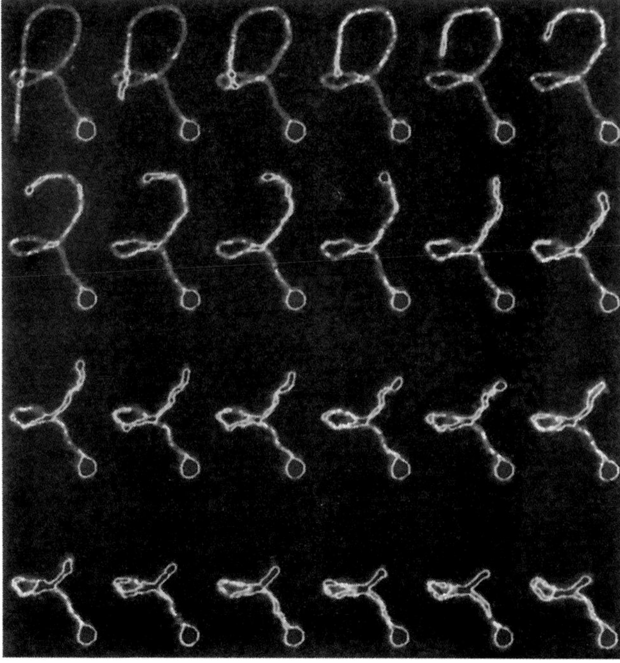

Fig. 11b: The relaxation of a stained DNA molecule in an entangled solution of unstained DNA. The molecule, initially pulled through the polymer solution with an optical tweezers, is seen to relax along a path defined by its contour. This work graphically shows that polymers in an entangled solution exhibit "tube-like" motion. [131] This result and a separate measurement of the diffusion of the DNA in a similar polymer solution[134] verifies de Gennes' reptation theory used to explain a general scaling feature of viscoelastic materials.

entropy considerations: the configurations of a flexible polymer are enumerated by counting the possible ways of taking a random walk of a large, but finite number of steps. A stretched molecule is in an unlikely configuration, and the system will move towards the much more likely equilibrium configuration of a random coil. Our accidental observation of a single-molecule rubber band created yet another detour: DNA from a lambda-phage virus is large enough to visualize and manipulate, and yet small enough so that the basic equations of motion describing a polymer are still valid. We stumbled onto a new way of addressing long-standing questions in polymer dynamics and began a program in polymer physics that is continuing today.[130, 131, 132, 133, 134, 135, 136, 137]

CLOSING REMARKS

The techniques I have discussed, to borrow an advertising slogan of AT&T, have enabled us to "reach out and touch" atoms and other neutral particles in powerful new ways. Laser beams can now reach into a vacuum chamber, capture and cool atoms to micro-kelvin temperatures, and toss them upwards in an atomic fountain. With this technique, a new generation of atomic clocks is being developed. Atoms quantum mechanically split apart and brought back together in an interferometer have given us inertial sensors of exquisite precision and will allow us to measure fundamental constants with unprecedented accuracy. Atom trapping and cooling methods allow us to Bose-condense a gas of atoms. With this condensate, we have begun to examine many-body effects in a totally new regime. The condensates are beginning to provide a still brighter source of atoms which we can exploit. Laser traps allow us to hold onto living cells and organelles within cells without puncturing the cell membrane. Single molecules of DNA are being used to study fundamental questions in polymer dynamics. The force and displacement generated by a dynesin molecule as it burns *one* ATP molecule can now be measured. This proliferation of applications into physics, chemistry, biology and medicine has occurred in less than a decade and is continuing, and we will no doubt see further applications of this new-found control over matter.

In 1985, when my colleagues and I first demonstrated optical molasses, I never foresaw the wealth of applications that would follow in just a few years. Instead of working with a clear vision of the future, I followed my nose, head close to the ground where the scent is strongest.

All of my contributions cited in this lecture were the result of working with the numerous gifted collaborators mentioned in this lecture. Without them, I would have done far less.

On a larger scale, the field of cooling and trapping was built out of the interwoven contributions of many researchers. Just as my associates and I were inspired by the work of others, our worldwide colleagues have already added immensely to our contributions. I consider this Nobel Prize to be the recognition of our collective endeavors. As scientists, we hope that others take note of what we have done and use our work to go in directions we never imagined. In this way, we continue to add to the collective scientific legacy.

REFERENCES

1. R. Conti, P. Bucksbaum, S. Chu, E. Commins, and L. Hunter, Phys. Rev. Lett. **42,** 343 (1979). See also, S. Chu, E. Commins, and R. Conti, Phys. Lett. **60A**, 96 (1977).
2. S. Chu, H.M. Gibbs, S.L. McCall, and A. Passner), Phys. Rev. Lett. **45**, 1715 (1980).
3. S. Chu, H.M. Gibbs, and S.L. McCall, Phys. Rev. B **24**, 7162 (1981).
4. P. Hu, S. Chu, and H.M. Gibbs, in *Picosecond Phenomena II*, eds. R.M. Hochstrasser, W. Kaiser and C.V. Shank (Springer-Verlag, 1980), p. 308.
5. S. Chu and S. Wong, Phys. Rev. Lett. **48**, 738 (1982); Also, see Comments, Phys. Rev. Lett. **49**, 1293 (1982).
6. S. Chu and A.P. Mills, Phys. Rev. Lett. **48**, 1333 (1982); S. Chu, A.P. Mills, Jr. and J.L. Hall, Phys. Rev. Lett. **52**, 1689 (1984).
7. A more complete account of this early history can be found in *Light Pressure on Atoms*, V.G. Minogin and V.S. Letokhov, (Gordon Breach Science, New York, 1987).
8. J.C. Maxwell, *A Treatise on Electricity and Magnetism*, 3rd ed. (1897), Reprint by Dover Publications, New York, (1954).
9. P. Lebedev, Ann. Phys., Leipzig, **6**, 433 (1901).
10. E.F. Nichols and G.F. Hull, Phys. Rev. **17**, 26 (1903), ibid. p 91.
11. A. Einstein, Phys. Z. **18**, 121 (1917). English translation in *Sources of Quantum mechanics*, ed. B.L. Waeerden (North-Holland, Amsterdam, 1967), pp. 63–78.
12. O.R. Frisch, Zs. Phys. **86**, 42 (1933) pp. 63–78.
13. Detailed calculation of lensing in the Mie scattering range (where the wavelength of the light is less than the diameter of the particle) can be found in a number of publications. See, for example, A. Ashkin, Biophys. J. **61**, 569 (1992).
14. G.A. Askar'yan, Zh. Eskp. Teor. Fiz. **42**, 1567 (1962).
15. V.S. Letokhov, Pis'ma Zh. Eskp. Teor. Fiz. 7, 348 (1968).
16. A. Ashkin, Science **210**, 1081 (1980).
17. A. Ashkin, Phys. Rev. Lett. **40**, 729 (1978).
18. J.E. Bjorkholm, R.R. Freeman, A. Ashkin and D.B. Pearson, Phys. Rev. Lett. **41**, 1361 (1978).
19. J.P. Gordon and A. Ashkin, Phys. Rev. A **21**, 1606 (1980).
20. S. Chu, AT&T Internal Memo, 11311-840509-12TM (1984).
21. T.W. Hänsch and A.L. Schawlow, Opt. Commun. **13**, 68 (1975).
22. D.J. Wineland and W.M. Itano, Bull. Am. Phys. Soc. **20**, 637 (1975).
23. D.J. Wineland and W.M. Itano, Phys. Rev. A **20**, 1521 (1979).
24. I wanted our paper to be titled "Demonstration of Optical Molasses". John Bjorkholm was a purist and felt that the phrase was specialized jargon at its worst. We compromised and omitted the phrase from the title but introduced it in the text of the paper.
25. For a comprehensive discussion of the work up to 1985, see W.D. Phillips, J.V. Prodan and H.J. Metcalf, J. Opt. Soc. Am. B **2**, 1751 (1985).
26. J. Prodan, A. Migdall, W.D. Phillips, I. So, H. Metcalf, and J. Dalibard., Phys. Rev. Lett. **54**, 992 (1985).
27. W. Ertmer, R. Blatt, J.L. Hall, and M. Zhu, Phys. Rev. Lett. **54**, 996 (1985).
28. I.P. Kaminow, *An Introduction to Electro-optic Devices*, (Academic Press, New York, 1974), pp 228–233.
29. S. Chu, L. Hollberg, J.E. Bjorkholm, A. Cable and A. Ashkin, Phys. Rev. Lett. **55**, 48 (1985).
30. The components of the experiment were assembled from parts of previous experiments: the cw dye laser needed for the optical molasses and the pulsed YAG laser

were previously used in a dye laser oscillator/amplifier system in a positronium spectroscopy experiment. A surplus vacuum chamber in a development section of Bell Laboratories became our molasses chamber.

31. S. Chu, M.G. Prentiss, A. Cable, and J.E. Bjorkholm, in *Laser Spectroscopy VII,* W. Persson and S. Svanberg, eds., (Springer-Verlag, Berlin, 1988) pp 64–67; Y. Shevy, D.S. Weiss, and S. Chu, in *Spin Polarized Systems,* ed. S. Stringari (World Scientific, Singapore, 1989), pp 287–294.
32. A. L. Migdall, J.V. Prodan, W.D. Phillips, T.H. Bergeman, and H.J. Metcalf, Phys. Rev. Lett. **54**, 2596 (1985).
33. A. Ashkin and J.P. Gordon, Opt. Lett., **8** 511 (1983).
34. A summary of this theorem and other "no trapping" theorems can be found in: S. Chu, in *Laser Manipulations of Atoms and Ions,* Proceedings of the International School of Physics "Enrico Fermi", course CXVIII, eds. E. Arimondo, W.D. Phillips, and F. Strumia (North-Holland, Amsterdam, 1992) pp 239–288.
35. A. Ashkin, Opt. Lett. **9,** 454 (1984).
36. S. Chu, J.E. Bjorkholm, A. Ashkin, L. Hollberg, and A. Cable, *Methods of Laser Spectroscopy,* eds. Y. Prior, A. Ben-Reuven, and M. Rosenbluh, (Plenum, New York 1985), pp. 41–50. The conference proceedings gave a snapshot of our thinking in December of 1985.
37. S. Chu, J.E. Bjorkholm, A. Ashkin, and A. Cable, Phys. Rev. Lett. **57**, 314 (1986).
38. A. Ashkin, J.M. Dziedzic, J.E. Bjorkholm, and S. Chu, Opt. Lett. **11**, 288 (1986).
39. D.E. Pritchard, E.L. Raab, V.S. Bagnato, C.E. Weiman, and R.N. Watts, Phys. Rev. Lett. **57**, 310 (1986).
40. Allan Mills had persuaded me to participate in a muonium spectroscopy experiment and I had been working on that experiment in parallel with the laser cooling and trapping work since 1985.
41. E.L. Raab, M. Prentiss, A. Cable, S. Chu, and D.E. Pritchard, Phys. Rev. Lett. **59**, 2631 (1987).
42. I gave a brief history of these events in reference 34. See also D. Pritchard and W. Ketterle, in *Laser Manipulations of Atoms and Ions,* Proceedings of the International School of Physics "Enrico Fermi", course CXVIII, eds. E. Arimondo, W.D. Phillips, and F. Strumia (North-Holland, Amsterdam, 1992) pp 473–496.
43. C. Monroe, W. Swann, H. Robinson, and C.E. Wieman, Phys. Rev. Lett. **65,** 1571 (1990). Raab and Pritchard had tried to get the scattering force trap to work at MIT by trying to capture the atoms directly from vapor, but the vapor pressure turned out to be too high for efficient capture.
44. K. Gibble, S. Kasapi and S. Chu, Opt. Lett. **17**, 526 (1992).
45. W. Ketterle, K. Davis, M. Joffe, A. Martin and D. Pritchard, Phys. Rev. Lett. **70**, 2253 (1993).
46. H.J. Lee, C.S. Adams, M. Kasevich and S. Chu, Phys. Rev. Lett. **76**, 2658 (1996).
47. I tried to persuade Alex Cable to come with me and become my first graduate student. By this time he was blossoming into a first rate researcher and I knew he would do well as a graduate at Stanford. He turned down my offer and a year later resigned from Bell Labs. While he was my technician, he had started a company making optical mounts. His company, "Thor Labs" is now a major supplier of optical components.
48. D. Sesko, C. Fan and C. Weiman, J. Opt. Soc. Am. B **5**, 1225 (1988).
49. W.D. Phillips, private communication.
50. P. Lett, R.N. Watts, C. Westbrook, and W.D. Phillips, Phys. Rev. Lett. **61**, 169 (1988).
51. P. Gould, P. Lett. and W. Phillips, in *Laser Spectroscopy VIII,* eds. W. Person and S. Svanberg (Springer-Verlag, Berlin, 1987) p 64.
52. Y. Shevy, D. Weiss and S. Chu, in *Spin Polarized Quantum Systems,* ed S. Stingari, (World Scientific, Singapore, 1989), pp. 287–294.
53. S. Chu, in *The Hydrogen Atom;* eds. G.F. Bassani, M. Inguscio, and T.W. Hänsch, (Springer-Verlag,1989), p 144.

54. M.S. Fee, A.P. Mills, S. Chu, E.D. Shaw, K. Danzmann, R.J. Chichester, D.M. Zuckerman, Phys. Rev. Lett. **70,** 1397 (1993); M.S. Fee, S. Chu, A.P. Mills, E.D. Shaw, K. Danzmann, R.J. Chichester, D.M. Zuckerman, Phys. Rev. A **48**, 192 (1993).
55. S. Chu, A.P. Mills, Jr., A.G. Yodh, K. Nagamine, H. Miyake, and T. Kuga, Phys. Rev. Lett. **60**, 101 (1988).
56. V.G. Minogin and O.T. Serimaa, Opt. Commun. **30,** 373 (1979). Also see V.G. Minogin Opt. Commun. **37**, 442 (1981).
57. J. Dalibard and C. Cohen-Tannoudji, J. Opt Soc. Am. B **2**, 1707 (1985).
58. J. Dalibard, C. Solomon, A. Aspect, E. Arimondo, R. Kasier, N. Vansteenkiste, and C. Cohen-Tannoudji, in *Atomic Physics 11*, eds. S. Haroche, J.C. Gay and G. Grynberg, (World Scientific, Singapore, 1989) pp. 199–214.
59. S. Chu, D. Weiss, Y. Shevy and P. Ungar, in *Atomic Physics 11,* eds. S. Haroche, J.C. Gay and G. Grynberg, (World Scientific, Singapore, 1989) pp 636–638.
60. J. Dalibard and C. Cohen-Tannoudji, J. Opt. Soc. Am. B **6**, 2023 (1989).
61. P.J. Ungar, D.S. Weiss, E. Riis, and S. Chu, J. Opt. Soc. Am. B **6**, 2058 (1989).
62. See Fig. 13 of ref. 60.
63. D.S. Weiss, E. Riis, Y. Shevy, P.J. Ungar and S. Chu, J. Opt. Soc. Am. B **6**, 2072 (1989).
64. We showed in Y. Shevy, D.S. Weiss, P.J. Ungar and S. Chu, Phys. Rev. Lett. 62, 1118, (1988) that a linear force vs. velocity dependence $\mathbf{F}(\upsilon) = -\alpha\upsilon$ would result in a Maxwell-Boltzmann distribution.
65. S-Q. Shang, B. Sheehy, P. van der Straten and H. Metcalf, Phys. Rev. Lett. **65**, 317 (1990).
66. See, for example, H. Dehmelt, Science **247**, 539, (1990).
67. F. Diedrich, J. C. Berquist, W. Itano, and D.J. Wineland, Phys. Rev. Lett. **62**, 403 (1989).
68. A. Aspect, E. Arimondo, R. Kaiser, N. Vansteenkiste, and C. Cohen-Tannoudji, Phys. Rev. Lett. **621**, 826 (1988); see also J. Opt. Soc. B **6,** 2112, (1989) by the same authors.
69. J. Lawall, S. Kulin, B. Saubamea, N. Bigelow, M. Leduc, and C. Cohen-Tannoudji, Phys. Rev. Lett. **75**, 4194 (1995).
70. E. Riis, D.S. Weiss, K.A. Moler, and S. Chu, Phys. Rev. Lett. **64**, 1658 (1990).
71. J. Nellessen, J. Werner and W. Ertmer, Opt. Commun. **78**, 300 (1990).
72. See, for example N.F. Ramsey, *Molecular Beams* (Oxford University Press, Oxford, 1956).
73. See, for example, A. De Marchi, G.D. Rovera and A. Premoli, Metrologia 20, 37 (1984). I had heard of this attempt while still a graduate student at Berkeley. We discussed the advantages of a cesium fountain atomic clock in K. Gibble and S. Chu, Metrologia **29**, 201, (1992).
74. M.A. Kasevich, E. Riis, S. Chu, and R.G. DeVoe, Phys. Rev. Lett. **63**, 612 (1989).
75. D.S. Weiss, E. Riis, M. Kasevich, K.A. Moler, and S. Chu, in *Light Induced Kinetic Effects on Atoms, Ions, and Molecules,* eds. L. Moi, S. Gozzini, C. Gabbanini, E. Arimondo, F. Strumia, (ETS Editrice, Pisa 1991) pp. 35–44.
76. A. Clarion, C. Salomon, S. Guellati and W.D. Phillips, Europhys. Lett. **16**, 165 (1991).
77. K. Gibble and S. Chu, Metrologia **29**, 201, (1992).
78. K. Gibble and S. Chu, Phys. Rev. Lett. **70**, 1771 (1993).
79. S. Ghezali, Ph. Laurent, S.N. Lea and A. Clairon, Europhys. Lett. **36**, 25 (1996).
80. V.I. Balykin, V.S. Letokhov, Yu.B. Ovchinnikov and A.I. Sidorov, Phys. Rev. Lett. **60**, 2137 (1988).
81. M.A. Kasevich, D.S. Weiss, and S. Chu, Opt. Lett., **15**, 667 (1990).
82. C.G. Aminoff, A.M. Steane, P. Bouyer, P. Desbiolles, J. Dalibard and C. Cohen-Tannoudji. Phys. Rev. Lett. **71**, 3083 (1993).
83. M. Kasevich, D. Weiss, E. Riis, K. Moler, S. Kasapi and S. Chu, Phys. Rev. Lett. **66**, 2297 (1991).
84. Ch. Bordé, Phys. Lett. A **140,** 10 (1989).

85. F. Riehle, Th. Kisters, A. Witte, S. Helmcke and Ch. Borde, Phys. Rev. Lett. **67,** 177 (1991).
86. O. Carnal and J. Mlynek, Phys. Rev. Lett. **66,** 2689 (1991).
87. D. Keith, C. Eksstrom, O. Turchette and D. Pritchard, Phys. Rev. Lett. **66,** 2693 (1991).
88. M. Kasevich and S. Chu, Phys. Rev. Lett. **67,** 181 (1991).
89. M. Kasevich and S. Chu, Applied Physics B **54,** 321 (1992).
90. K. Moler, D.S. Weiss, M. Kasevich, and S. Chu, Phys. Rev. A **45,** 342 (1991).
91. B. Young, M. Kasevich and S. Chu, in *Atom Interferometry,* ed. P. Berman (Academic Press, New York, 1997) pp 363–406.
92. A. Peters, K.Y. Chung, B. Young, J. Hensley and S. Chu, Philosophical Trans. A **355,** 2223 (1997).
93. P. Storey and C. Cohen-Tannoudji, J. Phys. **4,** 1999 (1994).
94. J. Hensley, A. Peters and S. Chu, Review of Scientific Instruments, in press (1998).
95. M. Kasevich and S. Chu, Phys. Rev. Lett. **69,** 1741 (1992).
96. N. Davidson, H.J. Lee, M. Kasevich, and S. Chu, Phys. Rev. Lett. **72,** 3158 (1994).
97. S. Rolston, C. Gerz, K. Helmerson, P.S. Jessen, P.D. Lett, W.D. Phillips, R.J.C. Spreeuw and C.I. Westbrook, Proc. SPIE **1726,** 205 (1992).
98. J.D. Miller, R. Cline, D. Heinzen, Phys. Rev. A**47,** R4567 (1993).
99. N. Davidson, H.J. Lee, C.S. Adams, M. Kasevich and S. Chu, Phys. Rev. Lett. **74,** 1311 (1995).
100. C.S. Adams, H.J. Lee, N. Davidson, M. Kasevich, and S. Chu, Phys. Rev. Lett. **74,** 3577 (1995).
101. Improved results are reported in H.J. Lee, C.S. Adams, N. Davidson, B. Young, M. Weitz, M. Kasevich, and S. Chu, in *Atomic Physics 14,* eds. C. Weiman and D. Wineland (AIP, New York) 1995, pp. 258–278.
102. M. Kasevich, H.J. Lee, C.A. Adams and S. Chu, in *Laser Spectroscopy 12,* eds. M. Inguscio, M. Allegrini, A. Sasso (World Scientific, Singapore, 1996) pp. 13–16.
103. D.M. Stamper-Kurn, M.R. Andrews, A.P. Chikkatur, S. Inouye, H.-J. Meisner, J. Strenger and W. Ketterle, Phys. Rev. Lett. **80,** 2072 (1998).
104. H. Feshbach, Ann. Phys. (NY) **19,** 287 (1962).
105. A.J. Moerdijk, B.J. Verhaar and A. Axelsson, Phys. Rev. A **51,** 4852 (1995).
106. The speed of light c is now a defined quantity. "2" is also known well.
107. D.S. Weiss, B.C. Young, and S. Chu, Phys. Rev. Lett. **70,** 2706 (1993).
108. D.S. Weiss, B.C. Young and S. Chu, Applied Physics B **59,** 217–256 (1994).
109. U. Gaubatz, P. Rudecker, M. Becker, S. Schiemann, M. Kültz, and K. Bergmann, Chem. Phys. Lett. **149,** 463 (1988).
110. M. Weitz, B. Young and S. Chu, Phys. Rev. A **50,** 2438 (1994).
111. P. Marte, P. Zoller and J.L. Hall, Phys. Rev. A **44,** R4118 (1991).
112. J. Lawall and M. Prentiss, Phys. Rev. Lett. **72,** 993 (1994).
113. L.S. Goldner, C. Gerz, R.J.C. Spreew, S.L. Rolstom, C.I. Westbrook, W. Phillips, P. Marte and P. Zoller, Phys. Rev. Lett. **72,** 997 (1994).
114. M. Weitz, B.C. Young, and S. Chu, Phys. Rev. Lett. **73,** 2563 (1994).
115. For a review of the current status of α, see T. Kinoshita, Rep. Prog. Phys. **59,** 1459 (1966).
116. M.H. Anderson, J.R. Ensher, M.R. Matthews, C.E. Wieman and E.A. Cornell, Science **269,** 198 (1995).
117. K.B. Davis, M-O Mewes, M. R. Anderson, N.J. van Druten, D.S. Durfee, D.M. Kurn and W. Ketterle, Phys. Rev. Lett. **75,** 3969 (1995).
118. C.C. Bradley, C.A. Sackett and R.G. Hulet, Phys. Rev. Lett. **78,** 985 (1997).
119. A. Ashkin and J.M. Dziedzic, Science **235,** 1517 (1987).
120. A. Ashkin, J.M. Dziedzic and T. Yamane, Nature (London) **330,** 769 (1987).
121. A. Ashkin and J.M. Dziedzic, Science **253,** 1517 (1987); Proc. Natl. Acad. Sci. USA **86,** 7914 (1989).
122. S. Block, D.F. Blair and H.C. Berg, Nature **338,** 514 (1989).

123. Y. Tadir, W. Wright, O. Vafa, T. Ord, R. Asch and M. Burns, Fertil. Steril. **52,** 870 (1989).
124. S. Block, L. Goldstein, and B. Schnapp, Nature **348,** 348 (1990); also see K. Svoboda and S. Block, Cell **77,** 773 (1994) and references contained within.
125. J.T. Finer, R.M. Simmons, J.A. Spudich, Nature **368,** 113 (1994).
126. H.M. Warrick, R.M. Simmons, J.F. Finer, T.Q.P. Uyeda, S. Chu, and J.A. Spudich, chapter 1 in *Methods in Cell Biology,* **39,** Academic, New York (1993) pp. 1–21; R.M. Simmons, J.T. Finer, S. Chu and J.A. Spudich, Biophys. J. **70,** 1813–1822 (1996).
127. S. Chu and S. Kron, Int. Quantum Electronics Conf. Tech Digest, (Optical Soc. of Am., Washington DC, 1990) p 202; M. Kasevich, K. Moler, E. Riis, E. Sunderman, D. Weiss, and S. Chu, *Atomic Physics 12,* eds. J.C. Zorn and R.R. Lewis, (Am. Inst. of Physics, New York 1990), pp. 47–57.
128. S. Chu, Science, **253,** 861, (1991).
129. Much of the activity has been reviewed by A. Ashkin, Proc. Natl. Acad. Sci. USA **94,** 4853 (1997).
130. Preliminary results were reported in S. Chu, Science, **253,** 861, (1991)
131. T. Perkins, D.E. Smith and S. Chu, Science **64,** 819 (1994).
132. T.T. Perkins, S.R. Quake, D.E. Smith and S. Chu, Science **264,** 822 (1994).
133. T.T. Perkins, D.E. Smith, R.G. Larson, and S. Chu, Science **268,** 83 (1994).
134. D.E. Smith, T.T. Perkins and S. Chu, Phys. Rev. Lett. **75,** 4146 (1995).
135. D.E. Smith, T.T. Perkins and S. Chu, Macromolecules **29,** 1372 (1996).
136 S.R. Quake and S. Chu, Nature, **388,** 151 (1997).
137. T.T. Perkins, D.E. Smith and S. Chu, Science **276,** 2016 (1997).

Claude Cohen-Tannoudji

CLAUDE N. COHEN-TANNOUDJI

I was born on April 1, 1933 in Constantine, Algeria, which was then part of France. My family, originally from Tangiers, settled in Tunisia and then in Algeria in the 16th century after having fled Spain during the Inquisition. In fact, our name, Cohen-Tannoudji, means simply the Cohen family from Tangiers. The Algerian Jews obtained the French citizenship in 1870 after Algeria became a French colony in 1830.

My parents lived a modest life and their main concern was the education of their children. My father was a self-taught man but had a great intellectual curiosity, not only for biblical and talmudic texts, but also for philosophy, psychoanalysis and history. He passed on to me his taste for studies, for discussion, for debate, and he taught me what I regard as being the fundamental features of the Jewish tradition–studying, learning and sharing knowledge with others.

As a child, I was very lucky to escape the tragic events which marked this century. The arrival of the Americans in Algeria, in November of 1942, saved us from the nazi persecutions that were spreading throughout Europe at the time. I completed my primary and secondary school education in Algiers. And I was also lucky enough to finish high school in very good conditions and to leave Algiers for Paris, in 1953, before the war in Algeria and the stormy period that preceded the independence.

I came to Paris because I was admitted to the Ecole Normale Supérieure. This French "grande école", founded during the French Revolution about 200 years ago, selects the top high school students who do well in the selective final examination. The four years at this school, from 1953 to 1957, were indeed a unique experience for me. During the first year, I attended a series of fascinating lectures in mathematics given by Henri Cartan and Laurent Schwartz, in physics by Alfred Kastler. Initially, I was more interested in mathematics but Kastler's lectures were so stimulating, and his personality so attractive, that I ended up changing to physics.

In 1955, when I joined Kastler's group to do my "diploma" work, the group was very small. One of Kastler's first students, Jean Brossel, who had returned four years before from M.I.T. where he had done research work with Francis Bitter, was supervising the thesis work of Jacques Emile Blamont and Jacques Michel Winter.

We were a small group, but the enthusiasm for research was exceptional and we worked hard. Brossel and Kastler were in the lab nearly day and night, even on weekends. We had endless discussions on how to interpret our experimental results. At the time, the equipment was rather poor and we did what we could without computers, recorders and signal averagers. We measured resonance curves point by point with a galvanometer, each curve five times, and then averaged by hand. We were, somehow, able to get nice curves and exciting results. I think that what I learned during that period was essential for my subsequent research work and key personalities such as Alfred Kastler and Jean Brossel certainly had a significant role in it.

We were going together, once a week, to attend the new lectures given in Saclay by Albert Messiah on quantum mechanics, by Anatole Abragam on NMR and by Claude Bloch on nuclear physics. I can still remember the stimulating atmosphere of these lectures.

During the summer of 1955, I also spent two months at the famous Les Houches summer school in the Alps. This school has contributed largely to the development of theoretical physics in France. At that time, the school offered an intense training in modern physics with about six lectures a day, for two months, and the lecturers were J. Schwinger, N. Ramsey, G. Uhlenbeck, W. Pauli, A. Abragam, A. Messiah, C. Bloch... to mention a few.

After finishing my "diploma" studies, I still had to get through the final examination "Agrégation" before leaving Ecole Normale as a student. The "Agrégation" is a competitive examination for teaching posts in high schools. The preparation consists of theoretical and experimental courses as well as some pedagogical training. You give a lecture attended by other students and a professor and after, there is a moment of general debate and constructive criticism in view of perfecting your lecture. Kastler, I remember, participated in the pedagogical training and he taught us how to organize and present our lecture.

Well, about this time I met Jacqueline who became my wife in 1958. She has shared with me all the difficult and happy times of life. She has been able to pursue her own career as a high school physics and chemistry teacher, to raise our three children Alain, Joëlle and Michel, to be part of the daily life of a researcher which can sometimes be very difficult and demanding. We have had, as many, our share of family tragedy and losing our oldest son Alain was a great misfortune to us all. Alain died in 1993, of a long illness, at the age of 34.

After the "Agrégation", I left the Ecole Normale and did my military service which was very long (28 months) because of the Algeria war. I was, though, assigned part of the time to a scientific department supervised by Jacques Emile Blamont. We were studying the upper atmosphere with rockets releasing sodium clouds at the sunset. By looking at the fluorescence light reemitted by the sodium atoms excited by the sunlight, it was possible to measure the variations with the altitude of various parameters such as the wind velocity or the temperature.

Then, in the beginning of 1960, I came back to the laboratory to do a Ph.D. under the supervision of Alfred Kastler and Jean Brossel with a research post at the CNRS (French National Center for Scientific Research). The lab had by then been expanded. Bernard Cagnac was finishing his thesis on the optical pumping of the odd isotopes of mercury and I was trying, with

Jean-Pierre Barrat, to derive a master equation for the optical pumping cycle and to understand the physics of the off-diagonal elements of the density matrix (the so-called atomic "coherences"). Our calculations predicted the existence of "light shifts" for the various Zeeman sublevels, a curious phenomenon we did not expect at all. I decided to try to see this effect. Cagnac left me his experimental set up during Christmas vacations and I remember getting the first experimental evidence on Christmas Eve of 1960. I was very excited and both Kastler and Brossel were very happy indeed. Kastler called the effect the "Lamp shift", since it is produced by the light coming from a discharge lamp. Nowadays, it is called light shift or a.c. Stark shift. I built a new experimental set up to check in detail several other predictions of our calculations, especially the conservation of Zeeman coherences during the optical pumping cycle. I submitted my Ph.D. in December of 1962. The members of the committee were Jean Brossel, Pierre Jacquinot, Alfred Kastler and Jacques Yvon.

Shortly after my Ph.D. Alfred Kastler urged me to accept a teaching position at the University of Paris. I followed his advice and started to teach at the undergraduate level. At about this time, there was a new reform in the University system : the so-called "troisième cycle" that consisted of teaching a graduate level with a flexible program. Jean Brossel asked me to teach quantum mechanics. He was teaching atomic physics, Alfred Kastler and Jacques Yvon statistical physics, Pierre Aigrain and Pierre-Gilles de Gennes solid state physics.

We had the best students of the Ecole Normale attending these lectures, so I set up a small group where every year a new student would join in and do a post-graduate thesis or a Ph.D. In 1967, I was asked to teach quantum mechanics at a lower level (second cycle). The book "Quantum Mechanics" originated from this teaching experience and was done in collaboration with Franck Laloë and Bernard Diu.

Understanding atom-photon interactions in the high intensity limit where perturbative treatments are no longer valid was one of the main goals of our research group. This led us to develop a new approach to these problems where one considers the "atom + photons" system as a global isolated system described by a time-independent Hamiltonian having true energy levels. We called such a system the "dressed atom". Although the quantum description of the electromagnetic field used in such an approach is not essential to interpret most physical effects encountered in atomic physics, it turned out that the dressed atom approach was very useful in providing new physical insights into atom-photon interactions. New physical effects, which were difficult to predict by standard semiclassical methods, were appearing clearly in the energy diagram of the dressed atom when examining how this energy diagram changes when the number of photons increases. We first introduced the dressed atom approach in the radio-frequency range while Nicole Polonsky, Serge Haroche, Jacques Dupont-Roc, Claire Landré, Gilbert Grynberg, Maryvonne Ledourneuf, Claude Fabre were working on their thesis. One of the new effects which were predicted and observed was the modification, and even the cancellation of the Landé factor of an atomic level by interaction with an intense, high frequency radio-frequency field. This effect presents some analogy with the g–2 anomaly of the electron spin except that it has the opposite sign: the g-factor of the atomic level is reduced by virtual absorption and re-emission of RF photons whereas the g-factor of the electron spin is enhanced by radiative corrections.

We devoted a lot of efforts to the interpretation of this change of sign and this led us, years later (with Jacques Dupont-Roc and Jean Dalibard), to propose new physical pictures involving the respective contributions of vacuum fluctuations and radiation reaction. And while this was going on, we had some very stimulating discussions with Victor Weisskopf who has always been interested in the physical interpretation of the g–2 anomaly.

The dressed atom approach has also been very useful in the optical domain. Spontaneous emission plays an important role as a damping mechanism and as a source of fluorescence photons. Serge Reynaud and I applied this approach to the interpretation of resonance fluorescence in intense resonant laser beams. New physical pictures were given for the Mollow triplet and for the absorption spectrum of a weak probe beam, with the prediction and the observation of new Doppler free lines resulting from a compensation of the Doppler effect by velocity dependent light shifts. The picture of the dressed atom radiative cascade also provided new insights into photon correlations and photon antibunching. New types of time correlations between the photons emitted in the two sidebands of the Mollow triplet were predicted in this way and observed experimentally at the Institut d'Optique in Orsay, in collaboration with Alain Aspect.

An important event in my scientific life has been my appointment as a Professor at the Collège de France in 1973. The Collège de France is a very special institution created in 1530, by King François I, to counterbalance the influence of the Sorbonne which was, at that time, too scholastic and where only latin and theology were taught. The first appointed by the King were 3 lecturers in Hebrew, 2 in Greek and 1 in Mathematics. This institution survived all revolutions and remains, to this day, reputed for its flexibility. Today there are 52 professors in all subjects, and lectures are open to all, for there is no registration and no degrees given. We professors are free to choose the topics of our lectures. The only rule is that these lectures must change and deal with different topics every year, which is very difficult and demanding. It is, however, very stimulating because this urges one to broaden one's knowledge, to explore new fields and to challenge oneself. No doubt that without such an effort I would not have started many of the research lines that have been explored by my research group. I am very grateful to Anatole Abragam who is at the origin of my appointment at the Collège de France. Part of this teaching experience incited the two books on quantum electrodynamics and quantum optics written with Jacques Dupont-Roc and Gilbert Grynberg.

In the early 1980s, I chose to lecture on radiative forces, a field which was very new at that time. I was also trying with Serge Reynaud, Christian Tanguy and Jean Dalibard to apply the dressed atom approach to the interpretation of atomic motion in a laser wave. New ideas were emerging from such an analysis related to, in particular, the interpretation of the mean value, the fluctuations and the velocity dependence of dipole forces in terms of spatial gradients of dressed state energies and of spontaneous transitions between these dressed states.

When in 1984 I was given the possibility to appoint someone to the position of Associate Director for my laboratory, at the Collège de France, I offered the post to Alain Aspect and then invited him to join me in forming, with Jean Dalibard, a new experimental group on laser cooling and trapping. A year later, Christophe Salomon who came back from a postdoctoral stay in JILA with Jan Hall, decided to join our group. This was a new very exciting scientific period for us. We began to investigate a new cooling mechanism suggested by the dressed atom approach and that resulted from correlations between the spatial modulations of the dressed state energies in a high intensity laser standing wave and the spatial modulations of the spontaneous rates between the dressed states. As a result of these correlations, the moving atom is running up potential hills more frequently than down. We first called such a scheme "stimulated blue molasses" because it appears for a blue detuning of the cooling lasers, contrary to what happens for Doppler molasses which require a red detuning. In fact, this new scheme was the first high intensity version of what is called now "Sisyphus cooling", a denomination that we introduced in 1986. We also observed, shortly after, the channeling of atoms at the nodes or antinodes of a standing wave. This was the first demonstration of laser confinement of neutral atoms in optical-wavelength-size regions.

A few years later, in 1988, when sub-Doppler temperatures were observed by Bill Phillips, who had been collaborating with us, we were prepared with our background in optical pumping, light shifts and dressed atoms, to find the explanation of such anomalous low temperatures. In fact, they were resulting from yet another (low intensity) version of Sisyphus cooling. Similar conclusions were reached by Steve Chu and his colleagues. At the same time, we were exploring, with Alain Aspect and Ennio Arimondo, the possibility of applying coherent population trapping to laser cooling. By making such a quantum interference effect velocity selective, we were able to demonstrate a new cooling scheme with no lower limit, which can notably cool atoms below the recoil limit corresponding to the recoil kinetic energy of an atom absorbing or emitting a single photon. These exciting developments opened the microKelvin and even the nanoKelvin range to laser cooling, and they allowed several new applications to be explored with success.

These applications will not be described here since they are the subject of the Nobel Lecture which follows this presentation. The purpose here was merely to give an idea of my scientific itinerary and to express my gratitude to all those who have helped me live such a great adventure: my family, my teachers, my students and my fellow colleagues all over the world.

I dedicate my Nobel Lecture to the memory of my son Alain.

MANIPULATING ATOMS WITH PHOTONS

Nobel Lecture, December 8, 1997

by

CLAUDE N. COHEN-TANNOUDJI

Collège de France et Laboratoire Kastler Brossel* de l'Ecole Normale Supérieure, 24 rue Lhomond, 75231 Paris Cedex 05, France

Electromagnetic interactions play a central role in low energy physics. They are responsible for the cohesion of atoms and molecules and they are at the origin of the emission and absorption of light by such systems. This light is not only a source of information on the structure of atoms. It can also be used to act on atoms, to manipulate them, to control their various degrees of freedom. With the development of laser sources, this research field has considerably expanded during the last few years. Methods have been developed to trap atoms and to cool them to very low temperatures. This has opened the way to a wealth of new investigations and applications.

Two types of degrees of freedom can be considered for an atom: the internal degrees of freedom, such as the electronic configuration or the spin polarization, in the center of mass system; the external degrees of freedom, which are essentially the position and the momentum of the center of mass. The manipulation of internal degrees of freedom goes back to optical pumping [1], which uses resonant exchanges of angular momentum between atoms and circularly polarized light for polarizing the spins of these atoms. These experiments predate the use of lasers in atomic physics. The manipulation of external degrees of freedom uses the concept of radiative forces resulting from the exchanges of linear momentum between atoms and light. Radiative forces exerted by the light coming from the sun were already invoked by J. Kepler to explain the tails of the comets. Although they are very small when one uses ordinary light sources, these forces were also investigated experimentally in the beginning of this century by P. Lebedev, E. F. Nichols and G. F. Hull, R. Frisch. For a historical survey of this research field, we refer the reader to review papers [2, 3, 4, 5] which also include a discussion of early theoretical work dealing with these problems by the groups of A. P. Kazantsev, V. S. Letokhov, in Russia, A. Ashkin at Bell Labs, S. Stenholm in Helsinki.

It turns out that there is a strong interplay between the dynamics of internal and external degrees of freedom. This is at the origin of efficient laser cooling mechanisms, such as "Sisyphus cooling" or "Velocity Selective Cohe-

* Laboratoire Kastler Brossel is a Laboratoire associé au CNRS et à l'Université Pierre et Marie Curie.

rent Population Trapping", which were discovered at the end of the 80's (for a historical survey of these developments, see for example [6]). These mechanisms have allowed laser cooling to overcome important fundamental limits, such as the Doppler limit and the single photon recoil limit, and to reach the microKelvin, and even the nanoKelvin range. We devote a large part of this paper (sections 2 and 3) to the discussion of these mechanisms and to the description of a few applications investigated by our group in Paris (section 4). There is a certain continuity between these recent developments in the field of laser cooling and trapping and early theoretical and experimental work performed in the 60's and the 70's, dealing with internal degrees of freedom. We illustrate this continuity by presenting in section 1 a brief survey of various physical processes, and by interpreting them in terms of two parameters, the radiative broadening and the light shift of the atomic ground state.

1. BRIEF REVIEW OF PHYSICAL PROCESSES

To classify the basic physical processes which are used for manipulating atoms by light, it is useful to distinguish two large categories of effects: dissipative (or absorptive) effects on the one hand, reactive (or dispersive) effects on the other hand. This partition is relevant for both internal and external degrees of freedom.

1.1 EXISTENCE OF TWO TYPES OF EFFECTS IN ATOM-PHOTON INTERACTIONS

Consider first a light beam with frequency ω_L propagating through a medium consisting of atoms with resonance frequency ω_A. The index of refraction describing this propagation has an imaginary part and a real part which are associated with two types of physical processes. The incident photons can be absorbed, more precisely scattered in all directions. The corresponding attenuation of the light beam is maximum at resonance. It is described by the imaginary part of the index of refraction which varies with $\omega_L - \omega_A$ as a Lorentz absorption curve. We will call such an effect a dissipative (or absorptive) effect. The speed of propagation of light is also modified. The corresponding dispersion is described by the real part n of the index of refraction whose difference from 1, $n-1$, varies with $\omega_L - \omega_A$ as a Lorentz dispersion curve. We will call such an effect a reactive (or dispersive) effect.

Dissipative effects and reactive effects also appear for the atoms, as a result of their interaction with photons. They correspond to a broadening and to a shift of the atomic energy levels, respectively. Such effects already appear when the atom interacts with the quantized radiation field in the vacuum state. It is well known that atomic excited states get a natural width Γ, which is also the rate at which a photon is spontaneously emitted from such states. Atomic energy levels are also shifted as a result of virtual emissions and reabsorptions of photons by the atom. Such a radiative correction is simply the Lamb shift [7].

Similar effects are associated with the interaction with an incident light beam. Atomic ground states get a radiative broadening Γ', which is also the rate at which photons are absorbed by the atom, or more precisely scattered from the incident beam. Atomic energy levels are also shifted as a result of virtual absorptions and reemissions of the incident photons by the atom. Such energy displacements $\hbar\Delta'$ are called light shifts, or ac Stark shifts [8, 9].

In view of their importance for the following discussions, we give now a brief derivation of the expressions of Γ' and Δ', using the so-called dressed-atom approach to atom-photon interactions (see for example [10], chapter VI). In the absence of coupling, the two dressed states $|g, N\rangle$ (atom in the ground state g in the presence of N photons) and $|e, N-1\rangle$ (atom in the excited state e in the presence of $N-1$ photons) are separated by a splitting $\hbar\delta$, where $\delta = \omega_L - \omega_A$ is the detuning between the light frequency ω_L and the atomic frequency ω_A. The atom-light interaction Hamiltonian V_{AL} couples these two states because the atom in g can absorb one photon and jump to e. The corresponding matrix element of V_{AL} can be written as $\hbar\Omega/2$, where the so-called Rabi frequency Ω is proportional to the transition dipole moment and to $\sqrt{N}$. Under the effect of such a coupling, the two states repel each other, and the state $|g, N\rangle$ is shifted by an amount $\hbar\Delta'$, which is the light shift of g. The contamination of $|g, N\rangle$ by the unstable state $|e, N-1\rangle$ (having a width Γ) also confers to the ground state a width Γ'. In the limit where $\Omega \ll \Gamma$ or $|\delta|$, a simple perturbative calculation gives:

$$\Gamma' = \Omega^2 \frac{\Gamma}{\Gamma^2 + 4\delta^2} \tag{1}$$

$$\Delta' = \Omega^2 \frac{\delta}{\Gamma^2 + 4\delta^2} \tag{2}$$

Both Γ' and Δ' are proportional to $\Omega^2 \propto N$, *i.e.* to the light intensity. They vary with the detuning $\delta = \omega_L - \omega_A$ as Lorentz absorption and dispersion curves, respectively, which justifies the denominations absorptive and dispersive used for these two types of effects. For large detunings ($|\delta| \gg \Gamma$), Γ' varies as $1/\delta^2$ and becomes negligible compared to Δ' which varies as $1/\delta$. On the other hand, for small detunings, ($|\delta| \ll \Gamma$), Γ' is much larger than Δ'. In the high intensity limit, when Ω is large compared to Γ and $|\delta|$, the two dressed states resulting from the coupling are the symmetric and antisymmetric linear combinations of $|g, N\rangle$ and $|e, N-1\rangle$. Their splitting is $\hbar\Omega$ and they share the instability Γ of e in equal parts, so that $\Gamma' = \Gamma/2$. One can explain in this way various physical effects such as the Rabi flopping or the Autler-Townes splittings of the spectral lines connecting e or g to a third level [11].

1.2 *MANIPULATION OF INTERNAL DEGREES OF FREEDOM*

1.2.1 *Optical pumping*

Optical pumping is one of the first examples of manipulation of atoms by light [1]. It uses resonant excitation of atoms by circularly polarized light for transferring to the atoms part of the angular momentum carried by the light beam. It is based on the fact that different Zeeman sublevels in the atomic ground state have in general different absorption rates for incoming polarized light. For example, for a $J_g = 1/2 \leftrightarrow J_e = 1/2$ transition, only atoms in the sublevel $M_g = -1/2$ can absorb σ^+-polarized light. They are excited in the sublevel $M_e = +1/2$ of e from which they can fall back in the sublevel $M_g = +1/2$ by spontaneous emission of a π-polarized photon. They then remain trapped in this state because no further σ^+-transition can take place. It is possible in this way to obtain high degrees of spin orientation in atomic ground states.

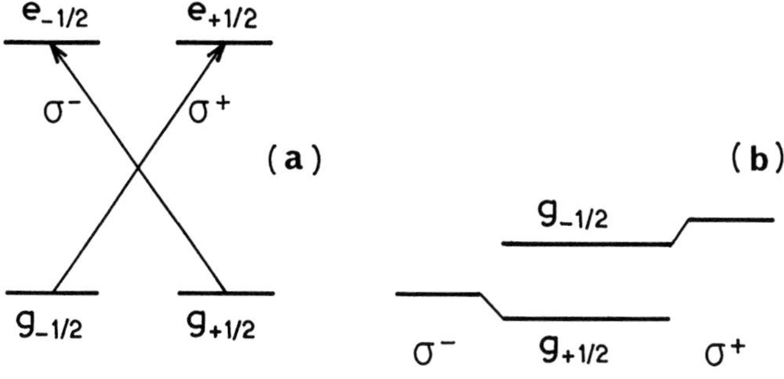

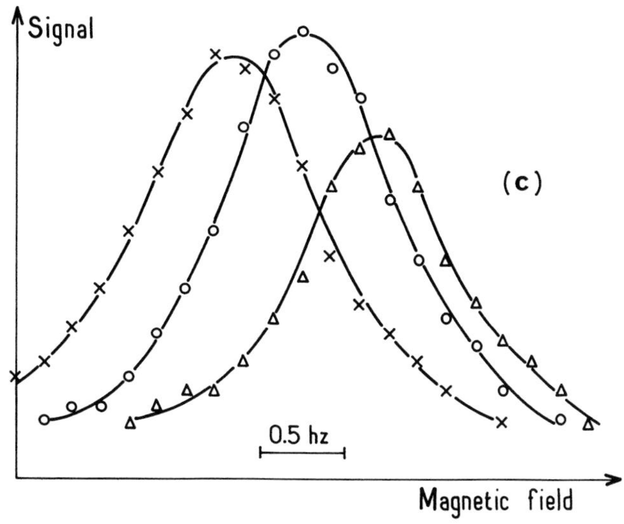

Figure 1. Experimental observation of light shifts (from reference [12]). For a $J_g = 1/2 \leftrightarrow J_e = 1/2$ transition (a), a σ^+-polarized non-resonant excitation shifts only the sublevel $g_{1/2}$ (right part of b), whereas a σ^--polarized excitation shifts only the sublevel $g_{+1/2}$ (left part of b). The detuning δ is positive and very large compared to the Zeeman splittings in e and g. The Zeeman splitting in the ground state is thus increased in the first case, decreased in the second one. (c) magnetic resonance signal versus magnetic field, in units of the corresponding Larmor frequency. The central curve (circles) is the resonance curve in the absence of light shift. When the non-resonant light beam is introduced, either σ^+-polarized (crosses) or σ^--polarized (triangles) the magnetic resonance curve is light shifted in opposite directions.

1.2.2 *Light shifts*

Optical pumping is a dissipative effect because it is associated with resonant absorption of photons by the atom. Non-resonant optical excitation produces light shifts of the ground state Zeeman sublevels. Because of the polarization selection rules, light shifts depend on the polarization of the exciting light and vary in general from one Zeeman sublevel to another. Consider for example a $J_g = 1/2 \leftrightarrow J_e = 1/2$ transition (Fig.1a). A σ^+-polarized excitation shifts only the Zeeman sublevel $M_g = -1/2$, whereas a σ^--polarized excitation shifts only the sublevel $M_g = +1/2$ (Fig.1b). Magnetic resonance curves in the ground state g which are very narrow because the relaxation time in g can be very long, are thus light shifted by a polarized non-resonant excitation, and the sign of this shift changes when one changes the polarization of the light beam from σ^+ to σ^-. It is in this way that light shifts were first observed [12]. Fig.1c gives an example of experimental results obtained by exciting the transition 6^1S_0, $F = 1/2 \leftrightarrow 6^3P_1$, $F = 1/2$ of ^{199}Hg atoms by the non-resonant light coming from a lamp filled with another isotope (^{201}Hg).

Light shifts can be considered from different points of view. First, they can be interpreted as a radiative correction, due to the interaction of the atom with an incident field rather than with the vacuum field. This is why Alfred Kastler called them "Lamp shifts". Secondly, they introduce perturbations to high precision measurements using optical methods, which must be taken into account before extracting spectroscopic data from these measurements. Finally, because of their variation from one sublevel to another, the effect of light shifts can be described in terms of fictitious magnetic or electric fields [13]. This explains why the light shifts produced by a non-resonant laser standing wave are being more and more frequently used to produce spatial modulations of the Zeeman splittings in the ground state on an optical wavelength scale. This would not be easily achieved with spatially varying real magnetic fields. We will see in section 2 interesting applications of such a situation.

1.3 *MANIPULATION OF EXTERNAL DEGREES OF FREEDOM*

1.3.1 *The two types of radiative forces*

There are two types of radiative forces, associated with dissipative and reactive effects respectively.

Dissipative forces, also called radiation pressure forces or scattering forces, are associated with the transfer of linear momentum from the incident light beam to the atom in resonant scattering processes. They are proportional to the scattering rate Γ'. Consider for example an atom in a laser plane wave with wave vector $\mathbf{k}$. Because photons are scattered with equal probabilities in two opposite directions, the mean momentum transferred to the atom in an absorption-spontaneous emission cycle is equal to the momentum $\hbar\mathbf{k}$ of the absorbed photon. The mean rate of momentum transfer, *i.e.* the mean force, is thus equal to $\hbar\mathbf{k}\Gamma'$. Since Γ' saturates to $\Gamma/2$ at high intensity (see section 1.1), the radiation pressure force saturates to $\hbar\mathbf{k}\Gamma/2$. The corresponding acceleration (or deceleration) which can be communicated to an atom with mass M, is equal to $a_{\max} = \hbar k\Gamma/2M = v_R/2\tau$, where $v_R = \hbar k/M$ is the recoil velocity of the atom absorbing or emitting a single photon, and $\tau = 1/\Gamma$ is the radiative lifetime of the excited state. For sodium atoms, $v_R = 3 \times 10^{-2}$ m/s and $\tau = 1.6 \times 10^{-8}$ s, so that $a_{\max}$ can reach values as large as 10^6 m/s^2, *i.e.* $10^5 g$ where g is the acceleration due to gravity. With such a force, one can stop a thermal atomic beam in a distance of the order of one meter, provided that one compensates for the Doppler shift of the decelerating atom, by using for example a spatially varying Zeeman shift [14, 15] or a chirped laser frequency [16].

Dispersive forces, also called dipole forces or gradient forces [2, 3, 17], can be interpreted in terms of position dependent light shifts $\hbar\Delta'(\mathbf{r})$ due to a spatially varying light intensity [18]. Consider for example a laser beam well detuned from resonance, so that one can neglect Γ' (no scattering process). The atom thus remains in the ground state and the light shift $\hbar\Delta'(\mathbf{r})$ of this state plays the role of a potential energy, giving rise to a force which is equal and opposite to its gradient : $\mathbf{F} = -\nabla[\hbar\Delta'(\mathbf{r})]$. Such a force can also be interpreted as resulting from a redistribution of photons between the various plane waves forming the laser wave in absorption-stimulated emission cycles. If the detuning is not large enough to allow Γ' to be neglected, spontaneous transitions occur between dressed states having opposite gradients, so that the instantaneous force oscillates back and forth between two opposite values in a random way. Such a dressed atom picture provides a simple interpretation of the mean value and of the fluctuations of dipole forces [19].

1.3.2 *Applications of dissipative forces – Doppler cooling and magneto-optical traps*

We have already mentioned in the previous subsection the possibility of decelerating an atomic beam by the radiation pressure force of a laser plane wave. Interesting effects can also be obtained by combining the effects of two counterpropagating laser waves.

A first example is Doppler cooling, first suggested for neutral atoms by T. W. Hänsch and A.L. Schawlow [20] and, independently for trapped ions, by D. Wineland and H. Dehmelt [21]. This cooling process results from a Doppler induced imbalance between two opposite radiation pressure forces. The two counterpropagating laser waves have the same (weak) intensity and the same frequency and they are slightly detuned to the red of the atomic frequency ($\omega_L < \omega_A$). For an atom at rest, the two radiation pressure forces exactly balance each other and the net force is equal to zero. For a moving atom, the apparent frequencies of the two laser waves are Doppler shifted. The counterpropagating wave gets closer to resonance and exerts a stronger radiation pressure force than the copropagating wave which gets farther from resonance. The net force is thus opposite to the atomic velocity v and can be written for small v as $F = -\alpha v$ where α is a friction coefficient. By using three pairs of counterpropagating laser waves along three orthogonal directions, one can damp the atomic velocity in a very short time, on the order of a few microseconds, achieving what is called an "optical molasses" [22].

The Doppler friction responsible for the cooling is necessarily accompanied by fluctuations due to the fluorescence photons which are spontaneously emitted in random directions and at random times. These photons communicate to the atom a random recoil momentum $\hbar k$, responsible for a momentum diffusion described by a diffusion coefficient D [3, 18, 25]. As in usual Brownian motion, competition between friction and diffusion usually leads to a steady-state, with an equilibrium temperature proportional to D/α. The theory of Doppler cooling [23, 24, 25] predicts that the equilibrium temperature obtained with such a scheme is always larger than a certain limit T_D, called the Doppler limit, and given by $k_B T_D = \hbar\Gamma/2$ where Γ is the natural width of the excited state and k_B the Boltzmann constant. This limit, which is reached for $\delta = \omega_L - \omega_A = -\Gamma/2$, is, for alkali atoms, on the order of 100 μK. In fact, when the measurements became precise enough, it appeared that the temperature in optical molasses was much lower than expected [26]. This indicates that other laser cooling mechanisms, more powerful than Doppler cooling, are operating. We will come back to this point in section 2.

The imbalance between two opposite radiation pressure forces can be also made position dependent though a spatially dependent Zeeman shift produced by a magnetic field gradient. In a one-dimensional configuration, first suggested by J. Dalibard in 1986, the two counterpropagating waves, which are detuned to the red ($\omega_L < \omega_A$) and which have opposite circular polarizations are in resonance with the atom at different places. This results in a restoring force towards the point where the magnetic field vanishes. Furthermore the non zero value of the detuning provides a Doppler cooling. In fact, such a scheme can be extended to three dimensions and leads to a robust, large and deep trap called "magneto-optical trap" or "MOT" [27]. It combines trapping and cooling, it has a large velocity capture range and it can be used for trapping atoms in a small cell filled with a low pressure vapour [28].

1.3.3 *Applications of dispersive forces : laser traps and atomic mirrors*

When the detuning is negative ($\omega_L - \omega_A < 0$), light shifts are negative. If the laser beam is focussed, the focal zone where the intensity is maximum appears as a minimum of potential energy, forming a potential well where sufficiently cold atoms can be trapped. This is a laser trap. Laser traps using a single focussed laser beam [29] or two crossed focussed laser beams [30, 31] have been realized. Early proposals [32] were considering trapping atoms at the antinodes or nodes of a non resonant laser standing wave. Channeling of atoms in a laser standing wave has been observed experimentally [33].

If the detuning is positive, light shifts are positive and can thus be used to produce potential barriers. For example an evanescent blue detuned wave at the surface of a piece of glass can prevent slow atoms impinging on the glass surface from touching the wall, making them bounce off a "carpet of light" [34]. This is the principle of mirrors for atoms. Plane atomic mirrors [35, 36] have been realized as well as concave mirrors [37].

2. SUB-DOPPLER COOLING

In the previous section, we discussed separately the manipulation of internal and external degrees of freedom, and we have described physical mechanisms involving only one type of physical effect, either dispersive or dissipative. In fact, there exist cooling mechanisms resulting from an interplay between spin and external degrees of freedom, and between dispersive and dissipative effects. We discuss in this section one of them, the so-called "Sisyphus cooling" or "polarization-gradient cooling" mechanism [38, 39] (see also [19]), which leads to temperatures much lower than Doppler cooling. One can understand in this way the sub-Doppler temperatures observed in optical molasses and mentioned above in section 1.3.2.

2.1 *SISYPHUS EFFECT*

Most atoms, in particular alkali atoms, have a Zeeman structure in the ground state. Since the detuning used in laser cooling experiments is not too large compared to Γ, both differential light shifts and optical pumping transitions exist for the various ground state Zeeman sublevels. Furthermore, the laser polarization varies in general in space so that light shifts and optical pumping rates are position-dependent. We show now, with a simple one-dimensional example, how the combination of these various effects can lead to a very efficient cooling mechanism.

Consider the laser configuration of Fig.2a, consisting of two counterpropagating plane waves along the z-axis, with orthogonal linear polarizations and with the same frequency and the same intensity. Because the phase shift between the two waves increases linearly with z, the polarization of the total field changes from σ^+ to σ^- and vice versa every $\lambda/4$. In between, it is elliptical or linear.

Consider now the simple case where the atomic ground state has an angular momentum $J_g = 1/2$. As shown in subsection (1.2), the two Zeeman sublevels $M_g = \pm 1/2$ undergo different light shifts, depending on the laser polarization, so that the Zeeman degeneracy in zero magnetic field is removed. This gives the energy diagram of Fig.2b showing spatial modulations of the Zeeman splitting between the two sublevels with a period $\lambda/2$.

If the detuning δ is not too large compared to Γ, there are also real absorptions of photons by the atom followed by spontaneous emission, which give rise to optical pumping transfers between the two sublevels, whose direction depends on the polarization: $M_g = -1/2 \rightarrow M_g = +1/2$ for a σ^+ polarization, $M_g = +1/2 \rightarrow M_g = -1/2$ for a σ^- polarization. Here also, the spatial modulation of the laser polarization results in a spatial modulation of the optical pumping rates with a period $\lambda/2$ (vertical arrows of Fig.2b).

The two spatial modulations of light shifts and optical pumping rates are of course correlated because they are due to the same cause, the spatial modulation of the light polarization. These correlations clearly appear in Fig.2b. With the proper sign of the detuning, optical pumping always transfers atoms from the higher Zeeman sublevel to the lower one. Suppose now that the

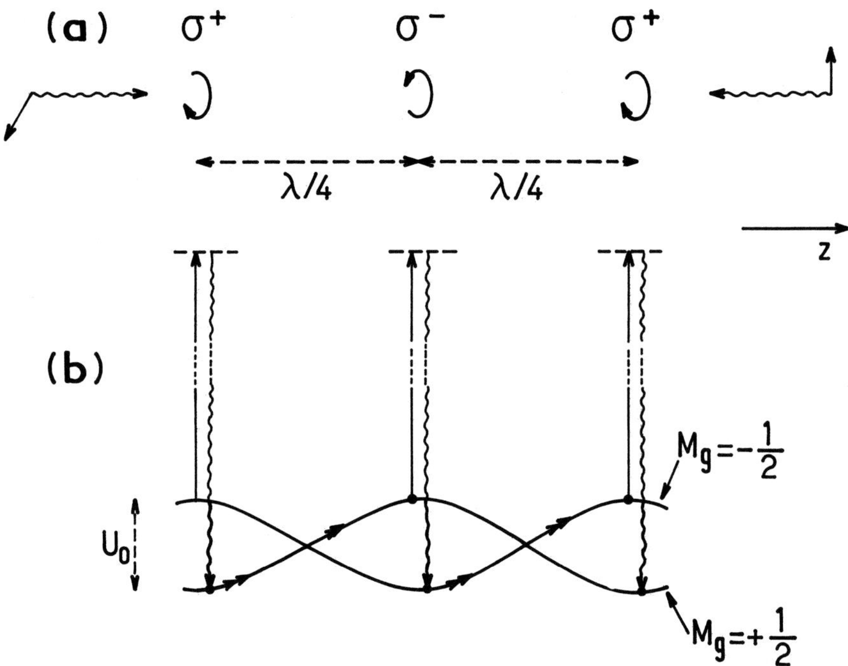

Figure 2. Sisyphus cooling. Laser configuration formed by two counterpropagating plane waves along the z-axis with orthogonal linear polarizations (a). The polarization of the resulting electric field is spatially modulated with a period $\lambda/2$. Every $\lambda/4$, it changes from σ^+ to σ^- and vice versa. For an atom with two ground state Zeeman sublevels $M_g = \pm 1/2$, the spatial modulation of the laser polarization results in correlated spatial modulations of the light shifts of these two sublevels and of the optical pumping rates between them (b). Because of these correlations, a moving atom runs up potential hills more frequently than down (double arrows of b).

atom is moving to the right, starting from the bottom of a valley, for example in the state $M_g = +1/2$ at a place where the polarization is σ^+. Because of the finite value of the optical pumping time, there is a time lag between internal and external variables and the atom can climb up the potential hill before absorbing a photon and reach the top of the hill where it has the maximum probability to be optically pumped in the other sublevel, i.e. in the bottom of a valley, and so on (double arrows of Fig.2b). Like Sisyphus in the Greek mythology, who was always rolling a stone up the slope, the atom is running up potential hills more frequently than down. When it climbs a potential hill, its kinetic energy is transformed into potential energy. Dissipation then occurs by light, since the spontaneously emitted photon has an energy higher than the absorbed laser photon (anti-Stokes Raman processes of Fig.2b). After each Sisyphus cycle, the total energy E of the atom decreases by an amount of the order of U_0, where U_0 is the depth of the optical potential wells of Fig.2b, until E becomes smaller than U_0, in which case the atom remains trapped in the potential wells.

The previous discussion shows that Sisyphus cooling leads to temperatures T_{Sis} such that $k_B\, T_{\text{Sis}} \simeq U_0$. According to Eq. (2), the light shift U_0 is proportional to $\hbar\Omega^2/\delta$ when $4|\delta| > \Gamma$. Such a dependence of T_{Sis} on the Rabi frequency Ω and on the detuning δ has been checked experimentally with cesium atoms [40]. Fig.3 presents the variations of the measured temperature T with the dimensionless parameter $\Omega^2/\Gamma|\delta|$. Measurements of T versus intensity for different values of δ show that T depends linearly, for low enough intensities, on a single parameter which is the light shift of the ground state Zeeman sublevels.

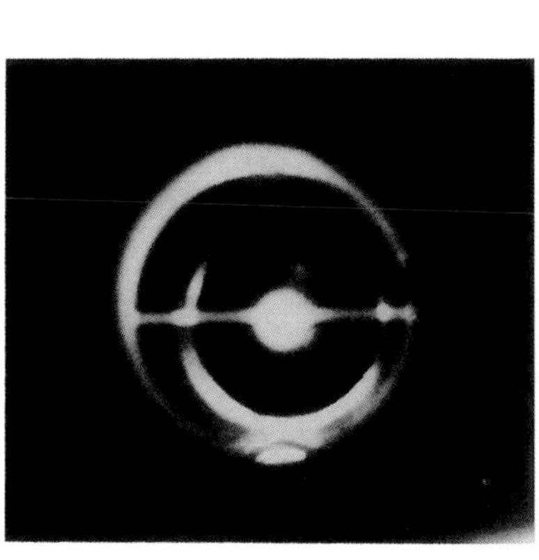

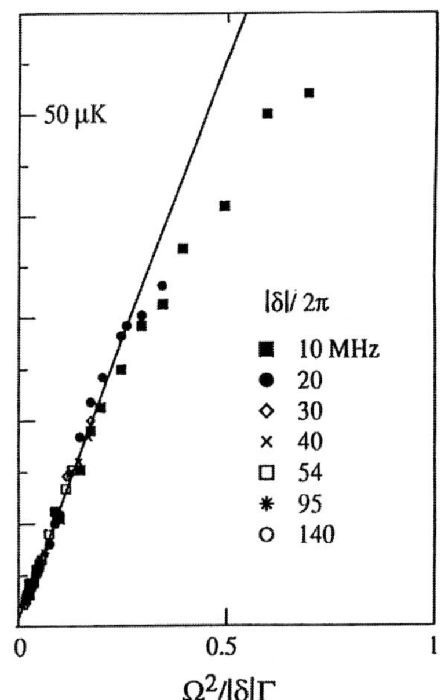

Figure 3. Temperature measurements in cesium optical molasses (from reference [40]). The left part of the figure shows the fluorescence light emitted by the molasses observed through a window of the vacuum chamber. The horizontal bright line is the fluorescence light emitted by the atomic beam which feeds the molasses and which is slowed down by a frequency chirped laser beam. Right part of the figure : temperature of the atoms measured by a time of flight technique versus the dimensionless parameter $\Omega^2/|\delta|\Gamma$ proportional to the light shift (Ω is the optical Rabi frequency, δ the detuning and Γ the natural width of the excited state).

2.2 *THE LIMITS OF SISYPHUS COOLING*

At low intensity, the light shift $U_0 \propto \hbar\Omega^2/\delta$ is much smaller than $\hbar\Gamma$. This explains why Sisyphus cooling leads to temperatures much lower than those achievable with Doppler cooling. One cannot however decrease indefinitely the laser intensity. The previous discussion ignores the recoil due to the spontaneously emitted photons which increases the kinetic energy of the atom by an amount on the order of E_R, where

$$E_R = \hbar^2 k^2 / 2M \tag{3}$$

is the recoil energy of an atom absorbing or emitting a single photon. When U_0 becomes on the order or smaller than E_R, the cooling due to Sisyphus cooling becomes weaker than the heating due to the recoil, and Sisyphus cooling no longer works. This shows that the lowest temperatures which can be achieved with such a scheme are on the order of a few E_R/k_B. This result is confirmed by a full quantum theory of Sisyphus cooling [41, 42] and is in good agreement with experimental results. The minimum temperature in Fig.3 is on the order of $10E_R/k_B$.

2.3 *OPTICAL LATTICES*

For the optimal conditions of Sisyphus cooling, atoms become so cold that they get trapped in the quantum vibrational levels of a potential well (see Fig. 4). More precisely, one must consider energy bands in this perodic structure [43]. Experimental observation of such a quantization of atomic motion

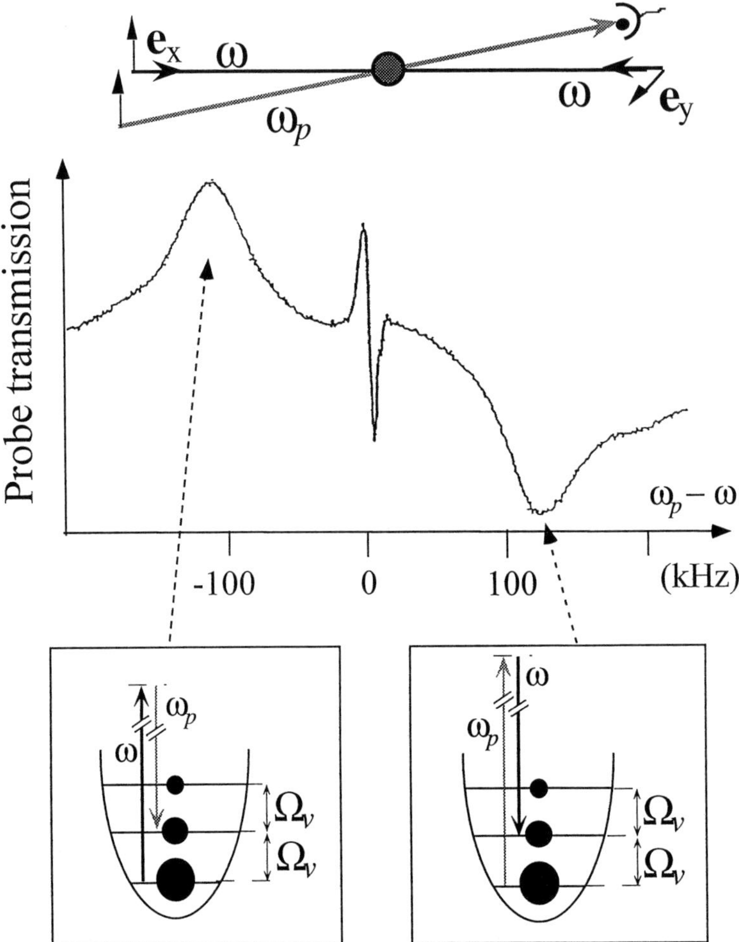

Figure 4. Probe absorption spectrum of a 1-D optical lattice (from reference [44]). The upper part of the figure shows the two counterpropagating laser beams with frequency ω and orthogonal linear polarizations forming the 1D-optical lattice, and the probe beam with frequency ω_p whose absorption is measured by a detector. The lower part of the figure shows the probe transmission versus $\omega_p-\omega$. The two lateral resonances corresponding to amplification or absorption of the probe are due to stimulated Raman processes between vibrational levels of the atoms trapped in the light field (see the two insets). The central narrow structure is a Rayleigh line due to the antiferromagnetic spatial order of the atoms.

in an optical potential was first achieved in one dimension [44] [45]. Atoms are trapped in a spatial periodic array of potential wells, called a "1D-optical lattice", with an antiferromagnetic order, since two adjacent potential wells correspond to opposite spin polarizations. 2D and 3D optical lattices have been realized subsequently (see the review papers [46] [47]).

3. SUBRECOIL LASER COOLING

3.1 *THE SINGLE PHOTON RECOIL LIMIT. HOW TO CIRCUMVENT IT*

In most laser cooling schemes, fluorescence cycles never cease. Since the random recoil $\hbar k$ communicated to the atom by the spontaneously emitted photons cannot be controlled, it seems impossible to reduce the atomic momentum spread δp below a value corresponding to the photon momentum $\hbar k$. Condition $\delta p = \hbar k$ defines the "single photon recoil limit". It is usual in laser cooling to define an effective temperature T in terms of the half-width δp (at $1/\sqrt{e}$) of the momentum distribution by $k_B T/2 = \delta p^2/2M$. In the temperature scale, condition $\delta p = \hbar k$ defines a "recoil temperature" T_R by:

$$\frac{k_B T_R}{2} = \frac{\hbar^2 k^2}{2M} = E_R \tag{4}$$

The value of T_R ranges from a few hundred nanoKelvin for alkalis to a few microKelvin for helium.

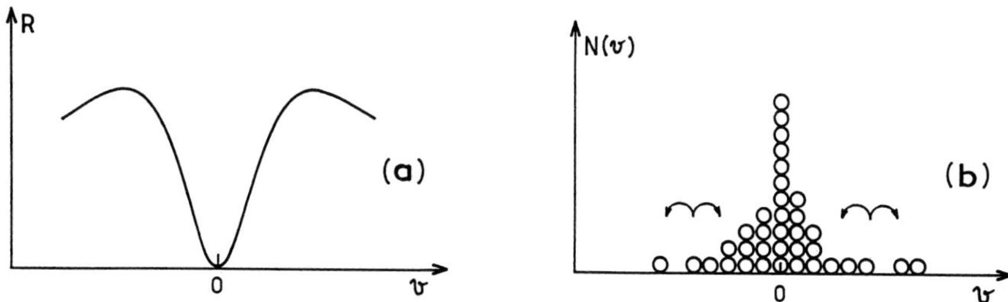

Figure 5. Subrecoil laser cooling. The random walk of the atom in velocity space is supposed to be characterized by a jump rate R which vanishes for $v = 0$ (a). As a result of this inhomogeneous random walk, atoms which fall in a small interval around $v = 0$ remain trapped there for a long time, on the order of $[R(v)]^{-1}$, and accumulate (b).

It is in fact possible to circumvent this limit and to reach temperatures T lower than T_R, a regime which is called "subrecoil" laser cooling. The basic idea is to create a situation where the photon absorption rate Γ', which is also the jump rate R of the atomic random walk in velocity space, depends on the atomic velocity $v = p/M$ and vanishes for $v = 0$ (Fig.5a). Consider then an atom with $v = 0$. For such an atom, the absorption of light is quenched. Consequently, there is no spontaneous reemission and no associated random recoil. One protects in this way ultracold atoms (with $v \simeq 0$) from the "bad" effects of the light. On the other hand, atoms with $v \neq 0$ can absorb and reemit light. In such absorption-spontaneous emission cycles, their velocities change in a random way and the corresponding random walk in v-space can transfer atoms from the $v \neq 0$ absorbing states into the $v \simeq 0$ dark states where they remain trapped and accumulate (see Fig. 5b). This reminds us of what happens in a Kundt's tube where sand grains vibrate in an acoustic standing wave and accumulate at the nodes of this wave where they no longer move. Note however that the random walk takes place in velocity space for the situation considered in Fig. 5b, whereas it takes place in position space in a Kundt's tube.

Up to now, two subrecoil cooling schemes have been proposed and demonstrated. In the first one, called "Velocity Selective Coherent Population Trapping" (VSCPT), the vanishing of $R(v)$ for $v = 0$ is achieved by using destructive quantum interference between different absorption amplitudes [48]. The second one, called Raman cooling, uses appropriate sequences of stimulated Raman and optical pumping pulses for tailoring the appropriate shape of $R(v)$ [49].

3.2 *BRIEF SURVEY OF VSCPT*

We first recall the principle of the quenching of absorption by "coherent population trapping", an effect which was discovered and studied in 1976 [50, 51]. Consider the 3-level system of Fig.6, with two ground state sublevels g_1 and g_2 and one excited sublevel e_0, driven by two laser fields with frequencies ω_{L1} and ω_{L2}, exciting the transitions $g_1 \leftrightarrow e_0$ and $g_2 \leftrightarrow e_0$, respectively. Let $\hbar\Delta$ be the detuning from resonance for the stimulated Raman process consisting of the absorption of one ω_{L1} photon and the stimulated emission of one ω_{L2} photon, the atom going from g_1 to g_2. One observes that the fluorescence

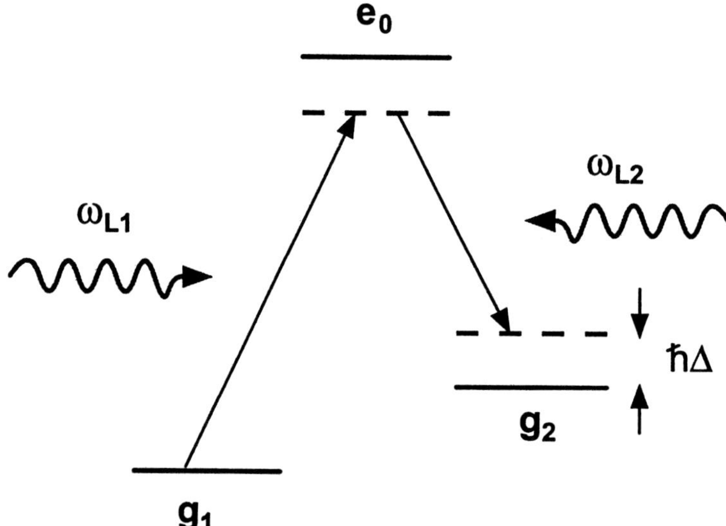

Figure 6. Coherent population trapping. A three-level atom g_1, g_2, e_0 is driven by two laser fields with frequences ω_{L1} and ω_{L2} exciting the transitions $g_1 \leftrightarrow e_0$ and $g_2 \leftrightarrow e_0$, respectively. $\hbar\Delta$ is the detuning from resonance for the stimulated Raman process induced between g_1 and g_2 by the two laser fields ω_{L1} and ω_{L2}. When $\Delta = 0$, atoms are optically pumped in a linear superposition of g_1 and g_2 which no longer absorbs light because of a destructive interference between the two absorption amplitudes $g_1 \to e_0$ and $g_2 \to e_0$.

rate R vanishes for $\Delta = 0$. Plotted versus Δ, the variations of R are similar to those of Fig.5a with v replaced by Δ. The interpretation of this effect is that atoms are optically pumped into a linear superposition of g_1 and g_2 which is not coupled to e_0 because of a destructive interference between the two absorption amplitudes $g_1 \rightarrow e_0$ and $g_2 \rightarrow e_0$.

The basic idea of VSCPT is to use the Doppler effect for making the detuning Δ of the stimulated Raman process of Fig.6 proportional to the atomic velocity v. The quenching of absorption by coherent population trapping is thus made velocity dependent and one achieves the situation of Fig. 5.a. This is obtained by taking the two laser waves ω_{L1} and ω_{L2} counterpropagating along the z-axis and by choosing their frequencies in such a way that $\Delta = 0$ for an atom at rest. Then, for an atom moving with a velocity v along the z-axis, the opposite Doppler shifts of the two laser waves result in a Raman detuning $\Delta = (k_1 + k_2)\, v$ proportional to v.

A more quantitative analysis of the cooling process [52] shows that the dark state, for which $R = 0$, is a linear superposition of two states which differ not only by the internal state (g_1 or g_2) but also by the momentum along the z-axis:

$$|\psi_D\rangle = c_1\,|g_1, -\hbar k_1\rangle + c_2\,|g_2, +\hbar k_2\rangle \tag{5}$$

This is due to the fact that g_1 and g_2 must be associated with different momenta, $-\hbar k_1$ and $+\hbar k_2$, in order to be coupled to the same excited state $|e_0, p = 0\rangle$ by absorption of photons with different momenta $+\hbar k_1$ and $-\hbar k_2$. Furthermore, when $\Delta = 0$, the state (5) is a stationary state of the total atom + laser photons system. As a result of the cooling by VSCPT, the atomic momentum distribution thus exhibits two sharp peaks, centered at $-\hbar k_1$ and

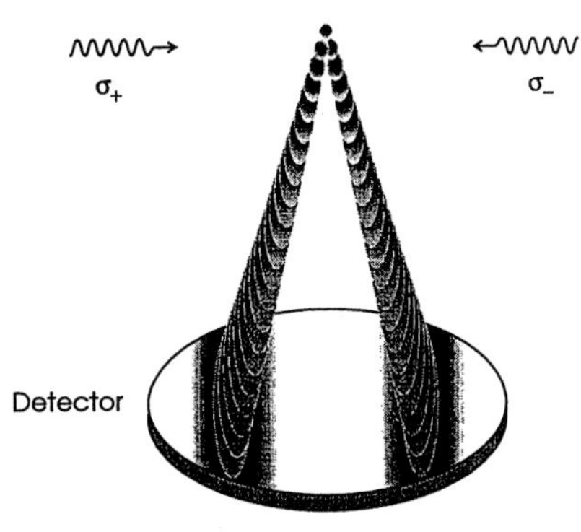

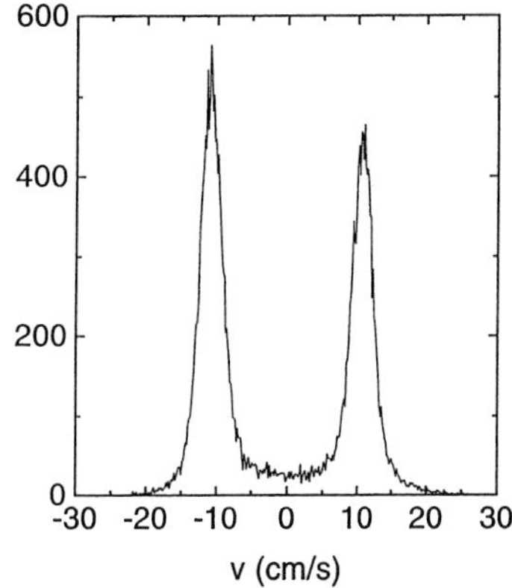

Figure 7. One-dimensional VSCPT experiment. The left part of the figure shows the experimental scheme. The cloud of precooled trapped atoms is released while the two counterpropagating VSCPT beams with orthogonal circular polarizations are applied during a time θ = 1ms. The atoms then fall freely and their positions are detected 6.5 cm below on a microchannel plate. The double band pattern is a signature of the 1D cooling process which accumulates the atoms in a state which is a linear superposition of two different momenta. The right part of the figure gives the velocity distribution of the atoms detected by the microchannel plate. The width δv of the two peaks is clearly smaller than their separation $2v_R$, where v_R = 9.2 cm/s is the recoil velocity. This is a clear signature of subrecoil cooling.

$+\hbar k_2$, with a width δp which tends to zero when the interaction time Θ tends to infinity.

The first VSCPT experiment [48] was performed on the 2^3S_1 metastable state of helium atoms. The two lower states g_1 and g_2 were the $M = -1$ and $M = +1$ Zeeman sublevels of the 2^3S_1 metastable state, e_0 was the $M = 0$ Zeeman sublevel of the excited 2^3P_1 state. The two counterpropagating laser waves had the same frequency $\omega_{L1} = \omega_{L2} = \omega_L$ and opposite circular polarizations. The two peaks of the momentum distribution were centered at $\pm\hbar k$, with a width corresponding to $T \simeq T_R/2$. The interaction time was then increased by starting from a cloud of trapped precooled helium atoms instead of using an atomic beam as in the first experiment [53]. This led to much lower temperatures (see Fig. 8). Very recently, temperatures as low as $T_R/800$ have been observed [54].

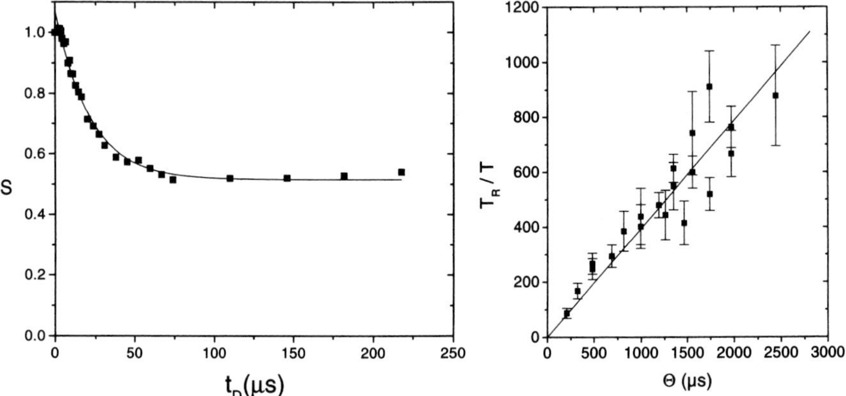

Figure 8. Measurement of the spatial correlation function of atoms cooled by VSCPT (from reference [54]). After a cooling period of duration θ, the two VSCPT beams are switched off during a dark period of duration t_D. The two coherent wave packets into which atoms are pumped fly apart with a relative velocity $2v_R$ and get separated by a distance $a = 2v_R t_D$. Reapplying the two VSCPT beams during a short probe pulse, one measures a signal S which can be shown to be equal to $[1 + G(a)]/2$ where $G(a)$ is the spatial overlap between the two identical wave packets separated by a. From $G(a)$, which is the spatial correlation function of each wave packet, one determines the atomic momentum distribution which is the Fourier transform of $G(a)$. The left part of the figure gives S versus a. The right part of the figure gives T_R/T versus the cooling time θ, where T_R is the recoil temperature and T the temperature of the cooled atoms determined from the width of $G(a)$. The straight line is a linear fit in agreement with the theoretical predictions of Lévy statistics. The lowest temperature, on the order of $T_R/800$, is equal to 5 nK.

In fact, it is not easy to measure such low temperatures by the usual time of flight techniques, because the resolution is then limited by the spatial extent of the initial atomic cloud. A new method has been developed [54] which consists of measuring directly the spatial correlation function of the atoms $G(a) = \int_{-\infty}^{+\infty} dz\phi^*(z + a)\phi(z)$, where $\phi(z)$ is the wave function of the atomic wave packet. This correlation function, which describes the degree of spatial coherence between two points separated by a distance a, is simply the Fourier transform of the momentum distribution $|\phi(p)|^2$. This method is analogous to Fourier spectroscopy in optics, where a narrow spectral line $I(\omega)$ is more easily inferred from the correlation function of the emitted electric field $G(\tau)$

$= \int_{-\infty}^{+\infty} dt E^*(t+\tau)E(t)$, which is the Fourier transform of $I(\omega)$. Experimentally, the measurement of $G(a)$ is achieved by letting the two coherent VSCPT wave packets fly apart with a relative velocity $2v_R = 2\hbar k/m$ during a dark period t_D, during which the VSCPT beams are switched off. During this dark period, the two wave packets get separated by a distance a = $2v_R t_D$, and one then measures with a probe pulse a signal proportional to their overlap. Fig.8a shows the variations with t_D of such a signal S (which is in fact equal to $[1 + G(a)]/2$). From such a curve, one deduces a temperature $T \simeq T_R/625$, corresponding to $\delta p \simeq \hbar k/25$. Fig.8b shows the variations of T_R/T with the VSCPT interaction time θ. As predicted by theory (see next subsection), T_R/T varies linearly with θ and can reach values as large as 800.

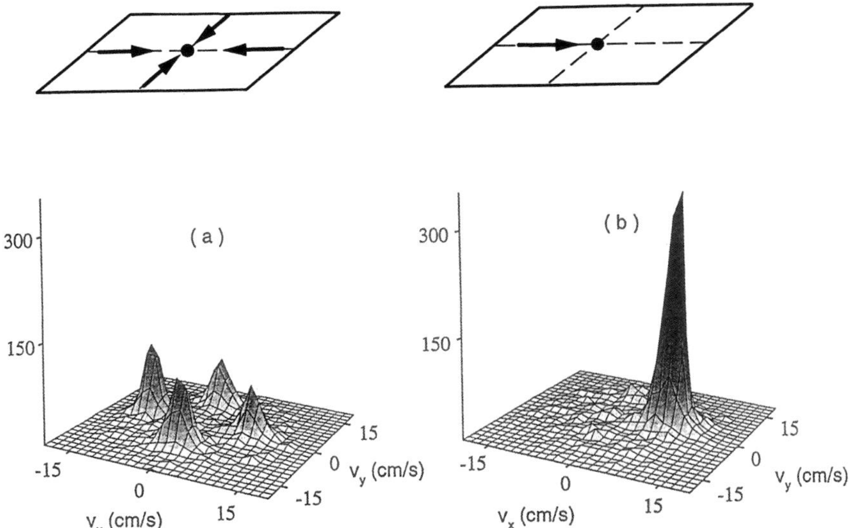

Figure 9. Two-dimensional VSCPT experiment (from reference [58]). The experimental scheme is the same as in Fig. 7, but one uses now four VSCPT beams in a horizontal plane and atoms are pumped into a linear superposition of four different momentum states giving rise to four peaks in the two-dimensional velocity distribution (a). When three of the four VSCPT laser beams are adiabatically switched off, the whole atomic population is transferred into a single wave packet (b).

VSCPT has been extended to two [55] and three [56] dimensions. For a $J_g = 1 \leftrightarrow J_e = 1$ transition, it has been shown [57] that there is a dark state which is described by the same vector field as the laser field. More precisely, if the laser field is formed by a linear superposition of N plane waves with wave vectors $\mathbf{k}_i$ ($i = 1,2,...N$) having the same modulus k, one finds that atoms are cooled in a coherent superposition of N wave packets with mean momenta $\hbar\mathbf{k}_i$ and with a momentum spread δp which becomes smaller and smaller as the interaction time θ increases. Furthermore, because of the isomorphism between the de Broglie dark state and the laser field, one can adiabatically change the laser configuration and transfer the whole atomic population into a single wave packet or two coherent wave packets chosen at will [58]. Fig. 9 shows an example of such a coherent manipulation of atomic wave packets in two dimensions. In Fig.9a, one sees the transverse velocity distribution associated with the four wave packets obtained with two pairs of counterpropagating laser waves along the *x* and *y*-axis in a horizontal plane; Fig.9b shows the single wave packet into which the whole atomic population is transferred by switching off adiabatically three of the four VSCPT beams. Similar results have been obtained in three dimensions.

3.3 *SUBRECOIL LASER COOLING AND LÉVY STATISTICS*

Quantum Monte Carlo simulations using the delay function [59, 60] have provided new physical insight into subrecoil laser cooling [61]. Fig.10 shows for example the random evolution of the momentum p of an atom in a 1*D*-VSCPT experiment. Each vertical discontinuity corresponds to a spontaneous emission process during which p changes in a random way. Between two successive jumps, p remains constant. It clearly appears that the random walk of the atom in velocity space is anomalous and dominated by a few rare events whose duration is a significant fraction of the total interaction time. A simple analysis shows that the distribution $P(\tau)$ of the trapping times τ in a small trapping zone near $v = 0$ is a broad distribution which falls as a power-law in the wings. These wings decrease so slowly that the average value $\langle\tau\rangle$ of τ (or the variance) can diverge. In such cases, the central limit theorem (CLT) can obviously no longer be used for studying the distribution of the total trapping time after N entries in the trapping zone separated by N exits.

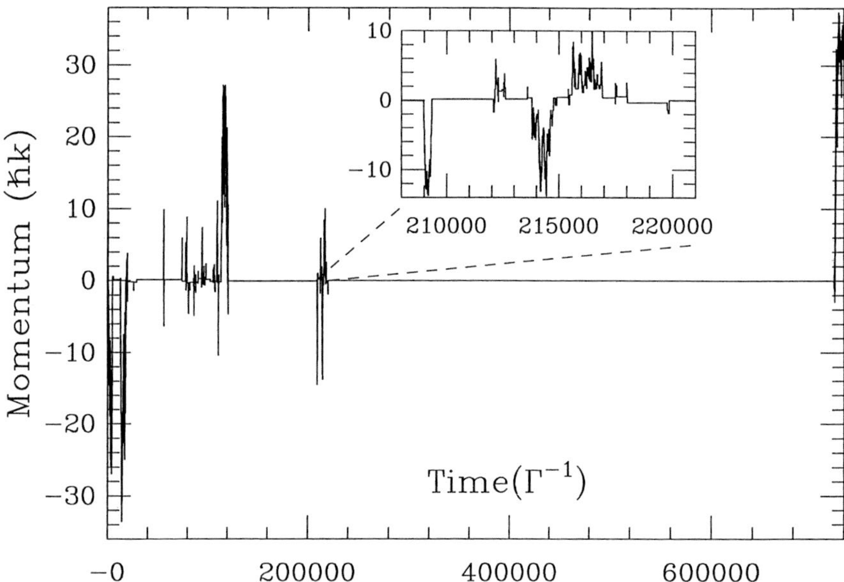

Figure 10. Monte Carlo wave function simulation of one-dimensional VSCPT (from reference [61]). Momentum p characterizing the cooled atoms versus time. Each vertical discontinuity corresponds to a spontaneous emission jump during which p changes in a random way. Between two successive jumps, p remains constant. The inset shows a zoomed part of the sequence.

It is possible to extend the CLT to broad distributions with power-law wings [62, 63]. We have applied the corresponding statistics, called "Lévy statistics", to subrecoil cooling and shown that one can obtain in this way a better understanding of the physical processes as well as quantitative analytical predictions for the asymptotic properties of the cooled atoms in the limit when the interaction time θ tends to infinity [61, 64]. For example, one predicts in this way that the temperature decreases as $1/\theta$ when $\theta \to \infty$, and that the wings of the momentum distribution decrease as $1/p^2$, which shows that the shape of the momentum distribution is closer to a Lorentzian than a Gaussian. This is in agreement with the experimental observations represented in Fig.8. (The fit in Fig.8a is an exponential, which is the Fourier transform of a Lorentzian).

One important feature revealed by this theoretical analysis is the non-ergodicity of the cooling process. Regardless of the interaction time θ, there are always atomic evolution times (trapping times in the small zone of Fig.5a around $v = 0$) which can be longer than θ. Another advantage of such a new approach is that it allows the parameters of the cooling lasers to be optimized for given experimental conditions. For example, by using different shapes for the laser pulses used in one-dimensional subrecoil Raman cooling, it has been possible to reach for Cesium atoms temperatures as low as 3 nK [65].

4. A FEW EXAMPLES OF APPLICATIONS

The possibility of trapping atoms and cooling them at very low temperatures, where their velocity can be as low as a few mm/s, has opened the way to a wealth of applications. Ultracold atoms can be observed during much longer times, which is important for high resolution spectroscopy and frequency standards. They also have very long de Broglie wavelengths, which has given rise to new research fields, such as atom optics, atom interferometry and Bose-Einstein condensation of dilute gases. It is impossible to discuss here all these developments. We refer the reader to recent reviews such as [5]. We will just describe in this section a few examples of applications which have been recently investigated by our group in Paris.

4.1 *CESIUM ATOMIC CLOCKS*

Cesium atoms cooled by Sisyphus cooling have an effective temperature on the order of 1 μK, corresponding to a r.m.s. velocity of 1 cm/s. This allows them to spend a longer time T in an observation zone where a microwave field induces resonant transitions between the two hyperfine levels g_1 and g_2 of the ground state. Increasing T decreases the width $\Delta\nu \sim 1/T$ of the microwave resonance line whose frequency is used to define the unit of time. The stability of atomic clocks can thus be considerably improved by using ultracold atoms [66. 67].

In usual atomic clocks, atoms from a thermal cesium beam cross two microwave cavities fed by the same oscillator. The average velocity of the atoms is several hundred m/s, the distance between the two cavities is on the order of 1 m. The microwave resonance between g_1 and g_2 is monitored and is used to lock the frequency of the oscillator to the center of the atomic line. The narrower the resonance line, the more stable the atomic clock. In fact, the microwave resonance line exhibits Ramsey interference fringes whose width $\Delta\nu$ is determined by the time of flight T of the atoms from one cavity to another. For the longest devices, T, which can be considered as the observation time, can reach 10 ms, leading to values of $\Delta\nu \sim 1/T$ on the order of 100 Hz.

Much narrower Ramsey fringes, with sub-Hertz linewidths can be obtained in the so-called "Zacharias atomic fountain" [68]. Atoms are captured in a magneto-optical trap and laser cooled before being launched upwards by a laser pulse through a microwave cavity. Because of gravity they are decelerated, they return and fall back, passing a second time through the cavity. Atoms therefore experience two coherent microwave pulses, when they pass through the cavity, the first time on their way up, the second time on their way down. The time interval between the two pulses can now be on the order of 1 sec, *i.e.* about two orders of magnitude longer than with usual clocks. Atomic fountains have been realized for sodium [69] and cesium [70]. A short-term relative frequency stability of $1.3 \times 10^{-13}\tau^{-1/2}$, where τ is the integration time, has been recently measured for a one meter high Cesium fountain [71, 72]. For $\tau = 10^4 s$, $\Delta\nu/\nu \sim 1.3 \times 10^{-15}$ and for $\tau = 3 \times 10^4 s$, $\Delta\nu/\nu \sim 8 \times 10^{-16}$ has been measured. In fact such a stability is most likely limited by the Hydrogen maser which is used as a reference source and the real stability, which could be more precisely determined by beating the signals of two fountain clocks, is expected to reach $\Delta\nu/\nu \sim 10^{-16}$ for a one day integration time. In addition to the stability, another very important property of a frequency standard is its accuracy. Because of the very low velocities in a fountain device, many systematic shifts are strongly reduced and can be evaluated with great precision. With an accuracy of 2×10^{-15}, the BNM-LPTF fountain is presently the most accurate primary standard [73]. A factor 10 improvement in this accuracy is expected in the near future.

To increase the observation time beyond one second, a possible solution consists of building a clock for operation in a reduced gravity environment. Such a microgravity clock has been recently tested in a jet plane making parabolic flights. A resonance signal with a width of 7 Hz has been recorded in a $10^{-2}g$ environment. This width is twice narrower than that produced on earth in the same apparatus. This clock prototype (see Fig.11) is a compact and transportable device which can be also used on earth for high precision frequency comparison.

Atomic clocks working with ultracold atoms could not only provide an improvement of positioning systems such as the GPS. They could be also used for fundamental studies. For example, one could build two fountains clocks, one with cesium and one with rubidium, in order to measure with a high accuracy the ratio between the hyperfine frequencies of these two atoms. Because of relativistic corrections, the hyperfine splitting is a function of $Z\alpha$ where α is the fine structure constant and Z is the atomic number [74]. Since Z is not the same for cesium and rubidium, the ratio of the two hyperfine

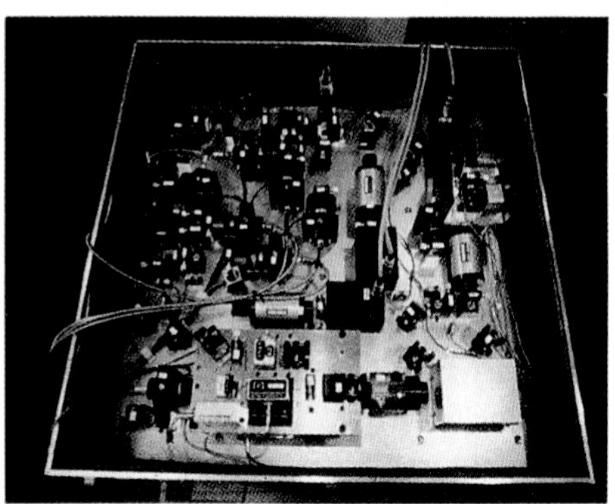

Figure 11. The microgravity clock prototype. The left part is the 60 cm × 60 cm × 15 cm optical bench containing the diode laser sources and the various optical components. The right part is the clock itself (about one meter long) containing the optical molasses, the microwave cavity and the detection region.

structures depends on α. By making several measurements of this ratio over long periods of time, one could check Dirac's suggestion concerning a possible variation of α with time. The present upper limit for $\dot{\alpha}/\alpha$ in laboratory tests [74] could be improved by two orders of magnitude.

Another interesting test would be to measure with a higher accuracy the gravitational red shift and the gravitational delay of an electromagnetic wave passing near a large mass (Shapiro effect [75]).

4.2 *GRAVITATIONAL CAVITIES FOR NEUTRAL ATOMS*

We have already mentioned in section 1.3 the possibility of making atomic mirrors for atoms by using blue detuned evanescent waves at the surface of a piece of glass. Concave mirrors (Fig.12a) are particularly interesting because the transverse atomic motion is then stable if atoms are released from a point located below the focus of the mirror. It has been possible in this way to observe several successive bounces of the atoms (Fig.12b) and such a system can be considered as a "trampoline for atoms" [37]. In such an experiment, it is a good approximation to consider atoms as classical particles bouncing off a concave mirror. In a quantum mechanical description of the experiment, one must consider the reflection of the atomic de Broglie waves by the mirror. Standing de Broglie waves can then be introduced for such a "gravitational cavity", which are quite analogous to the light standing waves for a Fabry-Perot cavity [76]. By modulating at frequency $\Omega/2\pi$ the intensity of the evanescent wave which forms the atomic mirror, one can produce the equivalent of a vibrating mirror for de Broglie waves. The reflected waves thus have a modulated Doppler shift. The corresponding frequency modulation of these waves has been recently demonstrated [77] by measuring the energy change ΔE of the bouncing atom, which is found to be equal to $n\hbar\Omega$, where $n = 0, \pm 1, \pm 2, \ldots$ (Figs.12c and d). The discrete nature of this energy spectrum is a pure quantum effect. For classical particles bouncing off a vibrating mirror, ΔE would vary continuously in a certain range.

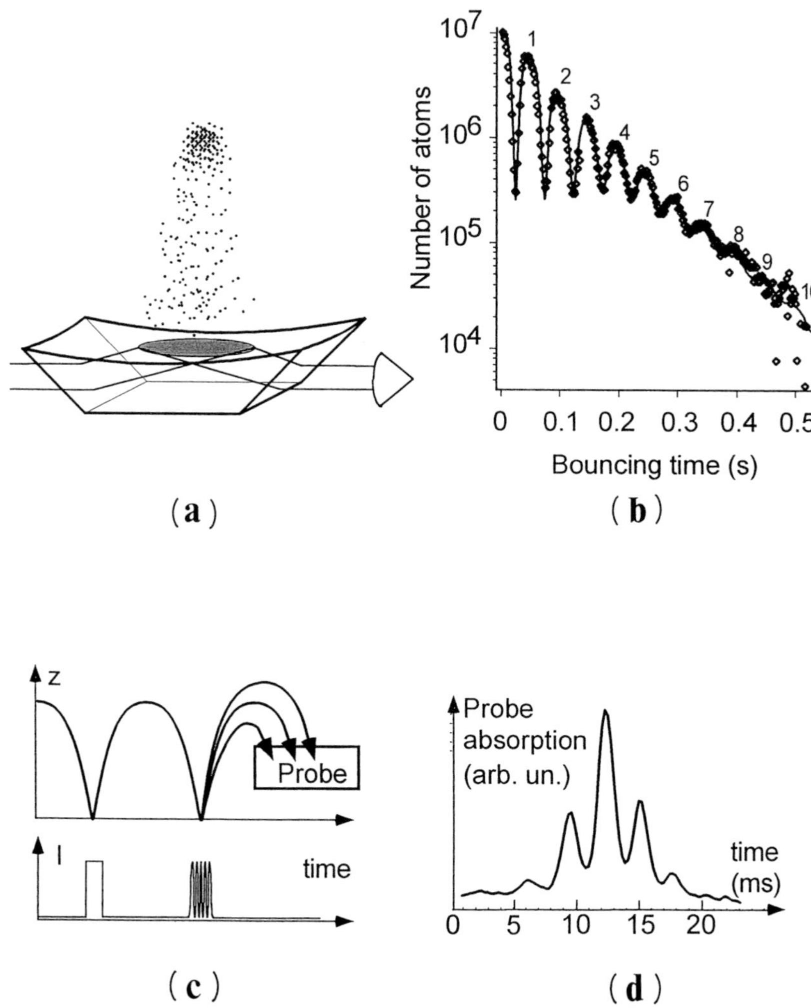

Figure 12. Gravitational cavity for neutral atoms (from references [37] and [77]). Trampoline for atoms (a) – Atoms released from a magneto-optical trap bounce off a concave mirror formed by a blue detuned evanescent wave at the surface of a curved glass prism. Number of atoms at the initial position of the trap versus time after the trap has been switched off (b). Ten successive bounces are visible in the figure. Principle of the experiment demonstrating the frequency modulation of de Broglie waves (c). The upper trace gives the atomic trajectories (vertical position z versus time). The lower trace gives the time dependence of the intensity I of the evanescent wave. The first pulse is used for making a velocity selection. The second pulse is modulated in intensity. This produces a vibrating mirror giving rise to a frequency modulated reflected de Broglie wave which consists in a carrier and sidebands at the modulation frequency. The energy spectrum of the reflected particles is thus discrete so that the trajectories of the reflected particles form a discrete set. This effect is detected by looking at the time dependence of the absorption of a probe beam located above the prism (d).

4.3 *BLOCH OSCILLATIONS*

In the subrecoil regime where δp becomes smaller than $\hbar k$, the atomic coherence length $h/\delta p$ becomes larger than the optical wavelength $\lambda = h/\hbar k = 2\pi/k$ of the lasers used to cool the atom. Consider then such an ultracold atom in the periodic light shift potential produced by a non resonant laser standing wave. The atomic de Broglie wave is delocalized over several periods of the periodic potential, which means that one can prepare in this way quasi-Bloch states. By chirping the frequency of the two counterpropagating laser waves forming the standing wave, one can produce an accelerated standing wave. In the rest frame of this wave, atoms thus feel a constant inertial force in addition to the periodic potential. They are accelerated and the de Broglie wavelength $\lambda_{dB} = h/M\langle v\rangle$ decreases. When $\lambda_{dB} = \lambda_{Laser}$, the de Broglie wave is Bragg-reflected by the periodic optical potential. Instead of increasing linearly with time, the mean velocity $\langle v\rangle$ of the atoms oscillates back and forth. Such Bloch oscillations, which are a textbook effect of solid-state physics, are more easily observed with ultracold atoms than with electrons in condensed matter because the Bloch period can be much shorter than the relaxation time for the coherence of de Broglie waves (in condensed matter, the relaxation processes due to collisions are very strong). Fig.13 shows an example of Bloch oscillations [78] observed on cesium atoms cooled by the improved subrecoil Raman cooling technique described in [65].

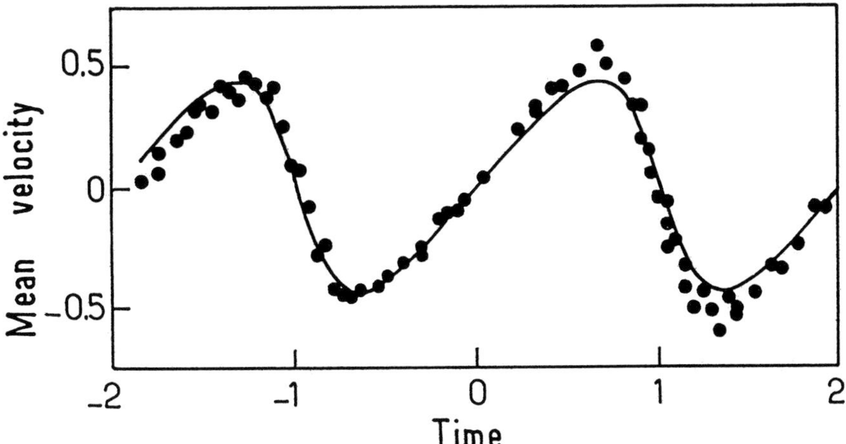

Figure 13. Bloch oscillations of atoms in a periodic optical potential (from reference [78]. Mean velocity (in units of the recoil velocity) versus time (in units of half the Bloch period) for ultracold cesium atoms moving in a periodic optical potential and submitted in addition to a constant force.

5. CONCLUSION

We have described in this paper a few physical mechanisms allowing one to manipulate neutral atoms with laser light. Several of these mechanisms can be simply interpreted in terms of resonant exchanges of energy, angular and linear momentum between atoms and photons. A few of them, among the most efficient ones, result from a new way of combining well known physical effects such as optical pumping, light shifts, coherent population trapping. We have given two examples of such cooling mechanisms, Sisyphus cooling and subrecoil cooling, which allow atoms to be cooled in the microKelvin and nanoKelvin ranges. A few possible applications of ultracold atoms have been also reviewed. They take advantage of the long interaction times and long de Broglie wavelengths which are now available with laser cooling and trapping techniques.

One can reasonably expect that further progress in this field will be made in the near future and that new applications will be found. Concerning fundamental problems, two directions of research at least look promising. First, a better control of "pure" situations involving a small number of atoms in well-defined states exhibiting quantum features such as very long spatial coherence lengths or entanglement. In that perspective, atomic, molecular and optical physics will continue to play an important role by providing a "testing bench" for improving our understanding of quantum phenomena. A second interesting direction is the investigation of new systems, such as Bose condensates involving a macroscopic number of atoms in the same quantum state. One can reasonably hope that new types of coherent atomic sources (sometimes called "atom lasers") will be realized, opening the way to interesting new possibilities.

It is clear finally that all the developments which have occurred in the field of laser cooling and trapping are strengthening the connections which can be established between atomic physics and other branches of physics, such as condensed matter or statistical physics. The use of Lévy statistics for analyzing subrecoil cooling is an example of such a fruitful dialogue. The interdisciplinary character of the present researches on the properties of Bose condensates is also a clear sign of the increase of these exchanges.

REFERENCES

[1] A. Kastler, J. Phys. Rad. **11,** 255 (1950).
[2] A. Ashkin, Science, **210,** 1081 (1980).
[3] V.S. Letokhov and V.G. Minogin, Phys.Reports, **73,** 1 (1981).
[4] S. Stenholm, Rev.Mod.Phys. **58,** 699 (1986).
[5] C.S. Adams and E. Riis, Prog. Quant. Electr. **21,** 1 (1997).
[6] C. Cohen-Tannoudji and W. Phillips, Physics Today **43,** No 10, 33 (1990).
[7] W. Heitler, "The quantum theory of radiation", 3rd ed. (Clarendon Press, Oxford, 1954).
[8] J-P. Barrat and C.Cohen-Tannoudji, J.Phys.Rad. **22,** 329 and 443 (1961).
[9] C.Cohen-Tannoudji , Ann. Phys. Paris **7,** 423 and 469 (1962).
[10] C. Cohen-Tannoudji, J. Dupont-Roc and G. Grynberg, "Atom-photon interactions– Basic processes and applications", (Wiley, New York, 1992).
[11] S.H. Autler and C.H. Townes, Phys.Rev. **100,** 703 (1955).
[12] C. Cohen-Tannoudji, C.R.Acad.Sci.(Fr) **252,** 394 (1961).
[13] C. Cohen-Tannoudji and J. Dupont-Roc, Phys.Rev. **A5,** 968 (1972).
[14] W.D.Phillips and H.Metcalf, Phys.Rev.Lett. **48,** 596 (1982).
[15] J.V.Prodan, W.D.Phillips and H.Metcalf, Phys.Rev.Lett. **49,** 1149 (1982).
[16] W. Ertmer, R. Blatt, J.L. Hall and M. Zhu, Phys.Rev.Lett. **54,** 996 (1985).

[17] G.A. Askarian, Zh.Eksp.Teor.Fiz. **42,** 1567 (1962) [Sov.Phys.JETP, 15, 1088 (1962)].
[18] C. Cohen-Tannoudji, "Atomic motion in laser light", in "Fundamental systems in quantum optics", J. Dalibard, J.-M. Raimond, and J. Zinn-Justin eds, Les Houches session LIII (1990), (North-Holland, Amsterdam 1992) p.1.
[19] J. Dalibard and C. Cohen-Tannoudji, J.Opt.Soc.Am. **B2,** 1707 (1985).
[20] T.W. Hänsch and A.L. Schawlow, Opt. Commun. **13,** 68 (1975).
[21] D.Wineland and H.Dehmelt, Bull.Am.Phys.Soc. **20,** 637 (1975).
[22] S. Chu, L. Hollberg, J.E. Bjorkholm, A. Cable and A. Ashkin, Phys. Rev. Lett. **55,** 48 (1985).
[23] V.S. Letokhov, V.G. Minogin and B.D. Pavlik, Zh.Eksp.Teor.Fiz. **72,** 1328 (1977) [Sov.Phys.JETP, **45,** 698 (1977)].
[24] D.J. Wineland and W. Itano, Phys.Rev. **A20,** 1521 (1979).
[25] J.P. Gordon and A. Ashkin, Phys.Rev. **A21,** 1606 (1980).
[26] P.D. Lett, R.N. Watts, C.I. Westbrook, W. Phillips, P.L. Gould and H.J. Metcalf, Phys. Rev. Lett. **61,** 169 (1988).
[27] E.L. Raab, M. Prentiss, A. Cable, S. Chu and D.E. Pritchard, Phys.Rev.Lett. **59,** 2631 (1987).
[28] C. Monroe, W. Swann, H. Robinson and C.E. Wieman, Phys.Rev.Lett. **65,** 1571 (1990).
[29] S. Chu, J.E. Bjorkholm, A. Ashkin and A. Cable, Phys.Rev.Lett. **57,** 314 (1986).
[30] C.S. Adams, H.J. Lee, N. Davidson, M. Kasevich and S. Chu, Phys. Rev. Lett. **74,** 3577 (1995).
[31] A. Kuhn, H. Perrin, W. Hänsel and C. Salomon, OSA TOPS on Ultracold Atoms and BEC, 1996, Vol. 7, p.58, Keith Burnett (ed.), (Optical Society of America, 1997).
[32] V.S. Letokhov, Pis'ma.Eksp.Teor.Fiz. **7,** 348 (1968) [JETP Lett., **7,** 272 (1968)].
[33] C. Salomon, J. Dalibard, A. Aspect, H. Metcalf and C. Cohen-Tannoudji, Phys. Rev. Lett. **59,** 1659 (1987).
[34] R.J. Cook and R.K. Hill, Opt. Commun. **43,** 258 (1982)
[35] V.I. Balykin, V.S. Letokhov, Yu. B. Ovchinnikov and A.I. Sidorov, Phys. Rev. Lett. **60,** 2137 (1988).
[36] M.A. Kasevich, D.S. Weiss and S. Chu, Opt. Lett. **15,** 607 (1990).
[37] C.G. Aminoff, A.M. Steane, P. Bouyer, P. Desbiolles, J. Dalibard and C. Cohen-Tannoudji, Phys. Rev. Lett. **71,** 3083 (1993).
[38] J. Dalibard and C. Cohen-Tannoudji, J. Opt. Soc. Am. **B6,** 2023 (1989).
[39] P.J. Ungar, D.S. Weiss, E. Riis and S. Chu, JOSA **B6,** 2058 (1989).
[40] C. Salomon, J. Dalibard, W. Phillips, A. Clairon and S. Guellati, Europhys. Lett. **12,** 683 (1990).
[41] Y. Castin, These de doctorat, Paris (1991).
[42] Y. Castin and K. Mølmer, Phys. Rev. Lett. **74,** 3772 (1995).
[43] Y. Castin and J. Dalibard, Europhys.Lett. **14,** 761 (1991).
[44] P. Verkerk, B. Lounis, C. Salomon, C. Cohen-Tannoudji, J.-Y. Courtois and G. Grynberg, Phys. Rev. Lett. **68,** 3861 (1992).
[45] P.S. Jessen, C. Gerz, P.D. Lett, W.D. Phillips, S.L. Rolston, R.J.C. Spreeuw and C.I. Westbrook, Phys. Rev. Lett. **69,** 49 (1992).
[46] G. Grynberg and C. Triché, in Proceedings of the International School of Physics "Enrico Fermi", Course CXXXI, A. Aspect, W. Barletta and R. Bonifacio (Eds), p.243, IOS Press, Amsterdam (1996); A. Hemmerich, M. Weidemüller and T.W. Hänsch, same Proceedings, p.503.
[47] P.S. Jessen and I.H. Deutsch, in Advances in Atomic, Molecular and Optical Physics, **37,** 95 (1996), ed. by B. Bederson and H. Walther.
[48] A. Aspect, E. Arimondo, R. Kaiser, N. Vansteenkiste, and C. Cohen-Tannoudji, Phys. Rev. Lett. **61,** 826 (1988).
[49] M. Kasevich and S. Chu, Phys. Rev. Lett. **69,** 1741 (1992).
[50] Alzetta G., Gozzini A., Moi L., Orriols G., Il Nuovo Cimento **36B,** 5 (1976).
[51] Arimondo E., Orriols G., Lett. Nuovo Cimento **17,** 333 (1976).

[52] A. Aspect, E. Arimondo, R. Kaiser, N. Vansteenkiste, and C. Cohen-Tannoudji, J. Opt. Soc. Am. **B6**, 2112 (1989).
[53] F. Bardou, B. Saubamea, J. Lawall, K. Shimizu, O. Emile, C. Westbrook, A. Aspect, C. Cohen-Tannoudji, C. R. Acad. Sci. Paris **318,** 877-885 (1994).
[54] B. Saubamea, T.W. Hijmans, S. Kulin, E. Rasel, E. Peik, M. Leduc and C. Cohen-Tannoudji, Phys.Rev.Lett. **79,** 3146 (1997).
[55] J. Lawall, F. Bardou, B. Saubamea, K. Shimizu, M. Leduc, A. Aspect and C. Cohen-Tannoudji, Phys. Rev. Lett. **73,** 1915 (1994).
[56] J. Lawall, S. Kulin, B. Saubamea, N. Bigelow, M. Leduc and C. Cohen-Tannoudji, Phys. Rev. Lett. **75,** 4194 (1995).
[57] M.A. Ol'shanii and V.G. Minogin, Opt. Commun. **89,** 393 (1992).
[58] S. Kulin, B. Saubamea, E. Peik, J. Lawall, T.W. Hijmans, M. Leduc and C .Cohen-Tannoudji, Phys. Rev. Lett. **78,** 4185 (1997).
[59] C. Cohen-Tannoudji and J. Dalibard, Europhys. Lett. **1,** 441 (1986).
[60] P. Zoller, M. Marte and D.F. Walls, Phys. Rev. **A35,** 198 (1987).
[61] **F. Bardou, J.-P. Bouchaud, O. Emile, A. Aspect and C. Cohen-Tannoudji, Phys. Rev. Lett. 72, 203 (1994).**
[62] B.V. Gnedenko and A.N. Kolmogorov, "Limit distributions for sum of independent random variables" (Addison Wesley, Reading, MA, 1954).
[63] J.P. Bouchaud and A. Georges, Phys. Rep. **195,** 127 (1990).
[64] F. Bardou, Ph. D. Thesis, University of Paris XI, Orsay (1995).
[65] J. Reichel, F. Bardou, M. Ben Dahan, E. Peik, S. Rand, C. Salomon and C. Cohen-Tannoudji, Phys. Rev. Lett. **75,** 4575 (1995).
[66] K. Gibble and S. Chu, Metrologia, **29,** 201 (1992).
[67] S.N. Lea, A. Clairon, C. Salomon, P. Laurent, B. Lounis, J. Reichel, A. Nadir, and G. Santarelli, Physica Scripta **T51,** 78 (1994).
[68] J. Zacharias, Phys.Rev. **94,** 751 (1954). See also : N. Ramsey, Molecular Beams, Oxford University Press, Oxford, 1956.
[69] M. Kasevich, E. Riis, S. Chu and R. de Voe, Phys. Rev. Lett. **63,** 612 (1989).
[70] A. Clairon, C. Salomon, S. Guellati and W.D. Phillips, Europhys. Lett. **16,** 165 (1991).
[71] S. Ghezali, Ph. Laurent, S.N. Lea and A. Clairon, Europhys. Lett. **36,** 25 (1996).
[72] S. Ghezali, Thèse de doctorat, Paris (1997).
[73] E. Simon, P. Laurent, C. Mandache and A. Clairon, Proceedings of EFTF 1997, Neuchatel, Switzerland.
[74] J. Prestage, R. Tjoelker and L. Maleki, Phys. Rev. Lett. **74,** 3511 (1995).
[75] I.I.Shapiro, Phys. Rev. Lett. **13,** 789 (1964).
[76] H. Wallis, J. Dalibard and C. Cohen-Tannoudji, Appl. Phys. **B54,** 407 (1992).
[77] A. Steane, P. Szriftgiser, P. Desbiolles and J. Dalibard, Phys. Rev. Lett. **74,** 4972 (1995).
[78] M. Ben Dahan, E. Peik, J. Reichel, Y. Castin and C. Salomon, Phys. Rev. Lett. **76,** 4508 (1996).

William D. Phillips

WILLIAM D. PHILLIPS

I was born on 5 November 1948 in Wilkes-Barre, Pennsylvania, just across the river from the town of Kingston, where my parents lived with my one and a half year old sister, Maxine. My parents had come to this small Pennsylvania town from places and backgrounds that were far apart and yet quite similar.

My mother, Mary Catherine Savino (later, Savine), was born in the southern Italian village of Ripacandida in 1913. Among her earliest memories are riding into her grandfather's vineyards in a horse-drawn cart. Her father emigrated to the US and brought the family to Altoona, Pennsylvania in 1920. Her new American schoolmates teased her for her inability to speak English and taunted her as a "Wop" for her Italian heritage. She resolved to excel, and so she did, graduating near the top of her class from Altoona High School.

My father, William (Bill) Cornelius Phillips, was born in Juniata, a community on the edge of Altoona, in 1907. His father was a carpenter and his mother operated a boarding house to augment the family income. His grandfather was a barrel-maker, who would demonstrate the quality of his product by jumping onto the finished barrel in front of the customer. Dad could trace his heritage to ancestors from Wales who fought in the American Revolution.

My father and mother were each the first in their families to go to college, each attending Juniata College, a small school in Huntingdon, Pennsylvania, founded and strongly influenced by the pacifist Church of the Brethren. My father and mother graduated from Juniata in 1930 and 1936, respectively, but never met until a Juniata professor who knew them both suggested to my father that he might call a young Juniata alumna and ask her out. This Italian Catholic young woman and this Welsh-American Methodist young man met, fell in love, got married, earned Masters degrees and became professional social workers in the hard coal country of Pennsylvania.

I grew up surrounded by family and friends, church and school, and physical and mental activity. I clearly remember the value my parents placed on reading and education. My parents read to us and encouraged us to read. As soon as I could read for myself, walking across town to the library became a regular activity. Almost as far back as I can remember, I was interested in science. I assembled a collection of bottles of household substances as my "chemistry set" and examined almost anything I could find with the microscope my parents gave me. Although they had no particular knowledge or special interest in science, they supported mine. Science was only one of the passions of my childhood, along with fishing, baseball, bike riding and tree climbing. But as time went on, Erector sets, microscopes, and chemistry sets captured more of my attention than baseball bats, fishing rods, and football helmets. In 1956, my family moved from Kingston to Butler, near Pittsburgh. I remember that during that time I decided that science was going to be my life work, and sometime during the late 1950s, I came to appreciate, in a very incomplete and naive way, the simplicity and beauty of physics.

My brother Tom was born in 1957–a concrete confirmation, my sister and I believed, of the power of prayer. We had been praying for a sibling, unaware that our parents could decide, and *had* decided, that two children were enough. Apparently our prayers were effective. The result was a thrill and a blessing for all of us. Another blessing was my being placed into an experimental "accelerated" class. There, dedicated and concerned teachers taught us things that were not part of the ordinary elementary school curriculum, like French and advanced mathematics. When my family moved to Camp Hill, near Harrisburg, in 1959, interested teachers continued to provide me with advanced instruction, and when I entered the 7th grade of Camp Hill High School in 1960, it was in another accelerated program.

During this time, I had a laboratory in the basement of our family home. Ignorant and heedless of the dangers of asbestos, electricity, and ultraviolet light, I spent many hours experimenting with fire, explosives, rockets and carbon arcs. But life was not all science. I ran for the track team and played for the tennis team at school. During the summer, I spent all day either on the tennis courts or in the community swimming pool, and considered the advantages of life as a tennis bum.

While my parents were not directly involved in my scientific interests, they tolerated my experiments, even when the circuit breakers all tripped because of my overloads. They were always encouraging, and there was never any lack of intellectual stimulation. Dinner table conversations included discussions of politics, history, sociology, and current events. We children were heard and respected, but we had to compete for the privilege of expressing our opinions. In these discussions our parents transmitted important values about respect for other people, for their cultures, their ethnic backgrounds, their faith and beliefs, even when very different from our own. We learned concern for others who were less fortunate than we were. These values were supported and strengthened by a maturing religious faith.

In high school, I enjoyed and profited from well-taught science and math classes, but in retrospect, I can see that the classes that emphasized language and writing skills were just as important for the development of my scientific career as were science and math. I certainly feel that my high school involvement in debating competitions helped me later to give better scientific talks, that the classes in writing style helped me to write better papers, and the study of French greatly enhanced the tremendously fruitful collaboration I was to have with Claude Cohen-Tannoudji's research group.

The summer after my junior year in high school, I worked at the University of Delaware doing sputtering experiments. It was a great experience and I learned an important truth from Jim Comas, the graduate student who supervised me. "An experimental physicist," he told me, "is someone who gets paid for working at his hobby."

Another important part of my high school experience was meeting Jane Van Wynen. Her family had moved from Maine when we were in ninth grade, but we largely ignored each other until our senior year when, during a school trip to the New York World's Fair during it's closing days in 1965, I became suddenly aware of her considerable charms. She was not so immediately convinced that I had any charms of interest to her, but my natural tenacity paid off, and we started dating.

In the fall of 1966 I started my studies at Juniata College, as my mother and father, my Aunt Betty, and my sister had before me, and as my younger brother, Tom would later. Juniata had a foreign language requirement, which could be satisfied by studying two years of a language or by passing a test. I passed the test in French, whereupon the chairman of the French department, who knew my sister, a French major in her senior year, suggested that I enroll in an advanced French literature class. Being a naive freshman, I did. The professor lectured in French, we read classic French literature and wrote our exams in French–not what I was used to in high school! I got a "C" on my first test and realized that college was not going to be as easy as high school. I finished the course with an "A", and learned an important lesson: I would have to work hard at Juniata.

Physics with calculus was a challenge as well, but a true joy. Ray Pfrogner, who taught that first course, revealed a beauty and a unity in physics and mathematics that, until then, I had lacked the tools to appreciate. Some evenings he invited us students to showings of films of Richard Feynman's classic public lectures on "The Character of Physical Law." These events included popcorn that Pfrogner popped himself. Feynman's breezy yet incisive style on occasional evenings and Pfrogner's clear expositions every other morning fueled my passion for physics.

My passion for Jane was also increasing during this time, fueled by daily letters, weekly phone calls and infrequent visits to her school, Penn State University. It is a passion that has matured and deepened but remained undiminished over the years. Our separation during our college years meant that I did not have a highly active social life, leaving lots of time for physics.

During my first year at Juniata, Wilfred Norris, the Physics Department chairman, invited me to start on the laboratory course normally taken by third-year students–a series of classic physics experiments, which I did under his supervision. Later, I started doing serious research under Norris's direction, rebuilding an X-band electron spin resonance (ESR) spectrometer and trying to resolve discrepancies in the literature about ESR linewidths.

In my senior year I spent a semester doing ESR at Argonne National Laboratories, working with Juan McMillan and Ted Halpern. There, I experienced full-time research, performed by a team of professionals who would discuss what the important problems were, decide what to do, how to do it, and then go into the lab and do it. I loved it!

Back at Juniata for my final semester, I was applying to graduate schools.

First on my list was Princeton–because I had heard its graduate program was superb and because a visitor to Juniata had told me that a physics student from my school would never be accepted to Princeton! I *was* accepted, but a visit to Princeton left me unconvinced that I wanted to go there. From the lobby of the Princeton physics building, I called Dan Kleppner at the Massachusetts Institute of Technology (MIT).

Dan had seen my application to MIT, including my experience in magnetic resonance, and had invited me to visit his group and consider working on a hydrogen maser experiment. So I visited MIT (and Harvard for good measure); I was struck by the pleasant camaraderie, and the friendly yet electric atmosphere that Dan had created in his group. That emotional reaction, and Jane's desire to return to New England, more than any purely scientific considerations, made me decide to go to MIT. I never regretted that decision, or any of the other decisions I made afterwards based on considerations of the heart.

During a hectic several weeks in 1970, Jane and I graduated from our respective colleges, married, honeymooned and moved to Boston. At MIT I started working with Fred Walther on the high-field hydrogen maser, another X-band magnetic resonance spectrometer. I learned how to do electronics, machining, plumbing and vacuum–all skills I have found essential in experimental research. I also learned from Dan, and from the others in his group, a way of thinking about physics intuitively, and a way of inquiring about a problem that has shaped the way I approach physics to this day. The style of open and lively discussion of physics problems that I found in Dan's group is one that I have tried to emulate in my own group at NIST. I also try to follow the principle Dan taught by example: that one can do physics at the frontiers, competing with the best in the world, and do it with openness, humanity and cooperation.

For my thesis research I measured the magnetic moment of the proton in H_2O. Through this project I met others in the community of precision measurements and fundamental constants–in particular, Barry Taylor and Ed Williams at the National Bureau of Standards. By the time I completed that measurement (which is, at least for the moment, still the best of its kind), tunable dye lasers had become commercially available and had found their way into our lab. I decided that I should learn more about these new toys and, with Dan's encouragement, embarked on an experiment to study the collisions of laser-excited atoms. I finally wrote up both experiments for my thesis and defended it in 1976.

I accepted a Chaim Weizmann fellowship to work on projects of my own choosing at MIT for another two years. During that time, I continued to work on collisions with Dave Pritchard and Jim Kinsey; I also started work on Bose-Einstein condensation (BEC) in spin-polarized hydrogen with Dan and Tom Greytak. We were filled with optimism in the early days of that experiment, but today, 22 years later, BEC of hydrogen is still "just around the corner." Nevertheless, the innovations achieved by that group, long after I left, along with the developments in laser cooling recognized by this year's Nobel Prize, were crucial in showing the way to the eventual success of BEC in alkali vapors.

At the party celebrating my thesis in 1976, Dan Kleppner said it was fortunate that I had done the second experiment, using lasers, because otherwise I would probably have ended up going to the National Bureau of Standards (NBS). In 1978 I accepted a position at NBS (later renamed the National Institute of Standards and Technology–NIST) in Barry Taylor's division, working with Ed Williams and Tom Olsen on precision measurements of the proton gyromagnetic ratio and of the Absolute Ampere. These were exciting projects, but my experience with lasers and atomic physics had also earned me the opportunity to devote part of my time to exploring ways of improving measurement capabilities using those tools. I used that opportunity to pursue laser cooling, and the story of how that went is told in the accompanying Nobel Lecture.

In 1979, shortly after Jane and I moved to Gaithersburg, we joined Fairhaven United Methodist Church. We had not been regular church-goers during our years at MIT, but Ed and Jean Williams invited us to Fairhaven and there we found a congregation whose ethnic and racial diversity offered an irresistible richness of worship experience. Later that year, our first daughter, Catherine, now known as Caitlin, was born. In 1981 Christine was born. Our children have been an unending source of blessing, adventure and challenge. Their arrival, at a time when both Jane and I were trying to establish ourselves in new jobs, required a delicate balancing of work, home, and church life. Somehow, our faith and our youthful energy got us through that period.

At NBS, with some borrowed equipment and some extra money that Barry Taylor, in his inimitable fashion, obtained from somewhere, I got started with laser cooling. Support from the Office of Naval Research allowed Hal Metcalf to spend time at NBS in those early days. I had worked with Hal a little at MIT, and I knew that his unbounded enthusiasm and his effervescent creativity were priceless qualities. My collaborating with Hal on laser cooling was the first and one of the most important among many valuable interactions with colleagues who came to NIST, or whom I met elsewhere. I have mentioned many of these in my Lecture, and I want to emphasize again how much they have contributed to the development of laser cooling, and particularly, how important the senior group members, Kris Helmerson, Paul Lett, Steve Rolston, and Chris Westbrook, have been. I also want to recall the words of Bengt Nagel in his formal remarks to Steve Chu, Claude Cohen-Tannoudji and myself on 10 December 1997 in Stockholm. He said that we were being recognized as leaders and *representatives* of our groups. The three of us feel very strongly that this Prize honors all of those wonderful colleagues who contributed so much to the development of laser cooling.

Since the announcement of the award of the 1997 Nobel Prize in Physics, I have been honored to receive greetings and congratulations from colleagues and friends all over the world, as well as from many people whom I did not know. One such greeting came, not to me but to my children, from Susan Hench Bowis. She had read newspaper accounts of the announcement and recalled to my teenage daughters that she had been 17 when in 1950 her father, Philip Hench, had been awarded the Nobel Prize in Physiology or Medicine. He had been far from home at the time of the announcement, as I had been, and, like Caitlin, Susan Hench had been away at school. Transatlantic telephone calls were not common in those days, and so when she eventually made contact and congratulated her father, it was by cable. He cabled back to her, "Prouder of you, my darling, than of any prize." Surely the Nobel Prize is the highest award a scientist could hope to receive, and I have received it with a sense of awe that I am in the company of those who have received it before. But no prize can compare in importance to the family and friends I count as my greatest treasures.

LASER COOLING AND TRAPPING OF NEUTRAL ATOMS

Nobel Lecture, December 8, 1997

by

WILLIAM D. PHILLIPS

National Institute of Standards and Technology, Physics Laboratory, Atomic Physics Division, Gaithersburg, MD 20899, USA

INTRODUCTION

In 1978, while I was a postdoctoral fellow at MIT, I read a paper [1] by Art Ashkin in which he described how one might slow down an atomic beam of sodium using the radiation pressure of a laser beam tuned to an atomic resonance. After being slowed, the atoms would be captured in a trap consisting of focused laser beams, with the atomic motion being damped until the temperature of the atoms reached the microkelvin range. That paper was my first introduction to laser cooling, although the idea of laser cooling (the reduction of random thermal velocities using radiative forces) had been proposed three years earlier in independent papers by Hänsch and Schawlow [2] and Wineland and Dehmelt [3]. Although the treatment in Ashkin's paper was necessarily over-simplified, it provided one of the important inspirations for what I tried to accomplish for about the next decade. Another inspiration appeared later that same year: Wineland, Drullinger and Walls published the first laser cooling experiment [4], in which they cooled a cloud of Mg ions held in a Penning trap. At essentially the same time, Neuhauser, Hohenstatt, Toschek and Dehmelt [5] also reported laser cooling of trapped Ba^+ ions.

Those laser cooling experiments of 1978 were a dramatic demonstration of the mechanical effects of light, but such effects have a much longer history. The understanding that electromagnetic radiation exerts a force became quantitative only with Maxwell's theory of electromagnetism, even though such a force had been conjectured much earlier, partly in response to the observation that comet tails point away from the sun. It was not until the turn of the century, however, that experiments by Lebedev [6] and Nichols and Hull [7, 8] gave a laboratory demonstration and quantitative measurement of radiation pressure on macroscopic objects. In 1933, Frisch [9] made the first demonstration of light pressure on atoms, deflecting an atomic sodium beam with resonance radiation from a lamp. With the advent of the laser, Ashkin [10] recognized the potential of intense, narrow-band light for manipulating atoms and in 1972 the first "modern" experiments demonstrated the deflection of atomic beams with lasers [11, 12]. All of this set the stage for the laser cooling proposals of 1975 and for the demonstrations in 1978 with ions.

Comet tails, deflection of atomic beams and the laser cooling proposed in

1975 are all manifestation of the radiative force that Ashkin has called the "scattering force," because it results when light strikes an object and is scattered in random directions. Another radiative force, the dipole force, can be thought of as arising from the interaction between an induced dipole moment and the gradient of the incident light field. The dipole force was recognized at least as early as 1962 by Askar'yan [13], and in 1968, Letokhov [14] proposed using it to trap atoms–even before the idea of laser cooling! The trap proposed by Ashkin in 1978 [1] relied on this "dipole" or "gradient" force as well. Nevertheless, in 1978, laser cooling, the reduction of random velocities, was understood to involve only the scattering force. Laser trapping, confinement in a potential created by light, which was still only a dream, involved both dipole and scattering forces. Within 10 years, however, the dipole force was seen to have a major impact on laser cooling as well.

Without understanding very much about what difficulties lay in store for me, or even appreciating the exciting possibilities of what one might do with laser cooled atoms, I decided to try to do for neutral atoms what the groups in Boulder and Heidelberg had done for ions: trap them and cool them. There was, however, a significant difficulty: we could not first trap and then cool neutral atoms. Ion traps were deep enough to easily trap ions having temperatures well above room temperature, but none of the proposed neutral atom traps had depths of more than a few kelvin. Significant cooling was required before trapping would be possible, as Ashkin had outlined in his paper [1], and it was with this idea that I began.

Before describing the first experiments on the deceleration of atomic beams, let me digress slightly and discuss why laser cooling is so exciting and why it has attracted so much attention in the scientific community: When one studies atoms in a gas, they are typically moving very rapidly. The molecules and atoms in air at room temperature are moving with speeds on the order of 300 m/s, the speed of sound. This thermal velocity can be reduced by refrigerating the gas, with the velocity varying as the square root of the temperature, but even at 77 K, the temperature at which N_2 condenses into a liquid, the nitrogen molecules are moving at about 150 m/s. At 4 K, the condensation temperature of helium, the He atoms have 90 m/s speeds. At temperatures for which atomic thermal velocities would be below 1 m/s, any gas in equilibrium (other than spin-polarized atomic hydrogen) would be condensed, with a vapor pressure so low that essentially no atoms would be in the gas phase. As a result, all studies of free atoms were done with fast atoms. The high speed of the atoms makes measurements difficult. The Doppler shift and the relativistic time dilation cause displacement and broadening of the spectral lines of thermal atoms, which have a wide spread of velocities. Furthermore, the high atomic velocities limit the observation time (and thus the spectral resolution) in any reasonably-sized apparatus. Atoms at 300 m/s pass through a meter-long apparatus in just 3 ms. These effects are a major limitation, for example, to the performance of conventional atomic clocks.

The desire to reduce motional effects in spectroscopy and atomic clocks was and remains a major motivation for the cooling of both neutral atoms and ions. In addition, some remarkable new phenomena appear when atoms are sufficiently cold. The wave, or quantum nature of particles with momentum p becomes apparent only when the deBroglie wavelength, given by $\lambda_{dB} = h/p$, becomes large, on the order of relevant distance scales like the atom-atom interaction distances, atom-atom separations, or the scale of confinement. Laser cooled atoms have allowed studies of collisions and of quantum collective behavior in regimes hitherto unattainable. Among the new phenomena seen with neutral atoms is Bose-Einstein condensation of an atomic gas [15, 16], which has been hailed as a new state of matter, and is already becoming a major new field of investigation. Equally impressive and exciting are the quantum phenomena seen with trapped ions, for example, quantum jumps [17–19], Schrödinger cats [20], and quantum logic gates [21].

LASER COOLING OF ATOMIC BEAMS

In 1978 I had only vague notions about the excitement that lay ahead with laser cooled atoms, but I concluded that slowing down an atomic beam was the first step. The atomic beam was to be slowed using the transfer of momentum that occurs when an atom absorbs a photon. Fig. 1 shows the basic

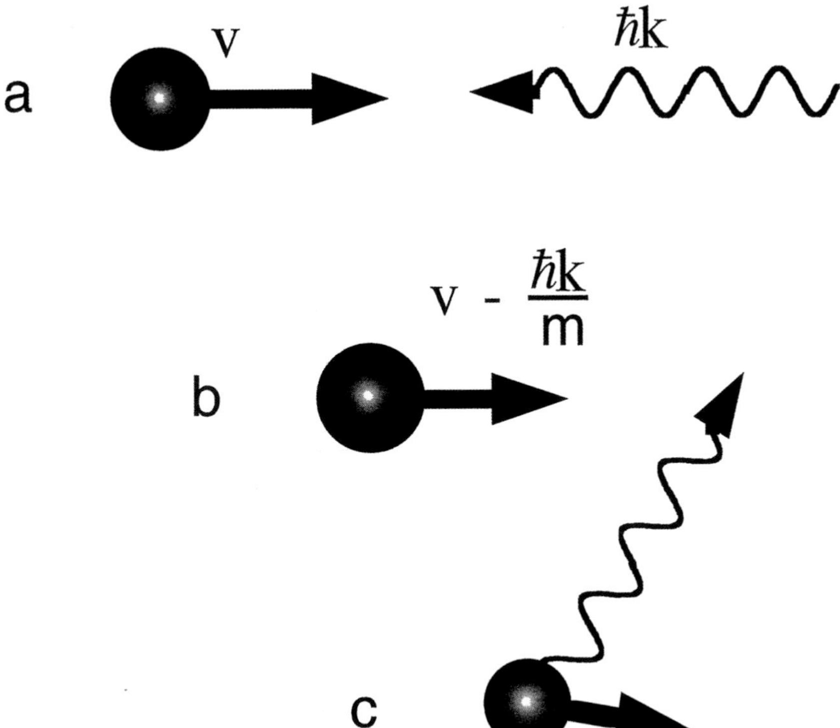

Figure 1. (a) An atom with velocity v encounters a photon with momentum $\hbar k = h/\lambda$; (b) after absorbing the photon, the atom is slowed by $\hbar k/m$; (c) after re-radiation in a random direction, on average the atom is slower than in 1a.

process underlying the "scattering force" that results. An atomic beam with velocity v is irradiated by an opposing laser beam. For each photon that a ground-state atom absorbs, it is slowed by $v_{rec} = \hbar k/m$ ($k = 2\pi/\lambda$ where λ is the wavelength of the light). In order to absorb again the atom must return to the ground state by emitting a photon. Photons are emitted in random directions, but with a symmetric average distribution, so their contribution to the atom's momentum averages to zero. The randomness results in a "heating" of the atom, discussed below.

For sodium atoms interacting with the familiar yellow resonance light, $v_{rec} = 3$ cm/s, while a typical beam velocity is about 10^5 cm/s, so the absorption-emission process must occur about 3×10^4 times to bring the Na atom to rest. In principle, an atom could radiate and absorb photons at half the radiative decay rate of the excited state (a 2-level atom in steady state can spend at most half of its time in the excited state). For Na, this implies that a photon could be radiated every 32 ns on average, bringing the atoms to rest in about 1 ms. Two problems, optical pumping and Doppler shifts, can prevent this from happening. I had an early indication of the difficulty of decelerating an atomic beam shortly after reading Ashkin's 1978 paper. I was then working with a sodium atomic beam at MIT, using tunable dye lasers to study the collisional properties of optically excited sodium. I tuned a laser to be resonant with the Na transition from $3S_{1/2} \rightarrow 3P_{3/2}$, the D2 line, and directed its beam opposite to the atomic beam. I saw that the atoms near the beam source were fluorescing brightly as they absorbed the laser light, while further away from the source, the atoms were relatively dim. The problem, I concluded, was optical pumping, illustrated in Fig. 2.

Sodium is not a two-level atom, but has two ground hyperfine levels (F = 1 and F = 2 in Fig. 2), each of which consists of several, normally degenerate,

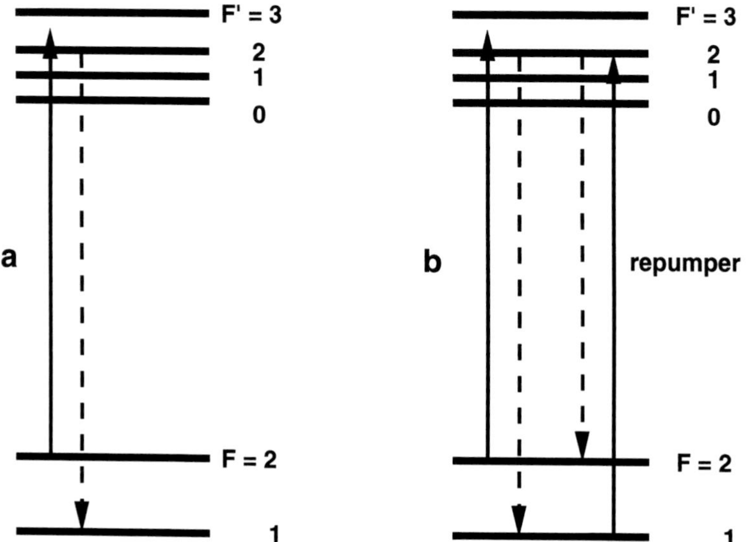

Figure 2. (a) The optical pumping process preventing cycling transitions in alkalis like Na; (b) Use of a repumping laser to allow many absorption-emission cycles.

states. Laser excitation out of one of the hyperfine levels to the excited state can result in the atom radiating to the other hyperfine level. This optical pumping essentially shuts off the absorption of laser light, because the linewidths of the transition and of the laser are much smaller than the separation between the ground state hyperfine components. Even for atoms excited on the $3S_{1/2}$ (F = 2) → $3P_{3/2}$ (F′ = 3) transition, where the only allowed decay channel is to F = 2, off-resonant excitation of F′ = 2 (the linewidth of the transition is 10 MHz, while the separation between F′ = 2 and F′ = 3 is 60 MHz) leads to optical pumping into F = 1 after only about a hundred absorptions. This optical pumping made the atoms "dark" to my laser after they traveled only a short distance from the source.

An obvious solution (Fig. 2b) is to use a second laser frequency, called a repumper, to excite the atoms out of the "wrong" (F = 1) hyperfine state so that they can decay to the "right" state (F = 2) where they can continue to cool. Given the repumper, another problem becomes apparent: the Doppler shift. In order for the laser light to be resonantly absorbed by a counterpropagating atom moving with velocity v, the frequency ω of the light must be lower by kv than the resonant frequency for an atom at rest. As the atom repeatedly absorbs photons, slowing down as desired, the Doppler shift changes and the atom goes out of resonance with the light. The natural linewidth $\Gamma/2\pi$ of the optical transition in Na is 10 MHz (full width at half maximum). A change in velocity of 6 m/s gives a Doppler shift this large, so after absorbing only 200 photons, the atom is far enough off resonance that the rate of absorption is significantly reduced. The result is that only atoms with the "proper" velocity to be resonant with the laser are slowed, and they are only slowed by a small amount.

Nevertheless, this process of atoms being slowed and pushed out of resonance results in a cooling or narrowing of the velocity distribution. In an atomic beam, there is typically a wide spread of velocities around $v_{th} = 3k_BT/m$. Those atoms with the proper velocity will absorb rapidly and decelerate. Those that are too fast will absorb more slowly, then more rapidly as they come into resonance, and finally more slowly as they continue to decelerate. Atoms that are too slow to begin with will absorb little and decelerate little. Thus atoms from a range of velocities around the resonant velocity are pushed into a narrower range centered on a lower velocity. This process was studied theoretically by Minogin [22] and in 1981, at Moscow's Institute for Spectroscopy, was used in the first experiment clearly demonstrating laser cooling of neutral atoms [23].

Fig. 3 shows the velocity distribution after such cooling of an atomic beam. The data was taken in our laboratory, but is equivalent to what had been done in Moscow. The characteristic of this kind of beam cooling is that only a small part of the total velocity distribution (the part near resonance with the laser beam) is slowed by only a small amount (until the atoms are no longer resonant). The narrow peak, while it represents true cooling in that its velocity distribution is narrow, consists of rather fast atoms.

One solution to this problem had already been outlined in 1976 by Letok-

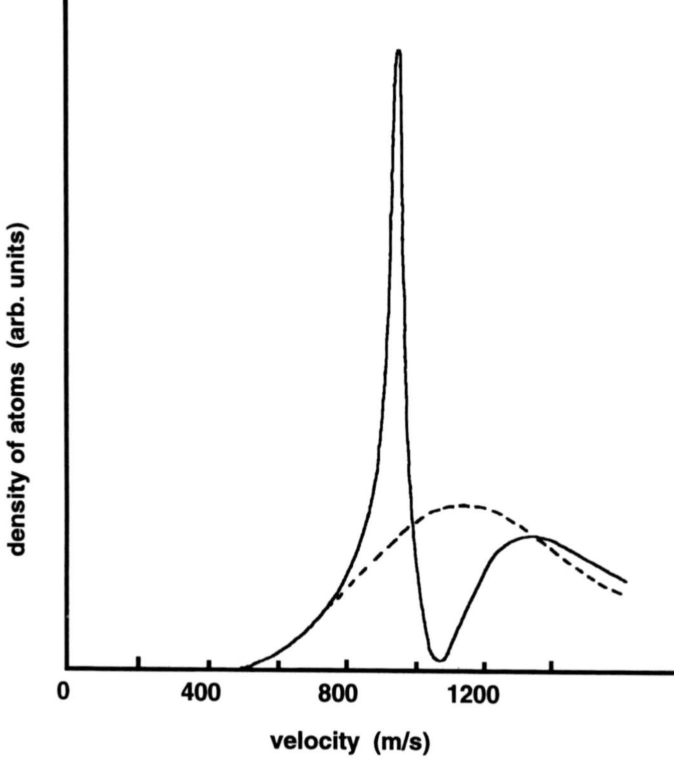

Figure. 3. Cooling an atomic beam with a fixed frequency laser. The dotted curve is the velocity distribution before cooling, and the solid curve is after cooling. Atoms from a narrow velocity range are transferred to a slightly narrower range centered on a lower velocity.

hov, Minogin and Pavlik [24] They suggested a general method of changing the frequency (chirping) of the cooling laser so as to interact with all the atoms in a wide distribution and to stay in resonance with the atoms as they are cooled. The Moscow group applied the technique to decelerating an atomic beam [25] but without clear success [26]. (Later, in 1983, John Prodan and I obtained the first clear deceleration and cooling of an atomic beam with this "chirp-cooling" technique [27–30]. Those first attempts failed to bring the atoms to rest, something that was finally achieved by Ertmer, Blatt, Hall and Zhu [31].) The chirp-cooling technique is now one of the two standard methods for decelerating beams. The other is "Zeeman cooling."

By late 1978, I had moved to the National Bureau of Standards (NBS), later named the National Institute of Standards and Technology (NIST), in Gaithersburg. I was considering how to slow an atomic beam, realizing that the optical pumping and Doppler shift problems would both need to be addressed. I understood how things would work using the Moscow chirp-cooling technique and a repumper. I also considered using a broadband laser, so that there would be light in resonance with the atoms, regardless of their velocity. (This idea was refined by Hoffnagle [32] and demonstrated by Hall's group [33]). Finally I considered that instead of changing the frequency of the laser to stay in resonance with the atoms (chirping), one could use a magnetic field to change the energy level separation in the atoms so as to keep them in resonance with the fixed-frequency laser (Zeeman cooling). All of these ideas for cooling an atomic beam, along with various schemes for avoiding optical pumping, were contained in a proposal [34] that I submitted to the Office of Naval Research in 1979. Around this time Hal Metcalf, from the State University of New York at Stony Brook, joined me in Gaithersburg and we began to consider what would be the best way to proceed. Hal contended that all the methods looked reasonable, but we should work on the Zeeman cooler because it would be the most fun! Not only was Hal right about the fun we would have, but his suggestion led us to develop a technique with particularly advantageous properties. The idea is illustrated in Fig. 4.

The atomic beam source directs atoms, which have a wide range of velocities, along the axis (*z*-direction) of a tapered solenoid. This magnet has more windings at its entrance end, near the source, so the field is higher at that end. The laser is tuned so that, given the field-induced Zeeman shift and the velocity-induced Doppler shift of the atomic transition frequency, atoms with velocity v_0 are resonant with the laser when they reach the point where the field is maximum. Those atoms then absorb light and begin to slow down. As their velocity changes, their Doppler shift changes, but is compensated by the change in Zeeman shift as the atoms move to a point where the field is weaker. At this point, atoms with initial velocities slightly lower than v_0 come into resonance and begin to slow down. The process continues with the initially fast atoms decelerating and staying in resonance while initially slower atoms come into resonance and begin to be slowed as they move further down the solenoid. Eventually all the atoms with velocities lower than v_0 are brought to a final velocity that depends on the details of the magnetic field and laser tuning.

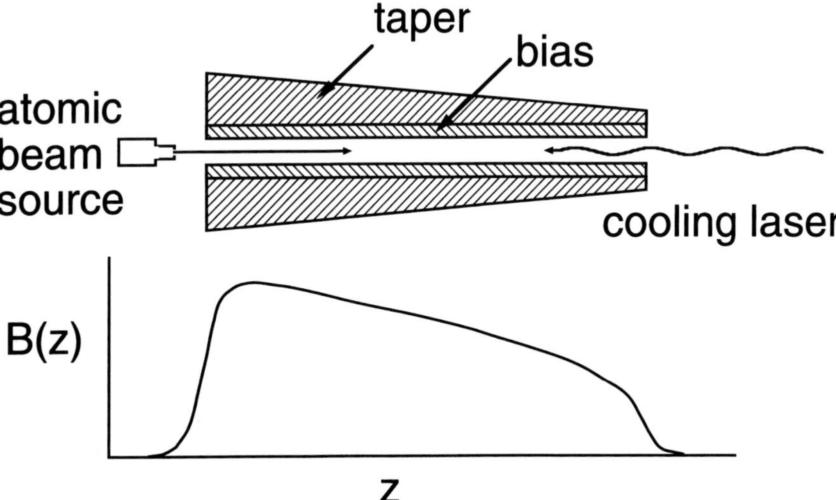

Figure 4. Upper: Schematic representation of a Zeeman slower. Lower: Variation of the axial field with position.

The first tapered solenoids that Hal Metcalf and I used for Zeeman cooling of atomic beams had only a few sections of windings and had to be cooled with air blown by fans or with wet towels wrapped around the coils. Shortly after our initial success in getting some substantial deceleration, we were joined by my first post-doc, John Prodan. We developed more sophisticated solenoids, wound with wires in many layers of different lengths, so as to produce a smoothly varying field that would allow the atoms to slow down to a stop while remaining in resonance with the cooling laser.

These later solenoids were cooled with water flowing over the coils. To improve the heat transfer, we filled the spaces between the wires with various heat-conducting substances. One was a white silicone grease that we put onto the wires with our hands as we wound the coil on a lathe. The grease was about the same color and consistency as the diaper rash ointment I was then using on my baby daughters, so there was a period of time when, whether at home or at work, I seemed to be up to my elbows in white grease.

The grease-covered, water-cooled solenoids had the annoying habit of burning out as electrolytic action attacked the wires during operation. Sometimes it seemed that we no sooner obtained some data than the solenoid would burn out and we were winding a new one.

On the bright side, the frequent burn-outs provided the opportunity for refinement and redesign. Soon we were potting the coils in a black, rubbery resin. While it was supposed to be impervious to water, it did not have good adhesion properties (except to clothing and human flesh) and the solenoids continued to burn out. Eventually, an epoxy coating sealed the solenoid against the water that allowed the electrolysis, and in more recent times we replaced water with a fluorocarbon liquid that does not conduct electricity or support electrolysis. Along the way to a reliable solenoid, we learned how to slow and stop atoms efficiently [27, 35–41].

The velocity distribution after deceleration is measured in a detection region some distance from the exit end of the solenoid. Here a separate detection laser beam produces fluorescence from atoms having the correct velocity to be resonant. By scanning the frequency of the detection laser, we were able to determine the velocity distribution in the atomic beam. Observations with the detection laser were made just after turning off the cooling laser, so as to avoid any difficulties with having both lasers on at the same time. Fig. 5 shows the velocity distribution resulting from Zeeman cooling: a large fraction of the initial distribution has been swept down into a narrow final velocity group.

One of the advantages of the Zeeman cooling technique is the ease with which the optical pumping problem is avoided. Because the atoms are always in a strong axial magnetic field (that is the reason for the "bias" windings in Fig. 4), there is a well-defined axis of quantization that allowed us to make use of the selection rules for radiative transitions and to avoid the undesirable optical pumping. Figure 6 shows the energy levels of Na in a magnetic field. Atoms in the $3S_{1/2}$ ($m_F = 2$) state, irradiated with circularly polarized σ^+ light, must increase their m_F by one unit, and so can go only to the $3P_{3/2}$ ($m_{F'} = 3$)

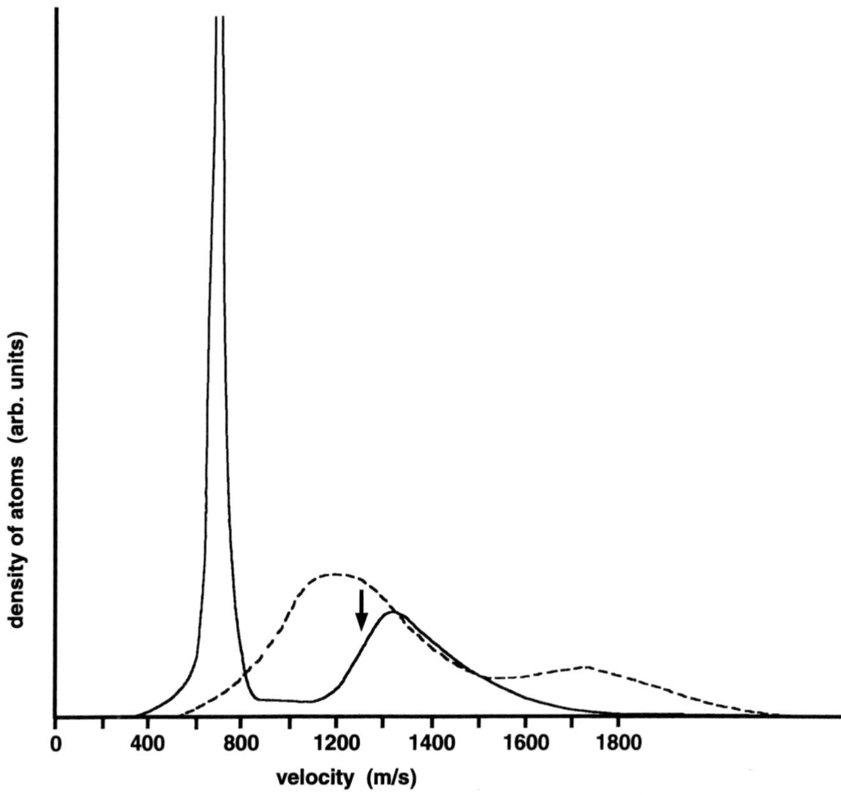

Figure 5. Velocity distribution before (dashed) and after (solid) Zeeman cooling. The arrow indicates the highest velocity resonant with the slowing laser. (The extra bump at 1700 m/s is from F = 1 atoms, which are optically pumped into F = 2 during the cooling process)

state. This state in turn can decay only to $3S_{1/2}$ ($m_F = 2$), and the excitation process can be repeated indefinitely. Of course, the circular polarization is not perfect, so other excitations are possible, and these may lead to decay to other states. Fortunately, in a high magnetic field, such transitions are highly unlikely [35] : either they involve a change in the nuclear spin projection m_I, which is forbidden in the high field limit, or they are far from resonance. These features, combined with high purity of the circular polarization, allowed us to achieve, without a "wrong transition," the 3×10^4 excitations required to stop the atoms. Furthermore, the circular polarization produced some "good" optical pumping: atoms not initially in the $3S_{1/2}$ ($m_F = 2$) state were pumped into this state, the "stretched" state of maximum projection of angular momentum, as they absorbed the angular momentum of the light. These various aspects of optical selection rules and optical pumping allowed the process of Zeeman cooling to be very efficient, decelerating a large fraction of the atoms in the beam.

In 1983 we discussed a number of these aspects of laser deceleration, including our early chirp-cooling results, at a two-day workshop on "Laser-Cooled and Trapped Atoms" held at NBS in Gaithersburg [42]. I view this as

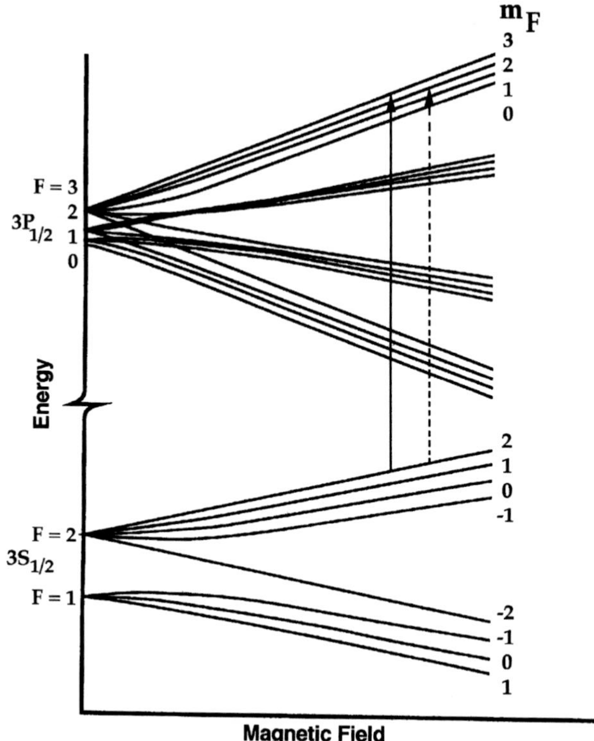

Figure 6. Energy levels of Na in a magnetic field. The cycling transition used for laser cooling is shown as a solid arrow, and one of the nearly forbidden excitation channels leading to undesirable optical pumping is shown dashed.

an important meeting in that it and its proceedings stimulated interest in laser cooling . In early 1984, Stig Stenholm, then of the University of Helsinki, organized an international meeting on laser cooling in Tvärminne, a remote peninsula in Finland. Fig. 7 shows the small group attending (I was the photographer), and in that group, only some of the participants were even active in laser cooling at the time. Among these were Stig Stenholm (who had done pioneering work in the theory of laser cooling and the mechanical effects of light on atoms [43–51]) along with some of his young colleagues; Victor Balykin and Vladimir Minogin from the Moscow group; and Claude Cohen-Tannoudji and Jean Dalibard from Ecole Normale Supérieure (ENS) in Paris, who had begun working on the theory of laser cooling and trapping. Also present were Jürgen Mlynek and Wolfgang Ertmer, both of whom now lead major research groups pursuing laser cooling and atom optics. At that time, however, only our group and the Moscow group had published any experiments on cooling of neutral atoms.

Much of the discussion at the Tvärminne meeting involved the techniques of beam deceleration and the problems with optical pumping. I took a lighthearted attitude toward our trials and tribulations with optical pumping, often joking that any unexplained features in our data could certainly be attributed to optical pumping. Of course, at the Ecole Normale, optical pump-

Figure 7. Stig Stenholm's "First International Conference on Laser Cooling" in Tvärminne, March 1984. Back row, left to right: Juha Javanainen, Markus Lindberg, Stig Stenholm, Matti Kaivola, Nis Bjerre, (unidentified), Erling Riis, Rainer Salomaa, Vladimir Minogin. Front row: Jürgen Mlynek, Angela Guzmann, Peter Jungner, Wolfgang Ertmer, Birger Ståhlberg, Olli Serimaa, Jean Dalibard, Claude Cohen-Tannoudji, Victor Balykin.

ing had a long and distinguished history. Having been pioneered by Alfred Kastler and Jean Brossel, optical pumping had been the backbone of many experiments in the Laboratoire de Spectroscopie Hertzienne (now the Laboratoire Kastler-Brossel). After one discussion in which I had joked about optical pumping, Jean Dalibard privately mentioned to me, "You know, Bill, at the Ecole Normale, optical pumping is not a joke." His gentle note of caution calmed me down a bit, but it turned out to be strangely prophetic as well. As we saw a few years later, optical pumping had an important, beautiful, and totally unanticipated role to play in laser cooling, and it was surely no joke.

STOPPING ATOMS

As successful as Zeeman cooling had been in producing large numbers of decelerated atoms as in Fig. 5, we had not actually observed the atoms at rest, nor had we trapped them. In fact, I recall a conversation with Steve Chu that took place during the International Conference on Laser Spectroscopy in Interlaken in 1983 in which I had presented our results on beam deceleration [27]. Steve was working on positronium spectroscopy but was wondering whether there still might be something interesting to be done with laser cooling of neutral atoms. I offered the opinion that there was still plenty to do, and in particular, that trapping of atoms was still an unrealized goal. It wasn't long before each of us achieved that goal, in very different ways.

Our approach was to first get some stopped atoms. The problem had been that, in a sense, Zeeman cooling worked too well. By adjusting the laser frequency and magnetic field, we could, up to a point, choose the final velocity of the atoms that had undergone laser deceleration. Unfortunately, if we chose too small a velocity, no slow atoms at all appeared in the detection region. Once brought below a certain velocity, about 200 m/s, the atoms always continued to absorb enough light while traveling from the solenoid to the detection region so as to stop before reaching the detector. By shutting off the cooling laser beam and delaying observation until the slow atoms arrived in the observation region, we were able to detect atoms as slow as 40 m/s with a spread of 10 m/s, corresponding to a temperature (in the atoms' rest frame) of 70 mK [36].

The next step was to get these atoms to come to rest in our observation region. We were joined by Alan Migdall, a new post-doc, Jean Dalibard, who was visiting from ENS, and Ivan So, Hal Metcalf's student. We decided that we needed to proceed as before, shutting off the cooling light, allowing the slow atoms to drift into the observation region, but then to apply a short pulse of additional cooling light to bring the atoms to rest. The sequence of laser pulses required to do this–a long pulse of several milliseconds for doing the initial deceleration, followed by a delay and then another pulse of a few hundred microseconds, followed by another delay before detection–was provided by a rotating wheel with a series of openings corresponding to the places where the laser was to be on. Today we accomplish such pulse sequences with acousto-optic modulators under computer control, but in those days it required careful construction and balancing of a rapidly rotating wheel.

The result of this sequence of laser pulses was that we had atoms at rest in our observation region with a velocity spread corresponding to <100 mK [52]. Just following our 1985 paper reporting this in Physical Review Letters was a report of the successful stopping of atoms by the chirp-cooling method in Jan Hall's group [31]. At last there were atoms slow enough to be trapped, and we decided to concentrate first on magnetostatic trapping.

MAGNETIC TRAPPING OF ATOMS

The idea for magnetic traps had first appeared in the literature as early as 1960 [53–55], although Wolfgang Paul had discussed them in lectures at the University of Bonn in the mid-1950s, as a natural extension of ideas about magnetic focusing of atomic beams [56–58]. Magnetic trapping had come to our attention particularly because of the successful trapping of cold neutrons [59]. We later learned that in unpublished experiments in Paul's laboratory, there were indications of confining sodium in a magnetic trap [60].

The idea of magnetic trapping is that in a magnetic field, an atom with a magnetic moment will have quantum states whose magnetic or Zeeman energy increases with increasing field and states whose energy decreases, depending on the orientation of the moment compared to the field. The increasing-energy states, or low-field-seekers, can be trapped in a magnetic field configuration having a point where the magnitude of the field is a relative minimum. (No dc field can have a relative maximum in free space [61], so high-field-seekers cannot be trapped.) The requirement for stable trapping, besides the kinetic energy of the atom being low enough, is that the magnetic moment move adiabatically in the field. That is, the orientation of the magnetic moment with respect to the field should not change.

We considered some of the published designs for trapping neutrons, including the spherical hexapole [62], a design comprising three current loops, but we found them less than ideal. Instead we decided upon a simpler design, with two loops, which we called a spherical quadrupole. The trap, its magnetic field lines and equipotentials are shown in Fig. 8. Although we thought that we had discovered an original trap design, we later learned that Wolfgang Paul had considered this many years ago, but had not given it much attention because atoms were not harmonically bound in such a trap. In fact, the potential for such a trap is linear in the displacement from the center and has a cusp there.

With a team consisting of Alan Migdall, John Prodan, Hal Metcalf and myself, and with the theoretical support of Tom Bergeman, we succeeded in trapping atoms in the apparatus shown in Fig. 9 [63]. As in the experiments that stopped atoms, we start with Zeeman slowing, decelerating the atoms to 100 m/s in the solenoid. The slowing laser beam is then extinguished, allowing the atoms to proceed unhindered for 4 ms to the magnetic trap. At this point, only one of the two trap coils has current; it produces a magnetic field that brings the atoms into resonance with the cooling laser when it is turned on again for 400 ms, bringing the atoms to rest. Once the atoms are stopped, the other coil is energized, producing the field shown in Fig. 8, and the trap is sprung. The atoms are held in the trap until released, or until collisions with the room-temperature background gas molecules in the imperfect vacuum knock them out. After the desired trapping time, we turn off the magnetic field, and turn on a probe laser, so as to see how many atoms remain in the trap. By varying the frequency of this probe on successive repetitions of the process, we could determine the velocity distribution of the atoms, via their Doppler shifts.

The depth of our trap was about 17 mK (25 mT), corresponding to Na atoms with a velocity of 3.5 m/s. In the absence of trapping fields, atoms that fast would escape from the region of the trap coils in a few milliseconds. Fig. 10 shows a section of chart paper with spectra of the atoms remaining after 35 ms of trapping time. If the trap had not been working, we would have seen essentially nothing after that length of time, but the signal, noisy as it was, was unmistakable. It went away when the trap was off, and it went away when we did not provide the second pulse of cooling light that stops the atoms before

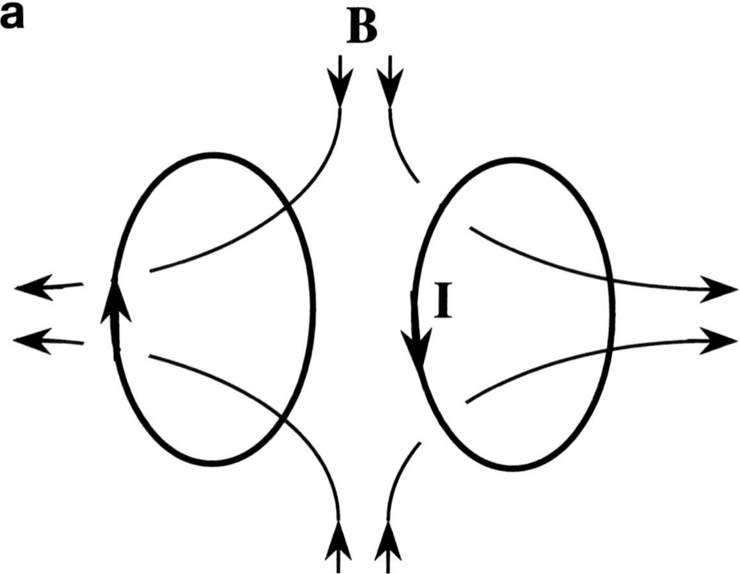

Figure 8. (a) Spherical quadrupole trap with lines of B-field.

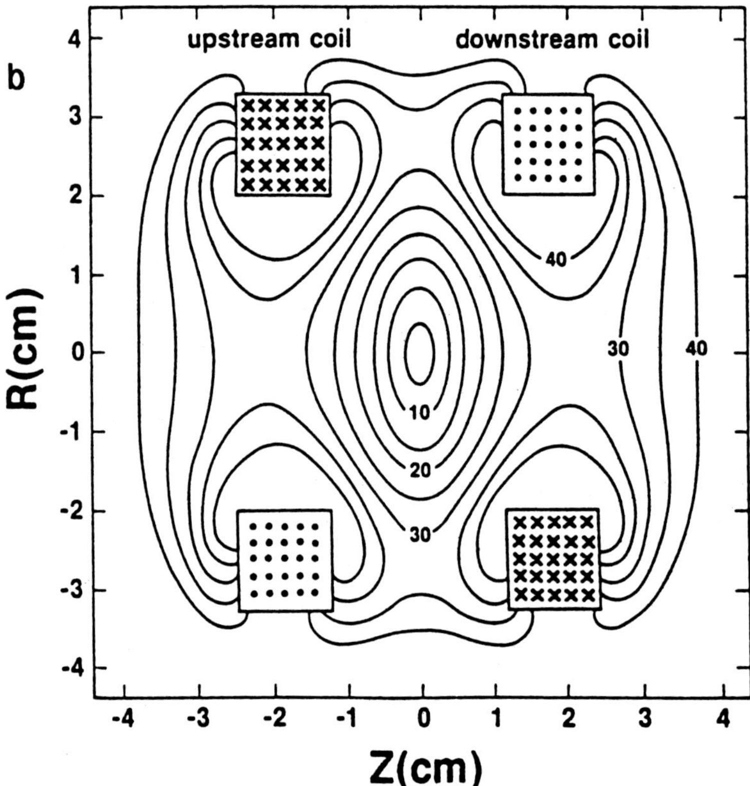

Figure 8. (b) Equipotentials of our trap (equal field magnitudes in millitesla), in a plane containing the symmetry (z) axis.

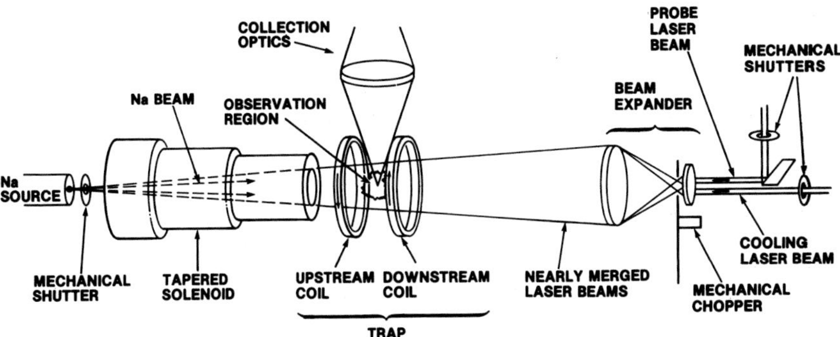

Figure 9: Schematic of the apparatus used to magnetically trap atoms.

trapping them. This was just the signature we were looking for, and Hal Metcalf expressed his characteristic elation at good results with his exuberant "WAHOO!!" at the top of the chart.

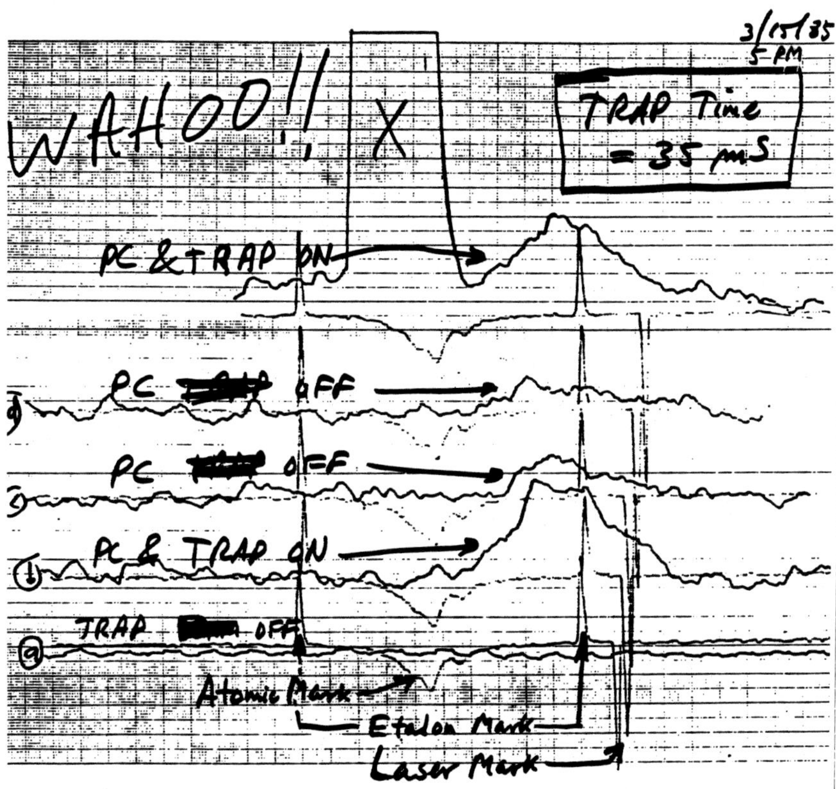

Figure 10. A section of chart paper from 15 March 1985. "PC" and "no PC" refer to presence or absence of the "post-cooling" pulse that brings the atoms to rest in the trapping region.

As the evening went on, we were able to improve the signal, but we found that the atoms did not stay very long in the trap, a feature we found a bit frustrating. Finally, late in the evening we decided to go out and get some fast food, talk about what was happening and attack the problem afresh. When we returned a little later that night, the signal had improved and we were able to trap atoms for much longer times. We soon realized that during our supper break the magnetic trap had cooled down, and stopped outgassing, so the vacuum just in the vicinity of the trap improved considerably. With this insight we knew to let the magnet cool off from time to time, and we were able to take a lot of useful data. We continued taking data until around 5:00 am, and it was probably close to 6:00 am when my wife Jane found Hal and me in our kitchen, eating ice cream as she prepared to leave for work. Her dismay at the lateness of our return and our choice of nourishment at that hour was partially assuaged by Hal's assurance that we had accomplished something pretty important that night.

Figure 11a presents the sequence of spectra taken after various trapping times, showing the decrease in signal as atoms are knocked out of the trap by collisions with the background gas molecules. Figure 11b shows that the loss of atoms from the trap is exponential, as expected, with a lifetime of a bit less than one second, in a vacuum of a few times 10^{-6} pascals. A point taken when the vacuum was allowed to get worse illustrates that poor vacuum made the signal decay faster. In more recent times, we and others have achieved much longer trapping times, mainly because of an improved vacuum. We now observe magnetic trap lifetimes of one minute or longer in our laboratory.

Since our demonstration [63] of magnetic trapping of atoms in 1985, many different kinds of magnetic atom traps have been used. At MIT, Dave Pritchard's group trapped [64] and cooled [65] Na atoms in a linear quadrupole magnetic field with an axial bias field, similar to the trap first discussed by Ioffe and collaborators [66] in 1962, and later by others [67, 68]. Similar traps were used by the Kleppner-Greytak group to trap [69] and evaporative-

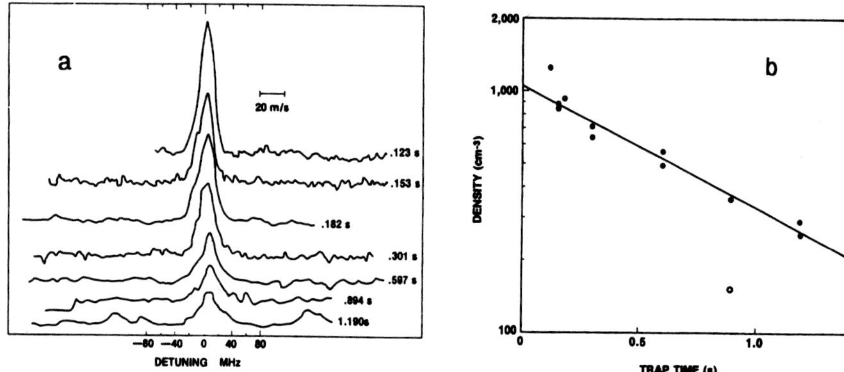

Figure 11. (a) Spectra of atoms remaining in the magnetic trap after various times; (b) Decay of number of trapped atoms with time. The open point was taken at twice the background pressure of the other points.

ly cool [70] atomic hydrogen, and by Walraven's group to trap [71] and laser-cool hydrogen [72]. The Ioffe trap has the advantage of having a non-zero magnetic field at the equilibrium point, in contrast to the spherical quadrupole, in which the field is zero at the equilibrium point. The zero field allows the magnetic moment of the atom to flip (often called Majorana flopping), so that the atom is in an untrapped spin state. While this problem did not cause difficulties in our 1985 demonstration, for colder atoms, which spend more time near the trap center, it can be a quite severe loss mechanism [73, 74]. In 1995, modifications to the simple quadrupole trap solved the problem of spins flips near the trap center, and allowed the achievement of Bose-Einstein condensation [15, 16].

OPTICAL MOLASSES

At the same time that we were doing the first magnetic trap experiments in Gaithersburg, the team at Bell Labs, led by Steve Chu, was working on a different and extremely important feature of laser cooling. After a beautiful demonstration in 1978 of the use of optical forces to focus an atomic beam [75], the Bell Labs team had made some preliminary attempts to decelerate an atom beam, and then moved on to other things. Encouraged by the beam deceleration experiments in Gaithersburg and in Boulder, Steve Chu reassembled much of that team and set out to demonstrate the kind of laser cooling suggested in 1975 by Hänsch and Schawlow [2]. (The physical principles behind the Hänsch and Schawlow proposal are, of course, identical to those expressed in the 1975 Wineland and Dehmelt laser cooling proposal [3]. These principles had already led to the laser cooling of trapped ions [4, 5]. The foci of Refs. [2, 3], however, has associated Hänsch and Schawlow with neutral atoms and Wineland and Dehmelt with ions.) In fact, the same physical principle of Doppler cooling results in the compression of the velocity distribution associated with laser deceleration of an atomic beam (see sections 2 and 3 of [76]). Nevertheless, in 1985, laser cooling of a gas of neutral atoms at rest, as proposed in [2] had yet to be demonstrated.

The idea behind the Hänsch and Schawlow proposal is illustrated in Fig. 12. A gas of atoms, represented here in one dimension, is irradiated from both sides by laser beams tuned slightly below the atomic resonance frequency. An atom moving toward the left sees that the laser beam opposing its motion is Doppler shifted toward the atomic resonance frequency. It sees that the laser beam directed along its motion is Doppler shifted further from its resonance. The atom therefore absorbs more strongly from the laser beam that opposes its motion, and it slows down. The same thing happens to an atom moving to the right, so all atoms are slowed by this arrangement of laser beams. With pairs of laser beams added along the other coordinate axes, one obtains cooling in three dimensions. Because of the role of the Doppler effect in the process, this is now called Doppler cooling.

Later treatments [4, 5, 43, 46, 77–79] recognized that this cooling process leads to a temperature whose lower limit is on the order of $\hbar\Gamma$, where Γ is the

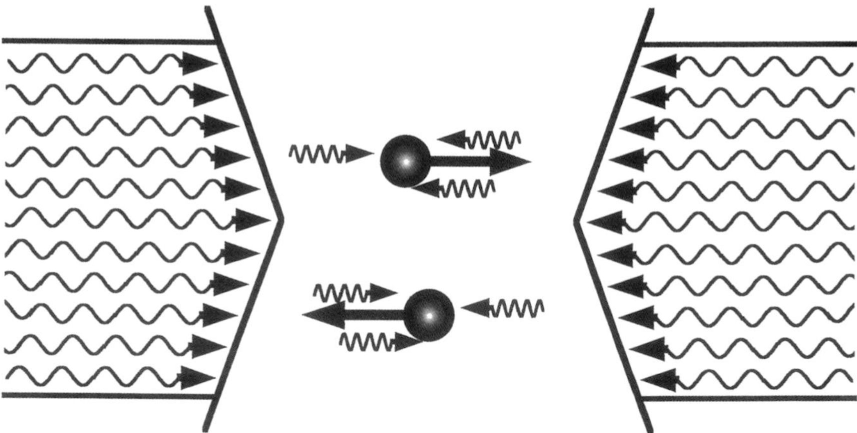

Figure 12. Doppler cooling in one dimension.

rate of spontaneous emission of the excited state (Γ^{-1} is the excited state lifetime). The temperature results from an equilibrium between laser cooling and the heating process arising from the random nature of both the absorption and emission of photons. The random addition to the average momentum transfer produces a random walk of the atomic momentum and an increase in the mean square atomic momentum. This heating is countered by the cooling force $\boldsymbol{F}$ opposing atomic motion. The force is proportional to the atomic velocity, as the Doppler shift is proportional to velocity. In this, the cooling force is similar to the friction force experienced by a body moving in a viscous fluid. The rate at which energy is removed by cooling is $F \cdot v$, which is proportional to v^2, so the cooling rate is proportional to the kinetic energy. By contrast the heating rate, proportional to the total photon scattering rate, is independent of atomic kinetic energy for low velocities. As a result, the heating and cooling come to equilibrium at a certain value of the average kinetic energy. This defines the temperature for Doppler cooling, which is

$$m < v_i^2 > = k_B T = \frac{\hbar\Gamma}{4}\left(\frac{\Gamma}{2\delta} + \frac{2\delta}{\Gamma}\right) \qquad (1)$$

where δ is the angular frequency of the detuning of the lasers from atomic resonance and v_i is the velocity along some axis. This expression is valid for 3D Doppler cooling in the limit of low intensity and when the recoil energy $\hbar^2 k^2/2m << \hbar\Gamma$. Interestingly, the equilibrium velocity distribution for Doppler cooling is the Maxwell-Boltzmann distribution. This follows from the fact that the Fokker-Planck equation describing the damping and heating in laser cooling is identical in form to the equation that describes collisional equilibrium of a gas [51]. Numerical simulations of real cases, where the recoil energy does not vanish, show that the distribution is still very close to Maxwellian [80]. The minimum value of this temperature called the Doppler cooling limit, occurring when $\delta = -\Gamma/2$, is

$$k_B T_{\text{Dopp}} = \frac{\hbar\Gamma}{2}. \qquad (2)$$

The first rigorous derivation of the cooling limit appears to be by Letokhov, Minogin and Pavilik [77] (although the reader should note that Eq. (32) is incorrectly identified with the rms velocity). Wineland and Itano [78] give derivations for a number of different situations involving trapped and free atoms and include the case where the recoil energy is not small but the atoms are in collisional equilibrium.

The Doppler cooling limit for sodium atoms cooled on the resonance transition at 589 nm where $\Gamma/2\pi = 10$ MHz, is 240 μK, and corresponds to an rms velocity of 30 cm/s along a given axis. The limits for other atoms and ions are similar, and such low temperatures were quite appealing. Before 1985, however, these limiting temperatures had not been obtained in either ions or neutral atoms.

A feature of laser cooling not appreciated in the first treatments was the fact that the spatial motion of atoms in any reasonably sized sample would be diffusive. For example, a simple calculation [80] shows that a sodium atom cooled to the Doppler limit has a "mean free path" (the mean distance it moves before its initial velocity is damped out and the atom is moving with a different, random velocity) of only 20 μm, while the size of the laser beams doing the cooling might easily be one centimeter. Thus, the atom undergoes diffusive, Brownian-like motion, and the time for a laser cooled atom to escape from the region where it is being cooled is much longer than the ballistic transit time across that region. This means that an atom is effectively "stuck" in the laser beams that cool it. This stickiness, and the similarity of laser cooling to viscous friction, prompted the Bell Labs group [81] to name the intersecting laser beams "optical molasses." At NBS [41], we independently used the term "molasses" to describe the cooling configuration, and the name "stuck." Note that an optical molasses is not a trap. There is no restoring force keeping the atoms in the molasses, only a viscous inhibition of their escape.

Using the techniques for chirp cooling an atomic beam developed at NBS-JILA [31] and a novel pulsed beam source, Chu's team at Bell Labs succeeded in loading cold sodium atoms into an optical molasses [81]. They observed the expected long "lifetime" (the time required for the atoms to diffuse out of the laser beams) of the molasses, and they developed a method, now called "release-and-recapture," for measuring the temperature of the atoms. The method is illustrated in Fig. 13. First, the atoms are captured and stored in the molasses, where for short periods of time they are essentially immobile due to the strong damping of atomic motion (13a). Then, the molasses laser beams are switched off, allowing the atoms to move ballistically away from the region to which they had originally been viscously confined (13b). Finally the laser beams are again turned on, recapturing the atoms that remain in the intersection (molasses) region (13c). From the fraction of atoms remaining after various periods of ballistic expansion one can determine the velocity distribution and therefore the temperature of the atoms at the time of release. The measured temperature at Bell Labs was 240^{+200}_{-60} μK [82]. The large uncertainty is due to the sensitive dependence of the analysis on the size and density distribution of atoms in the molasses, but the result was satisfyingly consistent with the predicted Doppler cooling limit.

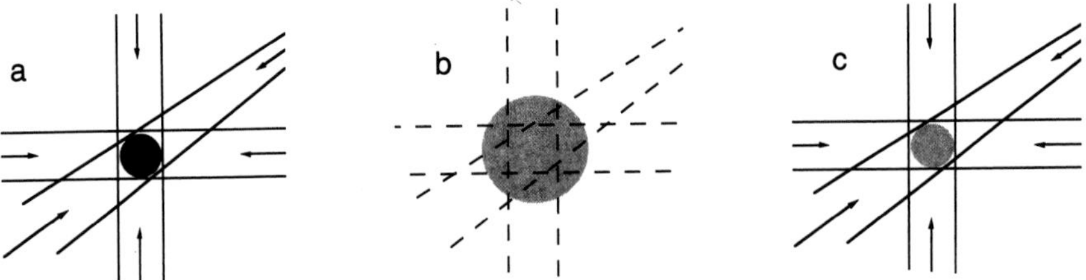

Figure 13. Release-and-recapture method for temperature measurement.

By the end of 1986, Phil Gould and Paul Lett had joined our group and we had achieved optical molasses in our laboratory at NBS, loading the molasses directly from a decelerated beam. (Today it is also routine to load atoms directly into a magneto-optical trap (MOT) [83] from an uncooled vapor [84, 85], and then into molasses.) We repeated the release-and-recapture temperature measurements, found them to be compatible with the reported measurements of the Bell Labs group, and we proceeded with other experiments. In particular, with Paul Julienne, Helen Thorsheim and John Wiener, we made a 2-focus laser trap and used it to perform the first measurements of a specific collision process (associative ionization) with laser cooled atoms [86]. (Earlier, Steve Chu and his colleagues had used optical molasses to load a single-focus laser trap–the first demonstration of an optical trap for atoms [87].) In a sense, our collision experiment represented a sort of closure for me because it realized the 2-focus trap proposed in Ashkin's 1978 paper, the paper that had started me thinking about laser cooling and trapping. It also was an important starting point for our group, because it began a new and highly productive line of research into cold collisions, producing some truly surprising and important results [88–94]. In another sense, though, that experiment was a detour from the road that was leading us to a new understanding of optical molasses and of how laser cooling worked.

SUB-DOPPLER LASER COOLING

During 1987 Gould, Lett and I investigated the behavior of optical molasses in more detail. Because the temperature was hard to measure and its measurement uncertainty was large, we concentrated instead on the molasses lifetime, the time for the atoms to diffuse out of the intersecting laser beams. We had calculated, on the basis of the Doppler cooling theory, how the lifetime would vary as a function of the laser frequency detuning and the laser intensity. We also calculated how the lifetime should change when we introduced a deliberate imbalance between the two beams of a counter-propagating pair. Now we wanted to compare experimental results with our calculations. The results took us somewhat by surprise.

Fig. 14 shows our measurements [80] of the molasses lifetime as a function of laser frequency along with the predicted behavior according to the Doppler cooling theory. The 1-D theory did not quantitatively reproduce the observed 3-D diffusion times, but that was expected. The surprise was the qualitative differences: the experimental lifetime peaked at a laser detuning above 3 linewidths, while the theory predicted a peak below one linewidth. We did not know how to reconcile this difficulty, and the results for the drift induced by beam imbalance were also in strong disagreement with the Doppler theory. In our 1987 paper, we described our failed attempts to bring the Doppler cooling theory into agreement with our data and ended saying [95] : "It remains to consider whether the multiple levels and sublevels of Na, multiple laser frequencies, or a consideration of the detailed motion of the atoms in 3-D can explain the surprising behavior of optical molasses." This was pure guesswork, of course, but it turned out to have an element of truth, as we shall see below.

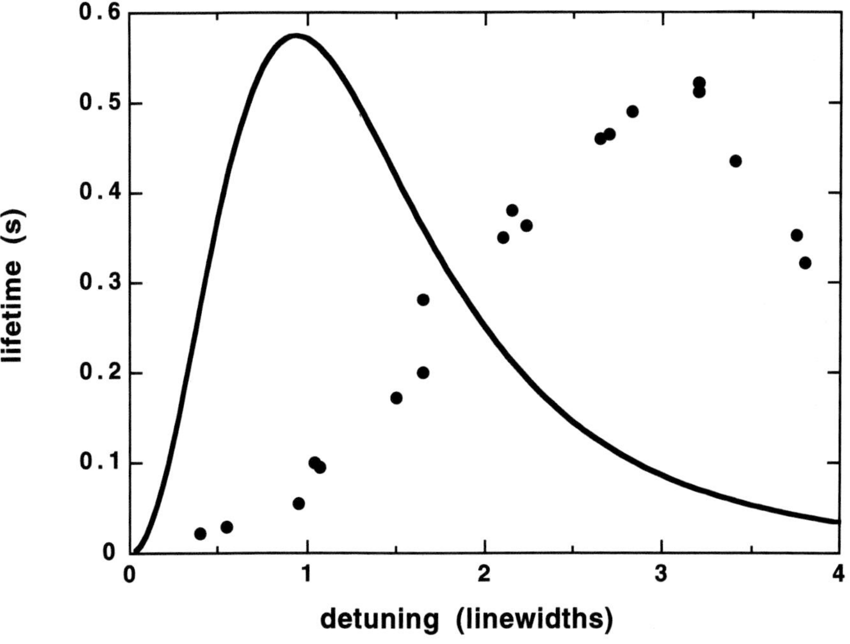

Figure. 14. Experimental molasses lifetime (points) and the theoretical decay time (curve) vs. detuning of molasses laser from resonance.

Having seen such a clear discrepancy between the Doppler cooling theory and the experimental results, with no resolution in sight, we, as experimentalists, decided to take more data. Paul Lett argued that we should measure the temperature again, this time as a function of the detuning, to see if it, too, would exhibit behavior different from that predicted by the theory. We felt, however, that the release-and-recapture method, given the large uncertainty associated with it in the past, would be unsuitable. Hal Metcalf suggested a different approach, illustrated in Fig. 15.

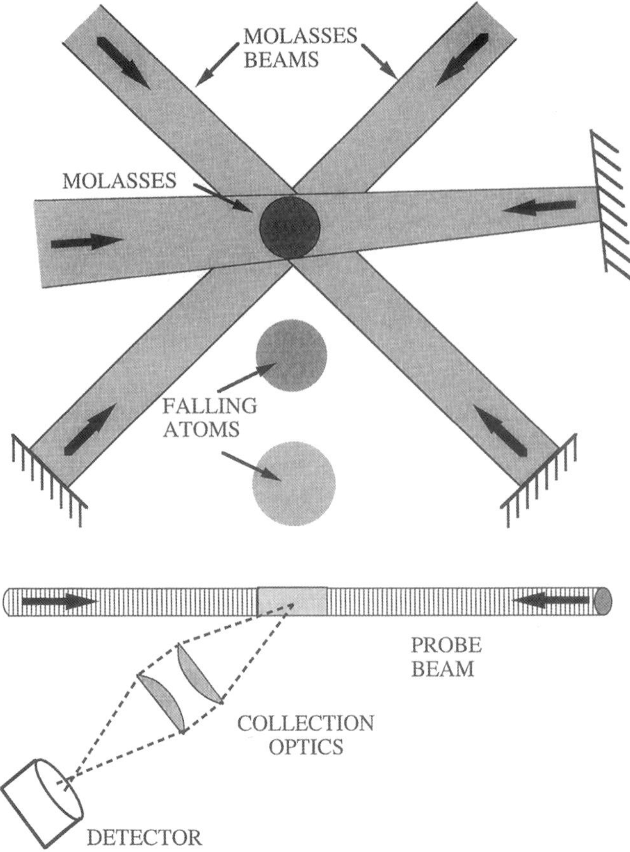

Figure 15. Time-of-flight method for measuring laser cooling temperatures.

In this time-of-flight (TOF) method, the atoms are first captured by the optical molasses, then released by switching off the molasses laser beams. The atom cloud expands ballistically, according to the distribution of atomic velocities. When atoms encounter the probe laser beam, they fluoresce, and the time distribution of fluorescence gives the time-of-flight distribution for atoms arriving at the probe. From this the temperature can be deduced. Now, with a team that included Paul Lett, Rich Watts, Chris Westbrook, Phil Gould, as well as Hal Metcalf and myself, we implemented the TOF temperature measurement. In our experiment, the probe was placed as close as 1 cm from the center of the molasses, which had a radius of about 4.5 mm. At the lowest expected temperature, the Doppler cooling limit of 240 μK for Na atoms, a significant fraction of the atoms would have been able to reach the probe, even with the probe above the molasses. For reasons of convenience, we did put the probe beam above the molasses, but we saw no fluorescence from atoms reaching the probe after the molasses was turned off. We spent a considerable time testing the detection system to be sure that everything was working properly. We deliberately "squirted" the atoms to the probe beam by heating them with a pair of laser beams in the horizontal plane, and verified that such heated atoms reached the probe and produced the expected time-of-flight signal.

Finally, we put the probe *under* the molasses. When we did, we immediately saw the TOF signals, but were reluctant to accept the conclusion that the atoms were colder than the Doppler cooling theory predicted, until we had completed a detailed modeling of the TOF signals. Figure 16 shows a typical TOF distribution for one of the colder observed temperatures, along with the model predictions. The conclusion was inescapable: Our atoms had a temperature of about 40 μK, much colder than the Doppler cooling limit of 240 μK. They had had insufficient kinetic energy to reach the probe when it

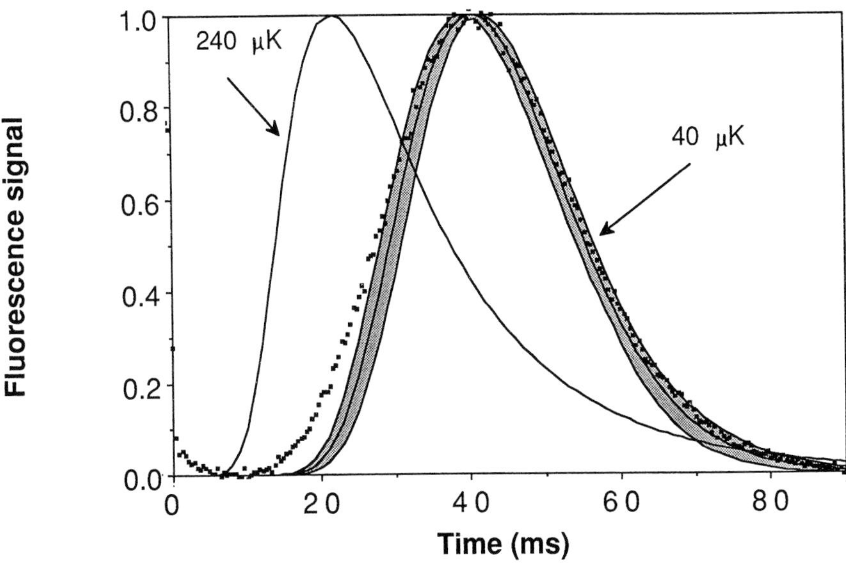

Figure. 16. The experimental TOF distribution (points) and the predicted distribution curves for 40 μK and 240 μK (the predicted lower limit of Doppler cooling). The band around the 40 μK curve reflects the uncertainty in the measurement of the geometry of the molasses and probe.

was placed above the molasses. As clear as this was, we were apprehensive. The theory of the Doppler limit was simple and compelling. In the limit of low intensity, one could derive the Doppler limit with a few lines of calculations (see for example, Ref. [80]); the most complete theory for cooling a two-level atom [96] did not predict a cooling limit any lower. Of course, everyone recognized that sodium was not a two-level atom, but it had seemed unlikely that it made any significant difference (our speculation in Ref. [95] notwithstanding). At low laser intensity the temperature depends on the laser detuning and the linewidth of the transition. Since the linewidth is identical for all possible transitions in the Na D2 manifold, and since the cooling transition ($3S_{1/2}$ (F = 2) → $3P_{3/2}$ (F = 3)) was well separated from nearby transitions, and all the Zeeman levels were degenerate, it seemed reasonable that the multilevel structure was unimportant in determining the cooling limit.

As it turned out, this was completely wrong. At the time, however, the

Doppler limit seemed to be on firm theoretical ground, and we were hesitant to claim that it was violated experimentally. Therefore, we sought to confirm our experimental results with other temperature measurement methods. One of these was to refine the "release and recapture" method described above. The large uncertainties in the earlier measurements [81] arose mainly from uncertainties in the size of the molasses and the recapture volume. We addressed that problem by sharply aperturing the molasses laser beams so the molasses and recapture volumes were well defined. We also found that it was essential to include the effect of gravity in the analysis (as we had done already for the TOF method). Because released atoms fall, the failure to recapture atoms could be interpreted as a higher temperature if gravity is not taken into account.

Another method was the "fountain" technique. Here we exploited our initial failure to observe a TOF signal with the probe above the molasses. By adjusting the height of the probe, we could measure how high the atoms could go before falling back under the influence of gravity. Essentially, this allowed us to measure the atoms' kinetic energy in terms of their gravitational potential energy, a principle very different from the TOF method. Finally, we used the "shower" method. This determined how far the atoms spread in the horizontal direction as they fell following release from the molasses. For this, we measured the fluorescence from atoms reaching the horizontal probe laser beam at different positions along that beam. From this transverse position distribution, we could get the transverse velocity distribution and therefore the temperature.

(The detailed modeling of the signals expected from the various temperature measurement methods was an essential element in establishing that the atomic temperature was well below the Doppler limit. Rich Watts, who had come to us from Hal Metcalf's lab and had done his doctoral dissertation with Carl Wieman, played a leading role in this modeling. Earlier, with Wieman, he had introduced the use of diode lasers in laser cooling. With Metcalf, he was the first to laser cool rubidium, the element with which Bose-Einstein condensation was first achieved. He was a pioneer of laser cooling and continued a distinguished scientific career at NIST after completing his postdoctoral studies in our group. Rich died in 1996 at the age of 39, and is greatly missed.)

While none of the additional methods proved to be as accurate as the TOF technique (which became a standard tool for studying laser cooling temperatures), each of them showed the temperature to be significantly below the Doppler limit. Sub-Doppler temperatures were not the only surprising results we obtained. We also (as Paul Lett had originally suggested) measured the temperature as a function of the detuning from resonance of the molasses laser. Figure 17 shows the results, along with the prediction of the Doppler cooling theory. The dependence of the temperature on detuning is strikingly different from the Doppler theory prediction, and recalls the discrepancy evident in Fig. 14. Our preliminary study indicated that the temperature did not depend on the laser intensity (although later measurements [80, 97, 98] showed that the temperature actually had a linear dependence on intensity). We observed that the temperature depended on the polarization of the molasses laser beams, and was highly sensitive to the ambient magnetic field. Changing the field by 0.2 mT increased the temperature from 40 μK to 120 μK when the laser was detuned 20 MHz from resonance (later experiments [80] showed even greater effects). This field dependence was particularly surprising, considering that transitions were being Zeeman shifted on the order of 14 MHz/mT, so the Zeeman shifts were much less than either the detuning or the 10 MHz transition linewidth.

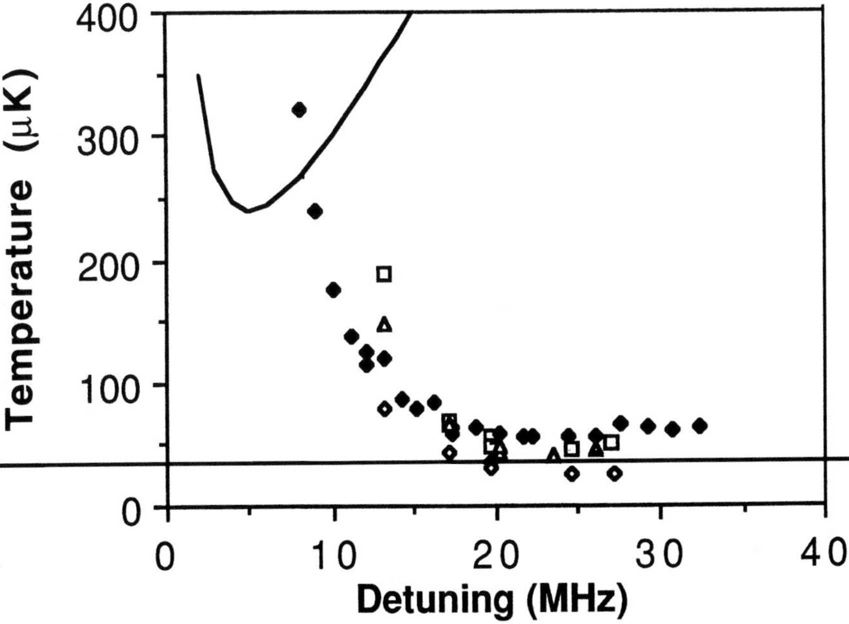

Figure 17. Dependence of molasses temperature on laser detuning (points) compared to the prediction of Doppler cooling theory (curve). The different symbols represent different molasses-to-probe separations.

Armed with these remarkable results, in the early spring of 1988 we sent a draft of the paper [99] describing our measurements to a number of experimental and theoretical groups working on laser cooling. I also traveled to a few of the leading laser cooling labs to describe the experiments in person and discuss them. Many of our colleagues were skeptical, as well they might have been, considering how surprising the results were. In the laboratories of Claude Cohen-Tannoudji and of Steve Chu, however, the response was: "Let's go into the lab and find out if it is true." Indeed, they soon confirmed sub-Doppler temperatures with their own measurements and they began to work on an understanding of how such low temperatures could come about. What emerged from these studies was a new concept of how laser cooling works, an understanding that is quite different from the original Hänsch-Schawlow and Wineland-Dehmelt picture.

During the spring and summer of 1988 our group was in close contact with Jean Dalibard and Claude Cohen-Tannoudji as they worked out the new theory of laser cooling and we continued our experiments. Their thinking centered on the multilevel character of the sodium atom, since the derivation of the Doppler limit was rigorous for a two-level atom. The sensitivity of temperature to magnetic field and to laser polarization suggested that the Zeeman sublevels were important, and this proved to be the case. Steve Chu (now at Stanford) and his colleagues followed a similar course, but the physical image that Dalibard and Cohen-Tannoudji developed has dominated the thinking about multi-level laser cooling. It involves a combination of multilevel atoms, polarization gradients, light shifts and optical pumping. How these work together to produce laser cooling is illustrated in simple form in Fig. 18, but the reader should see the Nobel Lectures of Cohen-Tannoudji and Chu along with the more detailed papers [100–103]

Figure 18a shows a 1-D set of counterpropagating beams with equal intensity and orthogonal, linear polarizations. The interference of these beams produces a standing wave whose polarization varies on a sub-wavelength distance scale. At points in space where the linear polarizations of the two beams are in phase with each other, the resultant polarization is linear, with an axis that bisects the polarization axes of the two individual beams. Where the phases are in quadrature, the resultant polarization is circular and at other places the polarization is elliptical. An atom in such a standing wave experiences a fortunate combination of light shifts and optical pumping processes.

Because of the differing Clebsch-Gordan coefficients governing the strength of coupling between the various ground and excited sublevels of the atom, the light shifts of the different sublevels are different, and they change with polarization (and therefore with position). Fig. 18b shows the sinusoidal variation of the ground-state energy levels (reflecting the varying light shifts or dipole forces) of a hypothetical $J_g = 1/2 \rightarrow J_e = 3/2$ atomic system. Now imagine an atom to be at rest at a place where the polarization is circular σ^- as at $z = \lambda/8$ in Fig. 18a. As the atom absorbs light with negative angular momentum and radiates back to the ground states, it will eventually be optically pumped into the $m_g = -1/2$ ground state, and simply cycle between this state and the excited $m_e = -3/2$ state. For low enough intensity and large enough detuning we can ignore the time the atom spends in the excited state and consider only the motion of the atom on the ground state potential. In the $m_g = -1/2$ state, the atom is in the lower energy level at $z = \lambda/8$, as shown in Fig. 18b. As the atom moves, it climbs the potential hill of the $m_g = -1/2$ state, but as it nears the top of the hill at $z = 3\lambda/8$, the polarization of the light becomes σ^+ and the optical pumping process tends to excite the atom in such a way that it decays to the $m_g = +1/2$ state. In the $m_g = +1/2$ state, the atom is now again at the bottom of a hill, and it again must climb, losing kinetic energy, as it moves. The continual climbing of hills recalls the Greek myth of Sisyphus, so this process, by which the atom rapidly slows down while passing through the polarization gradient, is called Sisyphus cooling. Dalibard and Cohen-Tannoudji had already described another kind of Sisyphus cooling, for 2-level

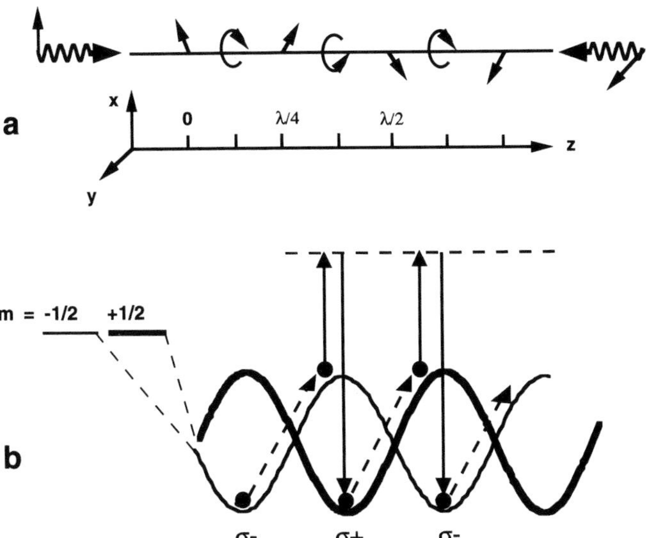

Figure 18. (a) Interfering, counterpropagating beams having orthogonal, linear polarizations create a polarization gradient. (b) The different Zeeman sublevels are shifted differently in light fields with different polarizations; optical pumping tends to put atomic population on the lowest energy level, but non-adiabatic motion results in "Sisyphus" cooling.

atoms [104] so the mechanism and the name were already familiar. In both kinds of Sisyphus cooling, the radiated photons, in comparison with the absorbed photons, have an excess energy equal to the light shift. By contrast, in Doppler cooling, the energy excess comes from the Doppler shift.

The details of this theory were still being worked out in the summer of 1988, the time of the International Conference on Atomic Physics, held that year in Paris. The sessions included talks about the experiments on sub-Doppler cooling and the new ideas to explain them. Beyond that, I had lively discussions with Dalibard and Cohen-Tannoudji about the new theory. One insight that emerged from those discussions was an understanding of why we had observed such high sensitivity of temperature to magnetic field: It was not the size of the Zeeman shift compared to the linewidth or the detuning that was important. Rather, when the Zeeman shift was comparable to the much smaller ($\approx$ 1 MHz) light shifts and optical pumping rates, the cooling mechanism, which depended on these phenomena, would be disturbed. We now suggested a crucial test: the effect of the magnetic field should be reduced if the light intensity were higher. From Paris, I telephoned back to the lab in Gaithersburg and urged my colleagues to perform the appropriate measurements.

The results were as we had hoped. Figure 19 shows temperature as a function of magnetic field for two different light intensities. At magnetic fields greater than 100 μT (1 gauss), the temperature was lower for higher light intensity, a reversal of the usual linear dependence of temperature and intensity [80, 98]. We considered this to be an important early confirmation of the qualitative correctness of the new theory, confirming the central role played

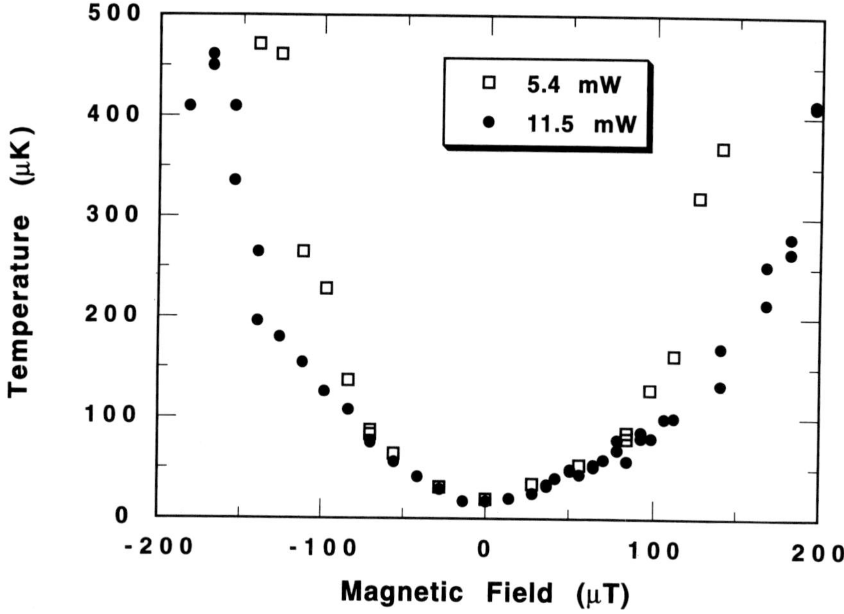

Figure 19. Temperature vs. magnetic field in a 3D optical molasses. Observation of lower temperature at higher intensity when the magnetic field was high provided an early confirmation of the new theory of sub-Doppler cooling.

by the light shift and the magnetic sublevels in the cooling mechanism. Joined by Steve Rolston and Carol Tanner we (Paul Lett, Rich Watts, Chris Westbrook and myself) carried out additional studies of the behavior of optical molasses, providing qualitative comparisons with the predictions of the new theory. Our 1989 paper [80], "Optical Molasses" summarized these results and contrasted the predictions of Doppler cooling with the new theory. Steve Chu's group also published additional measurements at the same time [105]. Other, even more detailed measurements in Paris [98] (where I was very privileged to spend the academic year of 1989–1990) left little doubt about the correctness of the new picture of laser cooling. In those experiments we cooled Cs atoms to 2.5 μK. It was a truly exciting time, when the developments in the theory and the experiments were pushing each other to better understanding and lower temperatures. Around this time, Jan Hall (whose pioneering work in chirp-cooling [31] had done so much to launch the explosive activity a few years before) commented that being in the field of laser cooling was an experience akin to being in Paris at the time of the Impressionists. Figure 20 symbolizes the truth of that comment.

OPTICAL LATTICES

In 1989 we began a different kind of measurement on laser cooled atoms, a measurement that was to lead us to a new and highly fruitful field of research. We had always been a bit concerned that all of our temperature measurements gave us information about the velocity distribution of atoms *after* their

Figure 20. Hal Metcalf, Claude Cohen-Tannoudji and the author on the famous bridge in Monet's garden at Giverny, ca. 1990.

release from the optical molasses and we wanted a way to measure the temperature *in situ*. Phil Gould suggested that we measure the spectrum of the light emitted from the atoms while they were being cooled. For continuous, single frequency irradiation at low intensity and large detuning, most of the fluorescence light scattered from the atoms should be "elastically" scattered, rather than belonging to the "Mollow triplet" of high-intensity resonance fluorescence [106]. This elastically scattered light will be Doppler shifted by the moving atoms and its spectrum should show a Doppler broadening characteristic of the temperature of the atomic sample. The spectrum will also contain the frequency fluctuations of the laser itself, but these are relatively slow for a dye laser, so Gould suggested a heterodyne method of detection, where the fluorescent light is mixed on a photodiode with local oscillator light derived from the molasses laser, producing a beat signal that is free of the laser frequency fluctuations.

The experiment was not easy, and it worked mainly because of the skill and perseverance of Chris Westbrook. An example of the surprising spectrum we obtained [107] is shown in Fig. 21. The broad pedestal corresponded well to

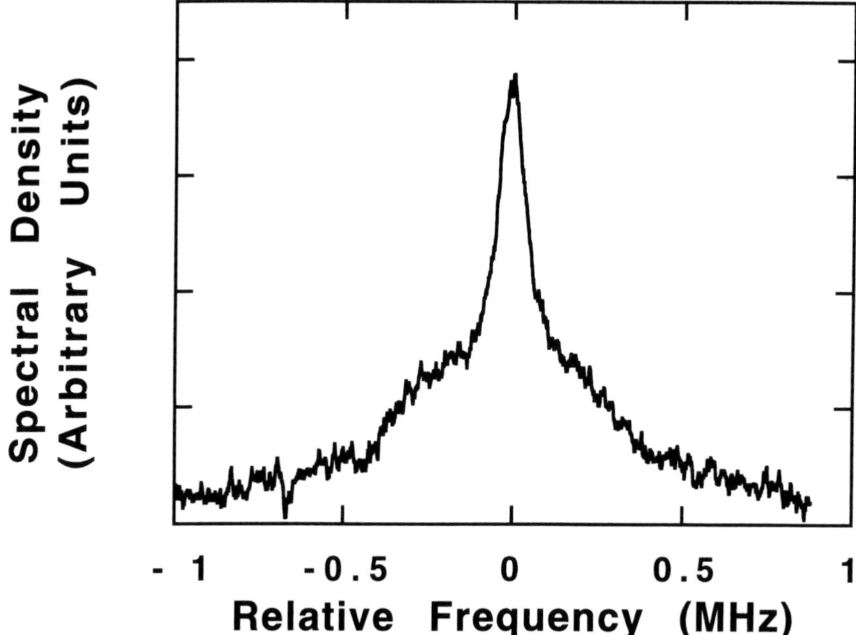

Figure 21. Heterodyne spectrum of fluorescence from Na atoms in optical molasses. The broad component corresponds to a temperature of 84 μK, which compares well with the temperature of 87 μK measured by TOF. The narrow component indicates a sub-wavelength localization of the atoms.

what we expected from the time-of-flight temperature measurement on a similar optical molasses, but the narrow central peak was a puzzle. After rejecting such wild possibilities as the achievement of Bose-Einstein condensation (Fig. 21 looks remarkably similar to velocity distributions in partially Bose-condensed atomic gases) we realized that the answer was quite simple: we were seeing line-narrowing from the Lamb-Dicke effect [108] of atoms localized to less than a wavelength of light. Atoms were being trapped by the dipole force in periodically spaced potential wells like those of Fig. 18b. We knew from both theory and experiments that the thermal energy of the atoms was less than the light shifts producing the potential wells, so it was quite reasonable that the atoms should be trapped. Confined within a region much less than a wavelength of light, the emitted spectrum shows a suppression of the Doppler width, the Lamb-Dicke effect, which is equivalent to the Mössbauer effect. This measurement [107] marked the start of our interest in what are now called optical lattices: spatially periodic patterns of light-shift-induced potential wells in which atoms are trapped and well localized. It also represents a realization of the 1968 proposal of Letokhov [14] to reduce the Doppler width by trapping atoms in a standing wave.

Joined by Poul Jessen, who was doing his Ph.D. research in our lab, we refined the heterodyne technique and measured the spectrum of Rb atoms in a 1-D laser field like that of Figure 18a. Figure 22 shows the results [109], which display well-resolved sidebands around a central, elastic peak. The si-

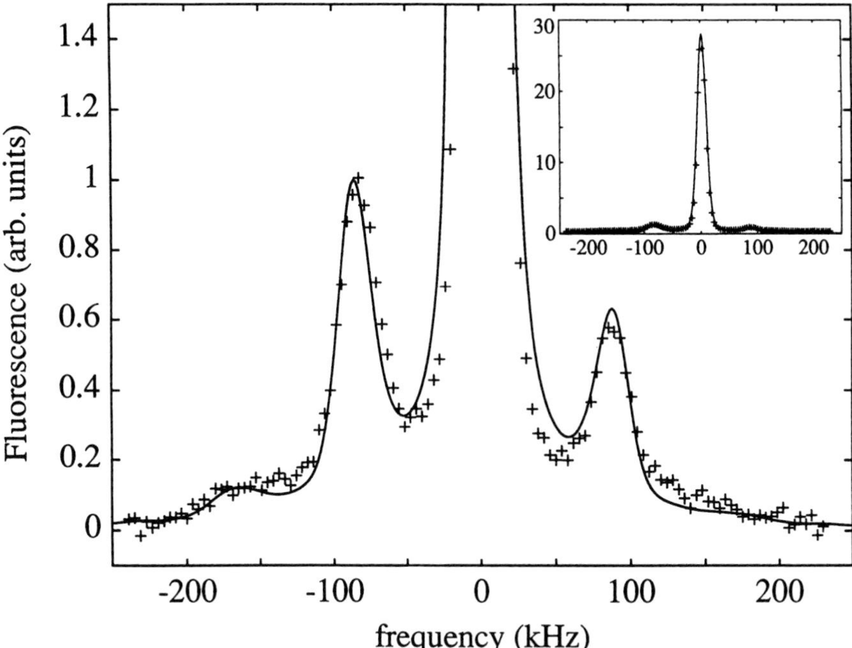

Figure 22. Vertical expansion of the spectrum emitted by Rb atoms in a 1-D optical lattice. The crosses are the data of ref. [109]; the curve is a first-principles calculation of the spectrum [110]. The calculation has no adjustable parameters other than an instrumental broadening. Inset: unexpanded spectrum.

debands are separated from the elastic peak by the frequency of vibration of atoms in the 1-D potential wells. The sideband spectrum can be interpreted as spontaneous Raman scattering, both Stokes and anti-Stokes, involving transitions that begin on a given quantized vibrational level for an atom bound in the optical potential and end on a higher vibrational level (the lower sideband), the same level (elastic peak) or a lower level (the higher sideband). We did not see sidebands in the earlier experiment in a 3D, 6-beam optical molasses [107] at least in part because of the lack of phase stability among the laser beams [111]. We have seen well-resolved sidebands in a 3D, 4-beam lattice [112].

The spectrum of Fig. 22 gives much information about the trapping of atoms in the potential wells. The ratio of sideband intensity to elastic peak intensity gives the degree of localization, the ratio of the two sideband intensities gives the temperature, and the spacing of the sidebands gives the potential well depth. Similar, but in many respects complementary, information can be obtained from the absorption spectrum of such an optical lattice, as illustrated by the experiments performed earlier in Paris [113]. The spectrum of Fig. 22 can be calculated from first principles [110] and the comparison of the experimental and theoretical spectra shown provides one of the most detailed confirmations of our ability to predict theoretically the behavior of laser cooled atoms.

In our laboratory, we have continued our studies of optical lattices, using adiabatic expansion to achieve temperatures as low as 700 nK [114], applying Bragg scattering to study the dynamics of atomic motion [115–118], and extending heterodyne spectral measurements to 3-D [112]. The Paris group has also continued to perform a wide range of experiments on optical lattices [119–122], as have a number of other groups all over the world.

The optical lattice work has emphasized that a typical atom is quite well localized within its potential well, implying a physical picture rather different from the Sisyphus cooling of Fig. 18, where atoms move from one well to the next. Although numerical calculations give results in excellent agreement with experiment in the case of lattice-trapped atoms, a physical picture with the simplicity and power of the original Sisyphus picture has not yet emerged. Nevertheless, the simplicity of the experimental behavior makes one think that such a picture should exist and remains to be found. The work of Refs. [123, 124] may point the way to such an understanding.

CONCLUSION

I have told only a part of the story of laser cooling and trapping at NIST in Gaithersburg, and I have left out most of the work that has been done in other laboratories throughout the world. I have told this story from my personal vantage point as an experimentalist in Gaithersburg, as I saw it unfold. The reader will get a much more complete picture by also reading the Nobel Lectures of Steve Chu and Claude Cohen-Tannoudji. For the work in my lab, I have tried to follow the thread that leads from laser deceleration and cooling of atomic beams [30, 35, 36, 52] to magnetic trapping [63], the discovery of sub-Doppler cooling [80, 99], and the beginnings of optical lattice studies [107, 109]. Topics such as later studies of lattices, led by Steve Rolston, and collisions of cold atoms, led by Paul Lett, have only been mentioned, and other areas such as the optical tweezer work [125, 126] led by Kris Helmerson have been left out completely.

The story of laser cooling and trapping is still rapidly unfolding, and one of the most active areas of progress is in applications. These include "practical" applications like atomic clocks, atom interferometers, atom lithography, and optical tweezers, as well as "scientific" applications such as collision studies, atomic parity non-conservation, and Bose-Einstein condensation (BEC). (The latter is a particularly beautiful and exciting outgrowth of laser cooling and trapping. Since the 1997 Nobel festivities, our laboratory has joined the growing number of groups having achieved BEC, as shown in Fig. 23.) Most of these applications were completely unanticipated when laser cooling started, and many would have been impossible without the unexpected occurrence of sub-Doppler cooling.

Laser cooling and trapping has from its beginnings been motivated by a blend of practical applications and basic curiosity. When I started doing laser cooling, I had firmly in mind that I wanted to make better atomic clocks. On the other hand, the discovery of sub-Doppler cooling came out of a desire to

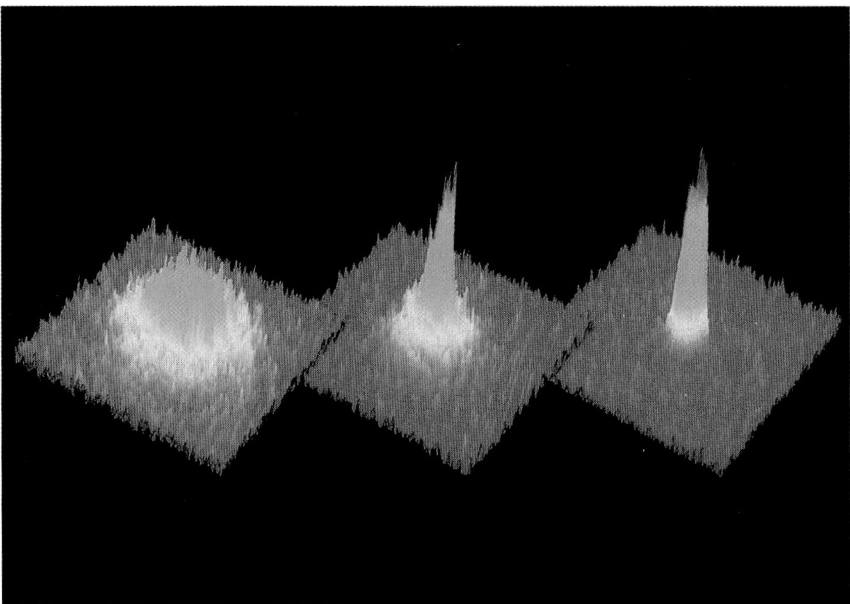

Figure 23. One of the most recent applications of laser-cooling and trapping is Bose-Einstein condensation in an atomic vapor. The figure shows a series of representations of the 2D velocity distribution of a gas of Na atoms at different stages of evaporative cooling through the BEC transition. The velocity distribution changes from a broad thermal one (left) to include a narrow, condensate peak (middle), and finally to be nearly pure condensate (right). The data were obtained in our laboratory in February of 1998, by L. Deng, E. Hagley, K. Helmerson, M. Kozuma, R. Lutwak, Y. Ovchinnikov, S. Rolston, J. Wen and the author. Our procedure was similar to that used in the first such observation of BEC, in Rb, at NIST/JILA in 1995 [15].

understand better the basic nature of the cooling process. Nevertheless, without sub-Doppler cooling, the present generation of atomic fountain clocks would not have been possible.

I hesitate to predict where the field of laser cooling and trapping will be even a few years from now. Such predictions have often been wrong in the past, and usually too pessimistic. But I firmly believe that progress, both in practical applications and in basic understanding, will be best achieved through research driven by both aims.

ACKNOWLEDGMENTS

I owe a great debt to all of the researchers in the many laboratories around the world who have contributed so much to the field of laser cooling and trapping of neutral atoms. Their friendly competition and generous sharing of understanding and insights has inspired me and educated me in an invaluable way. Very special thanks go to those researchers with whom I have been privileged to work here in Gaithersburg: to Hal Metcalf, who was part of the laser cooling experiments from the start, through most of the work described in this paper; to postdocs John Prodan, Alan Migdall, Phil Gould, Chris Westbrook, and Rich Watts, whose work led our group to the discovery of sub-Doppler cooling, and who moved on to distinguished careers elsewhere; to Paul Lett, Steve Rolston, and Kris Helmerson who also were pivotal figures in the development of laser cooling and trapping in Gaithersburg, who have formed the nucleus of the present Laser Cooling and Trapping Group (and who have graciously provided considerable help in the preparation of this manuscript); and to all the other postdocs, visitors and students who have so enriched our studies here. To all of these, I am thankful, not only for scientific riches but for shared friendship.

I know that I share with Claude Cohen-Tannoudji and with Steve Chu the firm belief that the 1997 Nobel Prize in Physics honors not only the three of us, but all those other researchers in this field who have made laser cooling and trapping such a rewarding and exciting subject.

I want to thank NIST for providing and sustaining the intellectual environment and the resources that have nurtured a new field of research and allowed it to grow from a few rudimentary ideas into a major branch of modern physics. I also thank the U.S. Office of Naval Research, which provided crucial support when I and my ideas were unproven, and which continues to provide invaluable support and encouragement.

There are many others, friends, family and teachers who have been of great importance. I thank especially my wife and daughters who have supported and encouraged me and provided that emotional and spiritual grounding that makes achievement worthwhile. Finally, I thank God for providing such a wonderful and intriguing world for us to explore, for allowing me to have the pleasure of learning some new things about it, and for allowing me to do so in the company of such good friends and colleagues.

REFERENCES

1. A. Ashkin, "Trapping of atoms by resonance radiation pressure," Phys. Rev. Lett. **40**, 729 (1978).
2. T. Hänsch and A. Schawlow, "Cooling of gases by laser radiation," Opt. Commun. **13**, 68 (1975).
3. D. Wineland and H. Dehmelt, "Proposed 10^{14} $\Delta\nu < \nu$ laser fluorescence spectroscopy on Tl^+ mono-ion oscillator III," Bull. Am. Phys. Soc. **20**, 637 (1975).
4. D. Wineland, R. Drullinger and F. Walls, "Radiation-pressure cooling of bound resonant absorbers," Phys. Rev. Lett. **40**, 1639 (1978).
5. W. Neuhauser, M. Hohenstatt, P. Toschek and H. Dehmelt, "Optical-sideband cooling of visible atom cloud confined in parabolic well," Phys. Rev. Lett. **41**, 233 (1978).
6. P. Lebedev, "Untersuchungen über die Druckkräfte des Lichtes," Ann. Phys. **6**, 433 (1901).
7. E. F. Nichols and G. F. Hull, "A Preliminary communication on the pressure of heat and light radiation," Phys. Rev. **13**, 307 (1901).
8. E. F. Nichols and G. F. Hull, "The pressure due to radiation," Phys. Rev. **17**, 26 (1903).
9. R. Frisch, "Experimenteller Nachweis des Einstenschen Strahlungsrückstosses," Z. Phys. **86**, 42 (1933).
10. A. Ashkin, "Atomic-beam deflection by resonance-radiation pressure," Phys. Rev. Lett. **25**, 1321 (1970).

11. R. Schieder, H. Walther and L. Wöste, "Atomic beam deflection by the light of a tunable dye laser," Opt. Commun. **5**, 337 (1972).
12. J.-L. Picqué and J.-L. Vialle, "Atomic-beam deflection and broadening by recoils due to photon absorption or emission," Opt. Commun. **5**, 402 (1972).
13. G. A. Askar'yan, "Effects of the gradient of a strong electromagnetic beam on electrons and atoms," Sov. Phys. JETP **15**, 1088 (1962).
14. V. Letokhov, "Narrowing of the Doppler width in a standing light wave," [Pisma Zh. Eksp. Teor. Fiz. **7**, 348 (1968)] JETP Lett. **7**, 272 (1968).
15. M. H. Anderson, J. R. Ensher, M. R. Matthews, C. E. Wieman and E. A. Cornell, "Observation of Bose-Einstein Condensation in a Dilute Atomic Vapor Below 200 Nanokelvin," Science **269**, 198 (1995).
16. K. B. Davis, M.-O. Mewes, M. R. Andrews, N. J. van Druten, D. S. Durfee, D. M. Kurn and W. Ketterle, "Bose-Einstein condensation in a gas of sodium atoms," Phys. Rev. Lett. **75**, 3969 (1995).
17. W. Nagourney, J. Sandberg and H. Dehmelt, "Shelved optical electron amplifier: observation of quantum jumps," Phys. Rev. Lett. **56**, 2797 (1986).
18. T. Sauter, W. Neuhauser, R. Blatt and P. Toschek, "Observation of quantum jumps," Phys. Rev. Lett. **57**, 1696 (1986).
19. J. C. Bergquist, R. G. Hulet, Wayne M. Itano and D. J. Wineland, "Observation of quantum jumps in a single atom," Phys. Rev. Lett. **57**, 1699 (1986).
20. C. Monroe, D. M. Meekhof, B. E. King and D. J. Wineland, "A "Schrödinger Cat" superposition state of an atom," Science **272**, 1131 (1996).
21. C. Monroe, D. M. Meekhof, B. E. King, W. M. Itano and D. J. Wineland, "Demonstration of a fundamental quantum logic gate," Phys. Rev. Lett. **75**, 4714 (1995).
22. V. G. Minogin, "Deceleration and monochromatization of atomic beams by radiation pressure," Opt. Commun. **34**, 265 (1980).
23. S. Andreev, V. Balykin, V. Letokhov and V. Minogin, "Radiative slowing and reduction of the energy spread of a beam of sodium atoms to 1.5K in an oppositely directed laser beam," [Pis'ma Zh. Eksp. Teor. Fiz. **34** 463 (1981)] JETP Lett. **34**, 442 (1981).
24. V. Letokhov, V. Minogin and B. Pavlik, "Cooling and trapping of atoms and molecules by a resonant laser field," Opt. Comm. **19**, 72 (1976).
25. V. Balykin, V. Letokhov and V. Mushin, "Observation of the cooling of free sodium atoms in a resonance laser field with a scanning frequency," [Pis'ma Zh. Eksp. Teor. Fiz. **29**, 614 (1979)] JETP Lett. **29**, 560 (1979).
26. V. Balykin, "Cyclic interaction of Na atoms with circularly polarized laser radiation," Opt. Comm. **33**, 31 (1980).
27. W. Phillips, J. Prodan and H. Metcalf, "Laser Cooling of Free Neutral Atoms in an Atomic Beam", in *Laser Spectroscopy VI*, H. Weber, W. Luthy, Ed. (Springer-Verlag, Berlin, 1983) p. 162.
28. W. D. Phillips and J. V. Prodan, "Chirping the Light - Fantastic?", in *Laser-cooled and Trapped Atoms*, W. D. Phillips, Ed. (Natl. Bur. Stand., Wash. DC, 1983), vol. Spec. Publ. 653, p. 137.
29. J. V. Prodan and W. D. Phillips, "Chirping the Light- Fantastic? Recent NBS Atom Cooling Experiments," Prog. Quant. Electr. **8**, 231 (1984).
30. W. D. Phillips and J. V. Prodan, "Cooling atoms with a frequency chirped laser", in *Coherence and Quantum Optics V*, L. Mandel, E. Wolf, Ed. (Plenum, New York, 1984) p. 15.
31. W. Ertmer, R. Blatt, J. Hall and M. Zhu, "Laser manipulation of atomic beam velocities: Demonstration of stopped atoms and velocity reversal," Phys. Rev. Lett. **54**, 996 (1985).
32. J. Hoffnagle, "Proposal for continuous white-light cooling of an atom beam," Opt. Lett. **13**, 102 (1988).
33. M. Zhu, C. W. Oates and J. L. Hall, "Continuous high-flux monovelocity atomic beam based on a broadband laser-cooling technique," Phys. Rev. Lett. **67**, 46 (1991).

34. W. D. Phillips, Proposal to the Office of Naval Research from the National Bureau of Standards, Laser cooling and trapping of neutral atoms (1979).
35. W. Phillips and H. Metcalf, "Laser deceleration of an atomic beam," Phys. Rev. Lett. **48**, 596 (1982).
36. J. Prodan, W. Phillips and H. Metcalf, "Laser production of a very slow monoenergetic atomic beam," Phys. Rev. Lett. **49**, 1149 (1982).
37. W. D. Phillips, J. V. Prodan and H. J. Metcalf, "Neutral Atomic Beam Cooling Experiments at NBS", in *Laser-cooled and Trapped Atoms*, W. D. Phillips, Ed. (Natl. Bur. Stand., Wash. DC, 1983), vol. Spec. Publ. 653, p. 1.
38. W. D. Phillips, J. V. Prodan and H. J. Metcalf, "Laser-cooled Atomic Beams", in *Atomic Physics IX*, R. S. Van Dyck, E. N. Fortson, Ed. (World Scientific, Singapore, 1984) p. 338.
39. W. D. Phillips, J. V. Prodan and H. J. Metcalf, "Neutral Atomic Beam Cooling Experiments at NBS," Prog. Quant. Electr. **8**, 119 (1984).
40. H. Metcalf and W. D. Phillips, "Laser Cooling of Atomic Beams," Comments At. Mol. Phys. **16**, 79 (1985).
41. W. D. Phillips, J. Prodan and H. Metcalf, "Laser cooling and electromagnetic trapping of neutral atoms," J. Opt. Soc. Am. B **2**, 1751 (1985).
42. W. D. Phillips, Ed., *Laser-Cooled and Trapped Atoms* , Natl. Bur. Stand. (U. S.), Spec. Publ. 653 (1993).
43. S. Stenholm, "Theoretical foundations of laser spectroscopy," Phys. Rep. **43**, 151 (1978).
44. S. Stenholm, "Redistribution of molecular velocities by optical processes," Appl. Phys. **15**, 287 (1978).
45. J. Javanainen and S. Stenholm, "Broad band resonant light pressure I: Basic equations," Appl. Phys. **21**, 35 (1980).
46. J. Javanainen and S. Stenholm, "Broad band resonant light pressure II: Cooling of gases," Appl. Phys. **21**, 163 (1980).
47. J. Javanainen and S. Stenholm, "Laser cooling of trapped particles I: The heavy particle limit," Appl. Phys. **21**, 283 (1980).
48. J. Javanainen and S. Stenholm, "Laser cooling of trapped particles II: The fast particle limit," Appl. Phys. **24**, 71 (1981).
49. J. Javanainen and S. Stenholm, "Laser cooling of trapped particles III: The Lamb-Dicke limit," Appl. Phys. **24**, 151 (1981).
50. S. Stenholm, "Dynamics of trapped particle cooling in the Lamb-Dicke limit," J. Opt. Soc. Am. B **2**, 1743 (1985).
51. S. Stenholm, "The semiclassical theory of laser cooling," Rev. Mod. Phys. **58**, 699 (1986).
52. J. Prodan, A. Migdall, W. D. Phillips, I. So, H. Metcalf and J. Dalibard, "Stopping atoms with laser light," Phys. Rev. Lett. **54**, 992 (1985).
53. C. V. Heer, "A low temperature atomic beam oscillator", in *Quantum Electronics*, C. H. Townes, Ed. (Columbia University Press, New York, 1960) p. 17.
54. V. V. Vladimirskii, "Magnetic mirror, channels and bottles for cold neutrons," Sov. Phys. JETP [Zh. Eksp. Teor. Fiz **39**, 1062 (1960)] **12**, 740 (1961).
55. C. V. Heer, "Feasibility of containment of quantum magnetic dipoles," Rev. Sci. Instrum. **34**, 532 (1963).
56. R. Vauthier, "Dispositif de focalisation pour particules électriquement neutres," C. R. Acad. Sci. (Paris) **228**, 1113 (1949).
57. H. Friedburg and W. Paul, "Optische Abbildung mit neutralen Atomen," Naturwissenschaften **38**, 159 (1951).
58. H. Friedburg, "Optische Abbildung mit neutralen Atomen," Z. Phys. **130**, 493 (1951).
59. K.-J. Kugler, W. Paul and U. Trinks, "A magnetic storage ring for neutrons," Phys. Lett. **72B**, 422 (1978).
60. B. Martin, thesis, Universität Bonn, Report No. Bonn-IR-75-8 (1975)

61. W. Wing, "On neutral particle trapping in quasistatic electromagnetic fields," Prog. Quant. Electr. **8**, 181 (1984).
62. R. Golub and J. Pendlebury, "Ultra-cold neutrons," Rep. Prog. Phys. **42**, 439 (1979).
63. A. Migdall, J. Prodan, W. Phillips, T. Bergeman and H. Metcalf, "First observation of magnetically trapped neutral atoms," Phys. Rev. Lett. **54**, 2596 (1985).
64. V. Bagnato, G. Lafyatis, A. Martin, E. Raab, R. Ahmad-Bitar and D. E. Pritchard, "Continuous slowing and trapping of neutral atoms," Phys. Rev. Lett. **58**, 2194 (1987).
65. K. Helmerson, A. Martin and D. Pritchard, "Laser cooling of magnetically trapped neutral atoms," J. Opt. Soc. Am. B **9**, 1988 (1992).
66. Y. V. Gott, M. S. Ioffe and V. G. Telkovsky, "Some new results on confining of plasmas in a magnetic trap", in *Nuclear Fusion*, Ed. (International Atomic Energy Agency, Vienna, 1962) p. 1045.
67. D. E. Pritchard, "Cooling neutral atoms in a magnetic trap for precision spectroscopy," Phys. Rev. Lett. **51**, 1336 (1983).
68. T. Bergeman, G. Erez and H. J. Metcalf, "Magnetostatic trapping fields for neutral atoms," Phys. Rev. A **35**, 1535 (1987).
69. H. F. Hess, G. P. Kochanski, J. M. Doyle, N. Masuhara, Daniel Kleppner and T. J. Greytak, "Magnetic Trapping of Spin-Polarized Atomic Hydrogen," Phys. Rev. Lett. **59**, 672 (1987).
70. N. Masuhara, J. M. Doyle, J. C. Sandberg, D. Kleppner, T. J. Greytak, H. F. Hess and G. P. Kochanski, "Evaporative cooling of spin-polarized atomic hydrogen," Phys. Rev. Lett. **61**, 935 (1988).
71. R. van Roijen, J. J. Berkhout, S. Jaakkol and J. T. M. Walraven, "Experiments with atomic hydrogen in a magnetic trapping field," Phys. Rev. Lett. **61**, 931 (1988).
72. I. D. Setija, H. G. C. Werij, O. J. Luiten, M. W. Reynolds, T. W. Hijmans and J. T. M. Walraven, "Optical Cooling of Atomic Hydrogen in a Magnetic Trap," Phy. Rev. Lett. **70**, 2257 (1994).
73. W. Petrich, M. H. Anderson, J. R. Ensher and E. A. Cornell, "Stable, Tightly Confining Magnetic Trap for Evaporative Cooling of Neutral Atoms," Phys. Rev. Lett. **74**, 3352 (1995).
74. K. B. Davis, M.-O. Mewes, M. A. Joffe, M. R. Andrews and W. Ketterle, "Evaporative cooling of sodium atoms," Phys. Rev. Lett. **74**, 5202 (1995).
75. J. Bjorkholm, R. Freeman, A. Ashkin and D. Pearson, "Observation of focusing of neutral atoms by the dipole forces of resonance-radiation pressure," Phys. Rev. Lett. **41**, 1361 (1978).
76. W. D. Phillips, "Laser cooling and trapping of neutral atoms", in *Laser Manipulation of Atoms and Ions (Proceedings of the International School of Physics "Enrico Fermi", Course CXVIII)*, E. Arimondo, W. Phillips, F. Strumia, Ed. (North Holland, Amsterdam, 1992) p. 289.
77. V. S. Letokhov, V. G. Minogin and B. D. Pavlik, "Cooling and capture of atoms and molecules by a resonant light field," Sov. Phys. JETP **45**, 698 (1977).
78. D. Wineland and W. Itano, "Laser cooling of atoms," Phys. Rev. A **20**, 1521 (1979).
79. J. Javanainen, "Light-pressure cooling of trapped ions in three dimensions," Appl. Phys. **23**, 175 (1980).
80. P. D. Lett, W. D. Phillips, S. L. Rolston, C. E. Tanner, R. N. Watts and C. I. Westbrook, "Optical molasses," J. Opt. Soc. Am. B **6**, 2084 (1989).
81. S. Chu, L. Hollberg, J. Bjorkholm, A. Cable and A. Ashkin, "Three-dimensional viscous confinement and cooling of atoms by resonance radiation pressure," Phys. Rev. Lett. **55**, 48 (1985).
82. The high temperature observed in this experiment has since been ascribed to the presence of a stray magnetic field from an ion pump. S. Chu, personal communication, (1997).
83. E. Raab, M. Prentiss, A. Cable, S. Chu and D. Pritchard, "Trapping of neutral sodium atoms with radiation pressure," Phys. Rev. Lett. **59**, 2631 (1987).

84. C. Monroe, W. Swann, H. Robinson and C. Wieman, "Very cold trapped atoms in a vapor cell," Phys. Rev. Lett. **65**, 1571 (1990).
85. A. Cable, M. Prentiss and N. P. Bigelow, "Observations of sodium atoms in a magnetic molasses trap loaded by a continuous uncooled source," Opt. Lett. **15**, 507 (1990).
86. P. L. Gould, P. D. Lett, P. S. Julienne, W. D. Phillips, H. R. Thorsheim and J. Weiner, "Observation of associative ionization of ultracold laser-trapped sodium atoms," Phys. Rev. Lett. **60**, 788 (1988).
87. S. Chu, J. Bjorkholm, A. Ashkin and A. Cable, "Experimental observation of optically trapped atoms," Phys. Rev. Lett. **57**, 314 (1986).
88. P. D. Lett, P. S. Jessen, W. D. Phillips, S. L. Rolston, C. I. Westbrook and P. L. Gould, "Laser modification of ultracold collisions: Experiment," Phys. Rev. Lett. **67**, 2139 (1991).
89. P. D. Lett, K. Helmerson, W. D. Phillips, L. P. Ratliff, S. L. Rolston and M. E. Wagshul, "Spectroscopy of Na_2 by photoassociation of ultracold Na," Phys. Rev. Lett. **71**, 2200 (1993).
90. L. P. Ratliff, M. E. Wagshul, P. D. Lett, S. L. Rolston and W. D. Phillips, "Photoassociative Spectroscopy of 1_g, 0_u^+ and 0_g^- States of Na_2," J. Chem. Phys. **101**, 2638 (1994).
91. P. D. Lett, P. S. Julienne and W. D. Phillips, "Photoassociative Spectroscopy of Laser-Cooled Atoms," Annu. Rev. Phys. Chem. **46**, 423 (1995).
92. K. Jones, P. Julienne, P. Lett, W. Phillips, E. Tiesinga and C. Williams, "Measurement of the atomic Na(3P) lifetime and of retardation in the interaction between two atoms bound in a molecule," Europhys. Lett. **35**, 85 (1996).
93. E. Tiesinga, C. J. Williams, P. S. Julienne, K. M. Jones, P. D. Lett and W. D. Phillips, "A spectroscopic determination of scattering lengths for sodium atom collisions," J. Res. Natl. Inst. Stand. Technol. **101**, 505 (1996).
94. M. Walhout, U. Sterr, C. Orzel, M. Hoogerland and S. L. Rolston, "Optical Control of Ultracold Collisions in Metastable Xenon," Phys. Rev. Lett. **74**, 506 (1995).
95. P. L. Gould, P. D. Lett and W. D. Phillips, "New Measurements with Optical Molasses", in *Laser Spectroscopy VIII*, W. Persson, S. Svanberg, Ed. (Springer-Verlag, Berlin, 1987) p. 64.
96. J. P. Gordon and A. Ashkin, "Motion of atoms in a radiation field," Phys. Rev. A **21**, 1606 (1980).
97. W. D. Phillips, C. I. Westbrook, P. D. Lett, R. N. Watts, P. L. Gould and H. J. Metcalf, "Observation of atoms laser-cooled below the Doppler limit", in *Atomic Physics 11*, S. Haroche, J. C. Gay, G. Grynberg, Ed. (World Scientific, Singapore, 1989) p. 633.
98. C. Salomon, J. Dalibard, W. D. Phillips, A. Clairon and S. Guellati, "Laser cooling of cesium atoms below 3 microkelvin," Europhys. Lett. **12**, 683 (1990).
99. P. D. Lett, R. N. Watts, C. I. Westbrook, W. D. Phillips, P. L. Gould and H. J. Metcalf, "Observation of atoms laser cooled below the Doppler limit," Phys. Rev. Lett. **61**, 169 (1988).
100. J. Dalibard and C. Cohen-Tannoudji, "Laser cooling below the Doppler limit by polarization gradients: simple theoretical models," J. Opt. Soc. Am. B **6**, 2023 (1989).
101. P. J. Ungar, D. S. Weiss, E. Riis and S. Chu, "Optical molasses and multilevel atoms: theory," J. Opt. Soc. Am. B **6**, 2058 (1989).
102. C. Cohen-Tannoudji and W. D. Phillips, Physics Today **43**, 33 (1990).
103. C. Cohen-Tannoudji, "Atomic Motion in Laser Light", in *Fundamental Systems in Quantum Optics*, J. Dalibard, J.-M. Raimond, J. Zinn-Justin, Ed. (North Holland, Amsterdam, 1992) p. 1.
104. J. Dalibard and C. Cohen-Tannoudji, "Dressed-atom approach to atomic motion in laser light: the dipole force revisited," J. Opt. Soc. Am. B **2**, 1707 (1985).
105. D. S. Weiss, E. Riis, Y. Shevy, P. J. Ungar and S. Chu, "Optical molasses and multilevel atoms: experiment," J. Opt. Soc. Am. B **6**, 2072 (1989).
106. B. R. Mollow, "Power spectrum of light scattered by two-level systems," Phys. Rev. **188**, 1969 (1969).

107. C. I. Westbrook, R. N. Watts, C. E. Tanner, S. L. Rolston, W. D. Phillips, P. D. Lett and P. L. Gould, "Localization of atoms in a three dimensional standing wave of light," Phys. Rev. Lett. **65**, 33 (1990).
108. R. H. Dicke, "The effect of collisions upon the Doppler width of spectral lines," Phys. Rev. **89**, 472 (1953).
109. P. S. Jessen, C. Gerz, P. D. Lett, W. D. Phillips, S. L. Rolston, R. J. C. Spreeuw and C. I. Westbrook, "Observation of quantized motion of Rb atoms in an optical field," Phys. Rev. Lett. **69**, 49 (1992).
110. P. Marte, R. Dum, R. Taïeb, P. Lett and P. Zoller, "Wave function calculation of the fluorescence spectrum of 1-D optical molasses," Phys. Rev. Lett. **71**, 1335 (1993).
111. G. Grynberg, B. Lounis, P. Verkerk, J.-Y. Courtois and C. Salomon, "Quantized motion of cold cesium atoms in two- and three-dimensional optical potentials," Phys. Rev. Lett. **70**, 2249 (1993).
112. M. Gatzke, G. Birkl, P. S. Jessen, A. Kastberg, S. L. Rolston and W. D. Phillips, "Temperature and localization of atoms in 3D optical lattices," Phys. Rev. A **55**, R3987 (1997).
113. P. Verkerk, B. Lounis, C. Salomon, C. Cohen-Tannoudji, J.-Y. Courtois and G. Grynberg, "Dynamics and Spatial Order of Cold Cesium Atoms in a Periodic Optical Potential," Phys. Rev. Lett. **68**, 3861 (1992).
114. A. Kastberg, W. D. Phillips, S. L. Rolston, R. J. C. Spreeuw and P. S. Jessen, "Adiabatic cooling of cesium to 700 nK in an optical lattice," Phys. Rev. Lett. **74**, 1542 (1995).
115. G. Birkl, M. Gatzke, I. H. Deutsch, S. L. Rolston and W. D. Phillips, "Bragg Scattering from Atoms in Optical Lattices," Phys. Rev. Lett. **75**, 2823 (1995).
116. G. Raithel, G. Birkl, A. Kastberg, W. D. Phillips and S. L. Rolston, "Cooling and localization dynamics in optical lattices," Phys. Rev. Lett. **78**, 630 (1997).
117. G. Raithel, G. Birkl, W. D. Phillips and S. L. Rolston, "Compression and parametric driving of atoms in optical lattices," Phys. Rev. Lett. **78**, 2928 (1997).
118. W. D. Phillips, "Quantum motion of atoms confined in an optical lattice," Materials Science and Engineering B **48**, 13 (1997).
119. B. Lounis, P. Verkerk, J.-Y. Courtois, C. Salomon and G. Grynberg, "Quantized atomic motion in 1D cesium molasses with magnetic field," Europhys. Lett. **21**, 13 (1993).
120. D. R. Meacher, D. Boiron, H. Metcalf, C. Salomon and G. Grynberg, "Method for velocimetry of cold atoms," Phys. Rev. **A50**, R1992 (1994).
121. P. Verkerk, D. R. Meacher, A. B. Coates, J.-Y. Courtois, S. Guibal, B. Lounis, C. Salomon and G. Grynberg, "Designing Optical Lattices: an Investigation with Cesium Atoms," Europhys. Lett. **26**, 171 (1994).
122. D. R. Meacher, S. Guibal, C. Mennerat, J.-Y. Courtois, K. I. Pestas and G. Grynberg, "Paramagnetism in a cesium lattice," Phys. Rev. Lett. **74**, 1958 (1995).
123. Y. Castin, Doctoral Dissertation, Ecole Normale Supèrieure (1992) (See section IV 3 e).
124. Y. Castin, K. Berg-Sorensen, J. Dalibard and K. Mølmer, "Two-dimensional Sisyphus cooling," Phys. Rev. A **50**, 5092 (1994).
125. K. Helmerson, R. Kishore, W. D. Phillips and H. H. Weetall, "Optical Tweezers-based immunosensor detects femtomolar concentration of antigens," Clinical Chemistry **43**, 379 (1997).
126. M. Mammen, K. Helmerson, R. Kishore, S.-K. Choi, W. D. Phillips and G. M. Whitesides, "Optically controlled collisions of biological Objects to evaluate potent polyvalent inhibitors of virus-cell adhesion," Chemistry & Biology **3**, 757 (1996).

Physics 1998

ROBERT D. LAUGHLIN, HORST L. STÖRMER and DANIEL C. TSUI

"for their discovery of a new form of quantum fluid with fractionally charged excitations"

THE NOBEL PRIZE IN PHYSICS

Speech by Professor Mats Jonson of the Royal Swedish Academy of Sciences. Translation of the Swedish text.

Your Majesties, Your Royal Highness, Ladies and Gentlemen,

For a long time, man has known how to use electricity. At first, he did so without having any knowledge of what an electric current actually is. This did not prevent the invention of the electrical motor, the telegraph and the telephone. But man also has a wonderful quality called curiosity. He wanted to know what it is that forms an electric current. The young physicist Edwin Hall had an idea. He assumed that a current consisted of some kind of particles that he could influence with a magnet. In 1879 he conducted an experiment which demonstrated that his thinking was correct. The Hall effect had been discovered.

But it was not until 1897 that Sir Joseph John Thomson discovered the electron. As it turned out, it is electrons – extremely tiny electrically charged particles – that pour through our electrical wires in large numbers.

We eventually learned how to control these electrons so well that with their aid, we were able to transmit sound and images. The transistor was invented in the late 1940s. Electronics replaced electricity as a force for societal change. The integrated circuit, satellite television, cellular telephones, the Internet – the world has shrunk and we humans have moved closer to each other because we managed to tame the electron.

To achieve this, physicists and engineers have had to work with materials containing many more electrons than there are stars in our Milky Way galaxy. Though all electrons affect each other because of their equal charges, it proved possible to bend the behavior of these electronic galaxies to the will of humans. Why? Nobel laureate Lev Landau provided us with an explanation. He showed that it is usually enough to understand electrons one by one, then put the parts together and create the whole. The exceptions to this 1 + 1 = 2 rule in electronic physics are so few and spectacular that they have often led to Nobel Prizes. This year's Prize is all about such an exception, in which all electrons cooperate in a new kind of carefully choreographed dance that must be viewed as something whole and indivisible.

"Zeig mir dein Handy" – show me your cell phone – replied Horst Störmer when a German journalist asked him about the usefulness of his research. This is because cellular phones employ a type of transistors that Störmer and Daniel Tsui helped develop. The miniaturization that is constantly underway in microelectronics has been pushed so far in this case that these transistors are being built atom by atom, with incredible precision, and in such a way that electrons are trapped between two layers of atoms and are unable to move sideways. This is advanced technology. But at the same time, it is a platform for brilliant basic science. If such a transistor is cooled to just above absolute zero, –273 degrees Celsius, and is exposed to a magnetic field a million times stronger than that of the earth, there is a good chance that the phenomenon discovered by Störmer and Tsui sixteen years ago will turn up again.

What did they discover? While Edwin Hall described the results of his measurements with a straight line, Tsui and Störmer found a stepwise curve when they made the same measurement. Their highest step was three times higher than previously observed, and they had the courage and insight to state that this indicated that particles with a charge of only one third that of an electron must be concealed here. But could such a revolutionary concept be taken seriously? The charge of an electron is indivisible, or is it not? Researchers were facing a totally unexpected discovery. How was it to be interpreted?

One person who took a serious look at the hints provided by this experiment was Robert Laughlin. After a year of thinking and calculating, he arrived at an explanation, based on the non-applicability of Lev Landau's method. He had to describe how perhaps 100 billion electrons interact all at once. It was something of a miracle that Laughlin managed to do this with a few formulas and four pages of text. This was possible because he succeeded in finding a new kind of order among electrons; they form a new type of quantum liquid.

Laughlin's new ordered electronic dance explained Störmer's and Tsui's experiments. It also implied that the swirls stirred up in an electronic sea by magnetic and electrical forces behave like particles that have only a fraction of the charge of one electron. Truly remarkable! Is there perhaps a connection with the fractionally charged quarks that hide deep inside atomic nuclei? Perhaps, perhaps not – but it is a sign of the profundity of the discovery now being awarded the Nobel Prize that the question about such an analogy can be asked.

Technology and science have gone hand in hand during this journey, each one stimulating the development of the other. So will the 1998 Physics Prize lead to new technology? We don't know, just as a century ago we could not have foreseen the usefulness of discovering the electron. Perhaps some of the young people who are able to watch this ceremony by means of today's electronics may some day find the answers to these questions.

Professors Daniel Tsui, Horst Störmer and Robert Laughlin,

Your wonderful discovery of a new type of quantum liquid with fractionally charged excitations has made us look at nature with new eyes. You have proved that there are indeed new secrets to uncover and new discoveries to be made, if one has the courage to look for them. In this way, you have set an example for new generations of scientists.

On behalf of the Royal Swedish Academy of Sciences, I warmly congratulate you and I ask you to step forward and receive the Prize from the hands of His Majesty the King.

ROBERT B. LAUGHLIN

EARLY YEARS

I was born on 1 November, 1950 in Visalia, California, a medium-sized town just south of Fresno in the San Joaquin Valley. It was at that time an agricultural community more like the Middle West or West Texas than Hollywood or Beverly Hills. The main highway into town was lined with magnificent walnut orchards and stands of valley oaks. My childhood home backed onto wheat and cotton fields. And when the navel orange crop was threatened by a freeze there was smudge in the air by day and talk of little else. A 10-minute drive in any direction brought one out of the town and into rows of tidy farms with peach orchards, olive orchards, avocado orchards, nuts of all sorts, row crops, and dairies. And above us stood the mighty Sierra Nevada, John Muir's Range of Light, the rivers of which irrigated the land and turned what would otherwise have been oak savannah into the richest farmland in the world. The mountains were obscured most of the time by the haze caused by irrigation and too many automobiles or the dense radiation fog that hides the sun most of the winter in that part of the world. My great Aunt recalled how they were visible most of the summer when she first came there after the great San Francisco earthquake of 1906. But on brilliant winter mornings just after a Pacific storm had blown through there they would be, a blazing wall of white stretching north and south as far as the eye could see, topped by the silhouettes of Sawtooth, Mineral King, and the Great Western Divide.

Both sides of my family landed in Visalia by accident. My mother was the daughter of a local doctor, Irvin Betts, who had come down from San Francisco after medical school "temporarily" and induced my grandmother to accompany him by promising her a return in a couple of years. She always laughed when she told this story. My father had grown up a widow's son in Chico, served as a naval officer in the war, followed his brother into the law, and had come to Visalia fresh out of law school to work in the Tulare County District Attorney's office. There he met and married my mother, and together they raised four children, of which I was the first. Like so many other American families mine had roots that were deep but temporary. We attended church, joined the Boy Scouts, contributed casseroles to PTA pot luck suppers, and celebrated many a Thanksgiving with family and friends, but in the end moved away. My father died in Visalia 18 years ago, and all of us, including my mother, now live elsewhere.

Early on in his career my father left the District Attorney's office and set up a private law practice in town. He worked very hard but was, as one of my uncles later put it, an "artist lawyer", meaning that he was more concerned with correctness than profits and often did work for needy clients for free. As a consequence while we had a roof over our heads, food on the table, and clothes to wear to school we were constantly conscious of being of modest means. Whether caused by this or our home environment generally it came to pass that all of us became quite self-reliant at an early age. I, for example, used to take appliances apart when they broke in an attempt to fix them, which I rarely did successfully, being a kid. I am better at this now. My sister Margaret, who is an attorney, still enjoys doing needlework from scratch. My brother John, a software engineer, prides himself in being able to fix any broken thing. It was through such creative play that I first learned about pump impellers, refrigerant cycles, material strength, corrosion, and the rudiments of electricity, and more importantly the idea that real understanding of a thing comes from taking it apart oneself, not reading about it in a book or hearing about it in a classroom. To this day I always insist on working out a problem from the beginning without reading up on it first, a habit that sometimes gets me into trouble but just as often helps me see things my predecessors have missed.

Another important aspect of our home was respect for ideas. At dinnertime one of my parents, usually my father, would lead a discussion about some controversial matter, such as racial integration of schools, whether John Lennon should have compared himself with Jesus Christ, support of Israel, or the morality of the Vietnam war, and all of us were expected to air and defend our views on these things, even if we did not want to. Over the course of time this gave us a deep respect for ideas, both our own and those of others, and an understanding that conflict through debate is a powerful means of revealing truth. This was, of course, before any of us understood rhetoric and how easily it can be misused. But the need for conflict to expose prejudice and unclear reasoning, which is deeply embedded in my philosophy of science, has its origin in these debates.

My mother, who was professional schoolteacher, was particularly concerned about our formal education and even went so far as to start a private school together with some other parents so that our intellectual needs would be met. They acquired an old two-room schoolhouse out in the country among the walnut groves at the foot of Venice Hill, added some indoor plumbing, and hired a small faculty to teach us a broad curriculum that included such things as Latin and French. I am afraid the money was largely wasted on me because I was not ready to learn French, or much of anything else, at that time, although I did rather enjoy watching the machines shaking walnuts off the trees in the fall. But it was impressed upon me that there was such a thing as good study habits and that I would have to acquire them if I wanted to be a scholar. My mother also had us take piano lessons, and this had a similar effect. I hated those lessons, but I now play regularly for pleasure and have even tried my hand at composing. So mothers everywhere take heart. The indoctrination you administer now may have unanticipated positive effects years later.

I was an extremely reclusive and introverted boy. It was to my parents' credit that they weathered the storm and encouraged my self-motivated study, even though it scared them to death, especially my mother. While still at Venice Hill, for example, I got very interested in how televisions worked, and electronics generally, so my parents bought me a Heathkit color TV, which I soldered together and eventually made work. It was a magnificent thing filled with vacuum tubes. One could probably have heated the living room with it. I found building this kit rather unsatisfactory, actually, because the manual did not explain how the circuits worked but only how to assemble them. So I went back to old discarded black-and-white models, which my father dutifully acquired for me, and began reading about what the various parts did and then testing the theory by removing them one at a time. It was in this way that I learned why it so bad to allow the 10 kilovolts stored on a cathode-ray tube to discharge through one's body. Thank God my mother never knew. I also taught myself how to blow glass using a propane torch from the hardware store and managed to make some elementary chemistry plumbing such as tees and small glass bulbs. The latter I filled with isopropyl alcohol and attached with a piece of surgical tubing to the intake of a cooling compressor I had scavenged from a broken refrigerator. This lowered the pressure sufficiently to boil the alcohol and lower the temperature well below the freezing point of water. I had ambitions of making liquid nitrogen, and could probably have done it with more compressors and some dewars. I also tried to make sodium metal by electrolysis of molten salts. I discovered that common wood lye had the lowest melting temperature of all the available materials, so I melted some in a orange juice can and electrolyzed it using an auto battery charger and an ice pick as the cathode. It worked, except that the sodium lived only a second or two before being oxidized by the surrounding air. It was at this time that I picked up the can to check for corrosion on the bottom and accidentally poured its contents all over my right hand, burning it severely. My father rushed me to the hospital, had it dressed, and then invented a story to tell my poor mother so that she would not have a heart attack. By good fortune the molten sodium hydroxide was so hot that it had vaporized the water in my skin and sloughed off without burning me chemically. My hand recovered fully. My parents would probably never have encouraged these things had they known how foolish and dangerous they were, but it is nonetheless a testament to their belief in the value of self-motivated exploration that they allowed me to cultivate such interests even though I got no credit from them toward college or employment.

In parallel with the development of my interests in technical gadgetry I began to acquire a profound love of and respect for the natural world which motivates my scientific thinking to this day. My maternal grandmother had a mountain cabin deep in the Tule river canyon just south of Sequoia Park, to which we were often invited as a family or as individuals. My grandfather had built it as a kind of hunting lodge before I was born, so it had a very masculine feeling despite being my grandmother's home. It had a big stone fireplace, knotty pine walls, a big cast iron chandelier for light, and a marvelous old Aeolean player piano with plenty of rolls. My grandmother was a complex person, but she loved the mountains and welcomed anyone else who did, including reclusive grandsons. So I spent time there whenever I was able, which was not very much because I had responsibilities at home, and over the course of time came to understand what a treasure it was – the house-sized boulders left in the riverbed by retreating glaciers, the massive ponderosa pines six feet in diameter at the base, delicate mosses and lichens of every imaginable color, the complex geometries of pine cones and oak boughs, the hundreds of fragrant herbs along the riverbank, the quiet rush of the river at night on a cool summer evening, and the vast tracts of wilderness beyond known to no one. I realized that nature is filled with a limitless number of wonderful things which have causes and reasons like anything else but nonetheless cannot be forseen but must be discovered, for their subtlety and complexity transcends the present state of science. The questions worth asking, in other words, come not from other people but from nature, and are for the most part delicate things easily drowned out by the noise of everyday life.

I owe my interest in mathematics to my father, or more precisely the sense that mathematics was something important and mysterious. He knew very little mathematics himself but was always reading about it and encouraging everybody else to do the same. He even mounted blackboards in the hall so that a person could write down a brilliant idea if he happened to be passing by. I remember particularly one day hearing a shout from my father's bedroom and rushing in to discover that he had just discovered Euler's theorem, the statement that $e^{i\pi} = -1$. He did not understand the proof completely, so it appeared to him more astonishing than it does to those of us with technical training, but he had correctly understood its significance and elegance. I did well in my mathematics courses in school but was not that challenged and, truthfully, not that interested either. But through my interests in electron motion in vacuum tubes I discovered a need to describe trajectories of moving particles with equations. So I taught myself calculus. I was terribly proud of this at the time, but I realize now that people at this age are simply developmentally ready to learn such things, which is why calculus is now taught in high school. But I was certainly the only person in my town to have done this, and my father's own interest in mathematics was the underlying cause.

BERKELEY

The experience that firmly placed me on a course toward a professional career in science was the four years I spent as an undergraduate at Berkeley. I entered in the fall of 1968 as an electrical engineer, my parents having prevailed upon me to take the economic facts of life seriously. I had applied to more elite schools but had not gotten in, presumably because my grades were not high enough, and also because I was what we now call an "angular" student, i.e. not well-rounded. My parents were not that disappointed, for they had themselves attended Berkeley, as eventually did my brother and two sisters. Berkeley was as different from the quiet country town of my youth as one could possibly imagine. It was full of coffee shops, politics, book stores, theaters, ethnic restaurants, stray dogs, junkies, street musicians, and fascinating people from every conceivable walk of life. As time passed I became more and more intoxicated with all this freedom and more and more convinced that the university was where my future lay. Here was the place ideas mattered, where everybody was eccentric, where originality was not only accepted but had actual market value. It was easy to get lost in the crowd at Berkeley, particularly in the great lecture courses, but this did not bother me because I had no intention of getting lost in the crowd, and anyway considered it a small price to pay for the freedom to think as I saw fit.

At Berkeley I had my first encounter with real professional scientists. I remember the Berkeley faculty as being particularly visionary and inspirational. In the physics department in particular there was a palpable sense of history going back to Heisenberg, Pauli, and Einstein. I later came to understand that Berkeley has always been a special place in American physics and that many of the greatest physicists in the world, perhaps even most of them, can trace their roots back to Berkeley in some way. It was this faculty that defined for me what physics was and should be, and thereby helped me make up my mind to pursue physics as a career. I came home in the middle of my sophomore year and announced, much to the horror of my parents, that I was switching to physics from engineering. After some discussion they gave in, as well-meaning parents tend to do in this situation, and I remember my father musing afterward that it would probably come out all right because these things usually did. Meanwhile at school I was experiencing such wonderful things as the surprise appearance of Charles Townes, winner of the Nobel Prize for invention of the laser, in one of my large lecture courses to explain simply and accurately how lasers work and how they came to be invented. I took quantum mechanics from Owen Chamberlain, who had won the Nobel Prize several years before for the discovery of the antiproton, and who was happy to discuss all sorts of unrelated things such as whether fusion would ever work and whether one should go East to graduate school. I learned electrodynamics from J. D. Jackson's wonderful book and had many occasions to ask him questions about the subject. I took introductory solid state physics from Charles Kittel, the acknowledged father of the field in which I was eventually to work. I took Goeffrey Chew's advanced quantum mechanics course and learned more about the S matrix than he probably intended. I also had many useful exchanges with Ray Sachs, who helped me learn differential geometry and general relativity on my own and guided me to a thesis. My work with Ray began with the question of whether a charged particle dropped in a gravitational field should radiate light, since the relativity principle said it was actually not accelerating. The correct answer is yes because the electromagnetic field knows about the curvature tensor. This line of thought led us to a calculation of the cross-section for scattering gravitational radiation off of a charged particle, the roles of the gravitational and electromagnetic fields in this case being exactly reversed. It was a wonderful time in my life. On commencement day we were addressed by Emilio Segré, sharer of the Nobel Prize for the antiproton discovery and author of a book on nuclear physics that is a delight to read to this day. He took the long view, told us all not to worry too much, and recounted how he and his fellow students in Rome had regularly scanned the obituaries in hopes that a job would become available soon. Many years later when I returned to Berkeley to talk about fractional quantization it was Professor Segré who rushed up after the lecture to ask if the particles we had identified in the fractional quantum hall effect might have something to do with quarks. It was his life's work to ask questions like that, and this was the reason I had found him and his colleagues so inspiring.

My years at Berkeley coincided almost exactly with the worst of the Vietnam war. It is not necessary to recount here the many terrible events of that time, but the political unrest at Berkeley caused ultimately by the war was a major constraint on student life, both intellectually and physically. It was also a real lesson in how people's perceptions of exactly the same facts can be profoundly different. I had no sympathy at all for the disrespect for property and formal education implicit in these demonstrations, but I did think long and hard about the issues raised and, more importantly, about what these events said about politics. Western society has many flaws, and it is good for an educated person to have thought some of these through, even at the expense of losing a lecture or two to tear gas. As to the war, I had no idea what to think about it, except that there were already scattered reports of people in my high school class having come back in body bags. So it came as quite a shock when President Nixon canceled student deferments arguing, correctly in my view, that they were unfair, held a lottery, and picked for me a draft number equal to my age – nineteen.

I remember vividly the day it was announced and the coldness I felt as the full implications slowly became clear. It was common knowledge that theoretical physicists do their best work before age 27, sometimes even earlier. I could not possibly meet this deadline now. There was also the moral question of whether to serve at all. Many people at that time were fleeing the country to avoid the draft, others were faking health problems, and still others were enlisting for long periods in exchange for safety. After stewing over this a long time I decided that I did not think defending one's country was wrong – although the Vietnam war had very little to do with defending one's country – that I could not lie about so important a matter, that I did not want to flee the country, and that I should obey its laws if I stayed. So that was that. I often question now whether this was the right decision, but in any event it is the one I made. But the weight of it bore down on me more and more heavily as my senior year progressed, and at the very end I lost focus, failed a laboratory, and graduated with only a degree in mathematics rather than with the double degrees in mathematics and physics I had actually earned. So I left Berkeley with everything I had come to value in ruins. The only thing I had left was the faith in myself instilled by my parents and the certainty that I had understood what theoretical physics was and was extremely good at it.

MILITARY

At the time I felt that my induction into the military was a giant step backward. It was certainly unfair taxation of my time, but then life is unfair, and getting reminded of this from time to time is perhaps not such a bad thing. I had decided not to become an officer because to do so would have required me to stay in a year longer, and time was critical. So I became an enlisted man and let the system do with me as it saw fit. Skill as a theoretical physicist matters very little in the lower ranks of the army – or perhaps has negative value. It is an interesting fact that during my tour I was never allowed access to computers, radios, or anything else that I might damage through curiosity, or perhaps something more sinister. What matters most is that one blend in. In basic training, which I had at Fort Ord near Monterey, one's identity and past are excised and a new one substituted. All one's clothes and possessions are removed and shipped home. All one's hair is shaved off so that one looks like a concentration camp victim. All people get the same hemorrhoid examination. All people get the same equipment. All people run with this equipment to the firing range. All people get the same cold. All people do the same chin-ups before meals. It was about as different from Berkeley as one could imagine, the suppression of individuality and freedom for the purpose of preventing mischief. In retrospect I consider my induction to have been not so much a step backward as an important lesson in civics, for it eventually became clear that these things I found so abhorrent were the very things required to make a large organization run well under stress. So I learned the hard way that freedom and efficiency conflict, that more of one means less of the other, and that this is fundamental. To this day I break out in a cold sweat every time I hear the term "programmatic science", for I know it really means tight bunks, shiny boots, and digging holes that will be filled back up by someone else the next day.

Some time near the end of basic training a computer somewhere decided that I was suited for missile school, so I was ordered to Fort Sill, Oklahoma, to learn how to fire Pershing missiles. This was a good deal less stressful than basic training, as the pace was slower, and this part of Oklahoma is laid back and rather beautiful, with rolling brown hills not unlike the ones in California. The Pershing missiles, on the other hand, were not beautiful. They were horrible weapons of war – solid-fuel rockets five feet in diameter at the base, long as a moving van, and capable of throwing a tactical nuclear warhead 500 miles. They were launched from trucks and required a team of 10 men to service and fire. The most interesting thing I learned during this time was how small a nuclear warhead was. The nose cone of a Pershing is only about 18 inches in diameter at the base. I had not been interested at all in nuclear weaponry as a student, and so I had never thought through carefully about their "efficiency". It is a sobering thought that these missiles were actually deployed in continental Europe in those days and that on at least one occasion, namely the 1973 Arab-Israeli war, there was an alert serious enough to leave the commanding officers trembling.

While at Fort Sill I met, or more precisely was grouped with, the people who were to be my companions for the rest of my tour in the military. They were a very personable bunch mostly from the upper Middle West, Pennsylvania, and Nebraska, and rather like a selection of the smarter students from my high school, except that the contingent from Detroit was rabidly racist, something that I had never encountered before and still have trouble understanding. Getting to know these people was my first of many reminders that the world is full of intelligent, well-meaning people who, for one reason or another, did not attend university but are nonetheless well-read and educated. Out there on the prairie lost opportunities of youth were the rule rather than the exception, and I slowly became disabused of the myth of the Bright Young Thing and have not believed in it since.

After missile school I was ordered to southern Germany, where I spent the remainder of my tour. This assignment was a welcome turn of events, but it was not a vacation, and it was in some respects extremely unpleasant. Most of the locals in my parents' generation were very accepting and helpful, for they were afraid of the Russians and remembered the many kindnesses done to them after the war. They were also prospering economically, which I know from personal experience helps one overlook indignities. But the people my age and younger hated the whole idea of a foreign army on their soil, especially one with nuclear weapons, felt little personal guilt for Germany's past, and felt that the Vietnam war had thoroughly discredited the alleged ethical superiority of English-speaking countries. So we were tolerated but not liked all that much. Also there were terrible morale problems in the unit to which I was assigned in Schwaebisch Gmuend, a small town near Stuttgart, which caused particularly heavy and widespread drug usage. These were largely corrected by a change in command about halfway through my tour, but they were nonetheless extremely scary.

During this time I tried to think about physics, and about university life generally, and I made a point of visiting the nearby technical universities and the great medieval university at Tübingen, but it was hopeless. I had a job to do, my time was too fragmented, and my unit discouraged much contact with university types, this being politically dangerous. Tübingen, in particular, was frowned upon because of the safe house for AWOL soldiers alleged to be there. So I decided to make the best of a bad situation and invest the time studying language. This turned out to be a better expenditure of time than reading physics books, for like most of my countrymen I had an incomplete understanding of how language is a vehicle for ideas rather than the other way around. While my language ability is still poor, I can still remember the day that radio stations began to sound clear, when newspapers began to inform more than frustrate, when I began to get jokes, and when I told my first joke. So in the light hindsight, I judge this time to have been well spent.

At the end of my tour I was released from duty in Europe, as I had elected to travel around a bit as a free man before going home. On the day of my emancipation I celebrated in traditional fashion by burning my boots – although in an especially thoughtful and creative way. I went downtown and bought 3 kilos of saltpetre, mixed it with sugar, and filled both boots up to the brim, fully laced, and lit one off. There was a tremendous pink flame, fierce heat, and dense smoke that began shooting straight up 30 feet as from a volcano as the fire ate down into the boot. Several of the battery officers came running up just as the experiment was ending to see what was left of the sole curling up like a shriveled bug. They had been playing baseball nearby and had thought that a radio unit was on fire. I assured them that the radio unit was not on fire and then proved it by lighting off the other boot.

MIT

I entered graduate school at MIT in the fall of 1974 with a sense of urgency sharpened by my two-year absence from academic life. I was behind all my friends and I was very impatient with any activity not leading directly to fundamental discovery, i.e. taking classes. However I soon found that things were not that simple. Physics graduate schools in America are for the most part set up as a first priority to service federal contracts, not to make fundamental discoveries, and a graduate student career makes no sense outside the context of one of these contracts. Indeed it was, and is, the practice at MIT to admit graduate students directly into research groups on an as-needed basis as a kind of labor pool. It took me a while to fully understand this depressing fact of life, but I eventually did and then proceeded to look for a home in a research group as a means of supporting myself while learning the things essential to achieving my larger ambitions. I had by this time become quite cynical about and suspicious of institutions of all kinds, and I felt that government-sponsored science was no more likely to be immune from economic pressures than business. So I directed my attention toward the branch of physics with the largest number of experiments, namely solid state physics, figuring that this was the best way to cut out the intellectual middleman and go directly to nature. I have since discovered that most good theorists think this way.

It was my good fortune at this time to fall in with John Joannopoulos, a young faculty member who had just come from Marvin Cohen's group at Berkeley. I had heard John talking at a research fair and had noticed that he was the only theorist who seemed genuinely interested in his own work, so I contacted him and asked for a job. Neither of us knew it at the time, but John was to become one a truly great trainer of graduate students, for the list of alumni from his group includes Prof. E. J. Mele at the University of Pennsylvania, Prof. A. D. Stone at Yale, Prof. Karen Rabe at Yale, Prof. D. Vanderbilt of Rutgers, Prof. T. Arias at MIT, Prof. E. Kaxiras at Harvard, Prof. D. H. Lee at Berkeley, and me. His main expertise was in using local exchange methods (c.f. Walter Kohn's 1998 Nobel lecture) to model electronic materials, which in those days meant defective silicon, silicate glasses, and amorphous selenium. I figured at the time that this was a good way to learn the basics, and I knew that William Shockley had started out doing similar things for John Slater at MIT. So I worked for John for a long time and published several papers with him that were not that memorable but kept dinner on the table while I was coming up the learning curve on the vast subject of solid state physics. John's strategy was to give students simple problems they could market right away and then invest enormous amounts of personal time making sure the research was on track. The physics training I got from John emphasized bread-and-butter things such as the basics of semiconductors, tight binding modeling methods, and pseudopotentials. The truly invaluable things I learned from him, however, were not technical at all, but organizational: how to mount a research campaign and execute it successfully, how to render a big body of work down to its essence, how to package work so that it is interesting and comprehensible to an audience, how to look for new physical content in old results, and how to think experimentally. John took as his highest priority that all his students have a professional niche to live in after graduation, something I now understand to be of paramount importance, for the science will come later if the person has what it takes, but it will never come if the student has no job in the critical years right after graduate school.

One of the terrific aspects of MIT in those days was the enormous variety of experimental work that either took place there or was talked about in seminars by outside speakers aggressively recruited by the faculty. It was motivated by questions that did not interest me that much, such as whether the Kosterlitz-Thouless transition could actually be observed in the laboratory or what renormalization group principles told one about scattering lineshapes. The important thing for me was the experiment itself, how it worked, and whether it might be saying something that the experimentalist himself had overlooked. So I learned about X-ray diffraction, neutron scattering, raman scattering, infrared absorption spectroscopy, heat capacity, transport, time-dependent transport, magnetic resonance, electron diffraction, electron energy loss spectroscopy – all the experimental techniques that constitute the eyes and ears of modern solid state physics. As this occurred I slowly became disillusioned with the reductionist ideal of physics, for it was completely clear that the outcome of these experiments was almost always impossible to predict from first principles, yet was right and meaningful and certainly regulated by the same microscopic laws that work in atoms. Only many years later did I finally understand that this truth, which seems so natural to solid state physicists because they confront experiments so frequently, is actually quite alien to other branches of physics and is vigorously repudiated by many scientists on the grounds that things not amenable to reductionist thinking are not physics.

It was at MIT that I met and married my wife Anita. We used to swim at the same time after work at the MIT swimming pool and were annoyed by the same guy in a leopard suit who obviously thought he was beautiful and talented. So one day I said, "That guy may look tough, but he keeps his suit on in the shower." It was absolutely true, of course, for I could not have made up such a good story. This broke the ice. Anita corrected many of my worst habits, in particular the one of returning to my office after swimming and working until midnight. We would instead go up to Harvard Square for a late-night snack or attend a movie or a poetry reading, the usual staples of student life. Also, Anita's family lived nearby and was quite close-knit and warm, so we used to escape from Cambridge regularly to visit them. They had a wonderful old saltbox house out in Concord with a huge fireplace heating an equally huge kitchen with low wooden beams and an old plank floor. In winter the hearth was always lit and there was always something interesting simmering on the stove. Her father, who was then Dean of Graduate Studies at Lesley College, is a yankee with a wicked sense of humor who had grown up on a dairy farm in Massachusetts and then gone on to a life of scholarship at Yale and Harvard. Her mother had grown up as a doctor's daughter in Palo Alto and graduated from Stanford. Thanks in part to this latter fact, Anita and I were married by candlelight at Memorial Church at Stanford, an interesting turn of events considering what was to happen later. Anita's mother got a bit carried away with this wedding and went so far as to get us onto the New York Times society page. But her father kept things in perspective. One afternoon, totally unprovoked, he held out a wedding gift that looked suspiciously like a dentist's bowl and said "spit please".

BELL LABS

I must have been doing something right at MIT, for at the end of my graduate career there the faculty got together and recommended me for a position in the Theory Group at Bell Labs, the best placement a young theoretical physicist could possibly have gotten. I had wanted to go to Xerox Palo Alto, but a job did not materialize, and in light of what happened later it was probably just as well. Bell Labs had been a kind of holy place of solid state physics since the 1950's when it was built up by Shockley after the invention of the transistor. I had no idea at the time of the significance of this placement, but I did notice during my job talk that everybody understood what I was saying immediately – this had never happened before – and that the audience had an irresistible urge to interrupt, heckle, and argue about the subject matter loudly among themselves during the talk so as to lob hand grenades into it, just like back-benchers do in the House of Commons. Being a combative person I rather liked this and lobbed a few grenades of my own to maintain control of my seminar. I later came to understand that this heckling was a sign of respect from these people, that the ability to handle it was a test of a person's worth, and that polite silence from them was an extremely bad sign, amounting to Pauli's famous criticism that the speaker was "not even wrong."

It was at Bell Labs that I first made direct contact with real semiconductor experts and thus began to fully understand what amazing materials they were and what they could do. I knew a little about semiconductors already, having worked on the theory of silicon-oxide interfaces at MIT and also having intimate familiarity with Marc Kastner's experimental amorphous silicon work there. But a thorough grasp of this great subject was not possible to acquire at MIT, or any other university, because no faculty could ever be big enough. I learned about cyclotron resonance measurements of electron masses and the associated disorder broadening from Jim Allen, defect-pair recombination luminescence from Michael Sturge, deep levels from John Poate and Dave Lang, silicide Schottky barriers from Marty Lepselter through Jim Phillips, infrared spectroscopy of shallow donors and acceptors from Gordon Thomas, and transport in the 2-dimensional electron gas from Dan Tsui. The theorists at Bell had all done work in semiconductors at some time or another and were very helpful in the learning process, particularly through their constant give-and-take with the experimentalists. While I was there, for example, Gordon Thomas and Tom Rosenbaum verified the continuous nature of the metal-insulator transition in phosphorus-doped silicon predicted by Phil Anderson and his "gang of four". Don Hamann and Michael Schlüter were doing ab-initio density functional computations for semiconductor surfaces, interfaces, and defects. Patrick Lee was working hard on the field theory of weak localization. I was also familiar with the cutting-edge work in gallium arsenide heterostructures being done at the time through seminars and informal conversations with Mike Schlüter, who was good friends with Horst Störmer. The fact that the two German expatriates at Bell were in the thick of this subject was no accident, for semiconductor physics had been particularly emphasized at that time in the German research establishment, and most of the careful, scholarly work on the subject, particularly the 2-dimensional electron gas, was being done in Germany.

It is a great irony that the work leading to the Nobel Prize this year began in a time of terrible defeat for me personally, as I had just learned that I would not get a permanent job at Bell. John Joannopoulos had recommended that I work closely with Mark Cardillo, who was diffracting neutral helium atoms from semiconductor surfaces as a means of diagnosing their structures, presumably on the theory that my proper niche at Bell would be as a modeler. Unfortunately, Mark was such a good experimentalist and so good at understanding the meaning of his results before I had even seen them that there was little left for me to do but confirm his insight after the fact. Also, there was no profound conceptual issue at stake. By about one year into my appointment I could see the inevitable but was unable to do anything about it. I had actually made a breakthrough in my helium diffraction work – I had discovered empirically by studying atomic beam experiments that the potential felt by the incoming helium atom was a universal constant times the electron density of the target – and was writing it up when Jim Phillips pointed out that the same idea had just been published by somebody else in Physical Review Letters. So it didn't count. The fateful vote on my promotion to permanent status came shortly thereafter, and rumor had it that I had only one supporter. While I had been expecting the axe to fall for some weeks my blood froze when it actually did. Once again my ambitions had been thwarted due to circumstances beyond my control, only this time the damage was much greater and almost certainly unrecoverable. I went home and told Anita, and together we began making plans for what to do. She was not all that unhappy, actually, for she did not like the New York metropolitan area all that much and had had ambitions to live in New England or out West.

It was at this moment that I wrote my first important paper in theoretical physics. I was 32 years old, 5 years beyond the alleged age of senility for theorists. Dan Tsui had come into the tearoom one afternoon with a copy of Klaus von Klitzing's famous paper on the integral quantum Hall effect to see what the theorists thought about it. Everybody was interested, for localization in the 2-dimensional electron gas was a timely topic. The version of it unique to two dimensions called weak localization had been discovered at Bell by Doug Osheroff and Gerry Dolan shortly before my arrival, and there was a raging controversy over the sign of magnetoresistance of these systems in weak fields. Von Klitzing's experiment was in the strong-field limit, for which there was no theory. I remember Phil Anderson's making a mumble about how there was probably a "gauge argument", by which he meant something like the physics of the Josephson effect, and this stuck in my mind. I knew that the enormous accuracy of Klitzing's effect precluded any complicated explanation, I knew from my work with John Joannopoulos what the Hamiltonian appropriate to the problem was, and I knew localization had to be occurring. I also knew how the experiment worked, in particular that the gate voltage on a field-effect transistor fixes the density and not the chemical potential, so that a gap in the density of states as proposed by Ando could not be the right answer. Within a few days I had hit on the idea of replacing a calculation of the current with a derivative of the energy with respect to vector potential, and shortly thereafter I made this physical by imagining an experiment in which the sample was wrapped into a ring. Thus was born what later became known as the "gauge argument" for accurate quantization of the Hall conductance. The upshot of this theory was that localization caused the effect and that the Hall conductance was accurately quantized because it was a measurement of the charge of the object being localized, in this case the electron.

The response at Bell to these events is a fascinating case study in how even well-informed people find a truly new idea difficult to understand and accept. Anderson complains regularly about this problem, and he often cites Planck's complaints about it, so I am in good company. A week later I gave a journal club presentation about von Klitzing's discovery, and finished off with my explanation, which could be given to that audience in two minutes. I got some questions about the experiment, but none about my ideas that were on the mark at all. I remember being challenged over, that well-known fact that all states were localized in two dimensions, something that made no sense at all in light of the experiments I had just shown. I remember giving the right answer, namely that the experiments showed the current theory of localization to be wrong in strong magnetic fields, and that there had to be a band of extended states below the fermi level carrying the current. But they were not convinced, and it was not until Bert Halperin wrote a paper repeating these arguments and elaborating upon them that they were accepted at Bell, by which time I was long gone.

LIVERMORE

By the time I began looking for jobs my fame for this work had begun to spread and I eventually was offered a job at Purdue which I accepted. A few weeks later I un-accepted and took a job at the Livermore lab as a post-doc, an act for which I feel guilty to this day, as Al Overhauser and Sergio Rodriguez had gone out on a limb on my behalf. I had received a call from Andy McMahan months earlier asking if I might want to come out, and on a whim I went for an interview and gave a talk on my quantum Hall theory. By good fortune one person in that audience, Dick More, understood its significance and caused an offer to be generated, even though my interests and training did not match the Laboratory's needs at all. Like most of the physicists at Livermore, Dick was an expert in atomic and plasma physics, but he had been trained as a solid state physicist and had even held Ted Holstein's old position at the University of Pittsburgh. I first turned the Livermore offer down, knowing full well that it was not an academic job. But after Anita and I had discussed it at length, I decided that I felt completely betrayed by the academic establishment and saw no reason to trust it a second time, especially for so little money. She felt the same way, noted that everybody switches careers nowadays, and suggested that we move to California where the economy was strong and just go out on the open market if things went bad at Livermore. So it was decided. We flew back to Newark, and that very evening I put Anita on a plane to San Francisco to look for a place to live. I remember watching her plane take off through a cyclone fence at North Terminal and standing there for a long time afterward with tears in my eyes wondering if I had done the right thing.

Livermore was and is a real industrial laboratory, by which I mean that it considered its job to be maintenance and development of technology for inertial-confinement fusion and nuclear weapons and not the generation of public-domain scientific knowledge. I was hired into what was then known as H-Division, the group responsible for generating equation-of-state and opacity tables for use in design codes. My job description said I was to work on modeling matter at a temperature of about 10 eV and a density of about 1/10 that of ordinary solids, a particularly difficult regime relevant to the X-ray laser program. However, it was the practice in those days to induce people to work on such applied problems by providing them with resources to work on real science as well. This is how Hugh DeWitt's famous Monte Carlo work on the 3-dimensional one-component plasma came to be done, for example, or Ceperley and Alder's excellent numerical work on the phase diagram of metals at zero temperature. So I was encouraged to continue thinking about the quantum Hall effect on the side and even given permission to use my computer account for any calculations that I might need to do. Also, for the first six months I was with Livermore I worked in a trailer known as the "cooler" outside the fence waiting for my clearance to come through. So at least in the short term my decision had been a a good one.

It was while I was in the cooler that I received the preprint from Horst and

Dan about their discovery of fractional quantum Hall effect. I remember flipping through to the figure at the back, staring at it for 10 seconds, and realizing that they must have found a many-body condensate with excitations carrying charge e/3, for there was no plausible explanation for the existence of a plateau other than localization of such a carrier. The temperature in the experiment was not that low by modern standards, so the plateau was not that flat and the parallel conductance not that small, but I knew Dan was very careful about localization physics and would have said something if this conductance had not been converging rapidly to zero with decreasing temperature. Also the bare eye could see that the quantization was at least 1% accurate, and there was no reason it should have been even that good unless it was von Klitzing's effect. I quickly telephoned Horst to make sure I had understood correctly and to find out all the little experimental details that never get published in formal papers, including in particular any evidence that the localization was incomplete. There was none. I told Horst what I thought it meant, and he told me they had had a similar idea. Dan had apparently seen the chart recorder plot, measured the field strength of the integral plateau with his fingers, displaced this 3 times to the right to land under the new plateau, and said "quarks".

So the task remaining was to find a prototype for this condensate simple enough to be convincing. I was familiar at the time with the theory of fractionally charged domain walls in 1-dimensional chains, an idea attributed by solid state physicists to Su and Schrieffer but actually going back further to a particle physics paper by Jackiw and Rebbi. I had learned about them from my fellow graduate student Gene Mele, who had worked on them at Xerox Webster. Gene had thought about these "solitons" deeply from every possible angle, knew as a result that they had to be real, and convinced me of it over the course of several meetings. It was actually a hot idea in those days, and there had been much experimental activity attempting to detect these solitons in polyacetylene, usually with light scattering. However, despite claims of success, the matter was never resolved because sample imperfection always corrupted the data and allowed the experiments to be interpreted more than one way. So my first attempt to write down a prototype for the fractional quantum Hall ground state borrowed heavily from the literature of solitons, in particular the idea of discrete broken symmetry on which it was based. I wrote up a theory and sent it in to Physical Review Letters. It was rejected, thank God. The referee, who I later discovered to be Steve Kivelson, observed that the discrete broken symmetry I had written down was actually a continuous broken symmetry and that its pinning on impurities would cause the sample to insulate. This was correct, and furthermore I had known about the problem before I sent in the paper and had deceived myself into believing that it did not matter.

It was at this time that I wrote the paper for which I have been awarded the Nobel Prize. Realizing that most people would require more than experimental phenomenology to be convinced I went back to the beginning and began computing the properties of the interacting 2-dimensional electron gas by the exact-diagonalization method. For most many-body problems this would have been a foolish thing to do, but I knew from the experiments that the system had an energy gap and that this would protect the calculation and give it meaning even when the number of particles was small. So I solved the problem for one and two particles, then powered up the computers to do three, four, five, and six. Each time the system locked in at particular densities as the pressure on it was increased, and thus exhibited the behavior seen in experiment. There was no sign of any tendency to crystallize, which would have shown up as a near degeneracy in the eigenvalue spectrum. So I knew that the right answer was indeed a uniform fluid with an energy gap. Having seen the behavior with small numbers of particles I began trying to guess the functional form of the wavefunction in hopes of then extrapolating to the thermodynamic limit. One particular functional form, a product of pair factors, caught my attention because it had occurred naturally as a basis element in the numerical calculations and had a particularly large weight in the correct ground state at filling factor 1/3, sometimes as much as 99.9 %. But it was not exact. Also there did not exist any standard mathematical machinery for computing the properties of such a state in the thermodynamic limit. Feeling rather discouraged I went to the library to read up on many-body physics, hoping to find some reason that the state I had proposed would be exact. I was looking through Eugene Feenberg's book on helium and chanced to open up the chapter on Jastrow ground states and there, in front of my eyes, was the functional form I had guessed! It was not exact at all, but rather a well-known variational technique for approximating the ground state of strongly-interacting many-body systems. I eagerly read about the analogy between such wavefunctions and the statistical mechanics of classical fluids and then realized that the fluid analogous to my proposed ground state was the very one-component plasma I had been learning about from Forrest Rogers and Huge DeWitt in the "cooler", albeit in one lower dimension. So I went to them to get guidance on how to compute the properties of this plasma. Once I had mastered the hypernetted chain and semiclassical Monte Carlo techniques and understood their error bars the rest was straightforward. The ground state energy was computed and found to be variationally superior to all known crystals. The charge-1/3 excitations were constructed from this ground state with an adiabatic thought experiment very similar to that used in the integral quantum hall effect. Wavefunctions for these excitations were proposed and their computed variational energies found to match the experimental activation energy for the parallel conductance in the plateau regions. It all fit. So I wrote the new theory up and sent it to Physical Review Letters. It was published there a few months later.

These rather heady events coincided almost exactly with the purchase of our first house and the birth of my first son, Nathaniel. Anita had been near the end of her pregnancy when we closed the deal, and I remember her scooting along the floor painting the baseboards white, the only chore she could do comfortably. Nat arrived a few weeks later in the dark of the morning on one of the rainiest day, I can ever remember. We somehow made it to the hospital, where he was delivered by Caesarean with me in attendance. It is quite something witnessing surgery on one's spouse while she is awake. My mother drove up from Visalia when she got the news and was almost annihilated by weather-induced traffic accidents. It was a wild and beautiful day. Thus I became a father and a homeowner at the same time and experienced all the changes in perceptions and priorities that happen to a person at this time in life. Our house was quite small, and in particular had one bathroom that needed its floor replaced twice due to dry rot caused by the children's shenanigans in the bathtub. But in the back was a small creek, a little grove of redwoods, and three wonderful apple trees which kept Anita busy canning apple sauce in the fall. Nat and my second child, Todd, who was born two years later, used to play endlessly back there. The larger environment was also quite rural. On occasion I would take one of the kids in a backpack across the street and up into Briones park to explore the wild oak woods and the herds of horses that roamed freely on the ridges.

We lived in this house many years, but the day finally came when we had to sell out and move to Stanford. I remember the last day very clearly. The moving van had left, Anita was just driving away with the last load of stuff, and Todd and I were left to vacuum up the last bits of dust. The house was echoing, as houses do when they have no people in them. I pointed out to him that this was the sound of ghosts. Here was where my children were born. Here was where we had run madly around the apple trees and smashed the violets. Here was where I had planted the Monterey pine which was now shading the patio. Todd stared at me for a moment and then said with obvious annoyance, "Let's go, Daddy."

STANFORD

My career at Livermore was effectively derailed by the fractional quantum hall theory, for I became so famous on account of it that I had to travel constantly and could not begin learning the classified parts of the Laboratory's business. I am to this day rather poorly educated about nuclear weapons, and I know virtually nothing about laser fusion capsules. The Lab had plenty of money in those days, and the head of H-Division, Hal Graboske, always supported my requests for travel, including once a trip to Denmark, Finland, and the Soviet Union. But the handwriting was on the wall. In 1984 offers began pouring in from universities, and there was a particularly good one from Stanford. Stanford at that time had a terrific solid state physics faculty and was quite a bit smaller, and therefore more malleable, than my old alma mater Berkeley. Also Anita and I were worried about the anti-intellectual attitudes toward the University of California expressed by the state legislature, worries that proved well-founded when a few years later salaries were capped, teaching loads were increased, and the state budget was not passed on time, causing faculty to be paid in IOU's. I had originally turned down all these university offers on the theory that the research environment would be better inside the fence at Livermore where somebody else took care of salaries and research support. But Stan Wojcicki, Sandy Fetter, and Mac Beasley were particularly persistent, and I got sound advice from both Berni Alder and Anita not to let this train go by, so in the end I relented and accepted their offer. Little did I know that in a few years the Cold War would end, the Department of Energy's budget would be squeezed, and the "expendable" public-domain science activities paid for by fat budgets would be the first thing to go. This is a well-known effect in industrial laboratories, as the recent histories of Bell Labs, Xerox, and IBM Research have sadly reminded us. But at this time I accepted mainly because of Berni's advice.

When I moved to Stanford I began to pursue the line of research I have been following ever since, namely trying understand the larger implications of fractional quantum hall discovery. The historical significance of the effect is a matter of some debate, but my own view is that it sets a precedent for completely trivial equations of motion to generate particles carrying fractional quantum numbers and a concomitant set of gauge forces between them, both of these also being postulates of the Standard Model. Thus I think the trail leads ultimately to big questions about the universe and cosmology. Progress in this direction has been painfully slow because the experiments have not cooperated. Most of the leads for finding other effects in nature analogous to the fractional quantum Hall effect turned out to be false, including particularly high-temperature superconductors, which have been my main materials physics research interest here for the last several years. However, Phil Anderson's idea that the phenomenology of the cuprates might be related to the known behavior of 1-dimensional antiferromagnets, which has a beautiful formal relationship to the fractional quantum hall ground state discovered by Duncan Haldane and Sriram Shastry, is still very intriguing, especially since the particles carrying fractional quantum numbers in the latter system, which we call spinons, are relativistic.

My experience with high-temperature superconductors has been very different from the fractional quantum Hall effect. The problem has been difficult to formulate clearly, and progress has been slow. I do not know whether this is due to an historical paradigm shift or just intellectual incompetence on our part, but I certainly find myself yearning for the good old days when the entirety of a problem could be understood in 10 seconds. It has been my good fortune to have an excellent group of experimentalists at Stanford with whom to work, notably Aharon Kapitulnik, Ted Geballe, and Mac Beasley, known collectively as the KGB, and Z.-X. Shen. I predicted optical rotary activity in bulk cuprate superconductors, which was unfortunately disproved by Aharon, although the symmetry breaking I predicted was eventually found by Laura Greene at the University of Illinois. I also predicted the so-called large pseudogap in the cuprates that was eventually discovered by Prof. Shen in photoemission from samples in extreme underdoping. I have also done some mathematical physics work for which I am very proud, most notably the invention of the Kalmeyer-Laughlin spin liquid vacuum and "anyon" superconductivity, although these things broke with my tradition of going directly to nature for inspiration and are in this sense flawed.

My job at Stanford is rather different from the ones I had held previously in that my own ambitions must take a back seat to the well-being of the students with whom I work. But this actually is not very difficult. My own sons are almost college age now, and my rule is simply to do for my students exactly what I hope someone else will do for my sons when the time comes: I teach them to have faith in themselves and in their own compass, to listen to nature to find truth, to love knowledge for the sake of itself, and to strive for greatness. A few days after the Nobel Prize announcement I got the following wonderful e-mail from Andrew Tikofsky, one of my best graduate students, who is now on Wall Street:

> Hi Bob,
> Ian McDonald, Steve Strong, and I are getting together for a beer near Grand Central Station this coming Tuesday in honor of your prize. You are cordially invited to attend.

FRACTIONAL QUANTIZATION

Nobel Lecture, December 8, 1998

by

ROBERT B. LAUGHLIN

Department of Physics, Stanford University, California 94305-4060, USA

One of my favorite times in the academic year occurs in early spring when I give my class of extremely bright graduate students, who have mastered quantum mechanics but are otherwise unsuspecting and innocent, a take-home exam in which they are asked to deduce superfluidity from first principles. There is no doubt a special place in hell being reserved for me at this very moment for this mean trick, for the task is impossible. Superfluidity, like the fractional quantum Hall effect, is an emergent phenomenon – a low-energy collective effect of huge numbers of particles that cannot be deduced from the microscopic equations of motion in a rigorous way and that disappears completely when the system is taken apart[A)]. There are prototypes for superfluids, of course, and students who memorize them have taken the first step down the long road to understanding the phenomenon, but these are all approximate and in the end not deductive at all, but fits to experiment. The students feel betrayed and hurt by this experience because they have been trained to think in reductionist terms and thus to believe that everything not amenable to such thinking is unimportant. But nature is much more heartless than I am, and those students who stay in physics long enough to seriously confront the experimental record eventually come to understand that the reductionist idea is wrong a great deal of the time, and perhaps always. One common response in the early stages of learning is that superconductivity and the quantum Hall effect are not fundamental and therefore not worth taking seriously. When this happens I just open up the AIP Handbook and show the disbeliever that the accepted values of e and h are defined by these effects, and that ends that. The world is full of things for which one's understanding, i.e. one's ability to predict what will happen in an experiment, is degraded by taking the system apart, including most delightfully the standard model of elementary particles itself. I myself have come to suspect that all the important outstanding problems in physics are emergent in nature, including particularly quantum gravity.

One of the things an emergent phenomenon can do is create new particles. When a large number of atoms condense into a crystal, the phonon, the elementary quantum of sound, becomes a perfectly legitimate particle at low energy scales. It propagates freely, does not decay, carries momentum and energy related to wavelength and frequency in the usual way, interacts by simple rules that may be verified experimentally, mediates the attractive interaction responsible for conventional superconductivity, and so forth, and none of these things depends in detail on the underlying equations of motion. They are generic properties of the crystalline state. The phonon ceases to have meaning when the crystal is taken apart, of course, because sound makes no sense in an isolated atom. A somewhat more esoteric example, although a more apt one, is the Landau quasiparticle of a metal[B)]. This is an excited quantum state that behaves like an extra electron added to a cold Fermi sea, but which is actually a complex motion of all the electrons in the metal. It is not possible to deduce the existence of quasiparticles from first principles. They exist instead as a generic feature of the metallic state; they cease to exist if the state does. This problem is not limited to solids. Even the humble electron, the most elementary particle imaginable, carries a polarization of the Dirac sea with it as it travels from place to place and is thus itself a complex motion of all the electrons in the sea. In quantum physics there is no logical way to distinguish a real particle from an excited state of the system that behaves like one. We therefore use the same word for both.

Whenever one is confronted with unpredictable emergent phenomena – which is to say almost always – there is need for a sound definition of sameness when comparing two states of matter. The one most of us prefer is the existence of a reversible adiabatic map. One imagines slowly changing the underlying equations of motion, checking at each stage to make sure the ground state and low-lying excitations have evolved in a one-to-one way. There is actually no need to check this if the system contains only a small number of particles, for the mapping is then guaranteed to be one-to-one by virtue of the adiabatic principle. But if the system contains a thermodynamically large number of particles it can happen that a small change to the equations of motion results in a violent rearrangement of the ground state and low-lying excitations, and a corresponding breakdown of the one-to-one mapping. This is a quantum phase transition. We say that two states are the same phase of matter if they can be slowly transformed into each other without encountering a quantum phase transition, and different phases of matter if they cannot. By this definition metals, insulators, and superconductors are all different phases of matter, but two metals with slightly different electron-electron repulsion strengths are the same. This definition of sameness is one of the most powerful ideas in physics, for it relieves us of the need to compute the properties of complex systems from first principles to understand them. Instead we can just find a prototype that is easy to solve and then map our solution backward adiabatically.

SOLITONS

The idea that particles carrying parts of an elementary quantum number might occur as an emergent phenomenon is not new. Already in the late 1970s there was a serious theory literature on this subject stemming from a key paper of Jackiw and Rebbi[1], but more generally from the rapid advances in field theory made in association with the solution of the strong inter-

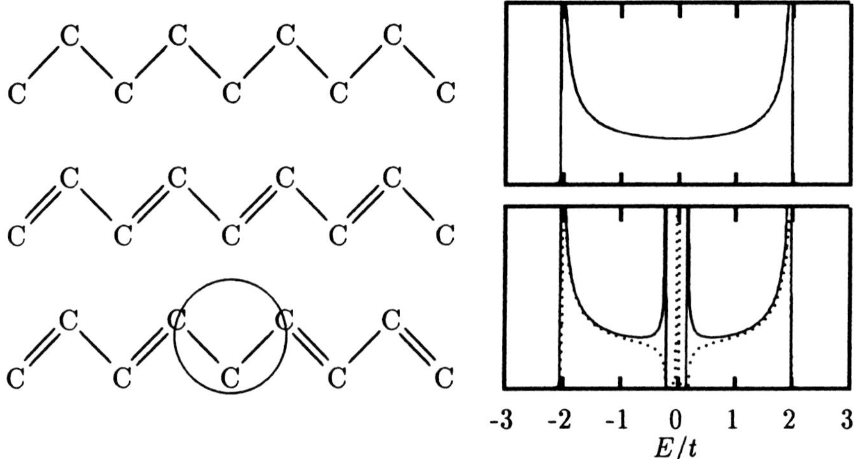

Figure 1. Physics of polyacetylene. The symmetric configuration of the atoms shown in the upper left is absolutely unstable to a distortion that contracts every other bond. This causes the 1-electron density of states, shown on the right, to acquire a gap. Formation of a domain wall results in the development of a state at mid gap (dotted curve on the right), and a net electric charge of $+e$ if this state is empty.

actions. The topological soliton or kink particle of Jackiw and Rebbi is conceptually similar to the 't Hooft-Polyakov monopole[2, 3] and the skyrmion[4], both of which were proposed as simplified models of real elementary particles. Solid state physicists first became aware of these ideas through a paper by Su, Schrieffer and Heeger[5] in which it was proposed that solitons might be the charge carriers of the conducting polymer polyacetylene. I first heard about this through my colleague Gene Mele, who was at the time working on the theory of polyacetylene in collaboration with Michael Rice at the Xerox Webster Research Center[6]. Gene was obsessed with solitons and had gone to such lengths to make realistic models of them that he could convince any reasonable person of their existence, and quickly convinced me. This is, of course, a theorist's statement. People who were not experts in quantum mechanics tended to find the idea outrageous, particularly since the experimental evidence for the existence of solitons was always indirect. It eventually came to pass that enthusiasm for the soliton idea among funding agents waned, the particle theorists diverted their energies to strings, and solitons were effectively forgotten. It is difficult nowadays to find anyone younger than I am who knows anything about them. I have always found the history of solitons to be a poignant comment on the effect of fashion on scientific thought, for there was no doubt among well-informed physicists that the idea was right and of potentially great importance.

The basic idea of the soliton is illustrated in Fig. 1. Polyacetylene is a zig-zag planar chain of CH units with every other bond contracted slightly. Because there are two equivalent ways of contracting the bonds, there is a possibility of domain walls between even-contracted regions and odd-contracted ones. These are the solitons. Realistic modeling reveals that these domain walls should be quite mobile, with an acceleration mass of about 10 times the electron mass, and that they should carry either a net electric charge of e and no net spin, or a spin of 1/2 and no charge. This breakup of the charge and spin quantum numbers of the electron is the important new effect, for an excitation carrying charge but no spin, or vice versa, cannot be adiabatically deformed into a free electron as a matter of principle. The excitations of a conventional insulator, on the other hand, are necessarily deformable into free electrons, and thus always have charge e and spin 1/2. The alleged properties of solitons were therefore quite unprecedented and extraordinary.

The property of polyacetylene that causes solitons to exist is its discrete broken symmetry. If one wraps a molecule with an even number of segments into a ring, one discovers that it has two equivalent quantum-mechanical ground states rather than one, and that these transform into each other under clockwise rotation by one segment, a symmetry of the underlying equations of motion. A conventional insulator would have only one ground state and would transform under this operation into itself. These two states acquire classical integrity in the thermodynamic limit. When the ring is small, local perturbations, such as a force applied to one atom only, can mix the two ground states in arbitrary ways, and in particular can tunnel the system from the even-doubled state to the odd-doubled one. But this tunneling becomes exponentially suppressed as the size of the ring grows, and eventually becomes insignificant. Classical integrity and two-fold degeneracy together make the broken-symmetry state fundamentally different from the conventional insulating state; one cannot be deformed into the other in the thermodynamic limit without encountering a quantum phase transition.

The peculiar quantum numbers of the soliton are caused by the formation of a mid-gap state in the electron spectrum. The N-electron model Hamiltonian[5]

$$\mathcal{H} = \sum_j^N \Big\{ t\Big(1 - \frac{x_{j+1} - x_j}{\ell}\Big) \sum_\sigma (c^\dagger_{j,\sigma} c_{j+1,\sigma} + c^\dagger_{j+1,\sigma} c_{j,\sigma}) - \frac{\hbar^2}{2M}\frac{\partial^2}{\partial x_j^2} + \frac{1}{2} k (x_{j+1} - x_j)^2 \Big\} \qquad (1)$$

is solved in the limit of large M by picking fixed displacements $x_{j+1} - x_j = \pm\, \delta l$ and then minimizing the expected energy per site;

$$\frac{<\mathcal{H}>}{N} = \frac{1}{2} k (\ell\delta)^2 - \frac{t}{2\pi}\int_0^{2\pi} \sqrt{(1+\delta^2) + (1-\delta^2)\cos(\theta)}\, d\theta \ . \qquad (2)$$

A nonzero value of δ is obtained for any value of k, as the chain is absolutely unstable to symmetry breaking by virtue of the Peierls effect. The gapped electron spectrum shown in Fig. 1 is the density of states for $\delta = 0.1$. When the equation is solved again for the soliton one finds an extra state in the center of the gap. The soliton has charge $+e$ when this state is unoccupied. The remaining charge and spin states just reflect the various ways the mid-gap state can be populated with electrons.

The fact that the soliton has charge $+e$ when unoccupied was first discovered numerically[7] and then subsequently explained by Schrieffer using his now-famous adiabatic winding argument[8]. This idea is critical to the theory of the fractional quantum hall effect, so I shall review it here briefly. One imagines adding a perturbation Hamiltonian that forces the order parameter to have a certain phase ϕ far from the origin. One then imagines locking the phase at the left end of the molecule to zero and then adiabatically advancing ϕ at the right end from 0 to π. While this is occurring the entire ground state at the right end is slowly sliding to the right, pumping electric charge out to infinity as it does so. The amount pumped by winding by π must be one electron, half the charge of the unit cell, since winding by twice this much just moves the unit cell over. But there is no net transfer of spin, as the system is a gapped singlet at every step in the operation. Since the operation also creates a soliton, we conclude that the soliton can have spin 0 and charge $+e$. It is deeply important that this argument is completely model-independent and relies only on the discrete broken symmetry of the bulk interior and the possibility of deforming the Hamiltonian into something simple at the ends of the sample without collapsing the gap.

There was an important variant of the soliton idea that foreshadowed the fractional quantum Hall discoveries, namely when the polyacetylene was imagined to be so severely p-type doped that the number of electrons in the π band is 2N/3 rather than N[8]. Then the instability is to contract every third bond, and there are two kinds of soliton – one winding forward by $2\pi/3$ and the other winding backward by the same amount – with charges $+2e/3$ and $+4e/3$ when all their mid-gap states are empty. So the separation of the spin and charge degrees of freedom of the electron in native polyacetylene is a special case of a more general effect in which the charge quantum number is fractionalized. The tendency of the Peierls instability to commensurate is actually so strong that fractionalization is expected to occur at other rational fillings as well, for example $3N/4$ and $3N/5$. Indeed the only reason it would not occur at all rational fillings is that the gap for most of these states is small and thus susceptible to being overwhelmed by the ion kinetic energy (which we have neglected), finite temperature, or dirt. These effects all favor fractions with small denominators.

The fact that these beautiful and reasonable ideas of Schrieffer and Su never became widely accepted may be traced in the end to one key difficulty: Long-range bond-contraction order was never found in polyacetylene. It is perfectly possible for solitons to exist if the discrete symmetry breaking has not set in globally – for example if it is interrupted every now and again by sample imperfections – and many experiments were done in highly defective and disordered polyacetylene with just this idea in mind. But the existence of solitons is *inescapable* only if the sample orders. This is a beautiful example of the special role ordering phenomena play in solid state physics, for while not always necessary they are often sufficient to demonstrate the truth of a thing. The problem with polymers in this context is that they owe most of their unique properties to noncrystallinity and are intentionally designed to tangle and disorder, i.e. not to do the one thing that would prove the existence of solitons. So in this sense the hope of demonstrating fractional quantization in real polyacetylene was doomed from the start.

LOCALIZATION

The 2-dimensional electron gas of a silicon field-effect transistor or a GaAs heterostructure, the venue of the integral and fractional quantum Hall effects, is notoriously imperfect. This is important because it immediately eliminates the possibility that microscopic details are responsible for Klaus von Klitzing's magnificent effect, the highly accurate quantization of the Hall conductance to integral multiples of e^2/h[9]. The field-effect transistor, for example, is made by oxidizing the surface of a piece of silicon, an operation that always results in microscopic strain and bond disorder at the surface because the silicon and SiO_2 lattice parameters do not match. This problem is so troublesome that it is customary to oxidize in the presence of a small partial pressure of water so that hydrogen is available to tying up the occasional dangling silicon bond. GaAs heterostructures are better in this regard, as the interface between the GaAs and the $Al_xGa_{1-x}As$ alloy is nominally epitaxial, but the Al atoms are still substituted at random in the GaAs lattice and are thus scattering centers. In either system there is the problem of the dopant ions, which are always strong scatterers because they are stripped of their carriers and thus Coulombic. The technique of modulation doping, invented by my co-winner Horst Störmer, mitigates this effect enormously but does not eliminate it. And of course there are chemical impurities gettered to the interface in unknown amounts. It is true that modern heterostructures have huge mobilities undreamed of in the old days, but they are not perfect.

Nor, as it turns out, would one want them to be. Imperfection is required for the quantum Hall effect to occur in the customary experimental configuration. Consider the sitution, illustrated in Fig. 2, of a translationally invariant strip of charge density ρ in a normal magnetic field B. Because of translational invariance, flowing current along the strip is the same as Lorentz boosting by speed v, which gives a current $j = v\rho$, an electric field $E = vB/c$, and a Hall conductance

$$\sigma_{xy} = \frac{J}{E} = \frac{\rho c}{B} \tag{3}$$

In a real field-effect transistor or heterostructure ρ is fixed by doping and the gate voltage and is *not* accurately quantized. Often it is the variable against which the Hall plateaus are plotted. Thus this result is inconsistent with all quantum Hall experiments. Sample imperfection is implicated in the formation of plateaus because it is the only agent in the problem, other than sample ends, capable of destroying translational invariance.

The most obvious thing for disorder to do is cause Anderson localization[10]. Localization is the underlying cause of the insulating state that results

Figure 2. When a disorder-free 2-d electron gas is placed in a magnetic field it can be Lorentz boosted to generate a current and a transverse electric field. The ratio of these then gives Eqn. (3), which is inconsistent with all quantum Hall experiments. This shows that disorder is essential for plateau formation.

when an idealized noninteracting metal is subjected to a sufficiently large random potential. It simply means that all the eigenstates of the 1-electron Hamiltonian below a certain energy have finite spatial extent, so that occupying them with electrons contributes nothing to the zero-frequency conductivity. Real metals, in which the electrons interact, have a similar metal-insulator transition, and it is believed that the two states of matter in question are adiabatically continuable to their noninteracting counterparts. Already at the time of the quantum Hall discovery there was a large literature on localization in 2-d metals, particularly those occurring in semiconductors, and it was known that localization effects were so strong in the absence of a magnetic field that only the insulating state should exist at zero temperature[11]-[14]. So there were many reasons to suspect the quantum Hall effect of being an emergent phenomenon that, like the Anderson insulator, could be understood solely in terms of 1-electron quantum mechanics and localization.

There are two important exact results that are particularly important for developing this idea. The first is the solution of the trivial model

$$\mathcal{H} = \sum_j^N \left\{ \frac{1}{2m} \left[\frac{\hbar}{i} \vec{\nabla}_j - \frac{e}{c} \vec{A}(\vec{r}_j) \right]^2 + E e y_j \right\} \quad , \tag{4}$$

where

$$\vec{A}(\vec{r}) = B y \hat{x} \quad . \tag{5}$$

With lengths measured in multiples of

$$\ell = \sqrt{\frac{\hbar}{m\omega_c}} = \sqrt{\frac{\hbar c}{eB}} \qquad (\omega_c = \frac{eB}{mc}) \quad , \tag{6}$$

one finds this to be a Slater determinant of the orbitals

$$\psi_{k,n}(x,y) = \frac{1}{\sqrt{2^n n! \sqrt{\pi} L_x}} e^{ikx} e^{(y+y_0-k)^2/2} (\frac{\partial}{\partial y})^n e^{-(y+y_0-k)^2} \quad , \tag{7}$$

where $y_0 = eEl/\hbar\omega_c$, the energies of which are

$$E_{k,n} = (n+1/2)\hbar\omega_c + \hbar ck(\frac{E}{B}) - \frac{mc^2}{2}(\frac{E}{B})^2 \quad . \tag{8}$$

If the electric field E is small, so there is a large gap between Landau levels n and $n + 1$, and the chemical potential is adjusted to lie in this gap, then the number of electrons in the sample is $N = n\, L_x L_y / 2\pi l^2$, the charge density is $\rho = ne/2\pi l^2$, and Eqn. (3) becomes $\sigma_{xy} = \rho c/B = ne^2/h$. As a check on this result we note that

$$\frac{1}{m} \int \int \psi^*_{k,n}(x,y)\left(\frac{\hbar}{i}\frac{\partial}{\partial x} - \frac{e}{c}A_x\right) \psi_{k,n}(x,y)\; dxdy = c\,\frac{E}{B} \quad . \tag{9}$$

Thus the current carried by each orbital is e times classical drift velocity cE/B. Adding these up and dividing by the sample area, we find that

$$J = \frac{ne^2}{\hbar} E \quad . \tag{10}$$

The second important result is the exact solution of the Hamiltonian

$$\mathcal{H}' = \mathcal{H} - V_0 \ell^2 \sum_j^N \delta^2(\vec{r}_j) \quad . \tag{11}$$

originally worked out by R. E. Prange[15]. The δ-function impurity potential is found to bind a single localized state down from each Landau level, and this state does not carry electric current, consistent with one's intuition. However, the remaining delocalized states turn out to be carrying too much, and the sum of the excess just exactly cancels the loss of a state! So the presence of the impurity has no effect on the Hall conductance. I have always found this result astonishing, for it is as though the remaining delocalized states understood that one of their comrades had been killed and were pulling harder to make up for its loss.

In the light of hindsight it is possible to prove that the Prange effect occurs for a broad class of impurity potentials. For this we must invoke some more general principles, for it is scarcely practical to diagonalize all possible impurity Hamiltonians and compute the current carried by their thermodynamically large number of orbitals one by one. Instead we notice that the current operator is formally the derivative of the Hamiltonian with respect to vector potential[16]. That is if we let

$$\vec{A} \rightarrow \vec{A} + A_0 \hat{x} \quad , \tag{12}$$

where A_0 is a constant, then

$$\frac{e}{m} \sum_j \left[\frac{\hbar}{i}\frac{\partial}{\partial x} - \frac{e}{c}A_x(\vec{r}_j)\right] = -c\,\frac{\partial \mathcal{H}}{\partial A_0} \tag{13}$$

This is actually true when the particles interact as well, a fact that will prove useful later. Normally this relation is of little help, for the addition of a constant vector potential is simply a gauge transformation, which has no physical meaning. But if the sample is wrapped into a loop, as shown in Fig. 3, then A_0 acquires the physical meaning of a magnetic flux $\phi = A_0 L_x$ forced through the

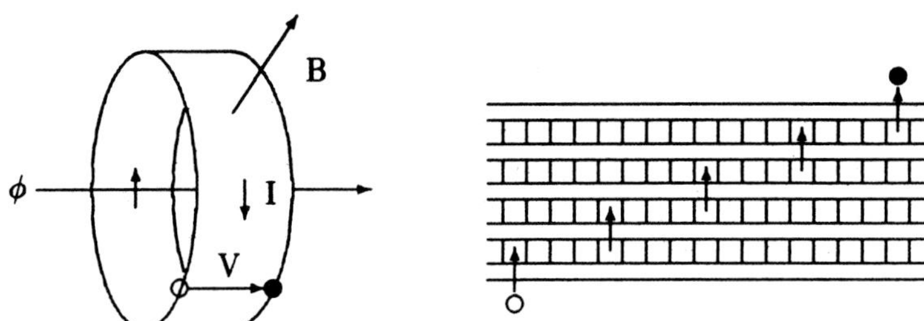

Figure 3. Illustration of thought experiment identifying quantum of Hall conductance with the electron charge. As magnetic flux $\Delta\phi = hc/e$ is adiabatically forced through the loop, one electron per Landau level is transferred from one edge to the other. In the absence of disorder the transfer is accomplished by a mechanical shift of the 1-body wavefunctions that evolves each wavefunction into its neighbor. When small amounts of disorder are present the number of electrons transferred must be exactly the same, but the mechanism of transfer is not, for the wavefunctions are violently distorted by even the smallest perturbation.

loop. Let us now imagine picking a sequence of flux values and solving the problem

$$\mathcal{H}_\phi|\Psi_\phi>= E_\phi|\Psi_\phi> \quad , \tag{14}$$

for each one, so that the many-body ground state $|\Psi_\phi>$ and corresponding energy eigenvalue E_ϕ are tabulated functions of ϕ. Then by Hellman-Feynman theorem we have

$$<\Psi_\phi|\frac{\partial \mathcal{H}_\phi}{\partial \phi}|\Psi_\phi>= \frac{\partial}{\partial \phi} <\Psi_\phi|\mathcal{H}_\phi|\Psi_\phi>= \frac{\partial E_\phi}{\partial \phi} \tag{15}$$

That is, the total current at any given value of of ϕ is just the adiabatic derivative of the total energy with respect to ϕ. This is actually not surprising, for slowly changing ϕ creates an electromotive force around the loop, which does work on the system if current is flowing. If there is no dissipation then the energy eigenvalue must increase accordingly. The effect is just Faraday's law of induction. Now if the loop is large, so that Aharonov-Bohm oscillations are suppressed and the current changes negligibly during the insertion process, the adiabatic derivative may be replaced by a differential

$$I = c\frac{\Delta E}{\Delta \phi} \quad , \tag{16}$$

where the denominator is the flux quantum hc/e. The advantage of this is that $\mathcal{H}_{\Delta\phi}$ is then exactly equal to $\mathcal{H}_0$ up to a gauge transformation. This means that the energy can have increased only through repopulation of the original states. How this comes about in the translationally-invariant case is shown in Fig. 3. The orbitals in the presence of nonzero ϕ are

$$\psi_{k,n}(x,y) = \frac{1}{\sqrt{2^n n! \sqrt{\pi} L_x}} e^{ikx} e^{(y+y_0-k-\alpha)^2/2} (\frac{\partial}{\partial y})^n e^{-(y+y_0-k-\alpha)^2} \quad , \tag{17}$$

where $\alpha = el\phi_0/\hbar cL_x$. As ϕ is advanced from 0 to $\Delta\phi$ they simply slide over like a shift register, the net result being to transfer one state per Landau level from the left side of the sample to the right, i.e.to transfer one electron across the sample per occupied Landau level. If the potential difference between the two sides is V
we thus have

$$I = c\frac{neV}{\Delta\phi} = n\frac{e^2}{h} \quad . \tag{18}$$

Now we can imagine turning on a small impurity potential in the interior of the ribbon. If the potential is sufficiently small that the gaps between Landau levels remain clean then there can be no change in the outcome of this thought experiment, for adiabatic evolution of ϕ stuffs exactly one state per Landau level into the disordered region on the left side and pulls exactly one out on the right. Conservation of states requires that they get through somehow. This applies not only to random potentials but non-random ones as well, including the δ-function used by Prange. Thus we have shown that any 1-electron Hamiltonian that can be adiabatically evolved into ideal Landau levels without having states cross the fermi level has an exactly quantized Hall conductance.

This previous argument can be significantly strengthened if the potential is random. In Fig. 4 we plot the 1-electron density of states of the Hamiltonian

$$\mathcal{H} = \sum_j^N \left\{ \frac{1}{2m}\left[\frac{\hbar}{i}\vec{\nabla}_j - \frac{e}{c}\vec{A}(\vec{r}_j)\right]^2 + V_{\rm random}(\vec{r}_j)\right\} \tag{19}$$

for various stengths of $V_{\rm random}$. The exact shape of this density of states is model-dependent, but the basic features are not. When $V_{\rm random}$ is small its primary effect is to break the degeneracy of the Landau level. The states in the tails of the distribution that results are localized. This is easiest to see in the limit that $V_{\rm random}$ is slowly-varying and smooth, for then the 1-body eigen-

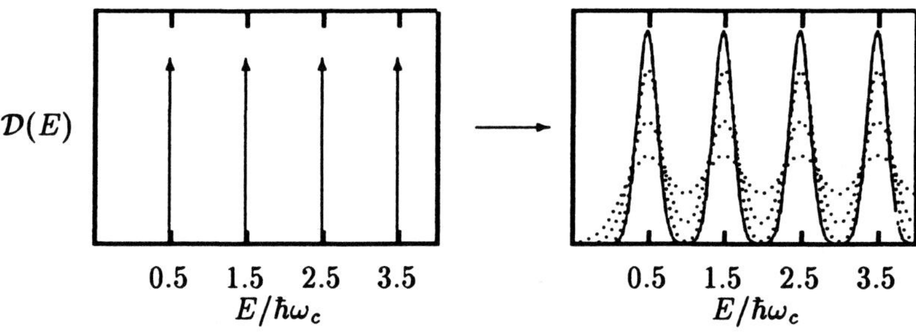

Figure 4. Effect of disorder on the 1-body density of states. Landau levels at $(n + 1/2)\hbar\omega_c$, which are highly degenerate, first broaden slightly without filling in the gap, thus maintaining the quantum Hall sum rule without the need to invoke localization. The states in the tails are localized, however, so the fermi level may be moved down into them with no ill effects. Additional disorder broadens the spectrum further until the tails grow together. The quantization is still exact, however, because all the states at the fermi surface are localized. This enables the quantum Hall effect to occur even when there are no true gaps.

states are well approximated as racetracks that travel along equipotentials. The analogy commonly drawn is with a mountainous landscape filled up to a certain height with water[17]. If the water level is low one gets small, isolated lakes, which become increasingly isolated as the water level drops because the deepest valleys occur rarely. If it is high one gets small, isolated islands, which become increasingly isolated as the water level rises because the tallest mountains occur rarely. Somewhere in the middle is a percolation point dominated at long length scales by a vast, crenylated shoreline which cannot distinguish islands from lakes. The significance of localization in the tails is that localized states are "not there" in the flux-winding thought experiment, in the sense that a state confined to one side of the loop cannot tell that the vector potential being added adiabatically is associated with flux ϕ through the loop. It thinks that we are doing a gauge transformation, so it simply changes its phase. It does not move in response to addition of ϕ, and its energy does not change. This means that it contributes nothing to the sum rule, and that whether it is occupied or not is irrelevant to the thought experiment. Thus the theorem of exact quantization is true not only if there are no states at the Fermi level, but also if all the states at the Fermi level are localized.

The ability of the exact quantization theorem to be extended to cases where the density of states does not have true gaps is crucial for accounting for real quantum Hall experiments. A field-effect transistor is basically a capacitor. It stores charge in response to the application of a gate voltage by the rule $Q = CV_g$, where C is determined by the oxide thickness, sample area, and little else. Accordingly, a sweep of the gate voltage is really a sweep of Q, not of chemical potential. The chemical potential in a real experiment simply adjusts itself to whatever is required to fix the charge at Q. Thus, if it were not for localization, the chemical potential would always be pinned in a Landau level, this Landau level would always be only partially occupied, and the conditions for observing the quantum Hall effect would never be achieved. But if the Fermi level lies in a region of localized states then the chemical potential can move about freely, populating or depopulating the localized states at will with no effect whatsoever on the Hall conductance. Another aspect of the experiments nicely accounted for by localization is lack of parallel resistance. In the limit that the electrons do not interact the parallel conductance σ_{xx} is due to electric dipole transitions from states just below the Fermi level to just above. If the states in question are localized, σ_{xx} must be zero. Localization causes insulation. However, the resistivity and conductivity tensors are inverses of each other, so that if $\sigma_{xy} = ne^2/h$ we also have

$$\begin{pmatrix} \rho_{xx} & \rho_{xy} \\ \rho_{yx} & \rho_{yy} \end{pmatrix} = \begin{pmatrix} \sigma_{xx} & \sigma_{xy} \\ \sigma_{yx} & \sigma_{yy} \end{pmatrix}^{-1} = \frac{h}{ne^2} \begin{pmatrix} 0 & -1 \\ 1 & 0 \end{pmatrix} \tag{20}$$

Note that experiments that tune the magnetic field strength instead of V_g are not fundamentally different, as they simply vary the length scale against which the charge density is measured. This still requires the chemical potential to adjust itself to keep Q fixed.

The flux-winding experiment also demonstrates that small amounts of disorder cannot completely localize all the states in a Landau level. There must be extended states, for if there were not, the whole Landau level would be "not there" and therefore incapable of carrying current. This is a very nontrivial result, for the arguments that all states should be localized in 2 dimensions in the absence of a magnetic field are sound. Just exactly where the extended states reside in the weak-disorder limit is a matter of some debate. Most experiments are consistent with the idea, originally proposed by Levine, Libby, and Pruisken[18], that there is a scaling theory of localization for this problem, although a different one from the field-free case, and that the extended states occur at one and only one energy somewhere near the center of the broadened Landau level. The strong-disorder limit is not controversial at all, however, for it is clear that complete localization must occur in this case. Another important implication of the flux-winding result therefore is that extended state bands cannot simply disappear but must "float" through the fermi surface as the disorder potential is increased, for only when extended states appear at the Fermi energy does the argument fail. The floating effect, which was predicted simultaneously by David Khmel'nitzkii and me[19, 20], was eventually observed experimentally[21].

The larger idea underlying these arguments is that the quantum Hall effect is an emergent phenomenon characterized by the ability of the matter in question to pump an integral number of electrons across the sample in a flux-winding experiment. The noninteracting models we have been discussing are nothing but prototypes; having understood them we identify the real experiments as their adiabatic continuations. This is not a very radical idea, for except for the presence of a magnetic field the ideal Landau level is not that different from a band insulator, a state of matter known to be continuable to its noninteracting prototype or, more precisely, defined by this property. The integrity of the low-lying excitations in either case is protected by the existence of the gap, which assures there are no other states into which a low-energy excitation can decay. This does not prove that other states are impossible – which is fortunate since the discovery of the fractional quantum Hall effect proves otherwise – but it does show that it makes sense, particularly since there are plenty of experimental instances of adiabatic continuability from an interacting system to a noninteracting one in solid state physics. If the state in question does map to a noninteracting prototype then a flux-winding experiment either dissipates or pumps an integral number of electrons across the sample. When the latter occurs, the Hall conductance is accurately quantized. The reason is that it measures the charge of the particle being pumped, in this case the charge of the electron.

FRACTIONAL QUANTUM HALL STATE

The fractional quantum Hall state is *not* adiabatically deformable to any noninteracting electron state. I am always astonished at how upset people get over this statement, for with a proper definition of a state of matter and a full understanding of the integral quantum Hall effect there is no other possible conclusion. The Hall conductance would necessarily be quantized to an integer because it is conserved by the adiabatic map and is an integer in the noninteracting limit by virtue of gauge invariance and the discreteness of the electron charge. So the fractional quantum Hall state is something unprecedented – a new state of matter.

Its phenomenology, however, is the same as that of the integral quantum Hall state in almost every detail[22]. There is a plateau. The Hall conductance in the plateau is accurately a pure number times e^2/h. The parallel resistance and conductance are both zero in the plateau. Finite-temperature deviations from exact quantization are activated or obey the Mott variable-range hopping law, depending on the temperature. The only qualitative difference between the two effects is the quantum of Hall conductance.

Given these facts the simplest and most obvious explanation, indeed the only conceivable one, is that the new state is adiabatically deformable into something physically similar to a filled Landau level except with fractionally-charged excitations. Adiabatic winding of a flux quantum – which returns the Hamiltonian back to itself exactly – must transfer an integral number of these objects across the sample. Localization of the objects must account for the existence of the plateau. All the arguments about deformability and exactness of the quantization must go through as before. As is commonly the case with new emergent phenomena, it is the experiments that tell us these things must be true, not theories. Theories can help us better understand the experiments, in particular by providing a tangible prototype vacuum, but the deeper reason to accept these conclusions is that the experiments give us no alternative.

My prototype ground state for the original 1/3 effect discovered by Tsui, Störmer, and Gossard is[23]

$$\Psi_m(z_1, ..., z_N) = \prod_{j<k>}^{N} (z_j - z_k)^m \exp\left[-\frac{1}{4\ell^2} \sum_j^N |z_j|^2\right] , \qquad (21)$$

where m is an odd integer, in this case 3, and $z_j = x_j + iy_j$ is the location of the jth particle expressed as a complex number. Horst likes to joke that his whole effect fits in one tiny equation, and I am deeply flattered every time he makes this joke in public, but the truth is that the equation is simple only because he and Dan had the good fortune to find the 1/3 state first. Most of the other 30-odd fractional quantum Hall vacua that have now been discovered do not have prototypes this simple, and I would be remiss in not pointing out that there are now reasonable alternatives[24]. These ground states are all adiabatically deformable into each other, however, and are in this sense fundamentally the same. My wavefunction was originally conceived as a variational ground state for the model Hamiltonian

$$\mathcal{H} = \sum_j^N \left\{ \frac{1}{2m}\left[\frac{\hbar}{i}\vec{\nabla}_j - \frac{e}{c}\vec{A}(\vec{r}_j)\right]^2 + V_{\text{ion}}(\vec{r}_j)\right\} + \sum_{j<k}^N v(\vec{r}_j - \vec{r}_k) , \qquad (22)$$

where

$$\vec{A}(\vec{r}) = \frac{B}{2}(x\hat{y} - y\hat{x}) \qquad V_{\text{ion}}(\vec{r}) = -\rho \int_{\text{sample}} v(\vec{r} - \vec{r}')\, d\vec{r}' \ . \quad (23)$$

The ion potential is present only because one of the electron-electron repulsions we wish to consider is the coulomb interaction $v(r) = e^2/r$, which must be neutralized by a background charge density ρ for the system to be stable. This wavefunction was subsequently shown by Duncan Haldane to be an exact ground state of a class of Hamiltonians with nonlocal potentials[25].

The most important feature of this wavefunction is that it locks the electron density at exactly $1/2\pi ml^2$ in the limit that N becomes thermodynamically large. We know this to be true because the square of the wavefunction is equivalent to the probability distribution function of a classical one-component plasma. Letting

$$|\Psi(z_1, ..., z_N)|^2 = e^{-\beta\Phi(z_1,...,z_N)} \quad , \qquad (24)$$

and choosing $\beta = 1/m$ to make the analogy transparent, we obtain

$$\Phi(z_1, ..., z_N) = -2m^2 \sum_{j<k}^{N} \ln|z_j - z_k| + \frac{m}{2\ell^2} \sum_{j}^{N} |z_j|^2 \qquad (25)$$

This is the potential energy of particles of "charge" m repelling each other logarithmically – the natural coulomb potential in 2 dimensions – and being attracted to the origin a uniform "charge" density $1/2\pi l^2$. In order to have local electrical neutrality, which is essential in a plasma, the particles must have density $\rho = 1/2\pi m\, l^2$.

It is also very important that the state is not crystalline when m is small. This is obvious for the case of $m = 1$, for then the wavefunction is just a full Landau level, but for other values of m one must appeal to the extensive literature on the classical one-component plasma[27]. Numerical studies have shown that crystallinity – or, more precisely, power-law correlations, for true crystallinity is impossible in 2 dimensions at finite temperature – occurs when the thermal coupling constant $\Gamma = 2m$ is about 140. Thus we are deeply in the liquid range at $m = 3$, and generally when m is a small odd integer. It was partially on the basis of this that I predicted the existence of a fractional quantum Hall state at 1/5, which was eventually found experimentally[28].

Another important feature of this state is the gap in its excitation spectrum, which it to say its preferred density. This gap is indicated by the experiments and also by the functional form of Ψ_m, which gives only densities $1/2\pi m\, l^2$, but was demonstrated rigorously to exist only when Haldane and Rezayi numerically diagonalized Eqn. (22) on a small sphere[29]. Exactly the same value for this gap – about 0.08 e^2/l for the coulombic case at $m = 3$ – was obtained by Girvin, MacDonald and Platzman[30] using hydrodynamic arguments. This is important because it identifies the lowest-energy excitation to be a quantum of compressional sound. In most quantum fluids the density operator ρ_q, appropriately projected, has a large amplitude to create a phonon. If one assumes this to be the only significant amplitude then one can exploit a sum rule to express the excitation energy E_q in terms of ground state properties solely[31]. Thus letting

$$\hat{\rho}_q = P\left[\sum_j^N e^{i\vec{q}\cdot\vec{r}_j}\right]P \quad , \tag{26}$$

where P is the projector onto the lowest Landau level, and denoting by $|\, x>$ an arbitrary excited state of the Hamiltonian of energy E_x above the ground state, we have

$$E_q = \frac{\sum_x E_x | <x|\hat{\rho}_q|\Psi_m> |^2}{\sum_x | <x|\hat{\rho}_q|\Psi_m> |^2} = \frac{1}{2}\frac{<\Psi_m|[\hat{\rho}_{-q},[\mathcal{H},\hat{\rho}_q]]|\Psi_m>}{<\Psi_m|\hat{\rho}_{-q}\hat{\rho}_q|\Psi_m>} \tag{27}$$

The rigidity of the classical plasma implicit in Ψ_m causes the denominator of this expression to vanish unusually rapidly as $\vec{q} \to 0$, the result being that E_q comes in at a finite constant. The dispersion relation has a shallow minimum at $q = 1.4/l$ for $m = 3$, a value close to the ordering wavevector $q = 1.56/l$ of the competing Wigner crystal. This minimum is aptly analogous to the roton of liquid ^{4}He. E_q is quite similar to the dispersion relation the conventional exciton in a filled Landau level[32].

The existence of an energy gap is sufficient to prove that the state has elementary excitations that carry a fraction of the electron charge. One imagines a thought experiment, illustrated in Fig. 5, in which the sample is poked with a thin magnetic solenoid through which magnetic flux ϕ is adiabatically inserted. What happens near the solenoid in this process is complicated and

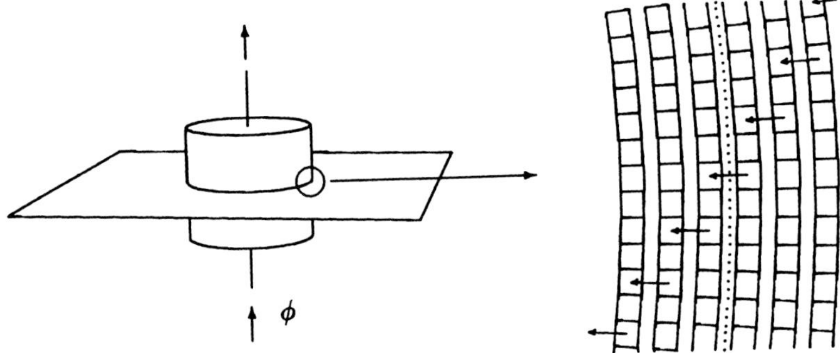

Figure 5. Illustration of thought experiment demonstrating the existence of fractionally-charged excitations. A fractional quantum Hall ground state is pierced with an infinitely thin magnetic solenoid, through which magnetic flux ρ is inserted adiabatically. The effect far away from the solenoid is to move states toward the solenoid. Advancing the phase all the way to $\Delta\rho = hc/e$ returns the Hamiltonian to its original state, up to an unimportant gauge transformation. This means that exactly one state must be transported through the surface of a large Gauss's law pillbox centered at the solenoid, and also that the solenoid may be removed at the end with no ill effects. The net result is the creation of an eigenstate of the original Hamiltonian with charge e/m, the average charge per state at infinity.

model-dependent, but far away from the solenoid the effect is simply to translate each Landau orbit inward as the phase is advanced, just as occurred in the integral quantum Hall loop experiment. Since the Hamiltonian is returned to its original state by the advance of ϕ from 0 to $\Delta\phi = hc/e$, up to an unimportant gauge transformation, the net effect of this advance must be to draw exactly one state per Landau level through the surface of an imaginary Gauss's law pillbox enclosing the solenoid. This draws in charge e/m, the average charge per state at infinity, and piles it up somewhere in the vicinity of the solenoid. The solenoid may then be removed, leaving behind an exact excited state of the original Hamiltonian carrying charge e/m, the charge *density* of the ground state.

A slight modification of this line of reasoning shows that the quantization of the fractional charge is exact. Let us imagine the situation illustrated in Fig. 6, in which the bare electron mass changes slowly across the sample, interpolating between a realistic value in region A to one in region C so small that Eqn. (21) becomes exact, all the while maintaining the integrity of the gap. We now perform the flux winding experiment as before, only this time we make the Gauss's law pillbox so big that it cuts region C. What happens in region A is complicated and impossible to predict accurately from first principles, but in region C exactly e/m of electric charge must be drawn in from infinity per the previous arguments. But there is no place for this charge to go other than the solenoid, at least if region A is large. Thus the operation must have made an excitation in region A with charge exactly e/m, regardless of microscopic details. We have actually proved something stronger in this example, for a second Gauss's law box placed inside region A shows that this same charge must be related in a fundamental way to the charge density in region A. Thus the latter is also quantized to the ideal value, even though Eqn. (21) is not exact in this region. These arguments are quite general and apply equally well to Hamiltonian parameters other than the bare mass we may wish to make non-ideal. The charge is conserved by any adiabatic modification of the Hamiltonian that maintains the gap, and is therefore characteristic of the entire phase of matter in question, not simply a particular prototype.

Wavefunctions that reasonably approximate these fractionally charged excitations, which I gave the unfortunate name "quasiparticles", are

$$\Psi^{+}_{z_0}(z_1, ..., z_N) = \exp\left[-\frac{1}{4\ell^2}\sum_j^N |z_j|^2\right] \prod_j^N (z_j - z_0) \prod_{j<k}^N (z_j - z_k)^m \tag{28}$$

for the positive excitation at location z_0 and

$$\Psi^{-}_{z_0}(z_1, ..., z_N) = \exp\left[-\frac{1}{4\ell^2}\sum_j^N |z_j|^2\right] \prod_j^N \left(2\ell^2 \frac{\partial}{\partial z_j} - z_0^*\right) \prod_{j<k}^N (z_j - z_k)^m \tag{29}$$

for the negative one. The excitation energies for these I originally estimated were 0.022 e^2/l and 0.025 e^2/l, respectively, for coulomb interactions at

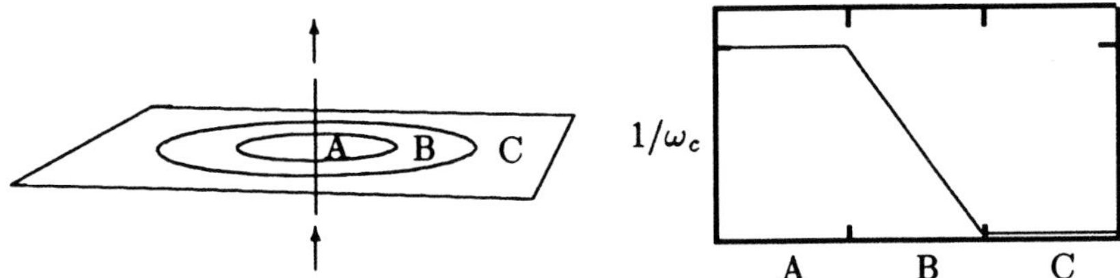

Figure 6. Illustration of thought experiment showing that the fractional charge is exact. One imagines a Hamiltonian parameter, such as the bare electron mass, that varies slowly in space so as to continuously deform a test system *A* into an ideal one at infinity *C*. A flux winding experiment performed in *A* then causes charge *e/m* to flow through a Gauss's law pillbox that cuts region *C*, regardless of the details, and this charge has no choice but to accumulate near the solenoid. In this way region *A* inherits exact properties from region *C* by virtue of integrity of the gap across region *B*.

$m = 3$[23]. Subsequent numerical work by Haldane and Rezayi[29] and Morf and Halperin[33] found the better numbers 0.026 e^2/l and 0.073 e^2/l. These improved energies sum to a value slightly higher than the roton gap, as expected if the collective mode is viewed as a quasiparticle-quasihole bound state, i.e. as an exciton. Charge transport gaps of this size are found experimentally[34] but are about half the theoretical value corrected for finite thickness of the 2-d electron gas. The remaining error is not of great concern because disorder is expected to severely lower the experimental gap. It is easy to see that the quasihole wavefunction describes a charge-1/3 excitation. Following the procedure of Eq. (24) we obtain

$$\Phi(z_1, ..., z_N) = -2\sum_j^N \ln|z_j - z_0| - 2m^2\sum_{j<k}^N \ln|z_j - z_k| + \frac{m}{2\ell^2}\sum_j^N |z_j|^2 \quad (30)$$

The plasma particles now see a phantom of "charge" 1 at location z_0 and arrange themselves on average to accumulate equal and opposite "charge" near z_0. It is somewhat more involved to demonstrate that the quasielectron wavefunction also describes a charge-1/3 excitation, but the reasoning is similar[35].

There are now a number of experimental papers which have reported the direct observation of charge e/3 quasiparticles. The most recent and widely cited are the shot noise measurements of Saminadayar et al[36]and de-Picciotto et al[37] which detect the charge through fluctuations in the current leaking across a narrow neck in a Hall bar. These very impressive experiments are more subtle than they appear at first glance because the relevant tunneling processes occur between sample edges, the excitation spectra of which are not gapped as they are in the bulk interior but rather the gapless spectra of chiral Luttinger liquids[38]. The carriers in this strange 1-dimensional metal carry a charge e/3 inherited from the bulk but are somewhat different physically from the quasiparticles in the interior and might even be construed as a different phenomenon. The shot noise expected from the tunneling of these excitations turns out to have the classical form with the electron charge reduced to $e/3$[39]. Somewhat more controversial, but in my view quite sound, are the resonant tunneling experiments of Goldman and Su[40], which were well-controlled versions of a an older experiment by Simmons et al[41]. These experiments also involve transport across a narrow neck of a Hall bar but measure zero-frequency transport as a function of carrier density and magnetic field instead of time-dependent current fluctuations[42]. I should also mention the famous experiments of Clark[43] which reported the observation of fractional charge in the high-temperature intercept of the activated parallel conductivity of the Hall plateau. These measurements were startling but controversial because no simple theoretical basis for the effect could be found. My own view, however, has continued to be that the best spectroscopy of this charge is the quantized Hall conductance itself, particularly in the limit that the sample is so dirty that arguments based on idealized edges make no sense, for the flux-winding sum rule measures the charge of the object transported across the sample, regardless of details.

FRACTIONAL STATISTICS

Fractional quantum Hall quasiparticles exert a long-range velocity-dependent force on each other – a gauge force – which is unique in the physics literature in having neither a progenitor in the underlying equations of motion or an associated continuous broken symmetry. It arises spontaneously along with the charge fractionalization, and is an essential part of the effect, in that the quantum states of the quasiparticles would not count up properly if it were absent. This force, which is called fractional statistics[44], has a measurable consequence, namely the values of the subsidiary fractions 2/5 and 2/7 and their daughters in the fractional quantum Hall hierarchy.

An isolated quasiparticle behaves physically like an ordinary electron or hole in a Landau level, except that its magnetic length is effectively $\sqrt{m}l$. This follows from its degeneracy in z_0, the functional form of the overlap matrix

$$\frac{<\Psi_{z_0}^{+}|\Psi_{z_0'}^{+}>}{\sqrt{<\Psi_{z_0}^{+}|\Psi_{z_0}^{+}><\Psi_{z_0'}^{+}|\Psi_{z_0'}^{+}>}} = \exp\left[-\frac{1}{4m\ell^2}(|z_0|^2 - 2z_0^* z_0' + |z_0'|^2)\right] \quad (31)$$

and the equivalence of the wavefunction at $m = 1$ to an ordinary hole in the orbital

$$\phi_{z_0}(z) = \exp\left[-\frac{1}{4\ell^2}(|z|^2 + |z_0|^2 + 2z_0^* z)\right] \quad (32)$$

This mapping is both one-to-one and physically apt.

A pair of quasiparticles, on the other hand, behaves like a pair of ordinary electrons or holes in a Landau level carrying magnetic solenoids containing a fraction of a flux quantum[45]. The 2-quasiparticle wavefunction

$$\Psi^{++}_{z_A z_B}(z_1, \ldots, z_N)$$

$$= \exp\left[-\frac{1}{4\ell^2}\sum_j^N |z_j|^2\right] \prod_j^N (z_j - z_A)(z_j - z_B) \prod_{j<k}^N (z_j - z_k)^m \tag{33}$$

is equivalent at $m = 1$ to a pair of ordinary holes in an otherwise full Landau level. It is obviously symmetric under interchange of the quantum numbers z_A and z_B and would thus appear to be the wavefunction of a pair of bosons. This is not correct, however, for it is uniquely the case in 2 dimensions that fermions have bose representations and vice versa. We must therefore use the more sophisticated concept of a Berry phase to determine whether the particles are physically fermions or bosons. Suppose the Hamiltonian is modified slightly so as to stabilize a pair of quasiparticles at locations z_A and z_B. This might be accomplished, for example, by adding shallow potential wells at these locations. If the Hamiltonian parameter z_A is then evolved around adiabatically in a closed loop P, the wavefunction returns to itself up to the phase

$$\phi = \oint_P \vec{A} \cdot d\vec{s} \qquad \vec{A}(z_A) = \lim_{z'_A \to z_A} \vec{\nabla}_A < \Psi^{++} z'_A z_B | \Psi^{++} z_A z_B > \tag{34}$$

This expression gives

$$\vec{A}(\vec{r}) = \frac{1}{2m}(\vec{r} \times \hat{z}) - \frac{1}{m}\frac{(\vec{r} - \vec{r}_B) \times \hat{z}}{|\vec{r} - \vec{r}_B|^2} \tag{35}$$

for $\vec{r}$ far from $\vec{r}_B$. The curl of this vector potential is just 2π times the local charge density, so the extra solenoidal component is simply a reflection of the missing charge associated with the presence of the quasihole at z_B. In the case of $m = 1$ the extra Berry phase incurred in moving A around B is just 2π. This means that the phase incurred in going around halfway, so as to exchange the particles, is π. The particles in this case are fermions. Had we picked a Fermi representation in which to do the calculation, for example by multiplying the wavefunction by $z_A - z_B$, this extra phase would have come out to be zero, but the end result would still have been that the wavefunction returned to minus itself when the particles were exchanged. When $m \neq 1$, however, neither the Fermi nor the bose representations get rid of this solenoidal component completely. In this case it is not an artifact of the choice of representation but a real velocity-dependent force.

The important experimental effect of the fractional statistics is to change the way quasiparticles pack. If quasiholes were fermions, for example, so that the analogy with a Landau level were exact, then occupying every available state would result in a uniform charge density of $e/2\pi m^2 l^2$. This is correct for $m = 1$ but for no other case, for the wavefunction

$$\Psi(z_1, \ldots, z_N) = \exp\left[-\frac{1}{4\ell^2}\sum_j^N |z_j|^2\right] \prod_j^N z_j^M \prod_{j<k}^N (z_j - z_k)^m \tag{36}$$

which describes M quasiholes packed as tightly together near the origin as possible, simply pushes the fluid back from a disc of area $2\pi Ml^2$, thereby creating a fluid of uniform charge density $e/2\pi ml^2$. Bosons, of course, would pack at any density they liked, so the actual behavior of quasiparticles is somewhere in between. It was on the basis of such observations that Halperin[46] first realized that the packing effect would account nicely for the observed subsidiary fractions 2/5 and 2/7 if the quasiparticles themselves were condensing into an analogue of $1/m$ state. The bose-representation wavefunction for condensing quasiparticles into the ± 1/3 fractional-statistics analogue of the 1/3 state is

$$\Psi_{\pm}(\eta_1, ..., \eta_N) = \prod_{j<k>}^{N} (\eta_j - \eta_k)^2 |\eta_j - \eta_k|^{\pm 1/3} \exp\left[-\frac{1}{4\ell^2} \sum_j^N |z_j|^2\right] \quad (37)$$

The corresponding charge densities are

$$\rho = \frac{e}{2\pi\ell^2}\left[\frac{1}{3} \mp \frac{1}{9(2 \pm 1/3)}\right] = \frac{1}{2\pi\ell^2}\left[\begin{array}{ll} 2/7 & \text{(quasiholes)} \\ 2/5 & \text{(quasielectrons)} \end{array}\right] \quad (38)$$

Repeating this argument hierarchically, Halperin was able to predict a sequence of fractional quantum Hall states which agreed with experiment and also with Haldane's more algebraic derivation of the sequence[25]. It was subsequently discovered by Jain[24] that the sequence of fractional quantum Hall ground states could be constructed by a method that did not employ quasiparticles at all, and thus the obvious conclusion that the occurrence of these **fractions *proves* the existence of fractional statistics was called into question.** However, it should not have been. The quasiparticles are quite far apart – about $3l$ – in the 2/7 and 2/5 states, and the gap to make them is large, so to assume that they simply vanish when these subsidiary condensates form makes no sense. Had the quasiparticles been fermions these densities would have been 10/27 = 0.370 rather than 2/5= 0.40 and 8/27 = 0.296 rather than 2/7 = 0.286. The effect of the fractional statistics is therefore small but measurable, about 5 % of the observed condensation fraction.

REMARKS

The fractional quantum Hall effect is fascinating for a long list of reasons, but it is important, in my view, primarily for one: It establishes experimentally that both particles carrying an exact fraction of the electron charge e and powerful gauge forces between these particles, two central postulates of the standard model of elementary particles, can arise spontaneously as emergent phenomena. Other important aspects of the standard model, such as free fermions, relativity, renormalizability, spontaneous symmetry breaking, and the Higgs mechanism, already have apt solid state analogues and in some cases were even modeled after them[C)], but fractional quantum numbers and gauge fields were thought to be fundamental, meaning that one had to postulate them. This is evidently not true. I have no idea whether the properties of the universe as we know it are fundamental or emergent, but I believe that the mere possibility of the latter should give the string theorists pause, for it would imply that more than one set of microscopic equations is consistent with experiment – so that we are blind to these equations until better experiments are designed – and also that the true nature of the microscopic equations is irrelevant to our world. So the challenge to conventional thinking about the universe posed by these small-science discoveries is actually troubling and very deep.

Fractional quantum Hall quasiparticles are the elementary excitations of a distinct state of matter that cannot be deformed into noninteracting electrons without crossing a phase boundary. That means they are different from electrons in the only sensible way we have of defining different, and in particular are not adiabatic images of electrons the way quasiparticle excitations of metals and band insulators are. Some composite fermion enthusiasts claim otherwise – that these particles are nothing more than screened electrons[24] – but this is incorrect. The alleged screening process always runs afoul of a phase boundary at some point, in the process doing some great violence to the ground state and low-lying excitations. I emphasize these things because there is a regrettable tendency in solid state physics to equate an understanding of nature with an ability to model, an attitude that sometimes leads to overlooking or misinterpreting the higher organizing principle actually responsible for an effect. In the case of the integral or fractional quantum Hall effects the essential thing is the accuracy of quantization. No amount of modeling done on any computer existing or contemplated will ever explain this accuracy by itself. Only a thermodynamic principle can do this. The idea that the quasiparticle is only a screened electron is unfortunately incompatible with the key principle at work in these experiments. If carefully analyzed it leads to the false conclusion that the Hall conductance is integrally quantized.

The work for which the three of us have been awarded the Nobel Prize was a collaborative effort of many excellent people in the most respected traditions of science. I join my colleagues in regretting that Art Gossard could not have shared in the Prize, as everyone in solid state physics understands that materials are the soul of our science and that no significant intellectual progress is ever possible without them. I gratefully acknowledge the numerical work of Duncan Haldane and Ed Rezayi[29], which was crucial in cementing the case that the energy gap existed and in calibrating the quasiparticle creation energies. I similarly acknowledge Bert Halperin's many outstanding contributions, including particularly his discovery that quasiparticles obey fractional statistics[46]. The list of fundamentally important contributions to the subject other than my own is so long that I cannot begin to do it justice. There are the numerous papers Steve Girvin and Alan MacDonald wrote, including particularly their obtaining, with Phil Platzman, the first accurate estimate of the energy gap[30]. There is Ad Pruisken's work on localization in a magnetic field and his proposition of the first appropriately modified scaling theory[18]. There is Xiao-Gang Wen's work on chiral edge excitations[38] and the follow-on work of Charles Kane and Matthew Fisher describing quasiparticle tunneling through mesoscopic necks[39]. There was the magnificent global phase diagram of the fractional quantum Hall effect proposed by Steve Kivelson, Dung-Hai Lee, and Shoucheng Zhang[47]. And of course there is the discovery of the strange Fermi surface at half-filling and its explanation in terms of composite fermions by Bert Halperin, Patrick Lee, and Nick Read[48] that is now defining the intellectual frontier in this field. I hope all my colleagues who have been involved with this subject over the years, both those I have mentioned and those I have not, will accept my gratitude and appreciation for all they have done, and my humble acknowledgement that the theory of the fractional quantum Hall effect, like all good science, is the work of many hands.

REFERENCES

A) P. W. Anderson Science **177,** 393 (1972).
B) D. Pines and P. Nozières, *The Theory of Fermi Liquids* (Benjamin, New York, 1966).
C) M. E. Peskin and D. V. Schroeder *Introduction to Quantum Field Theory* (Addison-Westey, Reading, MA; 1995).
[1] R. Jackiw and C. Rebbi, Phys. Rev. D **13**, 3398 (1976).
[2] G. 't Hooft, Nucl. Phys. B **79**, 276 (1974).
[3] A. M. Polyakov, Piz'ma Zh. Eksp. Teor. Fiz. 20, 430 (1974) [JETP Lett. **20**, 194 (1974)].
[4] T. H. R. Skyrme, Proc. Roy. Soc. A **262**, 233 (1961).
[5] W. P. Su, J. R. Schrieffer, and A. J. Heeger, Phys. Rev. Lett. **42**, 1698 (1979).
[6] M. J. Rice, A. R. Bishop, J. A. Krumhansl, and S. E. Trullinger, Phys. Rev. Lett. **36**, 432 (1976).
[7] J. A. Pople and J. H. Walmsley, Mol. Phys. **5**, 15 (1962).
[8] W. P. Su and J. R. Schrieffer, Phys. Rev. Lett. **46**, 738 (1981).
[9] K. von Klitzing, G. Dorda, and M. Pepper, Phys. Rev. Lett. **45**, 494 (1980).
[10] P. W. Anderson, Phys. Rev. **112**, 1900 (1958).
[11] E. Abrahams, P. W. Anderson, D. C. Licciardello, and T. V. Ramakrishnan, Phys. Rev. Lett. **42**, 673 (1979).
[12] G. J. Dolan and D. D. Osheroff, Phys. Rev. Lett. **43**, 721 (1979).
[13] D. J. Bishop, D. C. Tsui, and R. C. Dynes, Phys. Rev. Lett. **44**, 1153 (1980).
[14] G. Bergman, Phys. Rev. Lett. **48**, 1046 (1982).
[15] R. E. Prange, Phys. Rev. B **23**, 4802 (1981).
[16] R. B. Laughlin, Phys. Rev. B **23**, 5632 (1981).
[17] S. A. Trugman, Phys. Rev. B **27**, 7539 (1983).
[18] H. Levine, S. B. Libby, and A. M. M. Pruisken, Phys. Rev. Lett. **51**, 1915 (1983).
[19] R. B. Laughlin, Phys. Rev. Lett. **52**, 2034 (1984).
[20] D. E. Khmel'nitzkii, Phys. Lett. **106A**, 182 (1984).
[21] I. Glotzman et al, Phys. Rev. Lett. **74**, 594 (1995).
[22] D. C. Tsui, H. L. Störmer, and A. C. Gossard, Phys. Rev. Lett. **48**, 1559 (1982).
[23] R. B. Laughlin, Phys. Rev. Lett. **50**, 1395 (1983).
[24] J. K. Jain, Phys. Rev. Lett. **63**, 199 (1989).
[25] F. D. M. Haldane, Phys. Rev. Lett **51**, 605 (1983).
[26] R. B.Laughlin, Phys. Rev. B **27**, 3383 (1983).
[27] J. M Caillol, D. Levesque, J. J. Weis, and J. P. Hansen, J. Stat. Phys. **28**, 325 (1982).
[28] A. M. Chang *et al.*, Phys. Rev. Lett. **53**, 997 (1984).

[29] F. D. M. Haldane and E. H. Rezayi, Phys. Rev. Lett. **54**, 237 (1985).

[30] S. M. Girvin, A. H. MacDonald, and P. M. Platzman, Phys. Rev. Lett. **54**, 581 (1985).

[31] R. P. Feynman, *Statistical Mechanics* (Benjamin, Reading Mass, 1972).

[32] C. Kallin and B. I. Halperin, Phys. Rev. B. **30**, 5655 (1984).

[33] R. Morf and B. I. Halperin, Phys. Rev. B **33**, 1133 (1986).

[34] G. S. Boebinger *et al.*, Phys. Rev. Lett. **55**, 1606 (1985).

[35] R. B. Laughlin, in *The Quantum Hall Effect*, ed. by R. E. Prange and S. M. Girvin (Springer, Heidelberg, 1987), p. 233.

[36] L. Saminadayar *et al.*, Phys. Rev. Lett. **79**, 2526 (1997).

[37] R. de-Picciotto *et al.*, Nature **389**, 162 (1997).

[38] X.-G. Wen, Phys. Rev. Lett. **64**, 2206 (1990).

[39] C. L. Kane and M. P. A. Fisher, Phys. Rev. Lett. **72**, 724 (1994);

[40] V. J. Goldman and B. Su, Science **267**, 1010 (1995); V. J. Goldman, Surf. Sci. 361, **1** (1996).

[41] J. A. Simmons *et al.*, Phys. Rev. Lett. **63**, 1731 (1989).

[42] J. K. Jain, S. A. Kivelson, and D. J. Thouless, Phys. Rev. Lett. **71**, 3003 (1993).

[43] R. G. Clark *et al.*, Phys. Rev. Lett. **60**, 1747 (1988).

[44] J. M. Leinaas and J. Myrheim, Nuovo Cimento **37 B**, 1 (1977); F. Wilzcek, Phys. Rev. Lett. 48, 957 (1982).

[45] D. Arovas, F. Wilczek, and J. R. Schrieffer, Phys. Rev. Lett. **53**, 722 (1984).

[46] B. I. Halperin, Phys. Rev. Lett. **52**, 1583 (1984).

[47] S. Kivelson, D.-H. Lee, and S. Zhang, Phys. Rev. B **46**. 2223 (1992).

[48] B. I. Halperin, P. A. Lee, and N. Read, Phys. Rev. B **47**, 7312 (1993).

HORST L. STÖRMER

I was born on April 6, 1949 in a regional hospital in Frankfurt am Main in Germany. Having the umbilical cord wrapped twice tightly around my neck, my parents' fear for the mental health of their first-born son subsided only gradually.

My forefathers had been farmers, inn-keepers, blacksmiths, carpenters and shop keepers in the region. My mother, an elementary school teacher, and my father, having finished an apprenticeship, had been married during the previous year, shortly after a devastating war. Opening a store for interior decoration in my father's home town of Sprendlingen, they were trying to build an existence and start a family at the same time. Eighteen months later a brother, Heinz, was born without the umbilical complications.

Sprendlingen, today a part of Dreieich, just south of Frankfurt, was a town of some 15,000 inhabitants. I was raised in the circle of an extended family of four uncles and aunts, who, together with my parents, lived in two houses with barns and sheds and the store surrounding a large yard. It was an ideal playground for two boys growing up with their cousins – this group always extended by a horde of friends. Constructing huge sand castles with moats and bridges, cardboard tents from the shop's packing material, building elaborate knight's armour from scrap floor-covering and intricate race tracks for marbles from curtain rails remain fond memories of childhood.

I began kindergarten at age three and was soon after joined by my brother. The kindergarten's seemingly unlimited amount of toy building blocks must have fascinated me and I soon became somewhat of the establishment's chief architect. School, at six, was a happy time, complemented in the afternoons by playing soccer in our yard, roaming about the fields surrounding my home town, and building dozens of detailed cardboard model ships and airplanes from "Ausschneidebögen".

There was never a doubt in my parents' mind that their sons would receive the best possible education. Although none of my forefathers graduated from high school, my parents regarded highly the merits of a good education as a tool for social advancement. In their value system knowledge always ranked above wealth – although not rejecting a possible fortuitous marriage of both. To enter "Gymnasium", at ten, required the passing of a test. I was accepted and from then on commuted for eight years, five km each way, to the "Goethe Gymnasium" in the neighboring town of Neu Isenburg.

Gymnasium was hard. I was not a particularly good student. I loved mathematics and the sciences, but I barely scraped by in German and English and French. Receiving an "F" in either of these subjects always loomed over my head and kept me many a year at the brink of having to repeat a level. Luckily there was "Ausgleich", balancing a bad grade in one subject with a good grade in another. Mathematics and later physics got me through school without repeat performance. I also excelled in sports, particularly in track and field, where I won a school championship in the 50m dash. But sports could not be used for "Ausgleich".

One of my teachers stood out, Mr. Nick. He taught math and physics. A new teacher, basically straight out of college, young, open, articulate, fun, he represented what teachers could be like. His love and curiosity for the subjects he was teaching was contagious. As 15 or 16 year-olds, we read sections of Feynman's Lecture Notes in Physics in a voluntary afternoon course he offered.

Having mastered wooden building blocks and cardboard models, passed erector sets and toy trains, I had reached the level of "Elektro-Mann" and "Radio-Mann". Dozens of telephones and light boxes to communicate between the sheds at home were designed, constructed, improved, and mercilessly wrecked, possibly foreshadowing my later employment by a communications company. And then, of course, there was chemistry, a subject I did not appreciate in school, but it held the secrets for making explosives. I built a rocket that propelled a modified car of a toy train into the air. After several exhilarating launches, the rocket exploded in my hand and ripped off half my right thumb. I learned an important lesson: a rocket and a bomb differ only in the exhaust. Affecting me somewhat during adolescence, the missing thumb also relieved me from army duty. Today, it is only an unimportant, physical curiosity.

I always wanted to become a physicist. Supposedly, at age six, I had told just that to a technician, who was repairing a TV set in our home. Obviously, I had little clue as to what a physicist did. Nevertheless, the goal persisted all through high school, but suddenly got overthrown during the last year of "Gymnasium" when an art teacher discovered my talent for design. I passed my baccalaureate with average grades – quite good in the sciences but quite poor in the humanities – and started to study architecture at the Technical High School in Darmstadt, about 20 km south of my home town. Being too late at application time, I had to register for "Lehrfach für Bauwesen", a related subject, that consisted of similar freshmen courses as architecture. I turned out to be very good in making any technical drawing of a bird cage from any requested angle, but very poor in freehand drawing and decided that architecture was not for me. Instead I went on to pursue my true love – physics.

As with architecture in Darmstadt, I was too late for registration in physics at the Goethe University in Frankfurt and took up mathematics instead, transferring to physics the following year. The year was 1968. Student revolts swept the campuses from Berkeley to Berlin. Frankfurt was a major site for riots in the streets and in the lecture halls. For a young student, hardly familiar with university life, largely ignorant of the aim of the different protests, these were uncertain times. Legitimate educational reform requests became confused with larger political issues leading to absurd happenings around campus. Damage was done to the institution of the university and its teaching staff but, at the same time, 1968 marked the beginning of a gradual and rational reform.

Studying physics and mathematics was wonderful. It was a far cry from Gymnasium. I loved the rigor of mathematics. In physics we had fascinating beginners lectures by two descendents of the famous "Pohl School" of Göttingen, Prof. Martienssen and Prof. Queisser. I had joined a group of like-minded students that studied together and hung out in "Café Bauer" for relaxation. Life was good, until I took the "Vordiplom", the major exam in all courses at the end of the fourth semester.

All physics and math exams – some six to eight written or verbal tests – went very well. They went so well, that I thought I needn't study at all for the dreaded verbal chemistry test. With straight "A"s in physics and mathematics, what was the chemistry professor to do but let me pass? I was mistaken and flunked badly, requiring *all* tests to be taken again, six months, later. Thankfully, physics and math professors – some having had experiences of their own with chemistry tests – conspired and promised to maintain my grades in those subjects. It gave me six months, to study nothing but chemistry. I never felt more confident walking into an exam and succeeded getting an "A" in *chemistry*. I had been wary of the field of chemistry throughout high school and during much of my studies. Counting valences and bonds, memorizing dozens of exceptions to the rules and hundreds of arcane compounds never made much sense to me. I came to revise my attitude towards chemistry once I had grasped quantum mechanics and the origin of the chemical bond.

The thesis work for my Diploma – in Germany a required step towards the Ph.D. – was performed in Professor Werner Martienssen's Physical Institute under the supervision of a young assistant professor Eckhardt Hoenig. Professor Hoenig had just returned from the United States, where he had worked on highly-sensitive superconducting detectors, so-called SQUIDS. The aim was to use these new devices to study the magnetic properties of hemoglobin to derive the geometry of its bond with oxygen. It was a time of immense joy paired with intense learning of intricate low-temperature techniques. Hoenig was a wizard in inventing and building sophisticated instrumentation to attack physics questions. Gerd Binnig, who later shared the Nobel Prize for the invention of the Scanning Tunneling Microscope, was another student of a total of four working with Hoenig at this time in the same lab. It is probably coincidental, nevertheless, I believe our education in experimental physics down in this basement of the "Neubau" was second to none and strongly affected our experimental approaches throughout our careers. Hemoglobin did not bow to our instruments, at least over the course of a year, and I quickly performed some measurements on iron impurities in magnesium. I wrote an unimpressive diploma thesis on the magnetic anisotropy of their susceptibility and received the necessary license to start with a Ph.D. thesis.

At this time, my horizon unexpectedly widened. It had never occurred to me, nor to many of my town's youngsters, to go to university anywhere else but Frankfurt or Darmstadt. We went to the closest one and lived at home, where our families had been based for generations. However, in the fall of 1974, a former student from Frankfurt, Wolfgang Kottler, visited. He had since moved to Grenoble, France, where the Max-Planck-Institute for Solid State Research in Stuttgart was operating a high-magnetic field facility together with the French National Center for Scientific Research, CNRS. He was just finishing his Ph.D. thesis under Professor Hans-Joachim Queisser and was beating the bushes for his own replacement in Grenoble. Initially hesitant to make such a big step, moreover to a foreign country, the mastery of whose language I largely failed in school, I visited Grenoble and asked myself: Why not?

Going to Grenoble was the single most important step in my life. Leaving the familiar surroundings of home, diving into another culture, another language, meeting new people, making new friends was initially frightening, but eventually immensely educational and gratifying. Meeting my wife, Dominique Parchet, in Grenoble certainly added to the city's attractions.

Grenoble, at the edge of the Alps, not far from Switzerland was the French Science City. The magnet lab had been established only a few years back. Professor Klaus Dransfeld was the local director. There existed a frontier atmosphere with an exhilarating "can do" sentiment. It was an international place. Many famous scientists passed through and, due to the informality surrounding the lab, even the students were able to meet them on a very personal basis. This was quite different from other, more hierarchically structured research institutes. In a certain sense, students were kings at the magnet lab. They knew all the ins and outs of the magnets and the visiting collaborators were willing to share their scientific knowledge with them in return. It also was there, I first met Daniel Tsui from Bell Labs.

My thesis project was to work on the properties of electron hole droplets in high magnetic fields, a subject that had been proposed by Dieter Bimberg of the magnet lab. I was joined by Rolf Martin, who had just received his Ph.D. from the University of Stuttgart. Together we spent hundreds of immensely enjoyable and very productive research hours – daytime or nighttime – around the colossal magnets. Sharing a French "villa" with Ronald Ranvaud, where many distinguished visitors from abroad were often guests, life revolved totally around science. I finished my thesis in just over two years and received my Ph.D. from the University of Stuttgart, where my thesis advisor, Prof. Queisser, now a director at the Max-Planck-Institute in Stuttgart, held the position of an honorary professor. Instead of the usual dedication, my thesis had started with a cartoon. I learned only recently, that this had been a major cause of irritation and that removal of the cartoon as well as cutting my shoulder-length hair could barely be warded off.

All through my Ph.D. years, Prof. Queisser had urged me to finish my thesis swiftly and move on to the United States. He himself had been in the US, working at Bell Labs and later with Shockley, one of the inventors of the transistor. Bell Labs, the research arm of American Telephone and Telegraph (AT&T), was the "Mecca" of solid state research. Strongly encouraged and supported by my thesis advisor, I had visited Bell Labs and worked with John Hensel on electron hole droplets for several weeks during the spring of 1976. The visit was also intended to make contact with Raymond Dingle of Bell Labs. At the time, he was working on semiconductor quantum wells, an exciting new area of research made possible by the invention of molecular beam epitaxy (MBE) in the late '60s by Alfred Cho, also of Bell Labs. I had heard Dingle speaking on the topic at the 1975 March meeting of the German Physical Society and had decided that *this* was the subject I wanted to pursue. As it turned out Queisser knew Dingle personally and with partial financial support from the Max-Planck-Institute in Stuttgart I was accepted into a consultant position in Venky Narayanamurti's Department, working effectively as a postdoc with Ray Dingle. I moved to Bell Labs in June 1977.

Modulation-doping, the technique to generate ultra-high mobility two-dimensional electron systems, instrumental for practically all of my later research, was conceived about two weeks after my arrival at Bell Labs in a conversation with Ray Dingle. In his office, he had outlined their recent efforts to introduce free carriers into semiconductor superlattices and had sketched the positions of band edges, impurities and electrons on his white-board. It occurred to me that by placing impurities exclusively into the potential barriers, while keeping them out of the potential wells, the scattering of electrons by impurities should be reduced, thus increasing mobilities. It was a casual, almost trivial observation, which, however, turned out to have big impact.

Modifications to the MBE crystal growth instrumentation of Arthur Gossard and his assistant William Wiegmann to allow for such a selective doping were made over the course of a few months, and they demonstrated the anticipated gains in mobilities. Initially, mobilities improved by a mere factor two or three over conventionally doped superlattices, but they have since grown by another factor of ~1000. Loren Pfeiffer and Ken West, both from Bell Labs, have led this effort and have consistently provided the most exquisite samples for research. Much of our experimental success rests on our direct access to their "candy store".

Modulation-doping gained me a permanent position at Bell Labs in the fall of 1978, and I was soon joined by my long-time assistant, Kirk Baldwin. With such high-quality material available, many physics experiments – previously conducted on two-dimensional electron systems in silicon – became feasible in gallium arsenide. It also opened the door to many optical experiments on two-dimensional electron systems, largely performed by Aron Pinczuk and his colleagues at Bell Labs in Holmdel.

At the time, Dan Tsui of Bell Labs was already recognized as one of the world's leading experts on two-dimensional electron systems in silicon. He quickly recognized the potential of the new material for research and invited me to join him on his frequent trips to the MIT Francis Bitter High Magnetic Field Lab in Cambridge, Massachusetts. It was the beginning of a scientific collaboration and personal friendship, which has lasted now for almost 20 years.

The quantum Hall effect, having just been discovered in1980 by Klaus von Klitzing, was a major topic of our research. Another topic was the electron crystal, which was theoretically predicted to form in very low electron density samples in very high magnetic field. An exceptionally high quality, low electron density specimen had just been fabricated by Art Gossard and Willy Wiegmann. Dan Tsui had succeeded in contacting it electrically, and in October 1981 we took it to the Magnet Lab to look for signs of an electron crystal. What we discovered instead, during the evening of October 6, was the fractional quantum Hall effect.

Since this discovery, many outstanding graduate students (Gregory Boebinger, Robert Willett, Andrew Yeh, Wei Pan), postdocs (Albert Chang, Hong-Wen Jiang, Rui Du, Woowon Kang) and colleagues (James Eisenstein, Peter Berglund) joined us and made discoveries of their own in this fascinating research area. Other postdocs working with me (Edwin Batke, Rick Hall, Joe Spector, Ray Ashoori, and Amir Yacoby) have performed research in neighboring areas, but affected our thinking in lower-dimensional physics in general.

In 1983, I was promoted to head the department for Electronic and Optical Properties of Solids. Administration was a minor chore during those days, and I could continue to pursue my own research, practically full time. They were very exciting and intense research days during which the fractional quantum Hall effect and its implications were established in many laboratories around the world. Theoretical progress was rapid and exhilarating.

In 1991, I was promoted to director of the Physical Research Laboratory, heading some 100 researchers in eight departments in William Brinkman's Physics Research Division at Bell Labs. The time available for my own research dwindled, but I was compensated by becoming exposed to a wide range of exciting research topics. The initial satisfaction faded when the physical sciences at Bell Labs came under strong pressure from management to contract. These were difficult years, not just for me, but much more so for many of my friends and colleagues at Bell Labs. I was reminded of Gymnasium and the power of teachers. With the split-up of AT&T in 1996, the creation of Lucent Technologies, which subsumed Bell Labs, and a change of leadership, the physical sciences at Bell Labs are blossoming again today.

I always had thought of becoming a teacher one day. Being totally immersed in exciting research at Bell Labs, the idea had faded. It was resurfacing. I stepped down from my position in the Summer of 1997 and joined Columbia University in January of 1998 as a Professor of Physics and Applied Physics, while remaining Adjunct Physics Director at Bell Labs, part-time.

THE FRACTIONAL QUANTUM HALL EFFECT

Nobel Lecture, December 8, 1998

by

HORST L. STÖRMER

Department of Physics and Department of Applied Physics, Columbia University, New York, NY 10023, and Bell Labs, Lucent Technologies, Murray Hill, NJ 07974-0636, USA

INTRODUCTION

The fractional quantum Hall effect is a very counterintuitive physical phenomenon. It implies, that many electrons, acting in concert, can create new particles having a charge *smaller* than the charge of any individual electron. This is not the way things are supposed to be. A collection of objects may assemble to form a *bigger* object, or the parts may remain their size, but they don't create anything *smaller*. If the new particles were doubly-charged, it wouldn't be so paradoxical – electrons could "just stick together" and form pairs. But fractional charges are very bizarre, indeed. Not only are they smaller than the charge of any constituent electron, but they are exactly 1/3 or 1/5 or 1/7, etc. of an electronic charge, depending on the conditions under which they have been prepared. And yet we know with certainty, that none of these electrons has split up into pieces.

Fractional charge is the most puzzling of the observations, but there are others. Quantum numbers – usually integers or half-integers – turn out to be also fractional, such as 2/5, 4/9, and 11/7, or even 5/23. Moreover, bits of magnetic field can get attached to each electron, creating yet other objects. Such composite particles have properties very different from those of the electrons. They sometimes seem to be oblivious to huge magnetic fields and move in straight lines, although any bare electron would orbit on a very tight circle. Their mass is unrelated to the mass of the original electron but arises solely from interactions with their neighbors. More so, the attached magnetic field changes drastically the characteristics of the particles, from fermions to bosons and back to fermions, depending on the field strength. And finally, some of these composites are conjectured to coalesce and form pairs, vaguely similar to the formation of electron pairs in superconductivity. This would provide yet another astounding new state with weird properties.

All of these strange phenomena occur in two-dimensional electron systems at low temperatures exposed to a high magnetic field – only electrons and a magnetic field. The electrons reside within a solid, at the interface between two slightly different semiconductors. This is presently the smoothest plane we can fabricate to restrict the electrons' motion to two dimensions. Quantum mechanics does the rest.

Most of the experiments are very simple. Given a high magnetic field, typically from a commercial superconductive magnet, and given a temperature close to absolute zero, typically 1/100 to 1/10 of a degree Kelvin from a commercially available helium refrigerator, only a battery, a resistor, and a voltmeter are required. In reality one employs somewhat more sophisticated instrumentation to increase the data accumulation rate.

The samples are made from ultra-pure semiconductor materials. They are the essential ingredient for the experiments. Before diving into the mysterious caverns of two-dimensional many-particle physics one needs to get an appreciation for the sophisticated technologies that make the journey possible.

TWO-DIMENSIONAL ELECTRON SYSTEMS

In a three-dimensional world, the creation of a two-dimensional system usually requires a surface of an object or an interface between two substances and a force to keep things there. A game of billiards – on the surface of a table and held down by gravity – is a commonly cited model system. Electrons can be confined to the surface of liquid helium or to the surface of some insulator. They can be kept there by an electric field, which pushes them against a highly impenetrable barrier. The most successful method to create two-dimensional electron systems (2DES) is to confine them within a solid to the interface between a semiconductor and an insulator or to the interface between two different semiconductors. The first is the so-called silicon MOSFET (Metal Oxide Semiconductor Field Effect Transistor), in which the 2DES is confined to the interface between silicon and silicon oxide, see Fig. 1a.

In a silicon MOSFET electrons reside at the silicon side of the interface, pushed against the highly impenetrable, insulating silicon oxide glass by an electric field from a metal electrode atop the glass. The ability to vary the electron concentration in the silicon – and hence the electrical resistance – via the electrode (called gate) makes this structure an ideal transistor. Silicon MOSFETs are the workhorse of today's ~140B silicon industry – providing the central ingredient for everything from the PC to the digital watch.

In a MOSFET electrons can move along the plane of the interface but are bound to it in the perpendicular direction. In fact, due to quantum mechanics, they cannot move in this direction at all. The electric field from the electrode pushes the carriers so strongly against the glass and they become so strongly entrapped in this direction, that only a set of discrete states are quantum mechanically allowed in this dimension (see Fig. 1c). At low temperatures, much lower than the energetic spacing between these orbits, and at sufficiently low density, all electrons reside in the lowest of these states. Their behavior in this z-direction is rigidly confined. On the other hand, they are free to move in the x-y plane. The silicon MOSFET represents an almost ideal implementation of the concept of a 2DES and much of the physics of 2DES has relied on it.

As good and versatile as they are, such MOSFETs have their limitations.

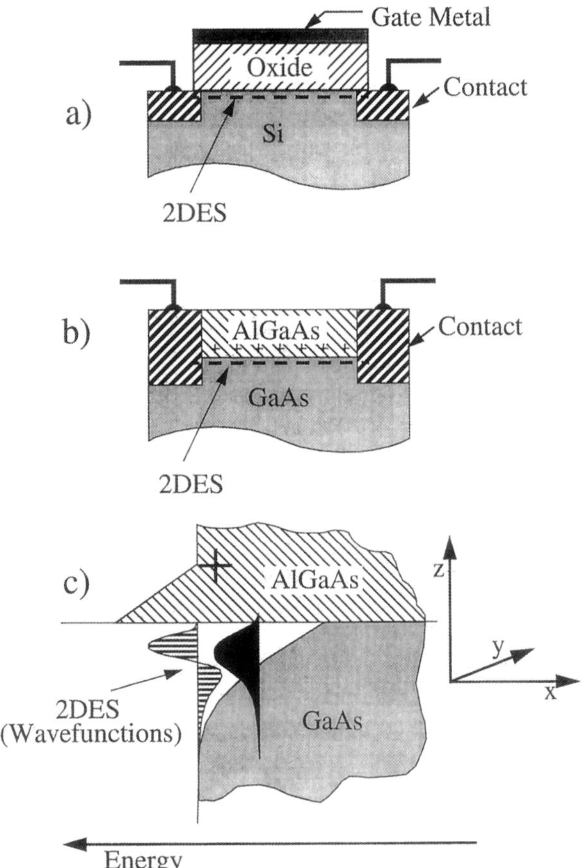

Figure 1 a). Schematic drawings of a silicon Metal Oxide Semiconductor Field Effect Transistor (MOSFET). The two-dimensional electron system (2DES) resides at the interface between silicon and silicon oxide. Electrons are held against the oxide by the electric field from the gate metal. b) Schematic drawings of a modulation-doped gallium arsenide/aluminum gallium arsenide (GaAs/AlGaAs) heterojunction. The 2DES resides at the interface between GaAs and AlGaAs. Electrons are held against the AlGaAs by the electric field from the charged silicon dopants (+) in the AlGaAs. c). Energetic condition in the modulation-doped structure (very similar to the condition in the MOSFET). Energy increases to the left. Electrons are trapped in the triangular-shaped quantum-well at the interface. They assume discrete energy states in the z-direction (black and horizontally striped). At low temperatures and low electron concentration only the lowest (black) electron state is occupied. The electrons are totally confined in the z-direction but can move freely in the x-y-plane.

Residing at the interface between a crystalline semiconductor and an insulating, random glass, electrons are often scattered by the roughness of the plane or by impurities that can penetrate the glass layer. Electron scattering is undesirable. It ejects electrons in a random fashion out of their trajectories, obscuring the observation of their "clean" behavior, governed solely by their mutual interactions and interactions with a magnetic field. Of course, electrons are also scattered by vibrations of the atoms, so-called phonons. Cooling the samples to temperatures near absolute zero reduces such vibrations to a level at which they become negligible as compared to scattering from any residual impurities.

Electrons, bound to the interface between two different *crystalline* semiconductors, should make for an even "better" 2DES than the one in a silicon MOSFET. Modulation-doped gallium-arsenide / aluminum-gallium-arsenide (GaAs / AlGaAs) hetero-structures have provided such a superior system for research and for some high-performance applications.

MODULATION-DOPING

Pure semiconductors do not conduct electricity at low temperatures. There are no free electrons that can move about the crystal. All of them have been consumed by the bonds that hold the solid together. To conduct electricity, semiconductors require the addition of a small number of impurities, known as doping. Doping entails somewhat of a physical "catch 22": without doping there are no free electrons, but doping introduces impurities, which strongly scatter the newly introduced free carriers. In a three-dimensional semiconductor this dilemma can practically not be circumnavigated. In two dimensions, however, there is a way. One can separate the mobile electrons from their parent impurities by confining them to different, neighboring planes. Such layers need to be in close proximity to each other for the impurities to transfer their electrons, but sufficiently far apart to prevent such electrons from scattering off the charged core of their bare parent impurities they leave behind. Molecular Beam Epitaxy (MBE) provides the tools for such an undertaking.

MBE is basically a high-vacuum evaporation technique, which allows one to evaporate high-quality, thin layers of semiconductors onto each other. Invented in the late 1960s by Al Cho at Bell Labs, it forms the basis of a large industry, manufacturing high-performance photonic and electronic devices, with an emphasis on communications. One standard materials combination used in MBE crystal growth is GaAs and GaAlAs. These are two semiconductors with practically identical atom-to-atom spacing (lattice constant) but they differ slightly in the energies of their free electrons (electron affinity). Electrons have a slight "preference" for GaAs over AlGaAs – about 300meV in a typical sandwich. An almost identical lattice constant guarantees a virtually defect-free, stress-free and hence high-quality interface. The difference in electron affinity allows to keep electrons at bay from their highly-scattering parent impurities.

In its most common implementation, the 2DES in an MBE-grown GaAs/AlGaAs sandwich resides at the GaAs side of a single interface with AlGaAs, see Fig.1b. A several mm thick GaAs layer is grown onto a 1/2 mm thick GaAs substrate. The substrate provides a template for the arriving atoms as well as mechanical support for the final structure. The GaAs layer is then covered by a ~0.5 μm thick layer of AlGaAs. During the high-quality, extremely clean, atomic-layer-by-atomic-layer growth process silicon impurities are introduced into the AlGaAs material at a distance of about 0.1 μm from the interface. Each silicon impurity has one more outer-shell electron than the gallium atom, which it replaces in the solid. It easily loses this additional electron, which wanders around the solid as a conduction electron. Seeking the energetically lowest state, the electron ventures over the energetic cliff and falls "down" into the GaAs material, only 0.1 μm away. In the highly pure GaAs layer such conduction electrons can move practically unimpeded by their parent silicon impurities, which remain in the AlGaAs layer, on the other side of the barrier. With modulation-doping you "can have your cake *and* eat it".

The attraction from all those positively charged (loss of one electron) stationary silicon ions pulls the mobile electrons against the AlGaAs barrier of the interface (see Fig. 1c). The conditions are completely analogous to the conditions in a Si MOSFET, in which the metal gate pulls the electrons against the silicon oxide barrier of its interface. The same quantization of the z-motion of the carriers arises and the carriers become quantum mechanically bound to the interface, but remain mobile within the x-y plane. The advantage that a modulation-doped GaAs/AlGaAs hetero-structure provides over a Si MOSFET originates from its atomically smooth interface between two crystalline semiconductors of very high purity. Transistors from such modulation-doped material (so-called HEMT transistors) represent today's lowest noise, highest frequency transistors and are extensively used in mobile telephony. Amazingly, much of the bizarre physics to be described below would occur in a transistor, not unlike those in many mobile phones, if cooled to low temperatures and placed in a high magnetic field.

Electron mobility is a common measure for the ease with which electrons move through a material. At low temperatures, where the scattering by phonons is negligible, mobilities in today's GaAs/AlGaAs hetero-structures exceed those in Si MOSFETs by almost a factor of 1000! Such modulation-doped specimens represent presently the best implementation of the concept of a two-dimensional metal, almost free of detrimental scattering from the host, see Fig. 2. This fact is best expressed as a mean free path of an electron before it scatters. It is ~1/5mm, meaning that a conduction electron passes by one million atoms of the semiconductor without scattering.

Modulation-doping was invented and implemented in 1977 by four researchers at Bell Labs. Fig. 3 shows a photograph taken around that time, in which they congregate around an early MBE machine. MBE technology has advanced immensely since these early days and MBE machines have grown in size and complexity. Fig. 4 shows a photograph of today's high mobility MBE system at Bell Labs and the researchers that employ it to fabricate the world's most exquisite modulation-doped specimens.

THE HALL EFFECT

The Hall effect was discovered in 1879 in a sheet of gold leaf by Edwin Hall, a graduate student at Johns Hopkins University in Baltimore. Running a current, I, through such a thin metal sheet, he measured two characteristic voltages, see Fig.5. The first, V, was the voltage *along* the current path, which, when divided by the current, represented the electrical resistance, R, of the

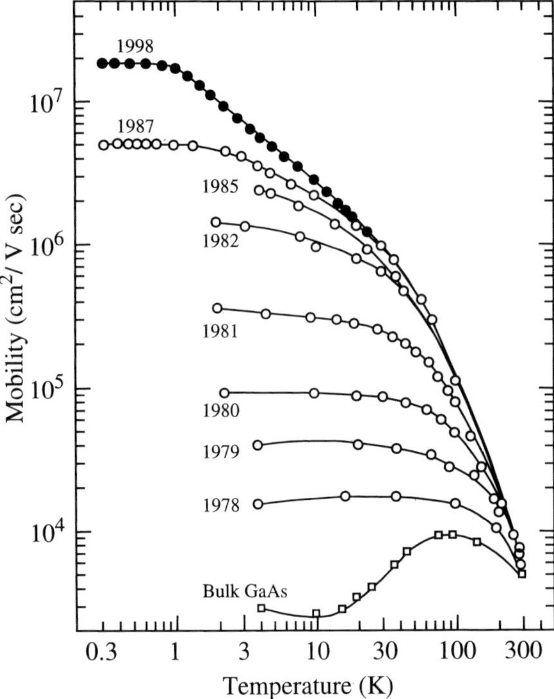

Figure 2. Progress made over the years in the mobility (μ) of electrons in two-dimensional electron systems in modulation-doped GaAs/AlGaAs as a function of temperature. At high temperature μ is limited by scattering with phonons of the solid. At the lowest temperatures μ is limited by impurities and defects in the material. "Bulk GaAs" represents a characteristic bulk sample. Since the inception of modulation-doping μ has risen by more than a factor of 1000. A mobility of 2×10^7 cm^2/Vsec corresponds roughly to 1/5 mm (!) ballistic flight of the electrons through the semiconductor before a collision takes place.

Figure 3. The inventors of the modulation-doping process congregating in 1978 around an early molecular beam machine at Bell Labs. From left: Willy Wiegmann, Art Gossard, Horst Störmer, and Ray Dingle.

Figure 4 .Today's Bell Labs ultra-high purity molecular beam epitaxy equipment with Loren Pfeiffer (center right) and Ken West (center left) who are synthesizing the worlds highest mobility material. They are joined by Kirk Baldwin (left) who is working with me since almost 20 years and Amir Yacoby (right), a postdoc, who worked on one-dimensional wires.

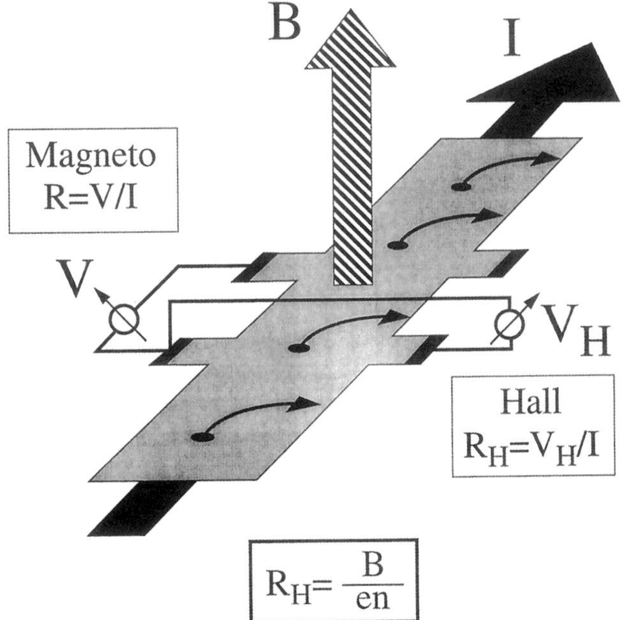

Figure 5. Geometry for measurement of the magneto resistance, R, and the Hall resistance, R_H, as a function of the current, I, and magnetic field, B. V represents the longitudinal voltage, which is dropping along the current path and V_H the Hall voltage, which is dropping perpendicular to the current path. The electron density per cm^2 is denoted as n and the charge of the electron as e. The black dots represent electrons that are forced towards one side of the bar following the Lorentz force from the magnetic field.

material. The second, V_H, was the voltage *across* the current path, which was expected to be zero since the current ran perpendicular to it. This was indeed Hall's observation until he applied a magnetic field, B, vertical to the metal sheet. It gave rise to a non-zero voltage, V_H, across the current path. From his different experiments, Hall deduced, that V_H was proportional to the current, I, and proportional to the magnetic field, B. Hence, denoting V_H/I as an electrical resistance, R_H, yielded $R_H \propto B$. Ever since this effect is known as the Hall effect. The associated voltage, V_H, is the Hall voltage, which, when divided by the current, I, becomes the Hall resistance, R_H.

The origin of the Hall effect is classical electrodynamics. The presence of the magnetic field exerts a sideward force (Lorentz force) onto the electrons which, on average, had been moving in the direction of the current. They are pushed toward one side of the specimen (depending on the direction of the magnetic field) giving rise to a charge accumulation on one side as compared to the other. This accumulation of charge ultimately results in the appearance of a voltage across the current path. Obviously, the higher the field the bigger the push, the bigger R_H. But also, the lower the density of electrons, the higher R_H. This sounds initially counterintuitive, but is rather simple, too. To generate the same current, less electrons need to travel *faster*. Faster electrons experience a stronger Lorenz force and create a bigger V_H and, hence, a bigger R_H.

In its final form $R_H=B/(ne)$, where n is the electron density per cm^2 (unit area) in the sample, which is equal to the electron density, N, per cm^3 (unit volume) times the thickness of the specimen and e is the elementary charge of an electron. Notice that no other electron parameter, such as its mass, nor any of the material parameters are entering – only the electron density. Most remarkably, R_H does not depend on the shape of the specimen. In fact, even a set of holes drilled into the specimen would not alter the result. A perforated metal sheet shows the same Hall resistance as a perfect sheet, as long as all electrical contacts remain mutually connected. Due to its independence from all intrinsic and extrinsic parameters, the Hall effect has become a standard tool for the determination of the density of free electrons in electrical conductors. In particular, the electron density of semiconductors, which can vary widely, depending on preparation, is measured via the Hall effect.

In 1879, Edwin Hall discovered, that in a normal conductor the resistance, R_H, depends linearly on the strength, B, of the magnetic field, see Fig.6. In 1980 Klaus von Klitzing discovered, that for the case of two-dimensional electron systems, the dependence is very different.

THE INTEGRAL QUANTUM HALL EFFECT

Perform a Hall experiment at the low temperature of liquid He (~4K) in a very high magnetic field (~10T) on the two-dimensional electron system of a Si-MOSFET and you will find a step-wise dependence of the Hall resistance on magnetic field, rather than Edwin Hall's linear relationship (see Fig.7). Yet more surprisingly, the value of R_H at the position of the plateaus of the

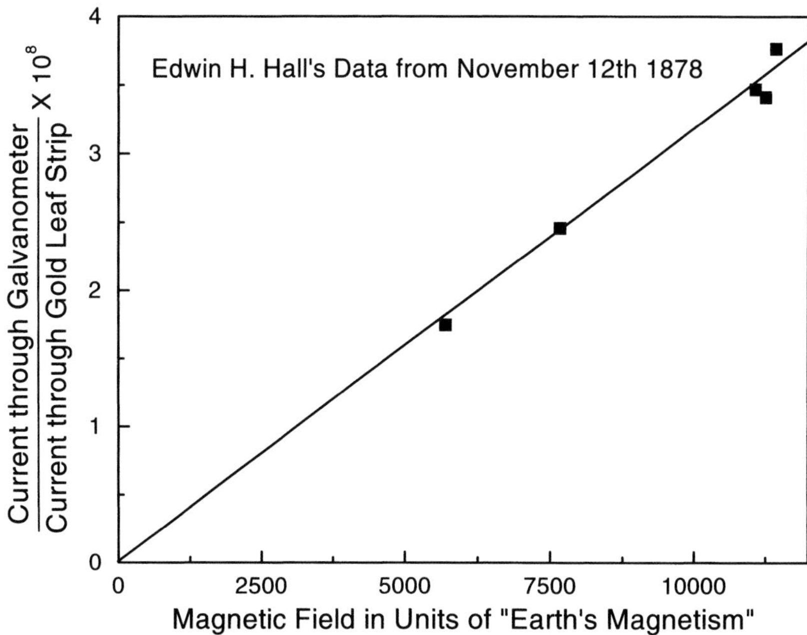

Figure 6. Edwin Hall's Hall data of 1878 as plotted from a table in his publication. The vertical axis is proportional to the Hall voltage, V_H of Fig. 5 and the horizontal axis is proportional to the magnetic field of Fig. 5. A linear relationship between V_H and B and hence between R_H and B is apparent. Since the days of Edwin Hall this strictly linear relationship has been confirmed by many, much more precise experiments.

steps is quantized to a few part per billion (!) to $R_H=h/(ie^2)$, where i is an integer and h is Planck's constant ($R_H \approx 25.812...k\Omega$ for i=1). In 1990, h/e^2, the quantum of resistance, as measured reproducibly to eight significant digits via this integral quantum Hall effect (IQHE), became the world's new resistance standard. Concomitant with the quantization of R_H the magneto resistance, R, drops to vanishingly small values. This is another hallmark of the IQHE and both are directly related.

Why are two-dimensional systems (2DES) so different? And what is the origin of the steps and minima? Classically, electrons in a high magnetic field are forced onto circular orbits, following the Lorentz force. Quantum mechanically, there exists only a discrete set of allowed orbits at a discrete set of energies. The situation is not unlike the discrete set of orbits that arise in an atom. Energetically, these so-called Landau levels represent an equally spaced ladder of states having energies, $E_i=(i-1/2)\ heB/(2\pi m)$, (i=1,2,3...), proportional to the magnetic field, B. m is the electron mass and h is Planck's constant. (Throughout the lecture we are neglecting the effects due to the electron spin. It simplifies the discussion without much loss of generality.) Electrons can only reside at these energies, but not in the large energy gaps in between. The existence of the gaps is crucial for the occurrence of the IQHE. Here 2DESs differ decisively from electrons in three dimensions. Motion in the third dimension, along the magnetic field, can add any amount of energy to the energy of the Landau levels. Therefore, in three dimensions,

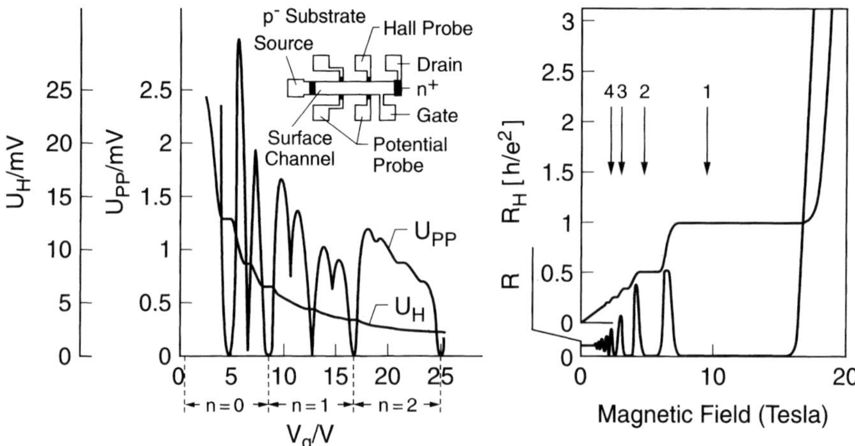

Figure 7. Left panel: original data of the discovery of the integral quantum Hall effect (IQHE) by Klaus von Klitzing in 1980 in the two-dimensional electron system of a silicon MOSFET transistor. Instead of a smooth curve he observed plateaus in the Hall voltage (U_H) and found concomitant deep minima in the magneto resistance (U_{PP}). The horizontal axis represents gate voltage (V_G) which varies the carrier density, n. The right panel shows equivalent data taken on a two-dimensional electron system in GaAs/AlGaAs. Since these data are plotted versus magnetic field they can directly be compared to Edwin Hall's data of Fig. 6. Rather than the linear dependence of the Hall resistance on magnetic field of Fig. 6, these data show wide plateaus in R_H and in addition deep minima in R.

the energy gaps are filled up and, hence, eliminated, preventing the quantum Hall effect from occurring. In 2DES, in addition to the existence of energy gaps, the number of electrons fitting into each Landau level is exactly quantized. It reflects the number, d, of orbits that can be packed per Landau level into each cm^2 of the specimen. It turns out to be d=eB/h. Notice that this capacity per Landau level, also called its degeneracy, apart from natural constants, depends only on the magnetic field, B. None of the materials parameters enters in any way. It is therefore a universal measure, independent of the material employed.

Let the sample have a fixed 2D electron density n. At low temperatures, where all electrons try to fall into the energetically lowest available states, and in a sufficiently high magnetic field, all electrons fit into the lowest Landau level, filling it only partially. As the field is lowered, the capacity of the Landau levels shrink according to d=eB/h. At B_1=nh/e the lowest Landau level is exactly full. Any further reduction of the field requires the first electron to leave the lowest Landau level and jump across the energy gap to the next higher Landau level at an energy cost of $heB_1/(2\pi m)$. Reducing the field to $B_2=(nh/e)/2=B_1/2$ fills two Landau levels and the first electron has to move to the third level, etc. This creates a sequence of fields, $B_i=(nh/e)/i$, at which all electrons fill up an exact number of Landau levels, keeping all higher Landau levels exactly empty. At these special points on the magnetic field axis, the magneto resistance, R, drops momentarily and the Hall resistance, R_H, assumes a set of very special values. Using $R_H=B/(ne)$ from the classical Hall resistance and inserting the values of the sequence of distinctive fields,

B_i, into the equation results in a quantized Hall resistance of R_H=h/(ie^2), i=1,2,3....While this is the desired result, it does not account for the true hallmark of the IQHE, which are wide plateaus in R_H and broad minima in R.

According to the above derivation, R_H would take on its quantized value only at very precise positions, B_i, of magnetic field. This would be a poor basis for a standard, since the precision to which R_H assumes one of the quantized values would depend on the precision to which one could determine B. In reality, in the IQHE the Hall resistance, R_H, assumes the quantized values over extended regions of B around B_i.

The origin for plateau formation and broad minima lies in electron localization. In spite of the extreme care with which the 2DES is prepared, there remain some energetic valleys and hillocks along the interface, be they due to residual defects, steps or impurities. Each Landau level is a reproduction of this uneven landscape. As a Landau level is being filled with electrons, some of the electrons get trapped (localized) and isolated. They no longer participate in the electrical conduction through the specimen and these patches of localized electrons become inert and act like a set of holes, cut out from the 2D sheet. As in a perforated metal sheet, such isolated patches do not affect the measurements of the density of mobile carriers in the flat part of the landscape, which are circumnavigating the hills and valleys. As long as filling and emptying of a Landau level fills or empties only the localized states at the energetic fringes, while keeping the Landau level in the extended flat regions full to capacity, the sample's Hall resistance, R_H, and magneto resistance, R, remain steady. Since, in the conducting regions, the Landau level is full, the Hall resistance remains fixed to its quantized value. Localized electrons provide a reservoir of carriers that keep the Landau levels in the energetically flat part of the sample exactly filled for finite stretches of magnetic field, giving rise to finite stretches of quantized Hall resistance and vanishing resistance in the IQHE.

The precision of quantization does not depend on the shape and size of the specimen, nor on the particular care taken to define its contact regions. (Fig. 8 shows a particularly egregious example.) In a quirk of nature, the existence and precision of the IQHE plateaus *requires* the existence of imperfections in the sample. Without such *dirt* there were no IQHE. Instead, even in a 2DES, one would revert to Edwin Hall's straight line.

In an ingenious thought experiment, Bob Laughlin was able to deduce the existence and precision of the IQHE from a set of very simple experimental ingredients (see his contribution to this volume). In his approach, the value of R_H=h/(ie^2)=(h/e)/(ie) emerges as a ratio of the magnetic flux quantum, ϕ_o=h/e and the electronic charge, e, together with the number of occupied Landau levels, i. Magnetic flux quanta are the elementary units in which a magnetic field interacts with a system of electrons. (The magnetic field itself is not quantized. This is different from charge, which usually comes in chunks of e. However, for the purposes of this lecture, which deals with magnetic fields in the presence of electrons, one may think of it as being quantized.) Being the ratio of ϕ_o to e, one can regard R_H as being a very precise measure

Figure 8. Photograph of a GaAs/AlGaAs sample. The size is about 6 x 1.5 *mm*. Black area (in reality mirror-like but reflecting the black camera) is the original surface above the 2DES. Gray areas have been scratched away to confine the current path to the center of the sample. White areas are indium blotches used to make contact to the 2DES. Gold wires are attached. Specimens like this one, prepared with little attention to exact dimension nor to tidiness, show quantization of the Hall resistance to an accuracy of a few 10 parts in a billion. The specimen shown is the sample in which the fractional quantum Hall effect (FQHE) was discovered in 1981.

of the electron charge when expressed as $e=\phi_o/(iR_H)$. From this purview, Klaus von Klitzing's experiment has provided a highly accurate electrometer to determine the charge of the current carrying particle in a 2DES.

THE FRACTIONAL QUANTUM HALL EFFECT

Discovery

In the beginning of October, 1981, Dan Tsui and I, both working at Bell Labs, had taken a specimen of a new sample made from modulation-doped GaAs/AlGaAs material to the Francis Bitter Magnet Lab at MIT, in Cambridge. The sample had been grown by Art Gossard, also of Bell Labs, and his assistant Willy Wiegmann. Having gained increasing experience with modulation-doping over the course of a couple of years, they had, for the first time, been able to fabricate a low electron density sample ($n=1.23\times10^{11}$ cm^{-2}) with an exceedingly high mobility of $\mu=90.000$ $cm^2/Vsec$. Fig. 8 is actually a photograph of this specimen. Given the high magnetic fields available at the magnet lab we foresaw being able to venture into the so-called extreme quantum limit, where the lowest Landau level is only partially occupied with electrons. The goal was to investigate this regime for signs of the so-called Wigner-solid, an electron crystal in two dimensions. The formation of such a regular array of electrons had been predicted theoretically, but remained unobserved.

On October 7, a Hall measurement on this specimen at the temperature of liquid He (4.2 K) produced the data of the top of Fig. 9. The largely linear relationship between Hall resistance, R_H, and magnetic field, B, is evident. Deviations at low field indicate the emergence of the IQHE. Knowledge of the electron density, as well as the values of the resistance steps (R_H=h/(ie^2), i=1,2,3..) clearly identify these features as the IQHE. With the last (i=1) step occurring at B≈5T (~7 cm on the mm-paper), for all fields beyond this point the electrons had to reside in the lowest Landau level, filling it to only a fraction n of its capacity. Cooling the sample to 1.5K the IQHE features firmed up, developing the familiar, flat plateaus (see top Fig.9). A remarkable feature occurred at B=15T: the Hall trace started to deviate from the originally

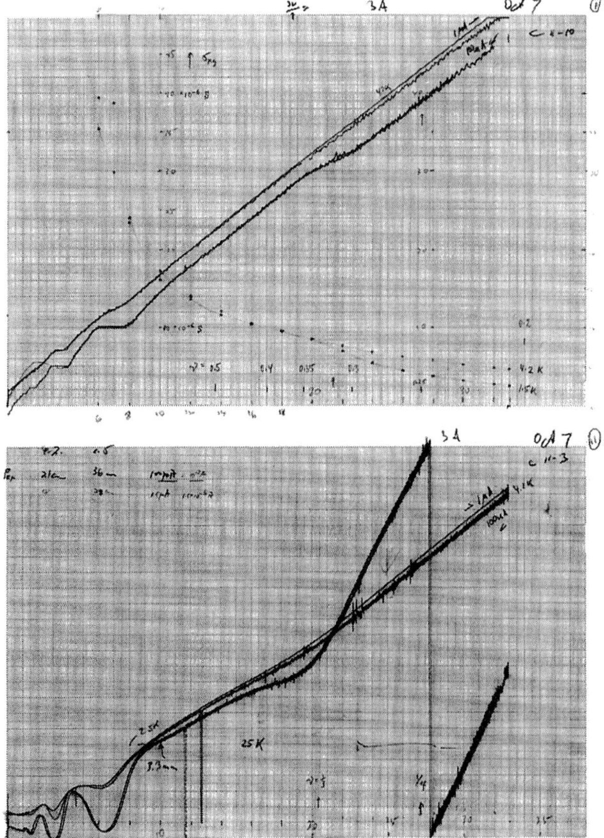

Figure 9. Data of Oct. 7, 1981 on the specimen # 6-19-81(3) (see Fig. 8) on millimeter paper. The top panel shows the Hall resistance, R_H, at temperatures 4.2 K and 1.5 K versus magnetic field, B. The bottom panel shows the magneto resistance, R, versus B at similar temperatures. 1T is equivalent to ~1.5 cm. Features at ~3 cm and ~7cm are due to the IQHE. Weaker features at ~21 cm are due to the FQHE. The scribbles in the top panel are from reuse of the millimeter-paper for data reduction from other traces.

straight line, showing a behavior not unlike that observed in the IQHE at the higher temperature of 4.2 K. This feature was totally unexpected. Beyond the emergence of a plateau in R_H, the magneto resistance, R, seemed to exhibit a concomitant minimum (see bottom Fig. 9).

The IQHE, arising from exact filling of Landau levels, could not have been at work, since above B≈5T the lowest level was only partially occupied. Furthermore, the Hall resistance in the vicinity of this change in slope far exceeded the largest possible of IQHE resistances of $R_H=h/e^2\approx25k\Omega$. Lightheartedly, Dan Tsui enclosed the distance between B=0 and the position of the last IQHE (~7 cm) between two fingers of one hand and measured the position of the new feature in this unit. He determined it to be three and exclaimed, "quarks!" Although obviously joking, with finely honed intuition, he had hit on the very essence of the data.

Following Laughlin's gedanken experiment and accepting quantization of the Hall resistance to measure the charge of the particle, a plateau three times as high as the last IQHE plateau meant the appearance of a charge $q=\phi_o/(3h/e^2)=e/3$. Obviously, our low-temperature, low-energy experiment (milli-eV, not Millions-eV) could not have generated anything even remotely related to quarks (sub-nuclear particles endowed with 1/3 charge) but, as it turned out, the implication of some kind of fractionally charged particle was dead right. At the time, we did not know what we had discovered. The paper on the findings (see Fig. 10), published in March 1982 in Physical Review Letter with Tsui, Störmer and Gossard as authors, speculated on it being a signature of a Wigner-solid or equivalent, but the paper also remarked on a fractional charge.

Origin

The IQHE can be understood solely on the basis of the quantized motion of individual 2D electrons in the presence of a magnetic field and random fluctuations of the interface potential which creates localized states. The existence of all fellow electrons enters only in the simplest of ways – as a filler of empty states of the Landau levels. The electrostatic interaction (so-called Coulomb interaction) between the like-charged carriers is irrelevant to the understanding of the IQHE. It is therefore called a single-particle effect.

The FQHE, on the other hand, can no longer be understood on the basis of the behavior of individual electrons in a magnetic field. The existence of an energy gap – so crucial for the exact quantization in the IQHE – is expected to be also essential for the occurrence of the FQHE. However, all magnetic field-induced energy gaps have been exhausted by the IQHE and have emerged as integral quantization of the Hall resistance to $R_H=h/(ie^2)$, i=1,2,3.... Other energy gaps, at fractional filling of a Landau level, must be of a different origin.

The origin of the FQHE is interaction between electrons. It is therefore termed a many-particle effect or an electron correlation effect, since the charged electrons are avoiding each other by correlating their relative motion in an intricate manner. In the IQHE, electrons have no freedom to

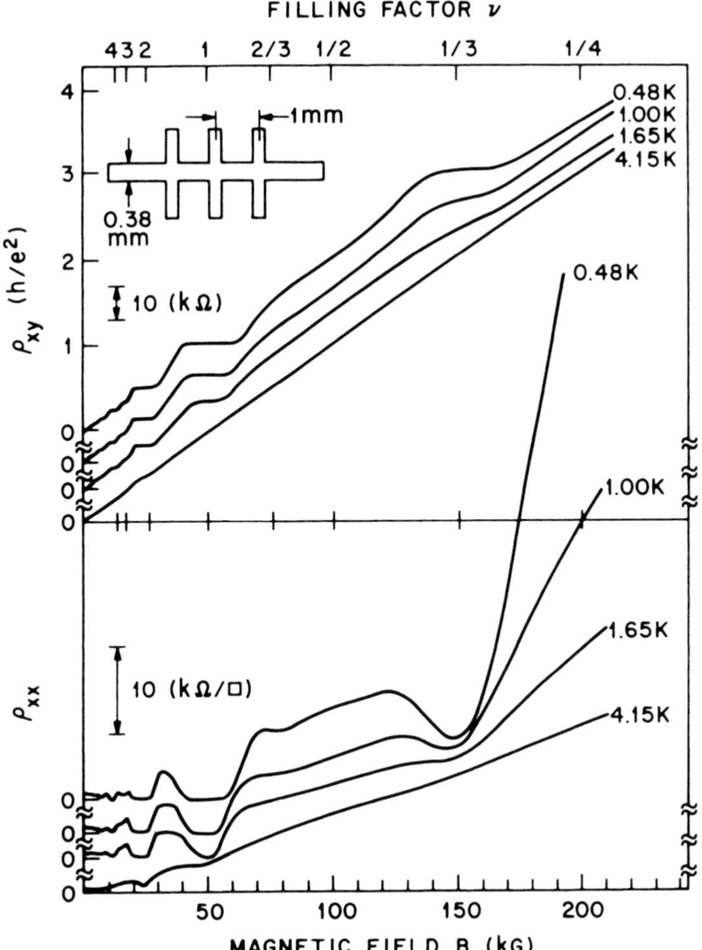

Figure 10. First publication on the FQHE. Hall resistance data (here ρ_{xy}) and magneto resistance data (here ρ_{xx}) are from the same specimen as in Fig. 9. The filling factor, ν, of the Landau level is indicated on the top. The features at ν=1,2,3.. are due to the IQHE. The features at ν=1/3 are due to the FQHE. Sample dimensions and sample temperatures are indicated.

avoid one another. Occurring at exact integral Landau level filling, electrons are already "close-packed" with no option for further avoidance. At fractional filling this is different. There is much "space" in a Landau level. Electrons have the freedom to avoid each other in the energetically most advantageous fashion. The electron-solid we had been searching for and in which electrons reside at fixed positions of maximum mutual distance, would have represented a static pattern that minimizes electron interaction. In the FQHE the electrons assume an even more favorable state, unforeseen by theory, by conducting an elaborate, mutual, quantum-mechanical dance.

Many-particle effects are extraordinarily challenging to address theoretically. In most situations they cause only a small adjustment of the behavior of the electrons and can be taken into account in an approximate manner.

Figure 11. First successful operation of our dilution refrigerator in high-magnetic field. The sign reads: 85mK, 280kG, Feb. 16, 84. The proud operators are clockwise from upper left: Albert Chang, Peter Berglund (who was largely responsible for the design and implementation of the instrument), Greg Boebinger, Dan Tsui and Horst Störmer.

Often such a treatment is quite adequate, but on occasion many-particle interactions becomes the essence of a physical effect. Superconductivity and superfluidity are of such intricate origin. To account for their occurrence one had to devise novel, sophisticated theoretical means. The emergence of the FQHE requires such a new kind of thinking.

Bob Laughlin had the correct theoretical insight and invented an elegant wavefunction which described the quantum-mechanical behavior of all those electrons in the 1/3 FQHE (as well as all other 1/q FQHE states) in a very succinct equation of some15 letters. It represents a triumph of many-particle theory. He also provided a reason for the existence of an energy gap and a derivation of this most mysterious charge of e/3 (see his contribution to this volume). In the following sections, I will attempt to give the reader an impression of the simple beauty of the physical concepts in the regime of the FQHE. Rather than addressing the expert, to whom several excellent monographs are available (see bibliography), my presentation aims at the scientifically knowledgeable layperson, who attempts to develop a sense for the origin of a phenomenon as strange as the FQHE. The discussion follows a non-historical path. It draws from the concept of the formation of composite particles between electrons and the magnetic field. From the vista of this model the serene beauty of electron correlation in 2D Landau levels manifests itself most clearly.

Of Electrons and Flux Quanta

In a classical model 2D electrons behave like charged billiard balls on a table, Fig.12a. They are distinguishable by virtue of their different history and they can be tracked individually. Quantum-mechanically, electrons are smeared out over the table. They are inherently indistinguishable and one can only cite a probability of finding an electron – any electron – at any particular location. In a perfect 2D system this probability is absolutely uniform over the whole plane. The electrons behave like a featureless liquid, Fig.12 b. That is not to say, that the motions of the electrons are not correlated. These like-charged carriers strongly avoid each other, as shown in Fig.12c in a classical representation. They also do this in the quantum-mechanical liquid of Fig.12b. It affects the probability of finding one electron *here* having detected another electron *there* (e.g. close by), but one cannot represent it in a graph as simple as Fig. 12c.

It was an important conceptual step to realize that an impinging magnetic

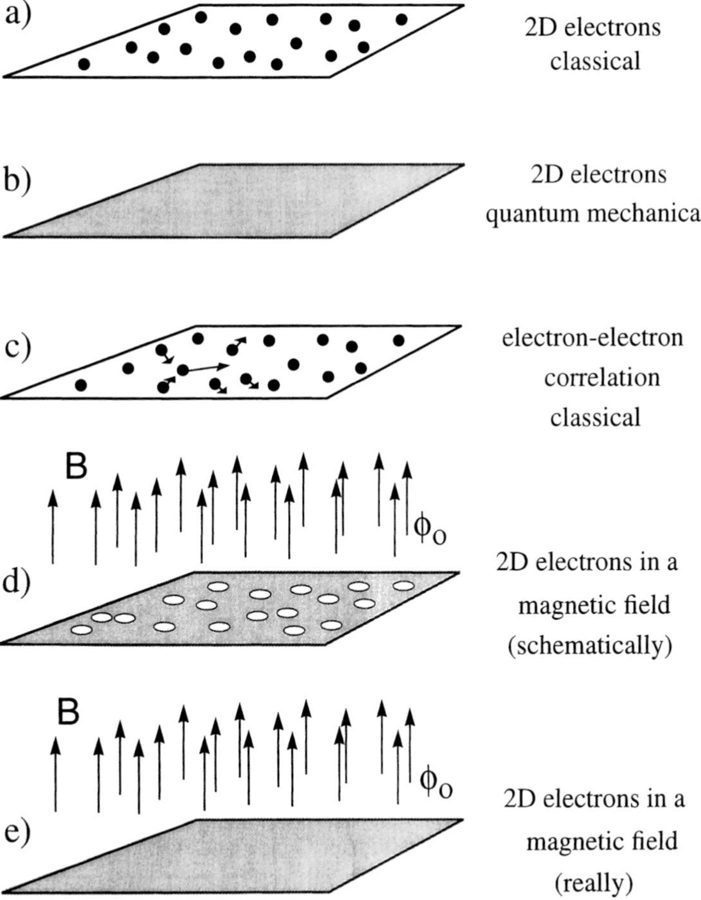

Figure 12. Schematic drawing of 2DES in various approximations. Black dots represent electrons. White holes represent vortices. Arrows represent magnetic flux quanta ϕ_o of the magnetic field, B.

field, B, could be viewed as creating tiny whirlpools, so-called vortices, in this lake of charge – one for each flux quantum, ϕ_o=h/e, of the magnetic field, Fig.12d. The notion of a whirlpool is quite appropriate, since such vortices have indeed a quantum-mechanical "swirl" – a phase twist – to them. Inside the vortex, electronic charge is displaced dropping to zero in the center and recovering to the average surrounding charge density at the edge of the vortex. The extent of a vortex is roughly the size of the area which contains one quantum of magnetic flux (area × B=ϕ_o). Therefore, each vortex can be thought of as carrying with it one flux quantum. Of course, just as the electrons are spread out uniformly over the plane, so are the vortices. As required by quantum mechanics, the probability of finding an electron – as well as a vortex – remains totally uniform, Fig.12e. However, the picture of electrons and vortices provides an intuitive way of looking at electron-electron correlation in the presence of a magnetic field.

Electrons and vortices are opposite objects, one representing a package of charge the other the *absence* of charge. Correlation of their mutual position can prove energetically very beneficial. Placing vortices directly onto electrons is particularly advantageous since the trough of the whirlpool, which represents the displacement of all fellow electrons, keeps their charges at bay and reduces mutual repulsion. This intuitive image requires some mental flexibility. Each electron is at the center of a vortex and at the same time is part of the pool of electrons generating vortices surrounding all those other electrons. Who says many-particle physics is easy?

Each electron *always needs* to be surrounded by *one* vortex. In the language of electrons and vortices, it is the system's way of satisfying the Pauli exclusion principle for electrons, which, in this situation, requires that no two electrons can be in the same position. The whirlpool provides the required place of respite. At complete filling of the lowest Landau level, where the number of electrons *equals* the number of flux quanta, the arrangement of electrons and vortices is totally controlled by the Pauli principle – one vortex per electron, no choices (see Fig.13). This is the condition of the i=1 IQHE. It can easily be extended to more Landau levels and also to include both electron spin directions and hence to i=2,3,4... The IQHE is driven by the Pauli exclusion principle for electrons. It is another way of expressing that the existence of other electrons enters the IQHE only in the simplest of ways – as a filler of empty states. When the number of vortices deviates from the number of electrons then there are choices.

At magnetic fields higher than the i=1 IQHE the stronger magnetic field provides more flux quanta and hence there are more vortices than there are electrons. The Pauli principle is readily satisfied by placing one vortex onto each electron (Fig.14a) – but there are more vortices available. The electron system can considerably reduce its electrostatic Coulomb energy by placing more vortices onto each electron (Fig.14b). More vortices on an electron generates a bigger surrounding whirlpool, pushing further away all fellow electrons, thereby reducing the repulsive energy. The so-established relative motion of electrons is no longer driven by the Pauli exclusion principle but

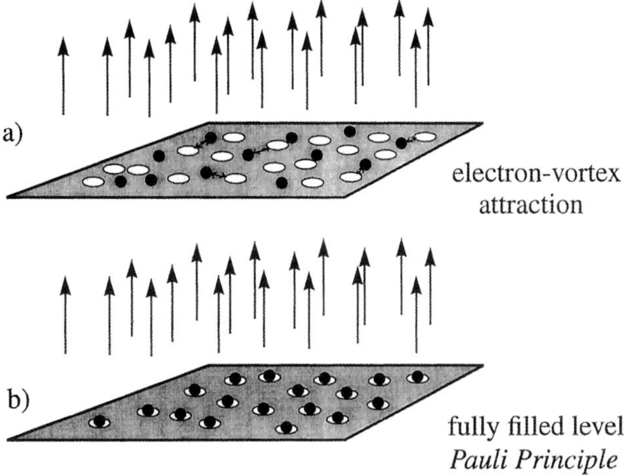

Figure 13. Schematic drawing of electron vortex attraction of a 2DES in the presence of a magnetic field. In the fully filled Landau level, ν=1, there are as many vortices as there are electrons and the Pauli exclusion principle forces the vortices onto the electrons. (The spin of the electron is neglected throughout.)

by reduction in Coulomb energy. This is the central principle underlying electron-electron correlation in 2DES in a magnetic field. Casting electron-electron correlation in terms of vortex attachment facilitates the comprehension of this intricate many-particle behavior. Regarding the vortices as little whirlpools, ultimately, remains a crutch for visualizing something that has no classical analog.

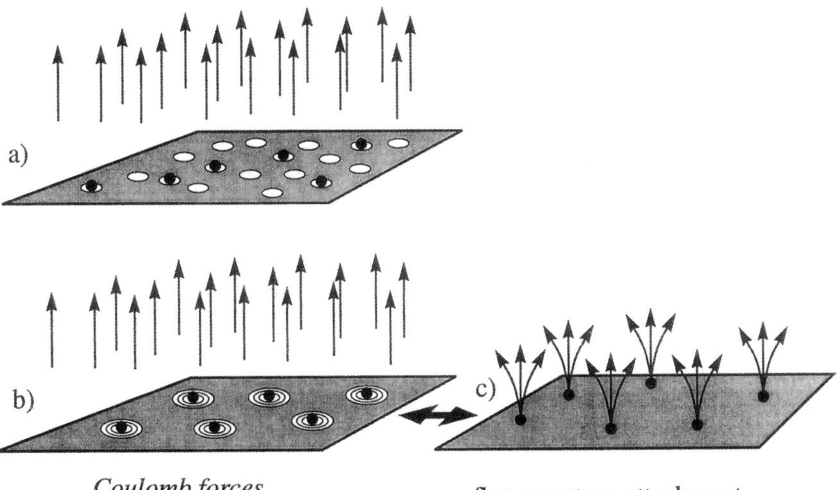

Figure 14. Schematic drawing of electron vortex attraction at fractional Landau level filling, ν=1/3. Now there are three times as many vortices as there are electrons. The Pauli principle is satisfied by placing one vortex onto each electron (a). Placing three vortices onto each electron reduces electron-electron (Coulomb) repulsion (b). Vortex attachment can be viewed as the attachment of magnetic flux quanta to the electrons transforming them to composite particles (c).

Composite Particles

Vortices are the expression of flux quanta in the 2D electron system and each vortex can be thought of as having been created by a flux quantum. Conceptually, it is advantageous to represent the vortices simply by their "generators", the flux quanta themselves. Then, the placement of vortices onto electrons becomes equivalent to the attachment of magnetic flux quanta to the carriers (Fig.14b,c). Electrons *plus* flux quanta can be viewed as new entities, which have come to be called composite particles, CPs. As these objects move through the liquid the flux quanta act as an invisible shield against other electrons. Replacing the system of highly interacting electrons by a system of electrons with such a "guard ring" – compliment of the magnetic field – removes most of the electron-electron interaction from the problem and leads to composite particles, which are almost void of mutual interactions. It is a minor miracle, that such a transformation from a very complex many-particle problem of well known objects (electrons in a magnetic field) to a much simpler single-particle problem of rather complex objects (electrons plus flux quanta) exits and that it was discovered.

CPs act differently from bare electrons. All of the external magnetic field has been incorporated into the particles via flux quantum attachment to the electrons. Therefore, from the perspective of CPs, the magnetic field has disappeared and they no longer are subject to it. They inhabit an apparently field-free 2D plane. Yet more importantly, the attached flux quanta change the character of the particles from fermions to bosons and back to fermions.

Fermions and Bosons

In physics one differentiates between two types of particles, bosons and fermions. Fermions, such as electrons or protons, have the property that all other such particles are *excluded* from being in the same quantum-mechanical state, e.g. in the same position. They are subject to the Pauli exclusion principle and fill sequentially one available state after the other. Bosons, such as photons or Helium atoms, have no such restriction and even have a preference for being in the *same* state. They follow Bose-Einstein statistics. In a very casual way, the exclusion principle for fermions is the reason for the world not collapsing (all identical fermions staying away from each other) and the second is the origin for phenomena such as lasing or superfluidity (all photons or Helium atoms condensing into the same state), usually referred to as bose-condensation. Fermions have half-integer spin while bosons have integer spin – spin being related to the "spinning" of the particle.

As the case of superfluidity in Helium shows, fermions, the elements of atoms (electrons, protons and neutrons), can be assembled to "make" bosons (Helium atoms). In a casual way, superconductivity, too, can be seen as the assembly of pairs of fermions (electrons) into bosons (Cooper pairs) which bose-condense into a superconducting state. One cannot assemble bosons to make "quasi" fermions. In a very simplistic way, the reason for the difference is that half-integers can be added to make integers but integers cannot be added to make half integers.

Systems of fermions and systems of bosons behave very differently under mutual exchange of the position of two of their constituent particles – it is said to change their *statistics*. Their wavefunction – the quantum-mechanical description of the behavior of all the particles – is multiplied by -1 in the case of fermions and by +1 in the case of bosons. It is one of the deeper mysteries of quantum mechanics and cannot be further illuminated here. In any case, one needs to accept nature's teachings.

Composite Particle Statistics
Electrons are fermions. As one slowly moves two electrons in a 2D electron system around each other and exchanges them, the wave function undergoes the sign reversal expected from fermions. It is different for CPs (Fig. 15). The attached flux quanta need to be taken into account and their presence changes the particles' statistics. As one slowly moves two CPs around each

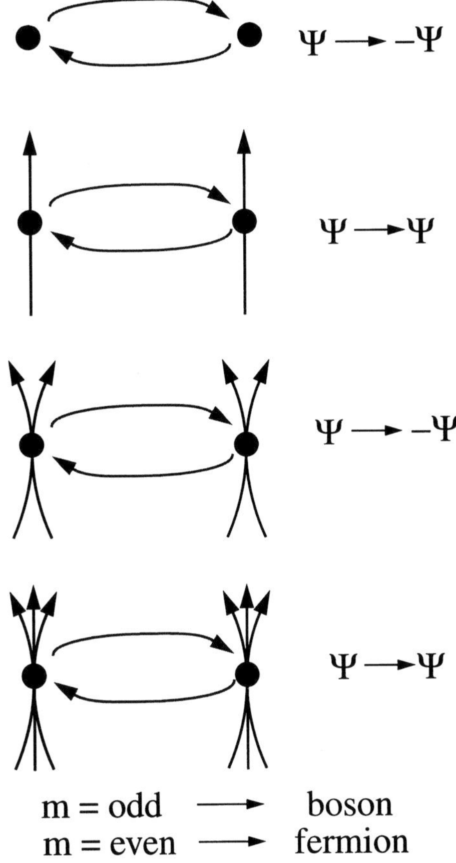

Figure 15. Statistics of electrons and composite particles. Exchange of two particles affects the wavefunction, Ψ, which described the quantum-mechanical behavior of the system. For electrons, Ψ is multiplied by –1, identifying the particles as fermions. With the attachment of an odd number of flux quanta Ψ remains *unchanged* under exchange (multiplication by +1), identifying these particles as bosons. Attachment of an even number of flux quanta returns the particles to fermions. m is the number of flux quanta.

other and exchanges them, the electrons by themselves reverse the sign of the wavefunction, but each attached flux quanta creates an extra "twist", multiplying it by an extra –1. As a result, CPs can be either fermions or bosons, depending on the number of attached flux quanta. An electron plus an *even* number of flux quanta becomes a composite *fermion* (CF), since the wavefunction is multiplied by -1 an *odd* number of times, i.e. by -1. An electron plus an *odd* number of flux quanta becomes a composite *boson* (CB), since the wavefunction is multiplied by –1 an *even* number of times, i.e. by +1. This so-called transmutation of the particle statistics through flux quantum attachment is deeply rooted in the two-dimensionality of the system. It represents a deep connection between space and particle statistics.

Accepting that CPs incorporate the external magnetic field and show either boson or fermion behavior, the perplexing properties of 2D electron systems in a high magnetic field can readily be appreciated.

1/3 Fractional Quantum Hall State

At 1/3 filling of the lowest Landau level (ν=1/3), the magnetic field contains three times as many flux quanta per unit area as there are electrons in the 2D system. Therefore, the electron liquid contains three times as many vortices as there are carriers. To minimize electron-electron interaction each electron accepts three vortices, which keeps fellow electrons optimally at bay. This is equivalent to the attachment of three magnetic flux quanta to each electron, which renders these objects CPs (Fig.14c). Since all the external magnetic field has been incorporated into the particles, they reside in an apparently magnetic field-free region. Consisting of an electron plus an *odd* number of flux quanta, the resulting composites are composite *bosons* (CBs). Being bosons and residing in apparently zero magnetic field, these CBs bose-condense into a new groundstate with an energy gap, characteristic for such bose-condensation. This is the sought after energy gap required for quantization of the Hall resistance and vanishing resistance to arise. It has been measured by various experimental techniques, most directly by light-scattering.

As the magnetic field deviates from exactly ν=1/3 filling, to higher fields, more vortices are being created (Fig. 16). They are not attached to any electrons since this would disturb the symmetry of the condensed state. The amount of charge deficit in any of these vortices amounts to exactly 1/3 of an electronic charge. These quasi-holes (whirlpool in the electron lake) are effectively positive charges as compared to the negatively charged electrons. An analogous argument can be made for magnetic fields slightly below ν=1/3 and the creation of quasi-electrons of negative charge e/3. Quasi-particles can move freely through the 2D plane and transport electrical current. They are the famous 1/3 charged particles of the FQHE that have been observed by various experimental means, most recently by measurement of the amount of electrical noise that they generate. Plateau formation in the FQHE arises, in analogy to plateau formation in the IQHE, from potential fluctuations and the resulting localization of carriers. In the case of the FQHE the carriers are not electrons, but, instead, the bizarre fractionally charged quasi-particles.

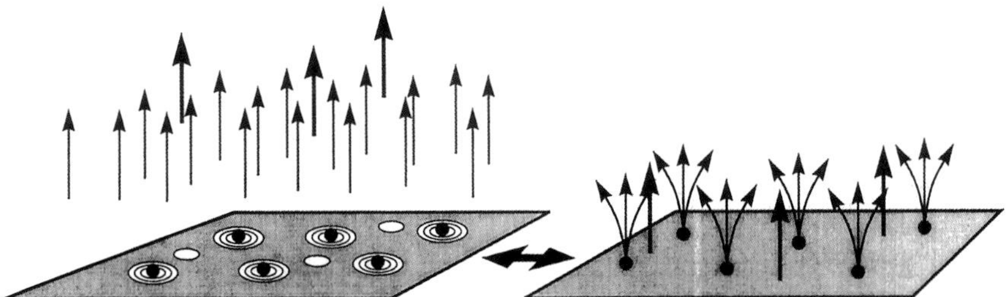

Figure 16. Schematic representation of 1/3 charged quasiparticles. At slightly higher B fields than at ν=1/3 additional vortices are created. They represent dimples in the electron lake. In the dimples exactly 1/3 of an electron charge is missing. These are the fractionally charged quasiparticles of the FQHE.

The FQHE at ν=1/5, 1/7, etc with quasi-particles of charge e/5, e/7, etc. can be accounted for in total analogy to the 1/3 FQHE by attaching 5, 7, etc. flux quanta to each electron. In fact, even states at ν=2/3, 4/5, 6/7, etc. and ν=1+1/3, 1+1/5 etc. can be covered by this procedure, regarding e.g. the ν=2/3 state as a full Landau level with 1/3 *missing* electrons. In this way all fractions at Landau level filling factor ν=i ± 1/q (often called the primary fractions) can be rationalized. But there are many others.

The State at ν=1/2

At first sight, the ν=1/2 state should be similar to the 1/3 state, yet it turns out to be very different. At half-filling of the lowest Landau level the magnetic field contains *two* times as many flux quanta per unit area and hence creates two times as many vortices as there are carriers. In analogy to the 1/3 state, each electron accepts now two vortices, which keeps the others at bay (Fig.17). However, the attachment of an *even* number of magnetic flux quanta to each electron renders these objects composite *fermions* (CFs) and not composite bosons. This drastically changes their behavior as compared to the 1/3 FQHE and its equivalents.

As at ν=1/3 so also at ν=1/2 all external magnetic field has been incorporated into the particles and they reside at apparently zero magnetic field. However, being fermions they are prevented from condensing into the lowest energy state. Instead, they fill up successively the sequence of lowest lying energy states, until a maximum is reached and all CFs have been accommo-

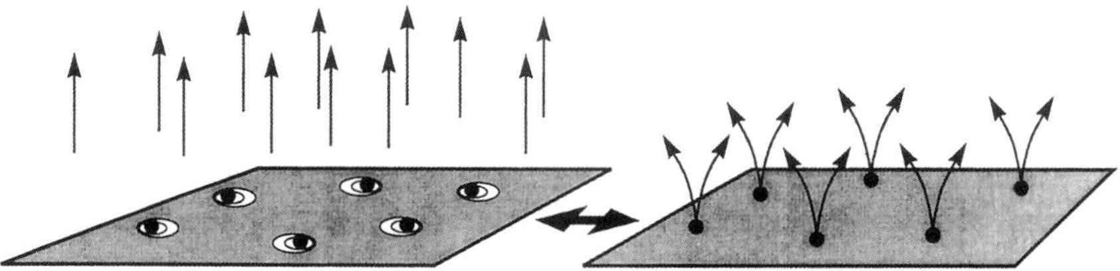

Figure 17. Schematic representation of the state at Landau level filling factor ν=1/2. Two vortices are bound to each electron, equivalent to the attachment of two flux quanta. The slight offset of the second vortex is meant to represent the formation of tiny in-plane electrical dipoles.

dated. The process is equivalent to the filling of states by electrons at B=0. Hence, from the point of view of CFs, the ν=1/2 state appears equivalent to the case for electrons at B=0. In spite of the *huge* external magnetic field at half-filling of the Landau level, CFs are moving in a similar fashion as electrons are moving in *zero* field. This has been directly observed in experiment. Flux quantum attachment has transformed these earlier electrons and they are propagating along straight trajectories in a high magnetic field, where normal electrons would orbit on very tight circles. The mass of a CF, usually considered to be a property of the particle, is unrelated to the mass of the underlying electron. Instead, the mass depends on the magnetic field and only on the magnetic field. In fact, it is a mass of purely many-particle origin, arising solely from interactions, rather than being a property of any individual particle. It is another one of these baffling implications of e-e interactions in high magnetic fields. The absence of condensation and the lack of an energy gap prevents the ν=1/2 state from showing a quantized Hall resistance. Instead the Hall line is featureless, just as it is for electrons around B=0, see Fig. 18.

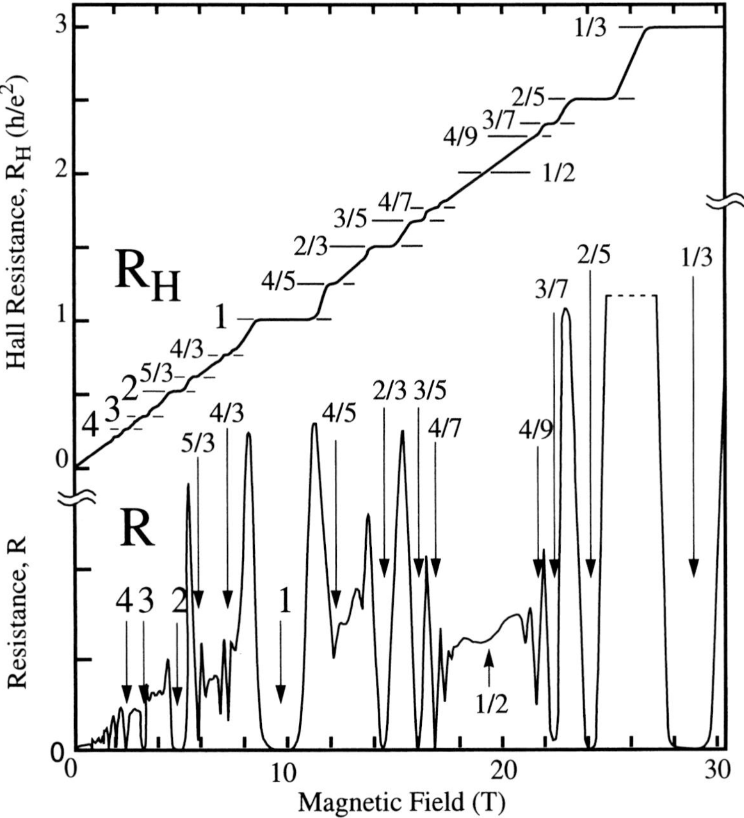

Figure 18. The FQHE as it appears today in ultra-high mobility modulation doped GaAs / AlGaAs 2DESs. Many fractions are visible. The most prominent sequence, ν=p/(2p±1), converges toward ν=1/2 and is discussed in the text.

The difference between ν=1/3 and ν=1/2 is striking. One is a bose-condensed many-particle state showing a quantized Hall effect and giving rise to fractionally charged particles. The other is a Fermi sea, in spite of the existence of a huge external field, and its particles have a mass that arises from interactions. One flux quantum per electron makes all the difference.

There are many fascinating open questions associated with the ν=1/2 state, such as: how does the mass vary with energy for CFs? and what is the microscopic structure of the particles? also, how does the electron spin (which we were neglecting throughout this lecture) affect CF formation? A beautiful picture of composite fermions being tiny dipoles is emerging. While one of the vortices is placed directly on the electron (Pauli principle) the position of the second vortex is a bit displaced from exact center, rendering the object an electric dipole in the 2D plane. There is great promise for future discovery and future theoretical insight.

All Those Other FQHE States

Bose condensation of CBs consisting of electrons and an odd number of flux quanta rationalizes the appearance of the FQHE at the primary fractions around Landau level filling factor ν=i ±1/q with quantized Hall resistances $R_H=h/(\nu e^2)$ and deep minima in the concomitant magneto resistance, R. However, a multitude of other FQHE states have been discovered over the years. Fig. 18 shows one of the best of today's experimental traces on a specimen with a multi-million cm^2/Vsec mobility. What is the origin of these other states? The composite fermion model offers an extraordinary lucid picture. We shall discuss it for the sequence of prominent fractions 2/5, 3/7, 4/9, 5/11... and 2/3, 3/5, 4/7, 5/9... (i.e. $\nu=p/(2p\pm1)$, p=2,3,4...) around ν=1/2.

At half-filling the electron system has been transformed into CFs consisting of electrons which carry two magnetic flux quanta. All of the external magnetic field has been incorporated into the particles and they reside in an apparently magnetic field-free 2D plane. Since they are fermions, the system of CFs at ν=1/2 resembles a system of electrons of the same density at B=0. What happens as the magnetic field deviates from B=0? For electrons their motion becomes quantized into electron-Landau orbits. They fill up their electron-Landau levels, encounter the energy gaps and exhibit the well know IQHE. CFs around ν=1/2 follow the same route. As the magnetic field deviates from exactly ν=1/2 the motion of CFs becomes quantized into CF-Landau orbits. They fill up their CF-Landau levels, encounter CF-energy gaps and exhibit an IQHE. However, this is not an IQHE of *electrons*, but an IQHE of *CFs*. This IQHE of CFs arises exactly at $\nu=p/(2p\pm1)$ which are the positions of the FQHE features. In fact, the oscillating features in the magneto resistance R of the FQHE around ν=1/2 closely resembles the oscillating features in R around B=0 and, once they have been shifted from B=0 to ν=1/2, they coincide with their position. This is very remarkable and in several ways.

CFs "survive" the additional (effective) magnetic field (away from ν=1/2) and the orbits of these composite particles mimic the orbits of electrons in the equivalent magnetic field in the vicinity of B=0. The CFs remain "good" particles. In this way, a complex electron many-particle problem at some rational fractional filling factor has been reduced to a single-particle problem at integer filling of CF-Landau levels in an effective magnetic field. Even the variation of the size of the energy gaps from one FQHE state to the next can be regarded as deriving from the ladder of Landau levels of CFs. More strikingly yet, excellent quantum-mechanical wavefunctions for these FQHE states can simply be derived from electron Landau levels. Therefore, the FQHE of electrons can be regarded as the IQHE of CFs.

The CFs model has been extraordinarily successful in conquering those other FQHE states. Even the 1/3 state can be viewed from the vantage point of this model. At $\nu=1/3$ the CFs emanating from $\nu=1/2$ have been quantized into CF-Landau levels and they are exactly filling the lowest of these levels. Hence, the $\nu=1/3$ FQHE state is the equivalent of the i=1 IQHE of CFs which had formed at $\nu=1/2$. In analogy to the electron case, the flux quanta – one per CF – create vortices in the CF liquid which are forced onto the CFs to satisfy the Pauli principle for CFs. Each electron, holding two flux quanta from being a CF in the first place, acquires a third one, exactly as required to become the $\nu=1/3$ state. Bose condensation of CBs reappears as Landau level formation in CFs. In the FQHE regime they represent two different sides of the same coin.

With all these similarities, is the FQHE then the same as the IQHE? Certainly not. From one point of view the FQHE is the result of a complex behavior (many-particle interactions) of simple particles (electrons) in the presence of a true external magnetic field. From another point of view it is the simple behavior (Landau quantization) of complex new multi-component particles (composite fermions) in the presence of an effective magnetic field. This view of the FQHE has developed over the past decade or so. Its extreme simplicity is testament to the beauty of nature as much as it demonstrates the ability and intuition of dozens of theorists that have shaped it over the years. Whatever model one constructs for the FQHE, its origin is an elegant quantum mechanical dance of electrons in a magnetic field.

CFs are "everywhere". All even-denominator fractions are candidates for CF formation. And all those CF states are capable of generating their own CF Landau levels leading to a panoply of FQHE states. Such other states are already visible in Fig.18. Yet more of such states appear in Fig.19. FQHE states are emanating from 1/2, 3/2, 1/4, 3/4, 5/4, and possibly from 3/8 and 3/14. There does not seem to be an end, although eventually the most fragile of states are destroyed by residual potential fluctuations, or by the formation of solids of electrons or CFs. Yet better 2DES should remove the curtain from those sensitive fractions. Their mere detection and classification may appear somewhat unimaginative. However, so far, whenever we dug a bit deeper into the rich soil of the 2DES in the presence of a magnetic field we were rewarded with more surprises. One of them has already occurred.

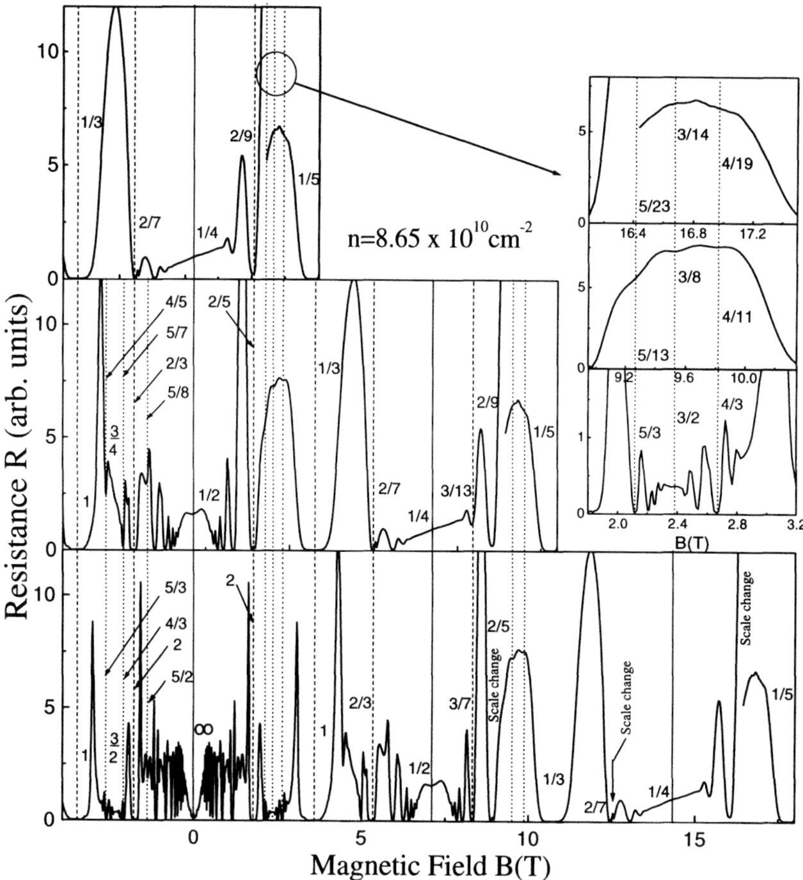

Figure 19. Multiple sequences of FQHE states are visible in this graph. Only magneto resistance data are shown for clarity. In the middle and top panel the bottom trace is shifted to the left by a magnetic field equivalent to the field at ν=1/2 and ν=1/4 respectively. The vertical lines show the self-similarity between different FQHE sequences. The right hand inset is a blow-up showing yet more developing FQHE states.

The Peculiar State at ν=5/2

Electrons with two attached flux quanta are fermions. They fill up sequentially the lowest energy states and are the starting point for multiple sequences of FQHE states. However, they themselves cannot be FQHE states. Yet the 5/2 state is exactly that. It has all the characteristics of a FQHE state, including energy gap and quantized Hall resistance, in spite of its even-denominator classification (see Fig.20). The ν=5/2 state resides in a higher Landau level (5/2=2+1/2) but this fact should not alter the simple reasoning. The Landau level below is energetically far removed and can be regarded as inert. Therefore the 5/2 state is really a 1/2 state in the next higher Landau level and should behave as such – but it does not. Discovered more than a decade ago, its true origin remains mysterious. With the advent of the CF model, the ν=5/2 state has recently been revisited and a most tantalizing possibility has arisen.

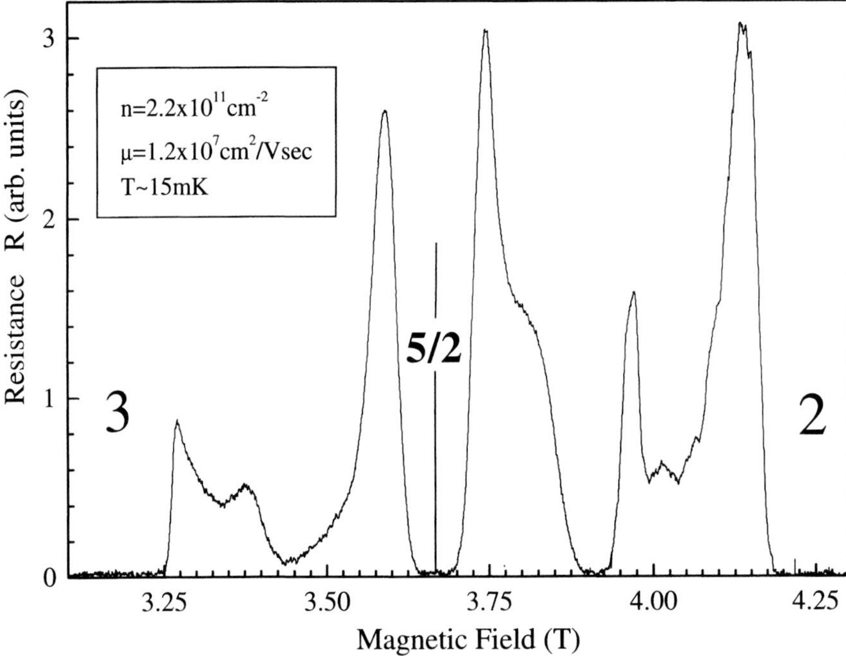

Figure 20. FQHE at ν=5/2. A FQHE state at such an even denominator fraction should not be allowed. The origin of the state remains unclear. An exciting possibility for the origin of this state is the formation of composite fermion pairs. These unpublished data were taken by J.-S. Xia and Wei Pan at the NHMFL in Gainesville, FL.

Driven by many-particle interactions the carriers at half-filling of the next Landau level indeed bind two flux quanta each – just like their ν=1/2 cousins. They form CFs and fill up the states, just as in Fig.17. However, many-particle physics pushes those CFs further to a yet lower energy ground state. In loose analogy to the formation of Cooper pairs in a normal electron systems at low temperatures and their subsequent condensation into a superconducting state with an energy gap, these CFs form new *CF-pairs* which condense into a novel many-particle groundstate. The resulting energy gap provides the essential ingredient for the observation of the characteristic FQHE features. This is a very exciting scenario, since it suggests, that yet other, higher-order electron-electron correlations than those of the CF/CB model can play a decisive role. The properties of the resulting particles are also expected to be very unusual (non-abelian).

At present, it remains unclear whether the 5/2 state is indeed of such an elaborate lineage, or whether some other, more mundane explanation will suffice. We will have to await future, more sophisticated experiments to tell us. If not the 5/2 state, there is a good chance that some other, yet to be unearthed FQHE state may be of such an intricate origin. And there may well be states that we have not even imagined.

CONCLUSIONS

Two-dimensional electron systems in high magnetic fields reveal to us totally new many-particle physics. Confined to a plane and exposed to a magnetic field such electrons display an enormously diverse spectrum of fascinating new properties: Totally unexpected new electron states with fractional quantum numbers, the attachment of magnetic flux to electrons, new particles obeying either bose or Fermi statistics, cancellation of exceedingly high magnetic fields, masses of purely electron-electron interaction origin, and possibly a strange, new process for particle pairing. These are but the most prominent of observations and implications. Most perplexing of all, such electrons create bizarre fractionally charged particles, without any individual electron splitting apart.

They are just electrons, although many of them. Indeed: "More is different!"

EPILOGUE

I am very honored having been chosen to share in this award of almost frightening proportions and I am grateful for receiving 1/3 of this very special prize. Unfortunately, the yet more delightful 1/4 version remained forbidden. I attribute it all to an immense amount of good fortune throughout my life and the truly outstanding colleagues with whom it allowed me to work.

As so often, this award is being given to a lucky few, but it truly honors the immense progress that has been made in many-particle physics over the years – in particular in the physics of two-dimensional systems – and the large group of experimentalists and theorists, that have brought it about. In this sense, I feel I share this award with so many of my colleagues and friends around the globe. To all of them I owe a great deal of gratitude.

As to our own contributions, the creators of materials remain the true heroes of the trade. Art Gossard and Willy Wiegmann fabricated the all-important sample in which the FQHE was discovered and many more after this event. Over the past decade or so, Loren Pfeiffer and Ken West brought the art and science of 2D material growth to new heights. It was in their samples, in which most of the exciting new discoveries in the FQHE were made. Kirk Baldwin's wizardry in the cleanroom and his screening of thousands of samples provided the underpinning to most of our experiments. Al Cho, John English, Jim Hwang, Mansour Shayegan, Charles Tu, Won Tsang, and Gunther Weiman also provided invaluable materials support.

I would not be here without the exceptional experimental skills and deep physical insights of postdocs, students and collaborators at Bell Labs, Princeton University, and other institutions. They include, Jim Allen, Ray Ashoori, Edwin Batke, Peter Berglund, Greg Boebinger, Albert Chang, Rui Du, Jim Eisenstein, Erich Gornik, Taisto Haavasoja, Rick Hall, Hong-Wen Jiang, Woowon Kang, Mikko Paalanen, Wei Pan, Aron Pinczuk, Zack

Schlesinger, Joe Spector, Werner Wegscheider, Claude Weisbuch, Bob Willett, Jian-Sheng Xia, Andrew Yeh, and Amir Yacoby.

None of what has been discovered in experiments by many in the field I would have appreciated without my theorist friends and colleagues patiently teaching me FQHE physics. Particular insights I received from: Nick d'Ambrumenil, Steve Girvin, Duncan Haldane, Bert Halperin, Song He, Jainendra Jain, Steve Kivelson, Bob Laughlin, Dung-Hai Lee, Peter Littlewood, Allan MacDonald, Rudolf Morf, Phil Platzman, Nick Read, Ed Rezayi, Ramamurti Shankar and Steve Simon. There are many more, too numerous to list.

My new colleagues at Columbia University I thank for a warm reception in their midst. Dominique, my wife, I thank for her unceasing support and cheerfulness. I also thank my producer.

Finally, I want to thank my long-time collaborator and friend, Dan Tsui, in his characteristically few words: "Thanks for taking me to the dance".

BIBLIOGRAPHY

Several authoritative monographs have been written on the different topics addressed in this lecture, which point back to the extensive original literature. The following represents a sampling of such books.

Physics in Silicon MOSFETs:

T. Ando, A. B. Fowler, and F. Stern, 1983, Electronic Properties of Two-Dimensional Systems, Rev. Mod. Phys. 54, 437 (1983).

Molecular Beam Epitaxy:

A. Cho, edt, 1994, Molecular Beam Epitaxy (AIP Press, Woodbury, NY).

M. A. Herman, and H. Sitter, 1998, Molecular Beam Epitaxy: Fundamentals and Current Status (Springer Verlag, Berlin, Heidelberg).

Modulation-Doping:

E. F. Schubert, 1993, Doping in III–V Semiconductors (Cambridge University Press).

Hall Effect:

H. Weiss, 1969, Structure and Application of Galvanomagnetic Devices (Pergamon Press).

Integral Quantum Hall Effect:

R. E. Prange, and S. M. Girvin, edts, 1990, The Quantum Hall Effect, 2nd ed. (Springer-Verlag, New York).

T. Chakraborty, and P. Pietilainen,1995, The Quantum Hall Effects (85 Springer Series in Solid State Sciences).

A. H. MacDonald, edt, 1990, Quantum Hall Effect: A Perspective (Kluwer Academic Publications).

Fractional Quantum Hall Effect and Composite Fermions:

R. E. Prange and S. M. Girvin, edts, 1990, The Quantum Hall Effect, 2nd ed. (Springer-Verlag, New York).

A. H. MacDonald, edt, 1990, Quantum Hall Effect : A Perspective (Kluwer Academic Publications).

T. Chakraborty and P. Pietilainen, 1995, The Quantum Hall Effects (85 Springer Series in Solid State Sciences).

S. Das Sarma and A. Pinczuk, edts, 1997, Perspectives in Quantum Hall Effects (Wiley and Sons).

O. Heinonen, edt, 1998, Composite Fermions: A Unified View of the Quantum Hall Regime (World Scientific, Singapore).

DANIEL C. TSUI

I tend to partition my life into three compartments: childhood years in a remote village in the province of Henan in central China, schooling years in Hong Kong, and the years since I came to attend college in the United States. The only thread connecting them is the kindness, generosity and friendship from the people around me that I have experienced all my life.

My childhood memories are filled with the years of drought, flood and war which were constantly on the consciousness of the inhabitants of my overpopulated village, but also with my parents' self-sacrificing love and the happy moments they created for me. Like most other villagers, my parents never had the opportunity to learn how to read and write. They suffered from their illiteracy and their suffering made them determined not to have their children follow the same path at any and whatever cost to them. In early 1951, my parents seized the first and perhaps the only opportunity to have me leave them and their village to pursue education in so far away a place that neither they nor I knew how far it truly was.

In Hong Kong, I began my formal schooling at the sixth grade level with fear and trembling, mixed with some pride and elation. I remember the difficulties that I encountered in not knowing the Cantonese dialect in the beginning, but, even more vividly, the overwhelming kindness of schoolmates who went out of their way to help by offering me their friendship, bringing me into their circle, and taking me to their out-of-class activities. In the middle of my second year in Hong Kong, I entered Pui Ching Middle School, which was known for being outstanding, especially in natural science subjects. Many of the teachers there were overqualified. They were the brightest graduates of the best universities in China and under normal circumstances would have been highly accomplished scholars and scientists. The upheaval of war in China, however, forced them to hibernate in Hong Kong teaching high school kids. They might not have been the best teachers pedagogically, but their intellects and their visions inspired us. Even their casual remarks and the stories from their romantic reminiscences of the glorious days at Peking University could leave indelible marks on us. It was they, I think, who in their unconscious ways dared us students, living in a most commercialized city, to look beyond the dollar sign and see the exploration of new frontiers in human knowledge as an intellectually rewarding and challenging pursuit.

I graduated from Pui Ching in 1957 and was admitted to the medical school of National Taiwan University in Taiwan. However, since it was unclear at the time how my parents were and whether I could return to them in China, I stayed in Hong Kong and entered a two-year special program run by the government to prepare Chinese high school graduates for the University of Hong Kong. In late spring the next year, I received the surprising good news from the United States that I was admitted with a full scholarship to my church pastor's Lutheran alma mater, Augustana College in Rock Island, Illinois. I arrived on campus right after Labor Day 1958, and there spent the best three years of my life. It was there that I had for the first time the leisure to wrestle with my Lutheran faith and to think through and make some sense out of my life experience. In Hong Kong, I was always extremely busy as a scholarship student, heavily involved with church activities and responsibilities, and worn-out from long distance daily commuting. Here, I was free to read, to learn and to think through things at my own pace. I knew from the start that I would go to graduate school, and the choice of subject and school was never a problem. C.N. Yang and T.D. Lee were awarded the Nobel Prize for Physics in 1957 and they both went to the University of Chicago. Yang and Lee were the role models for Chinese students of my generation and going to the University of Chicago for a graduate education was the ideal pilgrimage.

The University of Chicago was intense and intellectual. I liked its being in a major city, its cosmopolitan atmosphere, and even its grimy buildings and the austerity they appeared to convey. There, I luckily met and fell in love with Linda Varland, an undergraduate in the college, and we were married after her graduation. I was also fortunate that Royal Stark, who had just joined the physics faculty as a solid state experimentalist, took me on as a research assistant in the building-up of his laboratory. I realized quite early that I wanted to do experimental physics and that I lacked the aptitude for colossal experimental setups and also the taste for grandeur. I wanted to do table-top experiments and be allowed to tinker. Royal Stark trusted me and let me try my hands on everything in his laboratory. I was given the best opportunity to learn from the bottom up: from engineer drawing, soldering, machining, and design, to construction and building of our laboratory apparatus. By the time I received my Ph.D., I was confident that I could make a living using the technical skills I had learned there. Since I could always fall back on a job using my technical skills, I reasoned, why not then take a risk and try a research position doing something entirely novel and at the same time intellectually challenging.

I left Chicago in early spring 1968 and took a position in Bell Laboratories in Murray Hill, New Jersey to do research in solid state physics. I found myself a niche in semiconductor research, though I never got into the main stream either in semiconductor physics, which was mostly optics and high energy band-structures, or its use in device applications. I wandered into a new frontier, which was dubbed the physics of two-dimensional electrons. In February 1982, shortly after the discovery of the fractional quantum Hall effect, I moved to Princeton and started teaching.

Many of my friends and esteemed colleagues had asked me: "Why did you choose to leave Bell Laboratories and go to Princeton University?". Even today, I do not know the answer. Was it to do with the schooling I missed in my childhood? Maybe. Perhaps it was the Confucius in me, the faint voice I often heard when I was alone, that the only meaningful life is a life of learning. What better way is there to learn than through teaching!

INTERPLAY OF DISORDER AND INTERACTION IN TWO-DIMENSIONAL ELECTRON GAS IN INTENSE MAGNETIC FIELDS

Nobel Lecture, December 8, 1998

by

DANIEL C. TSUI

Department of Electrical Engineering, Princeton University, Princeton, NJ 08544-5263, USA

In this lecture, I would like to briefly go through the physics that I have learned in the years since I ventured into what we nowadays call research of semiconductor electronics in low dimensions, or in my case, more simply the electronic properties of two-dimensional systems.[1] To summarize, electrons confined to the interface of two different semiconductors normally behave like an ordinary gas of particles in two-dimensions. But, when taken to extreme conditions of low temperature and high magnetic field, they show new physics phenomena manifesting the interplay of electron-electron interaction and the interaction of the electrons with imperfections in the semiconductors. Let me first recall the events in my earlier research that led me to the journey that Art Gossard, Horst Störmer and I took in our adventure towards the discovery of the fractional quantum Hall effect (FQHE).[2]

PROLOGUE

I joined Bell Laboratories in the spring of 1968. I was sufficiently naive that I foolhardily convinced myself to leave behind the more familiar band structures and Fermi surfaces of metal physics, which I had become comfortable with through my years of research as a graduate student, and decided to try something different, e.g. surfaces or interfaces, and preferably some many-body interaction physics. I read about Anderson localization, Mott transition, and the notion that disorder and electron-electron interaction were the richest and most challenging problems in solid state physics. But, I did not have the foggiest idea on how to do what to get started.

Fortunately, I was advised to talk to John Rowell, who had at the time just completed the Rowell-McMillan electron-phonon interaction work using tunneling in superconductivity. John told me to look into point contact tunneling into the high T_c superconductors of those days and suggested that, instead of using tungsten wiskers, I should experiment with semiconductor tips which have built-in surface depletion layers as tunnel barriers. My effort to tunnel into the superconductors of niobium and vanadium compounds was unsuccessful, but the project forced me to learn some physics of semiconductor surfaces and interfaces. In fact, I was able to demonstrate experimentally the electric field quantization of the surface space-charge layer, first proposed by Schrieffer in the 1950s, by doing a tunneling experiment on InAs to observe directly the quantized energy levels and the Landau levels of the resulting two-dimensional (2D) electrons.[3] However, the most exciting part of this effort was my discovery, in writing the paper on this work, of the beautiful work on the Si metal-oxide-semiconductor field-effect transistor (Si-MOSFET) done by the IBM group in Yorktown Heights.[4] They laid a solid foundation for the development and growth of 2D electron physics in the subsequent decades.

Based on the IBM work, Jim Allen and I made a temperature dependence study of the inversion layer conductance in Si-MOSFET's to look for the 2D Anderson localization-delocalization transition. By varying the gate voltage on the device, we were able to move the Fermi level into the band-tail and observe directly a transition from the two-dimensional electron gas (2DEG) behavior to the behavior of an insulator. But, to our disappointment, we were unable to obtain quantitative agreement with theory. We thought interaction might be the cause and came to the conclusion that to enhance the interaction we should apply a magnetic field, which would change the individual electron's kinetic energy into the cyclotron energy. This should be especially effective in the extreme quantum limit when the cyclotron diameter was less than the average electron-electron separation.

In a 1976 paper, Kawaji and Wakabayashi[5] reported the observation of localized states in the energy gap between two Landau levels. This discovery was a most important milestone in the path to the quantized Hall effect. In response to their work, I studied at the Frances Bitter National Magnet Laboratory on MIT campus the conductance in the extreme quantum limit, when all electrons occupy the lowest Landau level, and saw some structures. Phil Anderson, after hearing this from John Rowell, asked to see the data. But by the time I showed them to him in the Bell Labs tearoom, I had repeated the experiment and found them to be sample specific. I told this to Phil and he made a cryptic remark under his breath that there should be some commensuration energy anyway. I reasoned: given that n is the 2D electron density and that the magnetic field B (applied perpendicular to the 2D plane) is expressed in terms of average flux density $n_\phi = B/\varphi_o$ (and $\varphi_o = {}^h\!/_e$ is the Dirac flux quantum), the ratio n/n_φ is the Landau level filling factor ν. For $n>n_\varphi$, an integer i number of Landau levels are filled at commensuration and the cyclotron energy, separating the filled from the empty levels, is the commensuration energy. I assumed he meant: In the $n<n_\varphi$ extreme quantum limit, at commensuration when $\nu = n/n_\varphi = {}^1\!/_i$, some interaction energy might become dominant to drive the 2D system into some new ground state. I was not brave enough to ask him: "What do you mean?" But I felt affirmed that I should continue to concentrate on the extreme quantum limit.

Indeed, with the advent of molecular beam epitaxy[6] and the invention of modulation doping to produce highly perfect 2D electron systems[7], it soon became quite clear to Art Gossard, Horst Störmer and me that where we wanted to go to look for new many-body interaction physics should be a highest mobility 2DEG sample placed in a most intense magnetic field.

TWO-DIMENSIONAL MAGNETO-TRANSPORT

In the presence of a perpendicular magnetic field the energy levels of a two-dimensional electron collapse, as a result of Landau quantization of its cyclotron orbits, into discrete Landau levels separated by the cyclotron energy quantum. Scattering broadens the Landau levels and gives rise to 2D magneto-transport described by the Ando-Uemura[8] theory. Fig.1 is an example showing the quantum oscillations in the diagonal resistivity ρ_{xx}, reflecting the broadened Landau level structure of the 2DEG, and the Hall resistance ρ_{xy}, well known from the Drude model. However, when the 2DEG is taken to the extreme condition of high B and low T, much more striking features appear, showing the interplay of disorder and electron-electron interaction in the system. More specifically, different physics phenomena are observed in three distinctly different physical regimes. The first is the disorder dominant regime, when the sample is dirty with low 2D electron mobility (e.g. $\mu < 10^5$ $cm^2/Vsec$ in the case of GaAs). The striking features in the data constitute the integral quantum Hall effect (IQHE)[9], which is understood in terms of the physics of independent electrons and their localization in the presence of

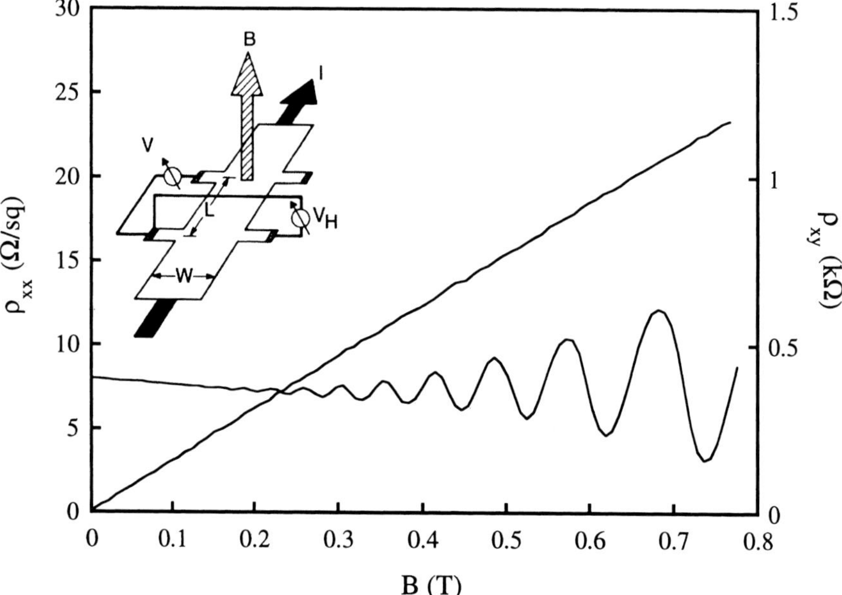

Figure 1. Magneto-transport coefficients ρ_{xx} and ρ_{xy} of a 2DEG in $GaAs/Al_xGa_{1-x}As$ at 0.35K in moderately low B. The insert shows the measurement geometry. The magnetic field B is perpendicular to the plane of the 2DEG and to the current I. The voltages V and V_H are respectively measured along and perpendicular to I. $\rho_{xx} = (^V/_L)/(^I/_W)$ is the resistance across a square, independent of the square size, and $\rho_{xy} = V_H/I$ is the Hall resistance independent of the sample width. Data taken by A. Majumdar.

random impurities in the semiconductors. The fractional quantum Hall effect (FQHE) is observed in high mobility samples in the second regime where the electron-electron interaction dominates. It manifests the many-body interaction physics of the 2DEG in the intense B field. Furthermore, even in the cleanest samples, the FQHE series terminates into an insulator in the high B limit. This insulator is believed to be an electron crystal pinned by defects to the semiconductor. The third regime is this high μ and high B field limit, where disorder and interaction play equally important roles and need to be treated on equal footing.

QUANTUM PHASE TRANSITIONS IN IQHE

Quantization of the Hall resistance in the natural conductance unit e^2/h is currently understood in terms of the existence of an energy gap, separating the excited states from the ground state, and localized states inside the gap. In the IQHE case, where the quantum numbers are integers identified with the number of completely filled Landau levels, the energy gap is the Landau gap of a cyclotron energy quantum. The accurate quantization was shown by Laughlin, using a gedanken experiment, as a consequence of charge quantization and that the experiment in effect measures the charge carried by the excited electron. The localized states arise from disorder in the 2D system, and the data, as shown in Fig.2, shows the localization-delocalization phase transitions. In other words, for B in the plateau regions, the states at E_F are localized, and in between, delocalized. As T is decreased, the range of B for the

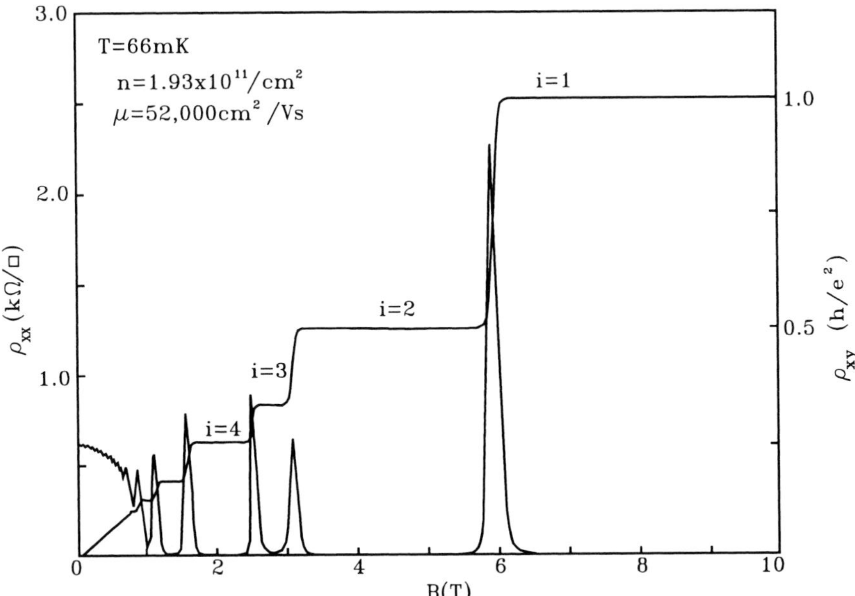

Figure 2. ρ_{xx} and ρ_{xy} of a relatively low mobility 2DEG in GaAs/$Al_xGa_{1-x}As$. The plateaus in ρ_{xy} are quantized in the natural conductance unit e^2/h with integer quantum numbers $i = 1, 2, \ldots$ Data taken by H. P. Wei.

existence of delocalized states decreases and the transition regions between the plateaus narrow. In the limit T→0, ρ_{xy} approaches a staircase. The underlying physics is the Anderson localization-delocalization quantum phase transition and the experiment is simply a magnificent display of 2D quantum critical phenomena, as first put forward by Pruisken.[10]

Quantum phase transitions take place at T=0. To relate them to experiments relies on the finite T behavior of the system at sufficiently close to the transition, which is governed by special rules, derived from simple scaling arguments. In our case, this boils down to the narrowing of the plateau to plateau transition regions following a power law dependence on T, with the temperature exponent a universal constant. Fig.3 is data from Wei *et al.*[11], where the narrowing of the transition regions is measured by the maximum in $d\rho_{xy}/dB$ and by the inverse half width $(\Delta B)^{-1}$ of the ρ_{xx} peak. The data shows a power law dependence $\sim T^{-\kappa}$ and the temperature exponent $\kappa=0.42 \pm 0.04$, independent of the Hall plateaus involved. Furthermore, the hallmark of a quantum phase transition is that the finite frequency f behavior of the system and its finite T behavior are similar; both follow power law dependence and are characterized by the same exponent. Crossover from one to the other oc-

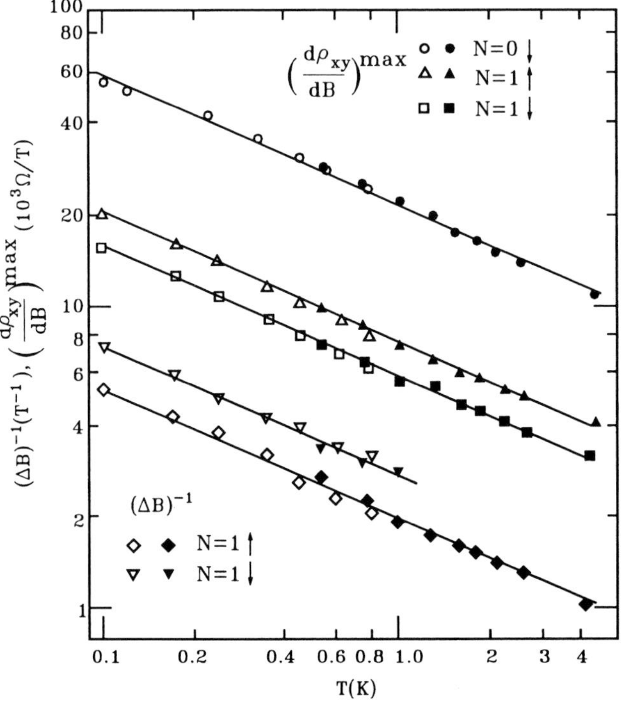

Figure 3. Data on narrowing of the plateau to plateau transition regions from an $In_xGa_{1-x}As/InP$ sample with $n=3.3 \times 10^{11}/cm^2$ and $\mu=3.4 \times 10^4\ cm^2/Vs$. The upper portion shows the maximum in $(d\rho_{xy}/dB)$ for the $i=1 \rightarrow 2$, $2 \rightarrow 3$, and $3 \rightarrow 4$ transitions (i.e. the N=0 ↓, 1 ↑, and 1↓ respectively); the lower portion shows the inverse half-width $(\Delta B)^{-1}$ of the ρ_{xx} peak for the $i=2 \rightarrow 3$ and $3 \rightarrow 4$ transitions. N is Landau level quantum number (From Ref. 11).

curs around $hf = kT$. Engel *et al.*[12] studied the microwave conductance in the frequency range from 0.2GHz to 16 GHz in a dilution refrigerator and were able to go from $f < kT/h$, where the scaling is dominated by T, to $f > kT/h$ where frequency scaling should hold. Their data is in Fig.4. They show power law dependence on f in the $f > kT/h$ limit with a frequency exponent equal to the T exponent within the experimental error bar, and are consistent with a crossover around $hf = kT$.

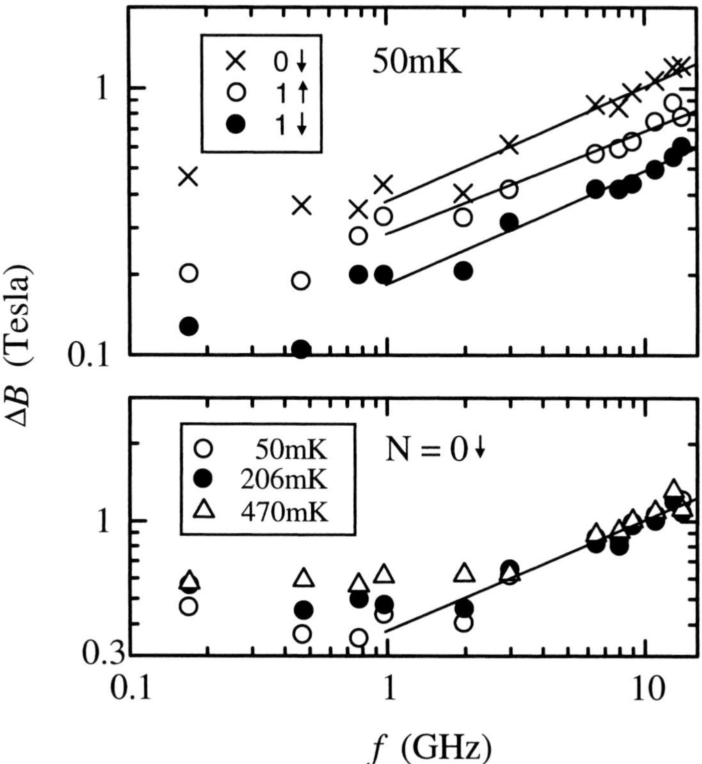

Figure 4. Microwave frequency dependence of the width of the plateau to plateau transition regions. Upper panel: Half width ΔB of $\sigma_{xx}(f)$ for the i=1 →2, 2 →3, and 3 →4 transitions (i.e. the N=0↓, N=1↑, and N=1↓, respectively). Lines are least squares fits of data for $f \geq 0.97$GHz to $\Delta B \sim f^{\gamma}$, with $\gamma = 0.43$, 0.38, and 0.42, respectively. Lower panel: ΔB of $\sigma_{xx}(f)$ for the i=1 →2 transition at three different T's. The line is a fit to $\Delta B \sim f^{0.43}$ (From Ref. 12).

THE FQHE

The second regime, where the electron-electron interaction dominates, is accessible using high mobility samples. In this regime, the FQHE becomes observable, and a large number of plateaus in ρ_{xy} and concomitant ρ_{xx} minima are apparent in the data (Figs. 5 and 6), even after the IQHE structures are exhausted in the $\nu<1$ extreme quantum limit. These plateaus, as determined

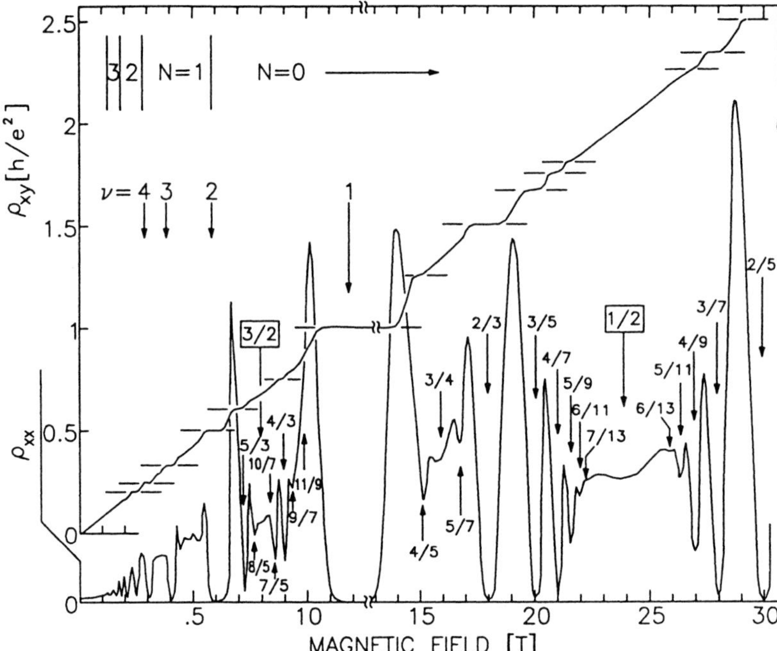

Figure 5. ρ_{xx} and ρ_{xy} of a 2DEG in $GaAs/Al_xGa_{1-x}As$ with n=3.0 x $10^{11}/cm^2$ and μ=1.3 x $10^6 cm^2/Vs$. The quantized Hall resistance plateaus are indicated by the horizontal bars and the odd denominator fractions marking the concomitant vanishing ρ_{xx}. The use of a hybrid magnet with fixed base field required composition of this figure from four different traces (breaks at ≃ 12T). Temperatures were ≈ 150 mK except for the high-field Hall trace at T=85 mK. The high-field ρ_{xx} trace is reduced in amplitude by a factor 2.5 for clarity. N is Landau level quantum number. Filling factor ν is indicated (From Ref. 13).

from their resistance values, are fractionally quantized and they occur around the same fraction of Landau level filling. At such fractional fillings, the single electron levels are highly degenerate and there is no energy gap across E_F to possibly give rise to Hall resistance quantization. Horst Störmer and Bob Laughlin will discuss the new many-body interaction physics manifested in the phenomenon and the broader implications of it in their lectures. Here, I simply want to mention the so-called odd denominator rule that all the fractional quantum numbers are odd denominator fractions, and to point out that there is now a firmly established exception at $\nu = {}^5/_2$. Over a decade ago, Willett *et al.*[13] reported the observation of a deep minimum in ρ_{xx} and a clear deviation of the Hall resistance from its classical line around $\nu = {}^5/_2$ filling, suggestive of an even denominator fraction ${}^5/_2$ FQHE. Very recently, Wei Pan and Jian-Sheng Xia, working with the University of Florida microkelvin group in Gainesville, cooled the 2DEG below 10 mK and were able to observe a Hall plateau quantized to better than 2 parts in 10^6, thus making the FQHE nature of the ground state at $\nu = {}^5/_2$ unequivocal.

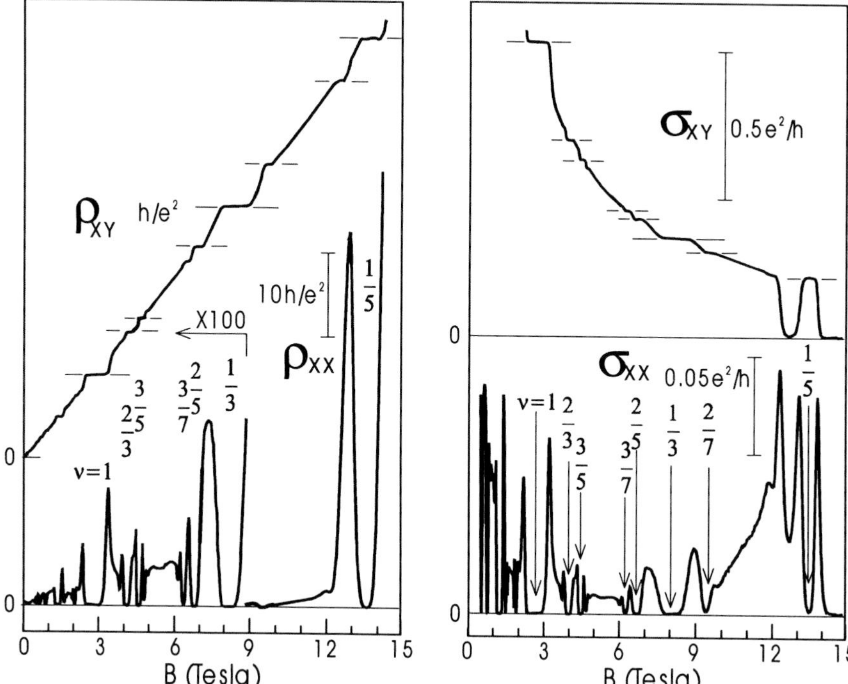

Figure 6. Magneto-transport coefficients of a 2DEG in GaAs/$Al_xGa_{1-x}As$ with n=6.5 x $10^{10}/cm^2$ and μ=1.5 x 10^6 cm^2/Vs. Left panel: ρ_{xx} and ρ_{xy} at 40 mK. Right panel: σ_{xx} and σ_{xy} obtained by inverting the ρ_{xx} and ρ_{xy} data. The vanishing of σ_{xy} together with σ_{xx} at B ~12.8T and B>14T indicates insulating behavior. Data taken by Y. P. Li.

THE MAGNETIC FIELD INDUCED CRYSTAL REGIME

Finally, there is the third regime, where disorder and interaction are equally important. This is the small filling limit after the FQHE series terminates into an insulating phase. In the absence of disorder, the ideal 2DEG is predicted to be a 2D electron crystal at sufficiently small fillings. But, in real physical systems, there is always disorder that can alter the ground state in fundamental ways. To date, insulating behavior is seen in the highest mobility 2D electrons for $\nu < 1/5$, and 2D holes for $\nu < 1/3$. This insulating phase in the cleanest 2D systems we have is attributed to crystallization of the 2D electron and 2D hole gases under intense B field. The crystal, being pinned to the semiconductor by defects in the semiconductor, cannot slide to conduct electricity.

Experimentally, this is a challenging regime. It requires all the best: the highest mobility samples, lowest T, and most intense B. Since it is an insulator, dc transport is limited. Microwave measurements, more appropriate at first sight, are notoriously hard and can be plagued with pitfalls. Consequently, unequivocal experiments are few and there is very little direct information on the crystalline nature of the ground state. A great deal of the properties of this insulating phase still remains unknown and unexplored. Chi-Chun Li and Lloyd Engel[14] recently improved their microwave absorption experiment and obtained data (Fig.7) showing a sharp conductance resonance in the in-

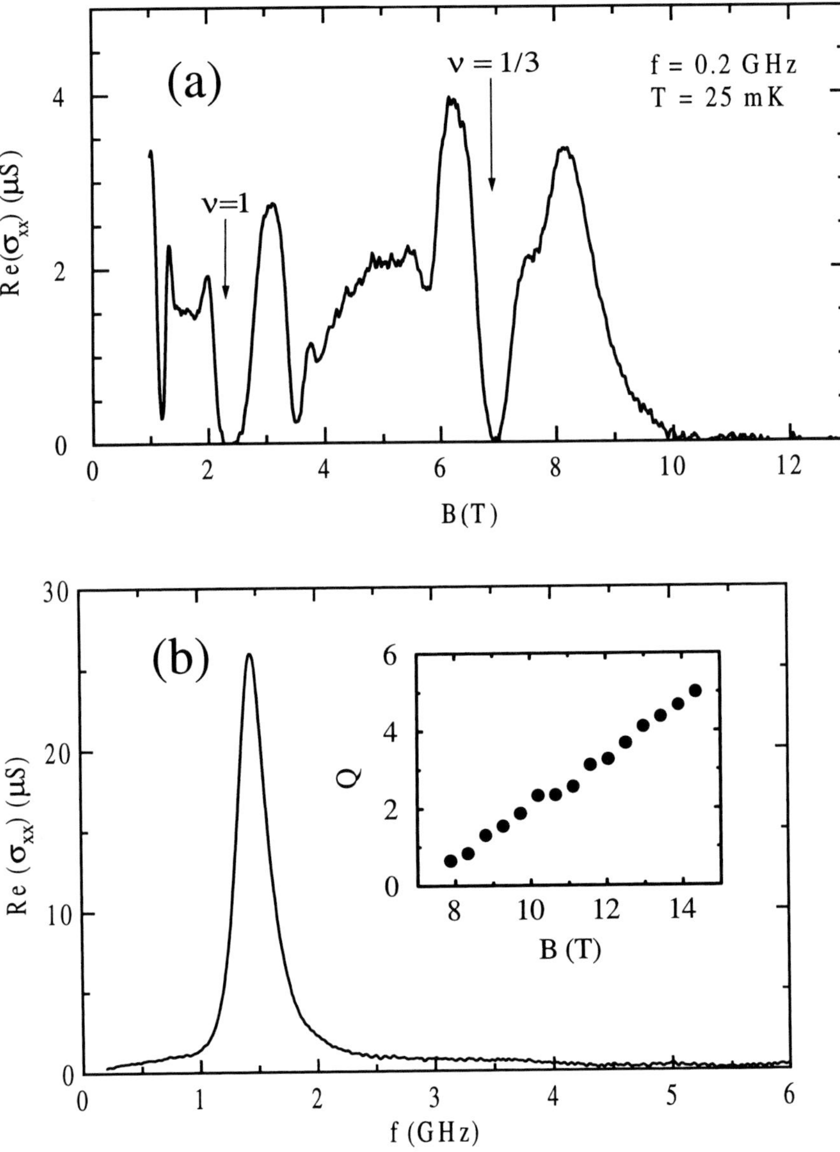

Figure 7. (a) Real part of diagonal conductivity vs. magnetic field of a two-dimensional hole gas (2DHG) in GaAs/$Al_xGa_{1-x}As$ with a hole density of 5.5 x $10^{10}/cm^2$ and μ=3.5 x $10^5 cm^2/Vs$. The 2DHG becomes insulating for B $\gtrsim$ 10T. (b) Resonant microwave absorption in the insulating phase of the 2DHG at B=13T and T=25mK. The insert shows the quality factor of the resonance Q vs. B (From Ref. 14).

sulating phase at ~1.5GHz, a frequency consistent with that expected of a pinning mode of the crystal. It is surprising, however, that the resonance is sharp, with a quality factor Q much larger than 1. The Q, as seen in the inset of Fig.7(b), increases with increasing B to the highest B studied. Thus, it appears that there is nontrivial fundamental physics hidden in this regime, and I look forward to further experimental efforts.

ACKNOWLEDGEMENTS

As a conclusion, I must emphasize that the FQHE journey that Horst Störmer and I have been on would never have gotten started without Art Gossard's participation in the early stage. To continue the journey to this day would also not have been possible without the hard work, the intellectual stimulation, and the youthful vitality of our graduate students and post-doctoral associates, and the contributions and collaborations of our colleagues in Bell Laboratories, Princeton University, and elsewhere in our community. They include K. W. Baldwin, P. Berglund, G. S. Boebinger, A. M. Chang, R. Du, J. P. Eisenstein, L. W. Engel, V. J. Goldman, M. Grayson, M. Hilke, H.W. Jiang, G. Kaminsky, C. Li, Y.P. Li, M. A. Paalanen, W. Pan, A. Pruisken, T. Sajoto, D. Shahar, J.A. Simmons, Y.W. Suen, H.P. Wei, R. Willett, and A. Yeh. Also, the most valuable support that made it possible for me to continue a research enterprise after moving to Princeton University was from the materials expert friends, K. Alavi, A. Y. Cho, J. E. Cunningham, J. H. English, A. C. Gossard, J. C. M. Hwang, J. F. Klem, R. A. Logan, L. N. Pfeiffer, M. Razeghi, M. B. Santos, M. Shayegan, W. T. Tsang, C. Tu, G. Weimann, K. W. West, and W. Wiegmann, who taught me the basics of semiconductor materials and technology and often provided samples for us to experiment with.

REFERENCES

1. T. Ando, A.B. Fowler, and F. Stern, Rev. Mod. Phys. **54**, 437 (1983).
2. D.C. Tsui, H.L. Störmer, and A.C. Gossard, Phys. Rev. Lett. **48**, 1559 (1982).
3. D.C. Tsui, Phys. Rev. Lett. **24**, 303 (1970).
4. F.F. Fang, A.B. Fowler, W.E. Howard, F. Stern, P.J. Stiles and their collaborators, see Ref. 1 for a review.
5. S. Kawaji and J. Wakabayashi, Surface Sci. **58**, 238 (1976).
6. For a recent review, see A.Y. Cho, MRS Bulletin, **20**, 21 (1995).
7. H.L. Störmer, R. Dingle, A.C. Gossard, W. Wiegmann, and M. Sturge, Solid State Commun. **29**, 705 (1979).
8. T. Ando and Y. Uemura, J. Phys. Soc. Japan **36**, 959 (1974).
9. K. von Klitzing, G. Dorda, and M. Pepper, Phys. Rev. Lett. **45**, 494 (1980).
10. A.M.M. Pruisken, Phys. Rev. B**32**, 2636 (1985).
11. H.P. Wei, D.C. Tsui, M.A. Paalanen, and A.M.M. Pruisken, Phys. Rev. Lett. **61**, 1294 (1988).
12 L.W. Engel, D. Shahar, C. Kurdak, and D.C. Tsui, Phys. Rev. Lett. **71**, 2638 (1993).
13. R. Willett, J.P. Eisenstein, H.L. Störmer, D.C. Tsui, A.C. Gossard, and J.H. English, Phys. Rev. Lett. **59**, 1776 (1987).
14. C.-C. Li, L.W. Engel, D. Shahar, D.C. Tsui, and M. Shayegan, Phys. Rev. Lett. **79**, 1353 (1997).

Physics 1999

GERARDUS 't HOOFT and MARTINUS J.G. VELTMAN

"for elucidating the quantum structure of electroweak interactions in physics"

THE NOBEL PRIZE IN PHYSICS

Speech by Professor Cecilia Jarlskog of the Royal Swedish Academy of Sciences.
Translation of the Swedish text.

Your Majesties, Your Royal Highness, Ladies and Gentlemen,

"Everything is made up of water," Thales told us 2,600 years ago. But he was speaking in the language of philosophers. The natural scientist of our time says instead: "Everything is made of elementary particles": the flowers in this hall, the bust of Alfred Nobel, indeed every one of us – even our distinguished Laureates! Elementary particles are nature's smallest building blocks, the roots of everything.

In order to describe the building blocks of nature we must have a language, a theory. This year's Laureates in Physics, Gerardus 't Hooft and Martinus Veltman, have made a decisive contribution to this language. Their contributions concern electromagnetic and weak interactions of the building blocks of matter.

Electromagnetism is the interaction which is responsible, for instance, for the existence of atoms. The principal actor here is the photon – light – because without light there would been no electromagnetism. Weak interactions also have their own agents: three particles that unfortunately have not been honored by beautiful and dignified names but are simply called W-plus, W-minus and Z. In spite of their dull names, these particles are of paramount importance. Consider, for example, W-plus. We know that the sun is like an oven. But who makes the fire there? The particle W-plus, of course! Without weak interactions the sun would not shine!

It turns out that the photon and the particles W and Z have a common origin. Electromagnetism and weak interactions are thus unified, and together they are called electroweak interactions.

This year's Laureates had their predecessors – prominent researchers who successively improved the description of these interactions. But weak interactions appeared to act like loose cannons. The calculations gave chaotic results – sometimes excellent and occasionally completely absurd. Nonsensical results appeared everywhere, as infinite probabilities and infinite quantum corrections.

The theory of weak interactions was undoubtedly very sick. Today, with hindsight, we see that it was in fact Veltman who indicated the correct direction to be taken. His guiding star was the concept of symmetry. The magic wand of symmetry converts a little fragment here, and a little one there, into a complete picture. The pattern that Veltman saw was that of a "non-abelian gauge theory," also called a "Yang-Mills" theory. Veltman embarked on studies of weak interactions within the framework of these theories and found encouraging results. A ray of light illuminated weak interactions!

Although 't Hooft joined Veltman, as his Ph.D. student, a year or so later, it was no easy task that awaited him. His contributions to the solution of their common problem were simply dazzling. 't Hooft and Veltman showed how the nasty infinities could be harnessed and interpreted. First the theory had to be modified, in order to be able to embark on calculations. This was done by introducing, among other things, a number of ghosts – particles that don't exist. These should, however, be well-mannered ghosts which, at the final stage, would say goodbye and disappear. And that is exactly what they do in 't Hooft's and Veltman's method. Albert Einstein taught us that we live in four dimensions – three spatial dimensions and time. 't Hooft and Veltman tell us instead: calculate as if the number of dimensions were slightly less than four, four minus epsilon, i.e., 3.99999... This approach proved to be highly effective. The nasty infinities became less frightening. They could be collected, harnessed and interpreted.

Although 't Hooft and Veltman did their – now prize-endowed – work around 1970, it has taken a long time for us to understand the extent of their ground-breaking efforts. We had to wait for the results from an accelerator called the Large Electron Positron (LEP) at the European Laboratory for Particle Physics (CERN) on the outskirts of Geneva. This accelerator, inaugurated in 1989 in the presence of – among others – His Majesty, has set the world record in precision measurements of electroweak interactions. 't Hooft's and Veltman's work has been a prerequisite for interpreting these results. The LEP results showed that a sixth quark, the top quark, was needed and that its mass could be determined even though there was not then sufficient energy to produce it. In order for this prediction to be confirmed we had to cross the Atlantic to the Fermi National Laboratory near Chicago where, in 1995, the top quark was discovered.

The theory of electroweak interactions predicts the existence of an extremely interesting particle, called the Higgs particle, whose mass can be determined with the help of 't Hooft's and Veltman's methods. Without the Higgs particle, says the theory, we would all be massless and thus doomed to perpetual motion at the speed of light. But then, we would not exist and thus not be able to think (nor indeed – perish the thought! – to enjoy this excellent evening). Will we discover the Higgs particle? The future can be summarized with a single word: Exciting!

Professor 't Hooft, Professor Veltman,

On behalf of the Royal Swedish Academy of Sciences I wish to congratulate you for your ground-breaking work. I now ask you to step forward to receive your Nobel Prizes from the hands of His Majesty the King.

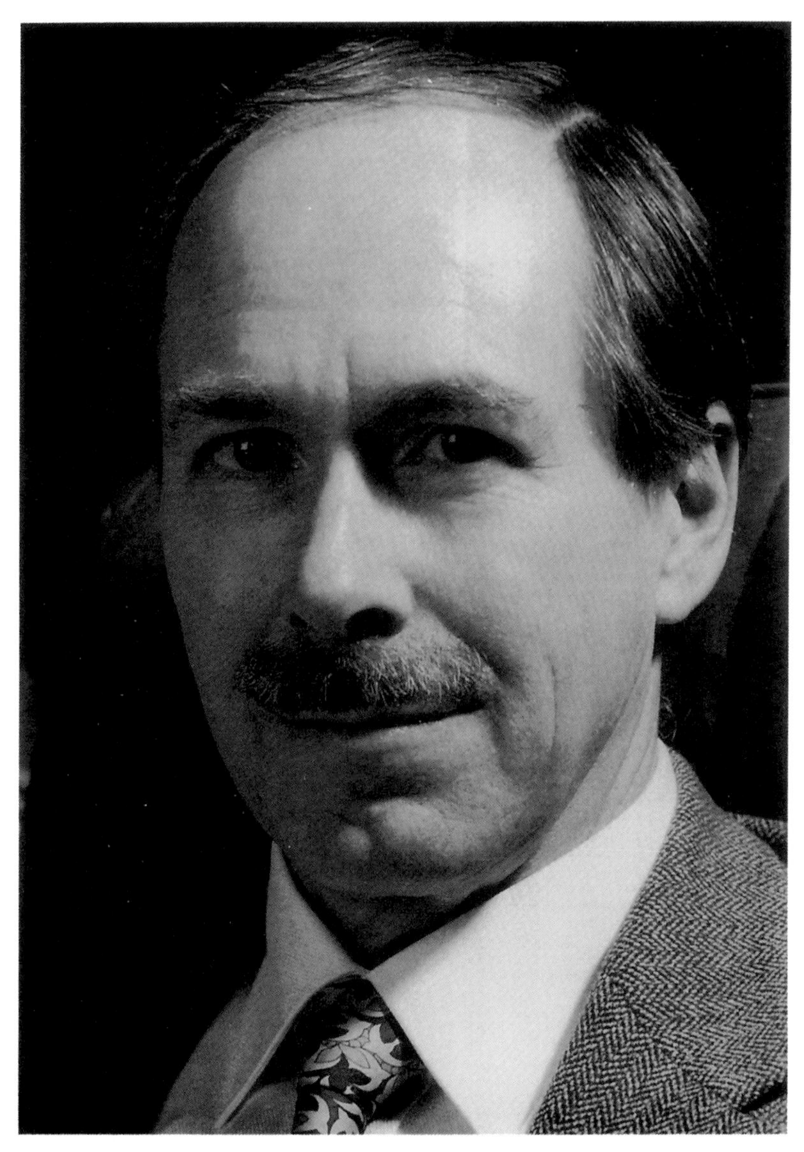

Gerard 't Hooft

GERARDUS 't HOOFT

"A man who knows everything". This, reportedly, was my reply to a school teacher asking me what I'd like to become when I grow up. I was eight years old, or thereabouts, and what I wanted to say was "professor", but, still not knowing everything, I had forgotten that word. And what I really meant was "scientist", someone who unravels the secrets of the fundamental Laws of Nature.

This perhaps was not such a strange wish. Science, after all, was in my family. Just about at that time, 1953, my grand-uncle, Frits Zernike had earned his Nobel Prize for work that had led him to the invention of the phase contrast microscope. He had worked out the theory and singlehandedly constructed his microscope, with which he had stunned biologists by showing them moving images of a living cell. My grandmother, Zernike's sister, used to tell us anecdotes about her brother when they were young. One day, for instance, he had purchased a telescope at a local market. That night, the police came at their door to warn her parents that there were "zinc thieves on their roof"; it was Frits however, trying out his new telescope and studying the heavens. She herself had married her professor, a well known zoologist, Pieter Nicolaas van Kampen at the university of Leyden. I never knew him, he passed away, after a long illness, when my mother was eighteen years old.

My uncle, Nicolaas Godfried van Kampen was appointed Professor of Theoretical Physics at the State University of Utrecht. My mother did not opt for a scientific career. "It never came up", she now says, adding that actually math and science were not particularly difficult for her at school, but being a girl, you wouldn't admit that you actually liked such subjects. She went to art school but later achieved a degree in French, and now she teaches that language in a private class.

Was it the environment or was it in my genes to choose to become a physicist? My grandmother adored scientists and by that she may have further determined my choice, but I think that my mind was made up long before I could talk. A picture was taken of me, at the age of two, studying a wheel. I do not remember the event, of course, but I do remember being fascinated by wheels when other kids were just running around, playing. My very earliest recollections are about being obsessed with phenomena I observed. I watched the ants crawling in the sand, and wondered what life would be like if you were an ant. You would be able to go into the tiniest spaces between the pebbles, and those would be as big as houses for you. But, I realized, an ant's life must be totally different from ours. Still being a toddler I saw one day how the wheels of two children's bikes, which were upside down, touched each other. If you turn one wheel, the other one would start rotating as well. You can make one wheel turn by rotating the other. The principle of transmission. How fascinating Nature is. I was well over two years old before I started to speak. Was it because there were so much more interesting things I wanted to understand than to communicate with people? I was also late in reading and writing. This, I remember, was because I thought reading meant being able to decipher my mother's handwriting.

Though born in Den Helder, I spent my childhood in The Hague, with my parents, my older sister who had changed her official name Elise into Ita as soon as she could talk, and my younger sister Agnes. My father had obtained a degree at Delft in naval engineering. He made his career at the dockyards of the big ocean cruisers of the Holland-America Line. He used to talk of the giants "Maasdam" and "Rijndam" as his ships. Then for a long time he worked at an oil company until he had enough of that. Like his father, he loved ships and all high-tech industry having to do with the sea. Noticing my interest in natural phenomena, he thought that it would be easy to get me interested in engineering as well. He bought me books about ships and car engines, which I never touched. "Those things have already been invented by someone else", I objected. "I want to investigate Nature, and discover new things."

When I was eight, my family moved for a ten month's period to London, England, where for the first time I was forced to master a foreign language, English. Too late, my parents discovered that sending their children to a private school would have required registration three years or more ahead. We went to a public school. School uniforms were not required, but there were strict regulations on clothing. One cold day I entered the school in long trousers. I was allowed in because I was a foreigner, and they always were very kind to me, but shorts, reaching until the knee, were the norm for the school. In summer time, during the week-ends, we would make long trips in the beautiful country-side. It seemed that all rain in England fell during the week-ends. I saw my first mountains, that is, hills taller than 100 meters, which hardly exist in the Netherlands. I was thrilled to notice that the tree trunks grow along the lines of gravity and ignore the direction of the slope. I also noticed some fundamental differences in English and Dutch architecture, so that, if you show me some houses, old or new, I can immediately tell the Dutch and English ones apart.

My father made more money than usual, and this afforded him to buy me some expensive boxes of Meccano. It was one of the great things he did for me. However, I had to make a deal with my father. Alternatingly, I would construct a model described in the book, and then construct something out of my own imagination. He thought the models in the book were more instructive, but I preferred my own imagination. The most fantastic things I constructed were robots, that I could persuade to pick up something, although infinite diligence was needed for that.

After primary school I went to the Dalton Lyceum, also in The Hague. It is a school system where students are given extra hours for studying homework material in the presence of teachers, and it worked well for me. After one year the choice was to be made between a non-classical and a classical continuation, the classical one including ancient Greek and Latin, which would take one year more, and it would be more demanding. My uncle said the choice would be immaterial. "You don't need Latin and Greek for physics", he said, "but it doesn't do any harm either." I chose to take the challenge. Why? I think I couldn't stand the idea that some kids would learn things I didn't know. I never regretted the choice.

My father bought me a book about radios, and that did interest me. "You know, Gerard", a schoolmate had once said to me, "nobody in the world understands how a radio works". This I found difficult to believe. "Look at all those things inside", I said, "the guy who designed that must have had some idea." But if there were any not understood secrets, I was going to find out about them, that I promised myself. The radio in the book had lamps in it, diodes, triodes, pentodes. Later I learned that transistors work the same way, and you could buy sets with complete instructions how to assemble a radio. I would never build a radio before I understood why it had to be assembled precisely this way. Why, for instance, would the designer always suppress the amplification power of a transistor by back coupling? I tried to make an amplifier with fewer transistors and no suppression. Can you make a radio with just one transistor for both the high frequency and the low frequency signal? I learned the answers to all these questions.

Of the modern languages, English, French and German, besides Dutch, were obligatory. I had difficulties with the logic of linguistic arguments and besides, the texts we had to translate were such that even in my own language I could hardly understand what they were about. But I managed, and now I am happy that I can communicate with the inhabitants of a major fraction of Europe.

So much easier were math (of which there was surprisingly much: algebra, analysis, trigonometry, stereometry), physics and chemistry. My physics teacher was a friendly, middle aged man with a small beard and a soft voice. He taught physics using a book that he and another teacher at our school had written, and which was being used throughout the country. It was sound and pedagogical, but not always equally accurate. Where fluids were discussed, it explained that the cross section of an airplane wing has "a droplet form" because "droplets take a shape of least resistance". Elsewhere, the rainbow is derived, and there droplets were spherical.

Being pedagogical was high on my teacher's priority list. But he also inspired us and made us think. "If there were any real geniuses in this class", he would say, "then they could have argued as follows, ...". But then, he assured us, there were of course no real geniuses in this class. Then, there was an interesting page in his book about photons. "A light bulb emits about 10^9 photons per second," it said. The argument was simple. "A single photon has a wave packet of about 10^{-9} seconds long. If there were much more than 10^9 photons, then for each photon vibrating in this way, you could find another photon vibrating in the opposite direction. You'd have destructive interference, and so there would not be any light." I had long arguments with him about this. Finally, with the help of my uncle, we could sort things out. This page does not appear in the later editions of the book.

Biology was taught by an elderly lady, too kind for this world. She would never give anyone failing marks, unless someone really asked for it, but high marks were also rare. My marks quickly dropped when the lessons became boring, such as the discussion of symmetry patterns in flowers (I thought the symmetry was never perfect anyway), or incomprehensible, when the human body was discussed (some parts were hardly mentioned, except outside hours among the pupils, and there were things that no-one could explain to me, and I didn't dare to ask).

Then, one of the teacher consultation days, my father noticed that none of the parents wished to talk to our biology teacher, since she never made anyone fail. He stepped towards her and said: "Did you know Professor P.N. van Kampen?" Of course she did, surely she did, she had attended all his lectures. He was such a scholar, he was so bright! Is Gerard really his grandson? If only had she known! The next day she started with zoology. I was given special attention. Van Kampen's grandson! My grades skyrocketed. She gave me the assignment to write a thesis. I chose to write about bacteria. Our local library had nothing about bacteria. One pre-war book was there, written in German in Gothic letters. I still don't know how I managed to produce a thesis using that. But it did not matter. My grade for it was superb.

I had the good fortune of having an enthusiastic art teacher. I suspect it was only because of my good geometric insight that I could make quite realistic drawings. But my mother spotted the weak points in my art. If you want to draw a human face or body, you have to know exactly how the bones and the muscles go, she said, otherwise you do it all wrong, and it doesn't look good. I was too shy to make a careful study of human bodies, and so I specialized in animals and landscapes. This will never make me a very good artist, I decided.

When I was ten I encountered my first piano. We were on a vacation in the hills in the south-east of Belgium. It was continuously raining during the entire two weeks. The cottage we had rented had an old piano in it. There were a few books with some songs in them. My father explained how the notes relate to the keys on the instrument. "The rest you can figure out by counting". Both my parents had suffered from compulsory piano lessons when they were young, and had intended not to subject any of their children to such a torment. But now that I wanted them, I could get my piano lessons. I had a private teacher. She was tough. She herself had had lessons from the well-known Dutch pianist Cor de Groot, and she wanted me to reach similar heights. I had to practice scales. It amazed her that the first time I tried to play a scale simultaneously with left and right hand, I nevertheless had the right idea to switch fingers left and right at different moments. "Most people first do this wrong", she said. She taught me Beethoven, Chopin, Debussy, Mendelssohn and many others. Much of it was too difficult for me, but I still play many of the pieces, and piano has become part of my life.

At age 16, the opportunity was offered to participate at the Dutch National Math Olympiad. It was the second time the olympiad was held. I passed easily the first round; only by being nervous I had misread the first exercise, which had been done correctly by most other participants. But I had done well with the others, and so I went with some 100 schoolkids to Utrecht for the next round. It was a tough one, and I had missed several questions. On hindsight, the questions had been very good ones, and I had only missed them because of lack of rigorous mathematical training. Today, math questions are phrased in such awkward ways, supersaturated with pedagogical nonsense, that I'd probably have missed them all.

Anyway, it came as a surprise when during a school break my younger sister came rushing towards me. "We searched for you everywhere," she said, "you're among the first ten!" The exact order was still kept secret. We came to Utrecht to learn that I had obtained the second prize. It consisted of two volumes of a book by Georg Pólya, "Mathematik und Plausibles Schliessen", and I devoured them. This was math of a kind that I liked very much. They must have seen by the way I had answered the questions that this was math to my liking. It contained, among other things, Euler's theorems for polygons in three-dimensional space, and this knowledge would turn out to be quite handy later in my career. I could have been number one in this Olympiad, if I hadn't flunked the first exercise in the first round, but then, probably, the others too had made avoidable mistakes.

The final examinations at high school, 1964, were tough. My only real problem was the languages, but what about biology? The high grades given by my teacher were ridiculous. Biology would be examined orally, and this time there would be a biology university professor who would independently judge the answers. When I entered the room, the first thing my teacher said to the university professor was: "Now this is Professor van Kampen's grandson!" His face brightened, "Really?", he said, he had followed all of Van Kampen's lectures. Such a brilliant zoologist. And here is his grandson. He must be very bright. They asked something about some obscure sponge. I vaguely remembered the text in the book, and tried to reproduce it. "Yes, yes!", they cried, "and sometimes it is said that ...", and then came the real text, which I had practically forgotten. They gave me a 10 out of 10. I gladly dedicate this result to the memory of my grandfather.

I passed the examination and went to the State University of Utrecht. Leyden was closer to The Hague, but my uncle was teaching at Utrecht, and his lectures I desired to follow. My father insisted that I become a member of the most elite student organization, the Utrecht Studenten Corps. Freshmen were shaven bold. This was actually one of the lesser humiliating things they did; the elderly students had developed a special skill at humiliating their freshmen. Some of the new students had already been in military service; for them, it was all only too familiar, and they had no problem. But I was easy to crack, and they could ridicule my lack of interest in anything but science. "So you wrote a thesis about bacteria? What kinds of bacteria are there?" It was an elderly medicine student who asked the question. When I mentioned the spirochetes, he asked: "and which disease is caused by them?" I knew what he wanted to hear. "Syphilis", I said. His opinion was that I should go into medicine, not physics.

But now I was near the Theoretical Physics Institute. I had rented a room just around the corner. Theoretical Physics occupied three adjacent houses opposite to a canal. One of the houses was owned by a lady who had introduced herself as a countess. There was some dispute as to whether she really was one. In summertime, when you opened the windows, chicken would hop in from the garden, and walk over the desks. Staff members would have coffee, lunch and discussions in a cellar. Through a narrow window you saw the legs of the pedestrians passing by. In earlier days the cellar had probably been in use by prostitutes. Of course, I was only a first year student, and I was not supposed to come in here. But more often than not, my uncle invited me in, and I adored the discussions, and the laughter.

The student organization forced me to spend time also on other things besides physics, which was exactly why my father had wanted me to become a member. I was coxswain in their celebrated Rowing Club, Triton, where I was appreciated because I could keep their boats coasting in straight lines. There was a student science discussion club, "Christiaan Huygens" where I have many fond memories, and together with some other students I organized a national congress for science students. But it was also at the student clubs where I learned to hate interminable meetings and pointless discussions. Especially the student revolts of the 60's I found silly and I kept at the greatest possible distance.

I wanted to go into what I saw as the heart of physics, the elementary particles. Unfortunately, my uncle had developed a dislike of the subject. People in that field are very aggressive, he warned. He had also investigated elementary particles, deriving what the mathematical consequences are of the fact that no information can go faster than light. You find equations, he explained, called dispersion relations, but they don't tell you everything about the particles. He had written a few articles, meticulously deriving these consequences. "And what happened? Others wrote dozens of papers, full of unwarranted assumptions, sloppy arguments and incredible results. But there were so many of those papers, that only they got all the citations." He thought that statistical physics was more to his liking.

There was a newly appointed Professor of Theoretical Physics who did specialize in subatomic particles, Martinus Veltman, or Tini, as he was normally called. When time came that I had to write an undergraduate thesis, somewhere in 1968, he was the person to advise me and judge me for it. Veltman naturally thought that those high grades of mine were just because of my family background, and if I were any good, he would first need some convincing. This never even bothered me, all I wanted was learn about elementary particles, and if he didn't think much of me, so be it. First things first, he said. Here is a paper by C.N. Yang and R.L. Mills. This stuff you must know.

Now this was a brilliant paper. It was beautiful, elegant and unique. But it was also considered to be useless. "It describes particles which do not exist in Nature", Veltman explained, "but in some modified form, they might". What modified form? To a fellow student, Veltman gave the assignment to study spontaneous symmetry breaking. There was a lot of confusion concerning the so-called Goldstone theorem. Jeffrey Goldstone had derived that spontaneous symmetry breaking implies the existence of massless particles. Spontaneous symmetry breaking could not be the resolution of the Yang-Mills problem because such massless particles do not exist. Later, this would be recognized as just one more example of too much adoration for abstract mathematical theorems; people did not bother to read the small-print, where Goldstone clearly said when his theorem does *not* apply. I am glad I ignored the problem; I did not understand why people thought there were massless particles if I could not see any in the equations.

My assignment was to study the so-called Adler-Bell-Jackiw anomaly. This was a subject in which Veltman was involved. He had a formal theorem saying that neutral pions cannot decay into photons. But when you actually calculate the decay, you find that it should occur. And the experimental data agree with that: neutral pions decay predominantly into photons. Something is wrong with the formal theorem. It was based upon flawed mathematics. The flaw was something highly interesting, and it would continue to play an interesting role later in particle physics. There were related problems with the eta particle. It decays into three pions while it shouldn't. The resolution to this problem was still entirely unknown.

They say that organizing a student congress causes one year delay in your studies. But I had never stopped thinking about physics, and I could begin my PhD studies in 1969. In Holland, the PhD is a very serious matter. I remembered my physics teacher being so proud of his thesis. My history teacher obtained his PhD late in his life, and he too had been telling us all about his defense of his lifetime work. Veltman was to be my advisor. He gave me the choice between various topics, but none could catch my imagination more than the subject he himself was working on: the renormalization of Yang-Mills fields. He explained to me that vector fields must be playing an elementary role in the weak interactions, but also in the strong interactions there were vector fields. All these fields were associated to spinning particles with mass. The mass was where the problem started. "These mass terms in the equations look quite innocent", he explained, "but in the end they impede all my attempts to obtain a finite, meaningful theory."

But he knew something else. He had studied the experimental data concerning the weak interactions. There, he had found very strong indications that the weak interactions have something to do with the theory of Yang and Mills. "But the matter becomes so complicated that you cannot do it by hand anymore", he said, and he had started designing a computer program to handle the complicated algebraic expressions. Computers were still in their infancy those days. Today's simplest hand-held calculators contain more electronic switches and are much faster than the bulky constructions that were called computers then. The monsters had to be fed with paper cards in which you had to punch your programs. His effort was an heroic one.

What I began thinking about was my own version of the Goldstone theorem, but I could not read those pompous mathematical theories. What I reconstructed in my own way was something that actually did exist already: it is now known as the Higgs mechanism, but important elements of it had also been derived by François Englert and Robert Brout. Unfortunately, these ideas were not along Veltman's line of thought. He wanted to derive everything just by looking at the experimental data, and by performing field transformations for which he could use his computer program. In his opinion, I clearly lacked insight in experimental subjects. Something had to be done about that. We sent my application to various summer schools in theoretical physics. My first choice was a school at Les-Houches, a ski resort high in the French Alps, near Chamonix. Famous French physicists would be teaching there. Presumably because my application was late, I was not admitted.

The next choice was Cargèse, and here I was admitted. Near this small town on the French island Corsica, right at the sea, the French physicist Maurice Lévy had established an Advanced Science Institute, ten years earlier. The story goes that Lévy had looked up in the atlas which French town has the maximal amount of sunshine during summer, and then he found this location. Now, Lévy had developed a model for the strongly interacting particles together with Murray Gell-Mann. Formally, the model could be renormalized, but in practice there were numerous problems, and they were going to be discussed. It was summer 1970. Lecturers were, besides Lévy and many others: the Korean Benjamin W. Lee, the German of Polish descent Kurt Symanzik, and many Frenchmen such as Jean-Loup Gervais.

The Gell-Mann-Lévy model is a model with spontaneous symmetry breaking. The pions are here interpreted as Goldstone particles. These lecturers were talking about renormalization in the presence of spontaneous symmetry breaking, and they were telling us that the mass terms that are generated (the mass of the proton) cause no problems whatsoever. As far as I remember, I only asked one question, both to Benjamin Lee and to Kurt Symanzik: "why can we not do the same for Yang-Mills theories?". They both gave the same answer: "if you are a student of Veltman's, ask him, we are no experts on Yang-Mills."

A general picture of how to deal with massive vector particles was forming in my mind, but I could not understand the negative attitude of all the experts towards such theories. Later, I would find out that they all had different reasons for rejecting such approaches: some people thought that there would be Goldstone bosons with physically unacceptable properties. Some thought that introducing fundamental scalar particles would not serve any fundamental physical principle such as local gauge invariance. To many people, a renormalization programme would seem to be so complicated that mathematical clashes would be unavoidable. Finally, there was the scaling problem. It was thought that scaling towards asymptotic freedom in the ultraviolet region would never happen in a field theory; this would imply that any relativistic quantum system with strongly interacting particles would explode nonperturbatively in the near ultraviolet, and therefore no perturbative quantum field theory would ever apply to such systems. Because of this universal agreement among the experts, no-one realized that all these arguments were wrong. Why had this faulty counter evidence not deterred me? Probably, Veltman's determination that there had to be something right about quantum field theory influenced me. But as a student I had also learned only to believe those arguments that I could truly understand.

What I did understand from the Cargèse lectures is that renormalization is complicated and delicate. At least at this point I could agree with my advisor, Veltman. When I returned to Utrecht, his assignment to me was that I should first study the pure Yang-Mills system, without anything resembling a Higgs mechanism for generating masses. There was not much literature on the subject, except some very elegant papers by Richard Feynman, Bryce DeWitt and by Ludwig D. Faddeev and his coworker Victor N. Popov at Leningrad. But some of the papers seemed to contradict one another, and so I began to collect the pieces of information that I could understand.

I learned how to formulate the Feynman rules for these Yang-Mills particles, and I learned that the discrepancy between the different papers was only an apparent one: you could perform gauge transformations to relate one to the other. I thought I was making tremendous progress towards formulating the exact renormalization procedure for this case, but Veltman had various objections. After long discussions, which again gave me many more insights, my first publication appeared. I had derived identities among amplitudes which were subsequently used by A.A. Slavnov and J.C. Taylor to derive more general identities, and their first references to my work made me feel very proud. The generally accepted name for these identities would be the "Slavnov-Taylor identities".

After having learned so much about renormalizing massless Yang-Mills fields, doing the same thing for theories with Higgs mechanism was relatively easy. But it was this second paper with which I caught world-wide attention. Veltman realized that now the problem that he had been working on for years had been solved, and he was enthusiastic. As he was one of the organizers of an international conference on particle physics at Amsterdam in 1971, he decided to use his new pawn (me) in his battle for the recognition of Yang-Mills theories, and gave me 10 minutes (but no place in the Proceedings) to explain our new results. A period of intensive cooperation followed. Together, we worked out the so-called dimensional renormalization technique. Certainly, the work I had done was considered to be good enough for a PhD degree, and I graduated in 1972.

This, by the way, was also the year of my marriage. While I was making my great discoveries in physics I had also discovered who I wanted to marry: Mrs Albertha A. Schik (Betteke). She had grown up in the town of Wageningen, and had studied medicine at Utrecht University.

We went to CERN, Geneva, where I had a fellowship, and Betteke could begin her work to obtain her certificate as a specialist in anesthesia, at the Hôpital Cantonal of the town of Geneva. The day before she was to meet her new superiors and colleagues there, we had made a trip to the Mont Blanc; on the way back we were in a minor car accident, and she fractured a bone in her foot. Her entry at the hospital will be remembered.

Veltman also came to CERN, and together we refined our methods for Yang-Mills theories. We were delighted with the great impact that our theories had. From 1971 onwards, all theories for the weak interactions that were proposed were Yang-Mills theories. Experiments were set up aimed at selecting out which of these Yang-Mills theories were correct. One of the simplest models continued to be successful; every now and then some particles were added to it, but its basic structure remained the same.

At CERN, I became interested in the quark confinement problem. I could not understand why none of the expert theoreticians would embrace quantum field theories for quarks. When I asked them, why not just a pure Yang-Mills theory?, they said that field theories were inconsistent with what J.D. Bjorken had found out about scaling in the strong interactions. This puzzled me, because when I computed the scaling properties of Yang-Mills fields, they seemed to be just what one needs. I simply could not believe that no-one besides me knew how Yang-Mills theories scale. I mentioned my result verbally at a small conference at Marseille, in 1972. The only person who listened to what I said was Kurt Symanzik. He urged me to publish my result about scaling. "If you don't, someone else will", he warned. I ignored his sensible advice. I had also made a remark about scaling in my 1971 paper on massive Yang-Mills fields. No-one had taken notice.

Veltman told me that my theory would be worthless if I could not explain why quarks cannot be isolated. He attached more importance to another project we had embarked upon: we had started a lengthy calculation concerning the renormalizability of quantum gravity models. Although complete renormalization would never be possible, it was still worth-while to study these theories at the one-loop level, and there were some important things to be learned. Our work would be continued by Stanley Deser and a fellow PhD student of Veltman's, Peter van Nieuwenhuizen, who discovered patterns in the renormalization counter terms that would lead to the discovery of supergravity theories.

But I also continued to think of gauge theories for the strong interaction. Quark confinement was indeed a problem, and I started to work on it. It was this question that led me to discover the magnetic monopole solutions in Higgs theories, the large N behaviour for theories with N colours (instead of 3, the physical number), and later the very important effects due to instantons. In the mean time, the scaling properties were rediscovered by H. David Politzer and by David Gross and Frank Wilczek in 1973, who now realized that this invalidated the age-old objections against simple, pure Yang-Mills theories for the strong interactions. The pure Yang-Mills theory with gauge group SU(3) was finally being accepted as the most likely explanation for the strong interactions, and it received the beautiful name "Quantum Chromodynamics" (QCD).

In 1974 we returned to Utrecht. I had been given an assistant professorship there. I was making progress understanding confinement as an effect due to Bose condensation of colour-magnetic monopoles. An important observation by Kenneth Wilson was that permanent quark confinement appears naturally if one performs the $1/g$ expansion instead of the g expansion in gauge theories, provided that a lattice cut-off is used. We were just beginning to see the extremely rich topological structure of gauge theories, and its consequences for the quantized system.

In 1976, I was invited for guest positions at Harvard (Morris Loeb lecturer) and Stanford. I worked on the question whether the delicate effects due to instantons – topologically twisted field configurations that should play a role in quantum chromodynamics – would survive when a renormalized perturbation expansion was applied. This led to one of the most complicated calculations I ever did: the one-loop corrections to instantons. It turned out that instantons in QCD give finite and well-defined contributions to the amplitudes. They give the symmetry structure a twist in such a way that many riddles in the experimental data concerning chiral symmetry were finally resolved, the most notable one being the problems with the eta particle, mentioned earlier. Several of my friends and colleagues at Harvard, MIT and Princeton such as Roman Jackiw, Sidney Coleman and David Gross but also physicists elsewhere (Moscow), students and postdocs joined the game of unraveling the secrets of instantons and monopoles. In the mean time my first daughter, Saskia Anne, was born, at Boston. When I returned to Utrecht I was appointed Full professor there. My second daughter, Ellen Marga, was born at Utrecht in 1978.

The years that followed I spent much energy and inventiveness to shed more light on the quark confinement problem. The neat and clean treatment of QCD that I hoped to find did not exactly materialize, but by the beginning of the 1980s the elementary mechanism for this phenomenon had become clear. QCD can be treated numerically when lattice cut-offs are used, and nowadays increasing accuracies are being reached by investigators using ever improving hardware and software. The problems remaining seem to be rather mathematical ones and not physical ones. QCD had become an integral ingredient of the Standard Model. I decided to turn towards the many open questions concerning the physics of this model.

I felt pain and sadness when for personal reasons Veltman left Utrecht in 1981. What about the deep, open problems in the Standard Model? Many of my colleagues agree that supersymmetry, a symmetry relation between particles with different spins, should play an essential role. I had seen how supersymmetry was born, back at CERN during the early 1970s. Bruno Zumino and Julius Wess were producing highly intriguing papers, while Van Nieuwenhuizen and Sergio Ferrara, and many others were making progress in supergravity. But what should a supersymmetric "parent theory" be like? How and why should supersymmetry be broken to explain the world as we observe it today? Do we really have to believe that there are dozens of particle types called "super partners", none of which have ever been seen? Such questions make me feel uncomfortable with supersymmetric theories.

The true answers must undoubtedly come from the incorporation of the gravitational force. At first sight it may seem difficult to believe that such an extremely weak force could cause so much havoc in a theoretical construction such as the Standard Model. The point is, however, that if gravity really corresponds to the curvature of space and time, as we must conclude from the successes of Einstein's theory of General Relativity, then Quantum Mechanics predicts quantum fluctuations in this curvature that, at the tiniest distance scales, grow out of control. This means that either gravity theory, or Quantum Mechanics, or both, must be replaced by some superior paradigm when we wish to describe physics at distance scales smaller than 10^{-33} cm. Whatever paradigm this would be, it is likely to entirely reform our understanding of the fundamental interactions, answering all our present questions at one stroke.

In 1984, the superstring revolution took place. Many of my colleagues were enchanted by the coherence of the mathematical structures they saw in this theory. Would this not be exactly what we are looking for, a new paradigm that naturally generates the gravitational force and an apparent complete unification of all interactions?

But to me, superstring theories presented as many new problems as they may solve; I still cannot quite fathom the fundamental logical coherence of these ideas. The short distance structure is as mysterious as it was before and the predictive power of these theories was disappointing, to put it mildly. I decided to try a different route. When Stephen Hawking discovered that black holes will radiate due to quantum field theoretical effects, this to me appeared to be a more solid starting point. Are black holes elementary particles? Are elementary particles black holes? I was stunned to learn that Hawking's result would put black holes in a category fundamentally different from any ordinary form of matter. If that were so, then what exactly are the laws of physics for black holes? The answer is that present theories are inconclusive. They clash. They lead to a paradox that may be as elementary as the paradox that, one century ago, led Max Planck to revise the black body radiation law, and which ultimately gave us Quantum Mechanics. By studying this paradox, I hoped to stumble upon something equally great. Needless to say, I was asking for more luck than in the average lottery system. The problem is a sturdy one, and it still has not been solved. To illustrate the paradoxical nature of our problem I formulated a feature of the quantum gravitational degrees of freedom which, in discussions with Leonard Susskind, was called the "Holographic Principle".

For a long time, I was among a small selected group of extravagants who studied quantum black holes. But superstring theory was catching on. As I had expected, superstring theory was not within a stone's throw of "the final theory", which had been what its addicts had prophesied, but it underwent fundamental changes. Membranes of various dimensionalities ("p-branes") were added, and now a door was opened for studying black holes in string theory. Suddenly, I found myself to be nearly back in the "mainstream" of physics: string theoreticians are now seeing the "holographic principle" everywhere. But the solution to our problems, bringing the gravitational force fully in agreement with Quantum Mechanics, has not yet been achieved. As long as this is the case, we will not be able to produce verifiable predictions concerning the enigmatic details of the Standard Model.

A CONFRONTATION WITH INFINITY

Nobel Lecture, December 8, 1999

by

GERARDUS 't HOOFT

Institute for Theoretical Physics, University of Utrecht, Princetonplein 5, 3584 CC Utrecht, the Netherlands

1. APPETIZER

Early attempts at constructing realistic models for the weak interaction were offset by the emergence of infinite, hence meaningless, expressions when one tried to derive the radiative corrections. When models based on gauge theories with Higgs mechanism were discovered to be renormalizable, the bothersome infinities disappeared – they cancelled out. If this success seemed to be due to mathematical sorcery, it may be of interest to explain the physical insights on which it is actually based.

2. INTRODUCTION

It is the highest possible honor for a scientist in my field to be standing here and give this lecture. It is difficult to express how thankful I am, not only to the Nobel Committee and to the Royal Swedish Academy of Sciences, but also to the numerous fellow physicists and friends who considered our work to be of such importance that we should be nominated to receive this Prize. In this lecture I intend to reflect on the efforts that were needed to tame the gauge theories, the reasons for our successes at this point, and the lessons to be learned. I realize the dangers of that. Often in the past, progress was made precisely because lessons from the past were being ignored. Be that as it may, I nevertheless think these lessons are of great importance, and if researchers in the future should choose to ignore them, they must know what they are doing.

When I entered the field of elementary particle physics, no precise theory for the weak interactions existed[1]. It was said that any theory one attempted to write down was non-renormalizable. What was meant by that? In practice, what it meant was that when one tried to compute corrections to scattering amplitudes, physically impossible expressions were encountered. The result of the computations appeared to imply that these amplitudes should be infinite. Typically, integrals of the following form were found:

$$\int \mathrm{d}^4k \, \frac{\mathrm{Pol}(k_\mu)}{(k^2+m^2)\big((k+q)^2+m^2\big)} = \infty \,, \qquad (2.1)$$

where Pol(k_μ) stands for some polynomial in the integration variables k_μ. Physically, this must be nonsense. If, in whatever model calculation, the effects due to some obscure secondary phenomenon appear to be infinitely strong, one knows what this means: the so-called secondary effect is not as innocent as it might have appeared – it must have been represented incorrectly in the model; one has to improve the model by paying special attention to the features that were at first thought to be negligible. The infinities in the weak-interaction theories were due to interactions from virtual particles at extremely high energies. High energy also means high momentum, and in quantum mechanics this means that the waves associated with these particles have very short wavelengths. One had to conclude that the short distance structure of the existing theories was too poorly understood.

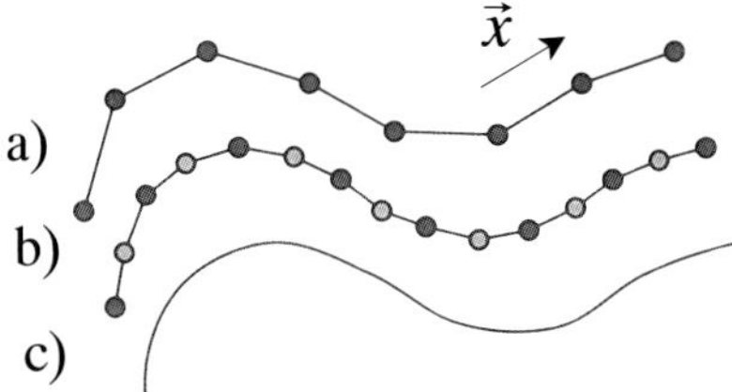

Fig. 1. Differentiation.

Short distance scales and short time intervals entered into theories of physics first when Newton and Leibniz introduced the notion of *differentiation*. In describing the motion of planets and moons, one had to consider some small time interval Δt and the displacement $\Delta\vec{x}$ of the object during this time interval (see Fig. 1a). The crucial observation was that, in the limit $\Delta t \to 0$, the ratio

$$\frac{\Delta\vec{x}}{\Delta t} = \vec{v} \tag{2.2}$$

makes sense, and we call it "velocity". In fact, one may again take the ratio of the velocity *change* $\Delta\vec{v}$ during such a small time interval Δt, and again the ratio

$$\frac{\Delta\vec{v}}{\Delta t} = \vec{a} \tag{2.3}$$

exists in the limit $\Delta t \to 0$; we call it "acceleration". Their big discovery was that it makes sense to write equations relating accelerations, velocities, and positions, and that in the limit where Δt goes to zero, you get good models describing the motion of celestial bodies (Fig. 1c). The mathematics of differential equations grew out of this, and nowadays it is such a central element in theoretical physics that we often do not realize how important and how nontrivial these observations actually were. In modern theories of physics we send distances and time intervals to zero all the time, also in multidimensional field theories, assuming that the philosophy of differential equations applies.

But occasionally it may happen that everything goes wrong. The limits that we thought to be familiar with, do not appear to exist. The behaviour of our model at the very tiniest time and distance scales then has to be reexamined.

Infinite integrals in particle theory were not new. They had been encountered many times before, and in some theories it was understood how to deal with them[2]. What had to be done was called "renormalization". Imagine a particle such as an electron to be something like a little sphere, of radius R and mass m_{bare}. Now attach an electric charge to this particle, of an amount Q. The electric-field energy would be

$$U = \frac{Q^2}{8\pi R} , \tag{2.4}$$

and, according to Einstein's special theory of relativity, this would represent an extra amount of mass, U/c^2, where c is the speed of light. Particle plus field would carry a mass equal to

$$m_{\text{phys}} = m_{\text{bare}} + \frac{Q^2}{8\pi c^2 R} . \tag{2.5}$$

It is this mass, called "physical mass", that an experimenter would measure if the particle were subject to Newton's law, $\vec{F} = m_{\text{phys}}\vec{a}$. What is alarming about this effect is that the mass correction diverges to infinity when the radius R of our particle is sent to zero. But we want R to be zero, because if R were finite it would be difficult to take into account that *forces* acting on the particle must be transmitted by a speed less than that of light, as is demanded by Einstein's theory of special relativity. If the particle was deformable, it would not be truly elementary. Therefore, finite-size particles cannot serve as a good basis for a theory of elementary objects.

In addition, there is an effect that alters the electric charge of the particle. This effect is called "vacuum polarization". During extremely short time intervals, quantum fluctuations cause the creation and subsequent annihilation of particle-antiparticle pairs. If these particles carry electric charges, the charges whose signs are opposite to our particle in question tend to move towards it, and this way they tend to neutralize it. Although this effect is usually quite small, there is a tendency of the vacuum to "screen" the charge of our particle. This screening effect implies that a particle whose charge is Q_{bare} looks like a particle with a smaller charge Q_{phys} when viewed at some distance. The relation between Q_{bare} and Q_{phys} again depends on R, and, as was the case for the mass of the particle, the charge renormalization also tends to infinity as the radius R is sent to zero (even though the effect is usually rather small at finite R).

It was already in the first half of the 20th century that physicists realized the following. The *only* properties of a particle such as an electron that we ever measure in an experiment are the physical mass m_{phys} and the physical charge Q_{phys}. So, the procedure we have to apply is that we should take the limit where R is sent to zero while m_{phys} and Q_{phys} are kept fixed. Whatever happens to the *bare* mass m_{bare} and the *bare* charge Q_{bare} in that limit is irrelevant, since these quantities can never be measured directly.

Of course, there is a danger in this argument. If, in Eq. (2.5), we send R to zero while keeping m_{phys} fixed, we notice that m_{bare} tends to *minus* infinity. Can theories in which particles have negative mass be nevertheless stable? The answer is no, but fortunately, Eq. (2.3) is replaced by a different equation in a quantized theory. m_{bare} tends to zero, not minus infinity.

3. THE RENORMALIZATION GROUP

The modern way to discuss the relevance of the small-distance structure is by performing *scale transformations*, using the *renormalization group*[3], and we can illustrate this again by considering the equation of motion of the planets. Assume that we took definite time intervals Δt, finding equations for the displacements Δx. Imagine that we wish to take the limit $\Delta t \to 0$ very carefully. We may decide first to divide all Δt's and all Δx's by 2 (see Fig. 1b). We observe that, if the original intervals are already sufficiently small, the new results of a calculation will be very nearly the same as the old ones. This is because during small time intervals, planets and moons move along small sections of their orbits, which are *very nearly straight lines.* If they had been moving exactly along straight lines, the division by 2 would have made no difference at all. Planets move along straight lines if *no force acts on them.* The reason why differential equations were at all successful for planets is that *we may ignore the effects of the forces* (the "interactions") when time and space intervals are taken to be very small.

In quantized field theories for elementary particles, we have learned how to do the same thing. We reconsider the system of interacting particles at very short time and distance scales. If at sufficiently tiny scales the interactions among the particles may be ignored, then we can understand how to take the limits where these scales go all the way to zero. Since then the interactions may be ignored, all particles move undisturbedly at these scales, and so the physics is then understood. Such theories can be based on a sound mathematical footing; we understand how to do calculations by approximating space and time as being divided into finite sections and intervals and taking the limits in the end.

So, how is the situation here? Do the mutual interactions among elementary particles vanish at sufficiently tiny scales? Here is the surprise that physicists had to learn to cope with: they do not.

Many theories indeed show very bad behaviour at short distances. A simple prototype of these is the so-called *chiral model*[4]. In such a model, a multicomponent scalar field is introduced which obeys a constraint: its total length is assumed to be fixed,

$$\sum_i |\phi_i|^2 = R^2 = \text{ fixed }. \tag{3.1}$$

At large distance scales, the effects of this constraint are mild, as the quantum fluctuations are small compared to R. At small distance scales, however, the quantum fluctuations are large compared to R, and hence the nonlinear effects of the constraint are felt much more strongly there. As a consequence, such a theory has large interactions at small distance scales and vice versa. Therefore, at infinitesimally small distance scales, such a theory is ill-defined, and the model is unsuitable for an accurate description of elementary particles. Other examples of models with bad small-distance behaviour are the old four-fermion interaction model for the weak interactions and most attempts at making a quantum version of Einstein's gravity theory.

But some specially designed models are not so bad. Examples are: a model with spinless particles whose fields ϕ interact only through a term of the form $\lambda\phi^4$ in the Lagrangian, and a model in which charged particles interact through Maxwell's equations (quantum electrodynamics, QED). In general, we choose the distance scale to be a parameter called $1/\mu$. A scale transformation by a factor of 2 amounts to adding ln 2 to ln μ and if the distance scale is Δx, then

$$\frac{\mu \mathrm{d}}{\mathrm{d}\mu}\Delta x = -\Delta x \ . \tag{3.2}$$

During the 1960's, it was found that in *all* theories existing at the time, the interaction parameters, being either the coefficient λ for $\lambda\phi^4$ theory, or the coefficient e^2 in quantum electrodynamics for electrons with charge e, the variation with μ is a positive function[5], called the β function:

$$\frac{\mu \mathrm{d}}{\mathrm{d}\mu}\lambda = \beta(\lambda) > 0 \ , \tag{3.3}$$

so, comparing this with Eq. (3.2), λ is seen to increase if Δx decreases.

In the very special models that we just mentioned, the function $\beta(\lambda)$ behaves as λ^2 when λ is small, which is so small that the coupling only varies very slightly as we go from one scale to the next. This implies that, although there are still interactions, no matter how small the scales at which we look, these interactions are not very harmful, and a consequence of this is that these theories are "renormalizable". If we apply the perturbation expansion

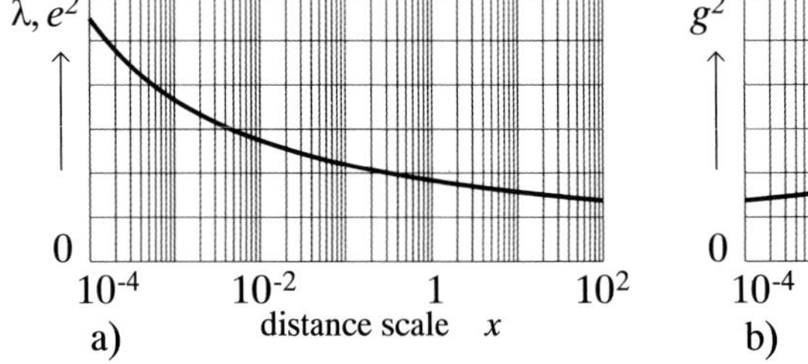

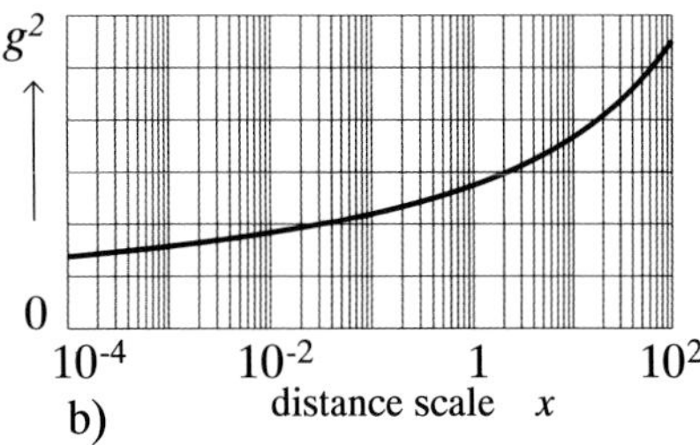

Fig. 2. Scaling of the coupling strength as the distance scale varies, a) for $\lambda\phi^4$ theories and QED, b) for Yang-Mills theories.

for small λ then, term by term, the expansion coefficients are uniquely defined, and we might be seduced into believing that there are no real problems with these theories.

However, many experts in these matters were worried indeed, and for good reason: If β is positive, then there will be a scale where the coupling strength among particles diverges. The solution to Eq. (3.3) is (see Fig. 2a):

$$\lambda(\mu) = 1/(C - \beta_2 \ln \mu) \ , \quad \text{if} \quad \beta(\lambda) = \beta_2 \lambda^2 \ , \tag{3.4}$$

where C is an integration constant, $C = 1/\lambda(1)$ if $\lambda(1)$ is λ measured at the scale $\mu = 1$. We see that at scales $\mu = \mathcal{O}\,[\exp\,(1/\beta_2\lambda(1))]$, the coupling explodes. Since for small $\lambda(1)$ this is exponentially far away, the problem is not noticed in the perturbative formulation of the theory, but it was recognized that if, as in physically realistic theories, λ is taken to be not very small, there is real trouble at some definite scale. And so it was not so crazy to conclude that these quantum field theories were sick and that other methods should be sought for describing particle theories.

I was never afflicted with such worries for a very simple reason. Back in 1971, I carried out my own calculations of the scaling properties of field theories, and the first theory I tried was Yang-Mills theory. My finding was, when phrased in modern notation, that for these theories,

$$\beta(g^2) = C\,g^4 + \mathcal{O}(g^6) \ , \quad \text{with} \quad C < 0 \ , \tag{3.5}$$

if the number of fermion species is less than 11 [for SU(2)] or $16\frac{1}{2}$ [for SU(3)]. The calculation, which was alluded to in my first paper on the massive Yang-Mills theory [6],was technically delicate but conceptually not very difficult. I could not possibly imagine what treasure I had here, or that none of the experts knew that β could be negative; they had always limited themselves to studying only scalar field theories and quantum electrodynamics.

4. THE STANDARD MODEL

If we were to confront the infinities in our calculations for the weak-interaction processes, we had to face the challenge of identifying a model for the weak interaction that shows the correct intertwining with the electromagnetic force at large distance scales but is sufficiently weakly interacting at small distances. The resolution here was to make use of spontaneous symmetry breaking*. We use a field with a quartic self-interaction but with a negative mass term, so that its energetically favored value is nonvanishing. The fact that such fields can be used to generate massive vector particles was known

* The mass generation mechanism discussed here should, strictly speaking, not be regarded as spontaneous symmetry breaking, since in these theories, the vacuum does not break the gauge symmetry. "Hidden symmetry" is a better phrase[7]. Let us simply refer to this mechanism as the "Higgs mechanism".

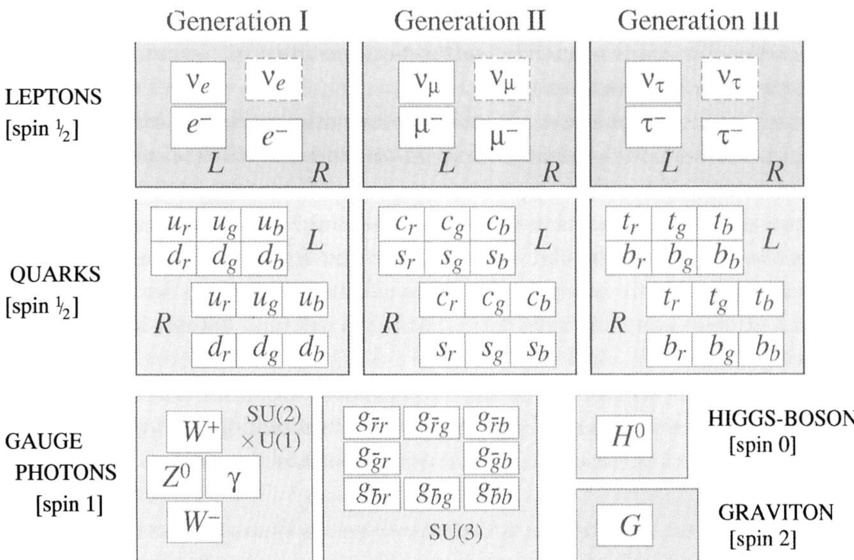

Fig. 3. The Standard Model, schematically. The leptons and the quarks form three nearly identical generations (only the masses of the entries in these three generations are different). Here, L stands for left-rotating, and R for right rotating (with respect to the direction of motion). Forces are mainly generated by the guage photons, SU(2) × U(1) for the electro-weak force, SU(3) for the strong force. The SU(2) forces act only on the left-rotating fermions.

but not used extensively in the literature. Also the fact that one could construct reasonable models for the weak interaction along these lines was known. These models, however, were thought to be inelegant, and the fact that they were the unique solution to our problems was not realized.

Not only did the newly revived models predict hitherto unknown channels for the weak interaction, they also predicted a new scalar particle, the Higgs boson[8]. The new weak interaction, the so-called neutral-current interaction, could be confirmed experimentally within a few years, but as of this writing, the Higgs boson is still fugitive. Some researchers suspect that it does not exist at all. Now if this were true then this would be tantamount to identifying the Higgs field with a chiral field – a field with a fixed length. We could also say that this corresponds to the limiting case in which the Higgs mass was sent to infinity. An infinite-mass particle cannot be produced, so it can be declared to be absent. But as we explained before, chiral theories have bad small-distance behaviour. We can also say that the interaction strength at small distances is proportional to the Higgs mass; if that would be taken to be infinite then we would have landed in a situation where the small-distance behaviour was out of control. Such models simply do not work. Perhaps experimentalists will not succeed in producing and detecting Higgs particles, but this then would imply that entirely new theories must be found to account for the small-distance structure. Candidates for such theories have been proposed. They seem to be inelegant at present, but of course that could be due to our present limited understanding, who knows? New theories would necessarily imply the existence of many presently unknown particle species, and experimenters would be delighted to detect and study such objects. We cannot lose here. Either the Higgs particle or other particles must be waiting there to be discovered, probably fairly soon.[9]

To the strong interactions, the same philosophy applies, but the outcome of our reasoning is very different. The good scaling behaviour of pure gauge theories (see Fig. 2b) allows us to construct a model in which the interactions at large distance scales is unboundedly strong, yet it decreases to zero (though only logarithmically) at small distances. Such a theory may describe the binding forces between quarks. It was found that these forces obtain a constant strength at arbitrarily large distances, where Coulomb forces would have decreased with an inverse square law. Quantum Chromodynamics, a Yang-Mills theory with gauge group SU(3), could therefore serve as a theory for the strong interactions. It is the only allowed model in which the coupling strength is large but nevertheless the small-distance structure is under control.

The weak force, in contrast, decreases *exponentially* as the distance between weakly interacting objects becomes large. Thus gauge theory allows us to construct models with physically acceptable behaviour at short distances, while the forces at large distances may vary in any of the following three distinctive ways:

(i) The force may drop exponentially fast, as in the weak interaction;
(ii) The force may drop according to an inverse-square law, as in electromagnetism, or
(iii) The force may tend towards a constant, as in the strong interactions.

The *Standard Model* is the most accurate model describing nature as it is known today. It is built exactly in accordance with the rules sketched above. Our philosophy is always that the experimentally obtained information about the elementary particles refers to their large distance behaviour. The small-distance structure of the theory is then postulated to be as regular as is possible without violating principles such as strict obedience of causality and Lorentz invariance. Not only do such models allow us to calculate their implications accurately, it appears that Nature really is built this way. In some sense, this result appears to be too good to be true. We shall shortly explain our reasons to suspect the existence of many kinds of particles and forces that could not yet be included in the Standard Model, and that the small-distance structure of the Standard Model does require modification.

5. FUTURE COLLIDERS

Theoreticians are most eager to derive all they want to know about the structures at smaller distances using pure thought and fundamental principles. Unfortunately, our present insights are hopelessly insufficient, and all we have are some wild speculations. Surely, the future of this field still largely depends on the insights to be obtained from new experiments.

The present experiments at the Large Electron Positron Collider (LEP) at CERN are coming to a close. They have provided us with impressive precision measurements that not only gave a beautiful confirmation of the Standard Model, but also allowed us to extrapolate to higher energies, which means that we were allowed a glimpse of structures at the smallest distance ranges yet accessible. The most remarkable result is that the structures there appear to be smooth; new interactions could not be detected, which indicates that the mass of the Higgs particle is not so large, a welcome stimulus for further experimental efforts to detect it.

In the immediate future we may expect interesting new experimental results first from the Tevatron Collider at Fermilab, near Chicago, and then from the Large Hadron Collider (LHC) at CERN, both of which will devote much effort to finding the still elusive Higgs particle. Who will be first depends on what the Higgs mass will turn out to be, as well as other not yet precisely known properties of the Higgs. Detailed analysis of what we know at present indicates that Fermilab has a sizeable chance at detecting the Higgs first, and the LHC almost certainly will not only detect these particles, but also measure many of their properties, such as their masses, with high precision. If supersymmetric particles exist, LHC will also be in a good position to be able to detect these, in measurements that are expected to begin shortly after 2005.

These machines, which will dissolve structures never seen before, however, also have their limits. They stop exactly at the point where our theories become highly interesting, and the need will be felt to proceed further. As before, the options are either to use hadrons such as protons colliding against antiprotons, which has the advantage that, due to their high mass, higher energies can be reached, or alternatively to use leptons, such as e^+ colliding against e^-, which has the advantage that these objects are much more point-like, and their signals are more suitable for precision experiments.[10] Of course, one should do both. A more ambitious plan is to collide muons, μ^+ against μ^-, since these are leptons with high masses, but this will require numerous technical hurdles to be overcome. Boosting the energies to ever increasing values requires such machines to be very large. In particular the high-energy electrons will be hard to force into circular orbits, which is why design studies of the future accelerators tend to take the form of straight lines, not circles. These linear accelerators have the interesting feature that they could be extended to larger sizes in the more distant future.

My hope is that efforts and enthusiasm to design and construct such machines in the future will not diminish. As much international cooperation as possible is called for. A sympathetic proposal[11] is called ELOISATRON, a machine in which the highest conceivable energies should be reached in a gigantically large circular tunnel. It could lead to a hundredfold improvement of our spatial resolution. What worries me, however, is that in practice one group, one nation, takes an initiative and then asks other groups and nations to join, not so much in the planning, but rather in financing the whole thing. It is clear to me that the best international collaborations arise when all partners are involved from the very earliest stages of the development onwards. The best successes will come from those institutions that are the closest approximations to what could be called "world machines". CERN claims to be a world machine, and indeed as such this laboratory has been, and hopefully will continue to be, extremely successful. Unfortunately, it still has an E in its name. This E should be made as meaningless as the N (after all, the physics studied at CERN has long ago ceased to be nuclear, it is subnuclear now). I would not propose to change the name, but to keep the name CERN only to commemorate its rich history.

6. BEYOND THE STANDARD MODEL

Other, equally interesting large scientific enterprises will be multinational by their very nature: plans are underway to construct neutrino beams that go right through the earth to be detected at the exit point, where it may be established how subtle oscillations due to their small mass values may have caused transitions from one type into another. Making world machines will not imply that competition will be eliminated; the competition, however, will not be between nations, but rather between the different collaborations who use different machines and different approaches towards physics questions.

The most interesting and important experiments are those of which we cannot guess the outcome reliably. This is exactly the case for the LHC experiments that are planned for the near future. What we do know is that the Standard Model, as it stands today, cannot be entirely correct, *in spite* of the fact that the interactions stay weak at ultra-short distance scales. Weakness of the interactions at short distances is not enough; we also insist that there be a certain amount of *stability*. Let us use the metaphor of the planets in their orbits once again. We insisted that, during extremely short time intervals, the forces acting on the planets have hardly any effect on their velocities, so that they move approximately in straight lines. In our present theories, it is as if at short time intervals several extremely strong forces act on the planets, but, for some reason, they all but balance out. The net force is so weak that only after long time intervals, days, weeks, months, the velocity change of the planets become apparent. In such a situation, however, a *reason* must be found as to why the forces at short time scales balance out. The way things are, at present, is that the forces balance out just by accident. It would be an inexplicable accident, and as no other examples of such accidents are known in Nature, at least not of this magnitude, it is reasonable to suspect that the true short-distance structure is not exactly as described in the Standard Model, but that there are more particles and forces involved, whose nature is as yet unclear. These particles and forces are arranged in a new symmetry pattern, and it is this symmetry that explains why the short-distance forces balance out.

It is generally agreed that the most attractive scenario is one involving "supersymmetry", a symmetry relating fermionic particles, whose spin is an integer plus one-half, and bosonic particles, which have integral spin.[12] It is the only symmetry that can be made to do the required job in the presence of the scalar fields that provide the Higgs mechanism, in an environment where all elementary particles interact weakly. However, when the interactions do eventually become strong then there are other scenarios. In that case, the objects playing the role of Higgs particles may be not elementary objects but composites, similar to the so-called Cooper pairs of bound electrons that perform a Higgs mechanism in ultracool solid substances, leading to superconductivity. Just because such phenomena are well known in physics, this is a scenario that cannot easily be dismissed. But, since there is no evidence at presence of a new strong interaction domain at the TeV scale, the bound-state Higgs theory is not favoured by most investigators.

One of the problems with the supersymmetry scenario is the supersymmetry breaking mechanism. Since at the distance scale where experiments are done at present no supersymmetry has been detected, the symmetry is broken. It is assumed that the breaking is "soft", which means that its effects are seen only at large distances, and only at the tiniest possible distance scales the symmetry is realized. Mathematically, this is a possibility, but there is as yet no plausible physical explanation of this situation. The only explanation can come from a theory at even smaller distance scales, where the gravitational force comes into play.

Until the early 1980's, the most promising model for the gravitational force was a supersymmetric variety of gravity: supergravity.[13] It appeared that the infinities that were insurmountable in a plain gravity theory, would be overcome in supergravity. Curiously, however, the infinities appeared to be controlled by the enhanced symmetry and not by an improved small-distance structure of the theory. Newton's constant, even if controlled by a dilaton field, still is dimensionful in such theories, with consequently uncontrolled strong interactions in the small-distance domain. As the small-distance structure of the theory was not understood, it appeared to be almost impossible to draw conclusions from the theory that could shed further light on empirical features of our world.

An era followed with even wilder speculations concerning the nature of the gravitational force. By far the most popular and potentially powerful theory is that of the superstrings.[14] The theory started out by presenting particles as made up of (either closed or open) pieces of string. Fermions living on the string provide it with a supersymmetric pattern, which may be the origin of the approximate supersymmetry that we need in our theories. It is now understood that only in a perturbative formulation do particles look like strings. In a nonperturbative formalism there seems to be a need not only of strings but also of higher-dimensional substances such as membranes. But what exactly is the perturbation expansion in question? It is not the approximation that can be used at the shortest infinitesimal distances. Instead, the shortest distances seem to be linked to the largest distances by means of duality relations. Just because superstrings are also held responsible for the gravitational force, they cause curvature of space and time to such an extent that it appears to be futile to consider distances *short* compared to the Planck scale.

According to superstring theory, it is a natural and inevitable aspect of the theory that distance scales shorter than the Planck scale cannot be properly addressed, and we should not worry about it. When outsiders or sometimes colleagues from unrelated branches of physics attack superstring theory, I come to its defense. The ideas are very powerful and promising. But when among friends, I have this critical note. As string theory makes heavy use of differential equations it is clear that some sort of continuity is counted on. We should attempt to find an improved short-distance formulation of theories of this sort, if only to justify the use of differential equations or even functional integrals.

Rather than regarding the above as criticism against existing theories, one should take our observations as indications of where to search for further improvements. Emphasizing the flaws of the existing constructions is the best way to find new and improved procedures. Only in this way can we hope to achieve theories that allow us to explain the observed structures of the Standard Model and to arrive at more new predictions, so that we can tell our experimental friends where to search for new particles and forces.

REFERENCES

1. For an account of the historical developments, see for instance: R. P. Crease and C. C. Mann, *The second creation: makers of the Revolution in Twentieth-century Physics* (New York, Macmillan, 1986), ISBN 0-02-521440-3.
2. See for instance A. Pais. *Inward bound: of matter and forces in the physical world,* Oxford University Press 1986, ISBN 0-19-851977-4.
3. K. G. Wilson and J. Kogut, *Phys. Reports* **12C** (1974) 75; H.D. Politzer, *Phys. Reports* **14C** (1974) 129.
4. See a description in: B. W. Lee, *Chiral Dynamics,* Gordon and Breach, New York, 1972, ISBN 0 677 01380 9 (cloth); 0 677 01385 X (paper), pp. 60–67.
5. D. J. Gross, in *The Rise of the Standard Model,* Cambridge Univ. Press (1997), ISBN 0-521-57816-7 (pbk), p. 199.
6. G. 't Hooft, *Nucl. Phys.* **B35** (1971) 167.
7. S. Coleman, *Secret Symmetries,* in *Laws of Hadronic Matter,* A. Zichichi, ed., (Academic Press, New York and London, 1975).
8. P. W. Higgs, *Phys. Lett.* **12**(1964) 132; *Phys. Rev. Lett.* **13** (1964) 508; *Phys. Rev.* **145** (1966) 1156; F. Englert and R. Brout, *Phys. Rev. Lett.* **13** (1964) 321.
9. See for instance: E. Accomando *et al, Phys. Reports* **299** (1998) 1, and P. M. Zerwas, *Physics with an e^+e^- Linear Collider at High Luminosity,* Cargèse lectures 1999, preprint DESY 99-178.
10. J. Ellis, *Possible Accelerators at CERN beyond the LHC,* preprint CERN-TH/99-350, hep-ph/9911440.
11. A. Zichichi, *Subnuclear Physics, the First Fifty Years, Highlights from Erice to ELN,* the Galvani Bicentenary Celebrations, Academia Delle Scienze and Bologna, 1998, pp. 117–135.
12. Supersymmetry has a vast literature. See for instance the collection of papers in: S. Ferrara, *Supersymmetry,* Vol. 1, North Holland, Amsterdam, etc., 1987.
13. S. Ferrara, *Supersymmetry,* Vol. 2, North Holland, Amsterdam, etc., 1987.
14. See for instance J. Polchinski, *String Theory, Vol. 1, An introduction to the Bosonic String,* Cambridge Monographs on Mathematical Physics, P. V. Landshoff *et al,* eds. Cambridge Univ. Press 1998, ISBN 0-521-63303-6.

MARTINUS J. G. VELTMAN

I was born on Saturday June 27, 1931, in a town called Waalwijk in the south of the Netherlands. My father was the head of the local primary school. One brother and two sisters of my father were primary school teachers as well, and in my family learning was held in high regard. My mother came from more practically oriented people: her father was a contractor and also ran a café. I have a bit from both sides. I was the fourth child in what became a six child family.

The town of Waalwijk had approximately 20 000 inhabitants. It was dominated by the shoe industry. For many life was not easy in the depression years but a head teacher was relatively well off. Consequently the life of my family was quiet and relatively uneventful.

In 1940 the Netherlands were overrun by the German army. I saw them marching in, and I heard people speak of the diffcult times coming. In this part of the Netherlands things were not as bad as in the big cities, and the main thing I remember is that my father's school was requisitioned by the Germans and troops were lodged therein. In the fall of 1944 we were liberated, contrary to the part of the Netherlands "above the rivers" that would have to endure till the capitulation in May 1945. This was due to the failure of the operation Market Garden; the last bridge near Arnhem was not taken. Thus we escaped what is called the hunger winter of 1944–1945, in which many Dutch people died of hunger.

The period of the war in which we were liberated, while the north was still occupied, was characterized by lots of artillery fire from our side, as Waalwijk was very near the front line. The allied troops were not very careful with their ammunition, and as a young boy of 14, I was very interested in playing with that. I remember how we would extract the powder from a failed tank grenade found in a ditch near where tanks had been firing. We would take the grenade by the point and then beat the lower end on the ground till the point came loose and we could then shake out the powder. I do not know how I survived this, but I did. We also survived the V1 flying bombs that came over on their way to Antwerp. Two of them actually fell on houses in Waalwijk, one of them at about 100 m of our house. With me is still the memory of dead bodies being extracted from the ruins.

In 1943, I went to high school (the Dutch HBS). The war period was marked by irregularities and at one point our class was in a horse stable. While I had been a very good student at the elementary school, I was quite mediocre at the HBS. Much of that was due to my bad aptitude for languages, which is a real handicap when they require you to learn three foreign languages. So it happened that I narrowly passed the final exam in 1948, at the age of 17.

In those years I had (and still have) electronics as a hobby. This is a somewhat exaggerated qualification, because there was practically no electronics material around. The Germans had confiscated all radios, and there was a great scarcity of anything that could catch a radio signal. I remember spending a whole day walking around in the nearby larger town of 's Hertogenbosch, trying to get a radio tube. Finally some kind person, having pity on me, gave me one.

I acquired my knowledge of electronics from the local plumber. I used to spend many evenings in his house, and during the holidays I would work for him. I thus also learned plumbing. When I started to understand radio's a bit better I became the local radio repairman. My only measuring instrument was my right index finger. If I touched a sensitive connection the radio would produce a hum. If a connection had the correct high voltage (about 200 V) I would get a shock. Commercially I was a failure, as I would usually not dare to ask money for my services.

When I passed the final high school exam the big question was: what now? Traditionally somebody like me would go to a medium level technical school in 's Hertogenbosch called MTS. However, given my low grades this did not go smoothly. At this point my physics high school teacher came to my home and suggested to my parents to send me to the University. This was a big thing, practically nobody did that in Waalwijk in those days. Universities were still very exclusive, and the south of the Netherlands was quite backwards in this respect. As the money situation was very tight the main point was to find a University were I could go to by train. This was possible with the University of Utrecht. For three years I commuted back and forth from Waalwijk to Utrecht, a 90 minute trip each way. I am still grateful to this high school teacher, Mr Beunes, as he did the extra thing, going to my parents house. Since then I have found out that many physicists owe their career to a good high school teacher.

Worse however was the state of the education at the University. The war had left the Netherlands ravaged, many good physicists had left the country or were killed. In retrospect, it is really a pity that Abraham Pais left Utrecht to go to the US. The teaching in Utrecht was uninspiring, and failed to awake much physics interest in me. After three quite mediocre years I left home and started to live in Utrecht. At that time I had no income as my father could not support that, and I was forced to work on the side. My main activity was typing lecture notes. Sometimes it was even diffcult to get decent meals. But by and large I lived a happy life, mainly bumming around.

I also got involved in another job, trying to sell some rather silly tool to the unsuspecting citizens. In this I was a complete failure. I cannot sell anything to anybody. In some sense that has remained true in my scientific life.

After five years (2 years longer than normal) I passed what was called the candidaats exam. Then I happened to stumble on a popular book on the theory of relativity. Mind you, up to then no teacher had ever mentioned this. This booklet really excited me, and I went to the Institute of Theoretical Physics to get a real book on the subject. After some nagging they gave me

Einstein's book "The Meaning of Relativity". Since then I was hooked. Also my financial situation improved slightly: given the very big shortage of high school teachers in the Netherlands it was not diffcult to get a job as parttime teacher. Actually, I started as a teacher at a lower technical school, teaching plumbers about physics. None of that was helpful in speeding up my studies.

After my candidaats exam I started initially as an experimentalist. I worked for some time studying medical physics, in particular the physical aspects of percussion sound (the sound that is produced when a doctor pounds your chest). Later I worked on a mass-spectrometer, mainly doing the electronics. I found out that this was not my real destiny, and switched to theoretical physics. However, I still have a considerable fondness for experimental physics.

In 1955, I landed a job as assistant to Prof. Michels, of the Van Der Waals laboratory in Amsterdam. Michels was an experimental physicist, involved in high pressure physics. My task was the upkeep of the library, remarkably well stocked, and occasionally preparing a talk for Michels. I remember that as a good period, and it brought me also into contact with the members of the theory institute in Amsterdam. Mostly they were interested in statistical mechanics, a subject that has never evoked the slightest enthusiasm in me. Sneeringly I used to say: you guys average out anything of interest.

Science wise my life improved greatly with the coming to Utrecht of Leon Van Hove, I believe in 1955. He was an excellent lecturer, and I volunteered to make official notes of his lectures. I finished my graduate studies in 1956, after which I had to go into military service for two years, coming out of that in February 1959. Van Hove was so kind to take me as a PhD student despite my relatively advanced age (27 years). Thus I started doing real theory.

As Van Hove did statistical mechanics like all other theorists in Utrecht there was some problem, because I wanted to do particle physics. At that time many European Universities did not have anybody doing research in that field, and the way to learn that was via physics schools. Typically, such schools would run over a period of two weeks, with internationally known speakers. In the spring of 1959, I went to such a school in Naples, were among others Kurt Symanzik and Bruno Zumino lectured. In August 1960, I went to yet another school, in Edinburgh, and that school has been of quite some importance to me. I met Shelly Glashow, at that time a student there, who was working on the subject for which he was to get the Nobel Prize. If someone had told him that, he would have been quite surprised. At these schools I became friends with several other students, among them Nicola Cabibbo and Derek Robinson (who went astray into statistical and mathematical directions). They have remained friends to this day. From the Edinburgh school I do remember fondly the lectures of Dave Jackson, now in Berkeley.

In 1960, Van Hove became director of the theory division at CERN, Geneva, Switzerland, the European High Energy laboratory. I followed him in 1961. Meanwhile, in 1960, I was married to my present wife Anneke, and before joining me in Geneva she delivered our daughter Hélène in the Netherlands, living in the house of my parents. Hélène followed in my footsteps and in due time completed her particle physics thesis with Mary Gaillard at Berkeley. She now works in the banking world in London. She is the one member of our family that understands what I have been doing.

At first I felt a bit lost at CERN, since my elementary particle physics knowledge was quite sketchy. For some time I had been working on a rather field theoretical problem, namely unstable particles. When that was finished I wanted to go into something closer to the experiments. This happened thanks to Sam Berman, an erstwhile student of Feynman. He was aware of the situation at the CERN theory division at the time, and he did put up an advertisement: "If you have nothing else to do and wish to be kept off the street please knock my door". So I did, and I am still very grateful to him. He suggested a calculation: Coulomb corrections to the production of vector bosons in the CERN neutrino experiment. This, after consultation with Van Hove, was then to be the second part of my thesis.

The problem consisted of two parts. One, a part very analogous to a previous calculation of Bethe and collaborators (Coulomb corrections to pair production), and secondly a part that was not solved by Bethe. I remember sitting in my offce for several months, staring at a single differential equation, trying to solve it using confluent hypergeometric functions. These are very disgusting functions, and after a while I felt that perhaps I should consult the world expert on that matter, Eyvind Wichmann. He happened to be in Copenhagen at the time, and I made a pilgrimage. Seldom have I made such a useless trip. Wichman tried to understand what I wanted, but he did not get it. He looked at me as if I was some strange animal.

Well, at some point I solved that problem, and that then completed my thesis. In due time (April 22, 1963) Van Hove and I went back to Utrecht for the ceremonies, in tails and white tie. I now understand that this is a preparation for the Nobel Prize. The thesis contained my work on unstable particles as well as the treatment of Coulomb corrections for vector boson production by neutrinos. Title of the thesis: Intermediate particles in S-matrix theory and calculation of higher order effects in the production of intermediate vector bosons.

At CERN, meanwhile, the experimentalists were gearing up for the CERN neutrino experiment. I was asked to speak at one of their meetings on vector boson production. I became almost instantly very good friends with Bernardini, the leader of the group. They then wanted me to do extensive calculations for them concerning vector boson production, as that would be needed for analysis of the data. Computer calculations of that type had been done already by Lee, Markstein and Yang, and when Lee came to CERN I took the occasion to ask for those programs, which he curtly refused. I then asked him if he could give me some advice, to which he answered: "don't make mistakes". I thought this funny, and started to laugh, but that was not appreciated by Lee, who took some moments to teach me the seriousness of this enterprise. Well, even if in the making of computer programs the not making of errors is usually the main problem, I still feel that I did not really need that advice!

The CERN neutrino experiment was a very big happening in my life. When they started I was more or less permanently around, looking at the pictures as they came out. When no spectacular events came out the enthusiasm of the experimentalists waned, and after a while the only ones to look at the pictures were Bernardini and myself. And so it came to pass that I became the spokesman for the group at the Brookhaven Conference in 1963. Somewhere in that period I acquired two lifelong friends, the experimentalists Mel Schwartz (Nobel Laureate 1988) and Val Telegdi.

The 1963 CERN neutrino experiment left me with an interest for experiments that never went away. I am a deep believer in the importance of experiments for the progress of physics. Also, the experiment left me with a feeling for these things, to recognize what is important and who are the good guys. All theorists ought to go through some such experience. These days, however, that is not really practical any more. The experiments have become gigantic enterprises, involving hundreds of physicists and a large number of engineers. The modern experimentalist is often more manager than physicist.

In 1963, I went to SLAC at Stanford, where Pief Panofsky was building a linear electron accelerator. Also Sam Berman was there, in fact he was much of the reason for going there. Meanwhile, at CERN, I had become good friends with John Bell who was one of the very few theorists that had any interest in the neutrino experiment. He also came to SLAC, and in fact we wrote a paper together that we however never published. He became quite involved with what is now known as the Bell inequalities, while I started constructing my symbolic computer program Schoonschip. That also had its origin in the neutrino experiment: in doing the necessary algebra for vector boson production I was often exasperated by the effort that it took to get an error free result, even if the work was quite mechanical. In a discussion on the CERN terrace, including among others Mike Levine, we concluded that somebody ought to write a program to do that type of work. I started doing that at SLAC, in the autumn of 1963. Many good things have been invented at the CERN terrace. Mike Levine later successfully completed the first QED sixth order calculation.

In the spring of 1964, I went back to CERN and worked there till my departure for Brookhaven in 1966. There I had the pleasure of getting to know Maurice Goldhaber, then the very successful director of Brookhaven National Laboratory. A man that impresses me to this day. He liked me as well, and in fact tried to get me to that laboratory, which I did not. I do remember getting a phone call from Brookhaven while sitting with my parents in law in the Dutch town of Leeuwarden. How they ever found me there is still a mystery to me.

In the meantime, just prior to going to the US, our son Hugo was born. He now runs a restaurant called Solstice in Los Angeles. If I want a really good dinner that is where I go. I hope the reader gets the hint.

September 1966, I went back to Utrecht, as successor of Van Hove, i.e. professor of theoretical physics. There was still nobody doing particle physics there, so I started to build that up. That took some time; it was really a big change after the rather hectic CERN life. I made a mistake: I thought that being relatively isolated in Utrecht it would be a good idea to become editor of Physics Letters. Indeed, it is a way to keep in touch with the action in the field; however, I received on the average 1 article a day, and I rejected about 90 % straight away. In other words, the big majority was junk, just cluttering up my mind. As far as I can see it has become worse, not better. Many a physicist has come to hate my "high handedness", as one of the victims called it. I was happy to get out of that job by the summer of 1968.

A turning point in my scientific life occurred during a one month visit (April 1968) to Rockefeller University. In the quietness of that institution I embarked on the scientific venture that has now been honoured with the Nobel Prize. I am still indebted to Bram Pais who got me there and counseled me in that period. Too bad that he left the Netherlands in 1945; I am sure that he would have kept Dutch particle physics on a high level. One man can make a big difference.

In the summer of 1968, I went to Orsay, near Paris, on the invitation of the French physicists Claude Bouchiat and Philippe Meyer. The stay lasted till September 1969, it was a sabbatical year (after two years in Utrecht...). As Utrecht partly paid me during that period, I told the French people that I did not need much in the way of a salary, and subsequently they did put me in some low job. This had an unpleasant consequence; Christmas 1968, I was fired, as de Gaulle had decided on some cost saving operation. Luckily they succeeded in patching it up, in some mysterious French way. Some well known French physicist told me: luckily that it happened to you and not to T.D. Lee. That made me aware of my place on the totem pole.

Back in Utrecht I continued my work, and had several students under my supervision. Among them Peter van Nieuwenhuizen, now director of the C.N. Yang institute at Stony Brook, Bernard de Wit, now holding my former position in Utrecht and Gerard 't Hooft, my co-laureate. Our group became known, especially so after the work of 't Hooft and myself that is cited in connection with this years prize. Besides my own research I was very busy in that period: reforming the physics educational system in Utrecht (see my complaints above), and trying to get a good computer system. The latter required endless meetings, mainly caused by some mathematician who insisted that the machine could run Algol, a by now largely forgotten computer language. I had literally to learn some 6 or 7 computers inside out to get to the final result: a CDC 6800 computer. During one of these meetings (January 1971) I received a phone call telling me that my wife was about to deliver another child; I went out of the meeting to the hospital and came back after about one hour. I gave everybody a cigar, celebrating the birth of my son Martijn and continued with the meeting. Martijn is now working in Hollywood, in the movie industry.

't Hooft and I worked together for a few more years, after which we drifted apart. I went my way doing calculations of radiative corrections, something that he was not interested in. The fame of Utrecht had spread, and two young Italian physicists came to work with me: Giam-Piero Passarino and Maurizio

Consoli. Some Dutch students at that time were Jochum van der Bij, now professor at Freiburg, Germany and Michel Lemoine, now a free lance Senior Petroleum Engineer. The latter has convinced me that theoretical physics is a good science to be educated in, it prepares for no job in particular but the scientific methods learned are of use in many positions in modern society. So never worry too much what kind of job you will get after finishing a theoretical physics education. For example, the first Prize in Economic Sciences in Memory of Alfred Nobel went to Tinbergen, a former theoretical physicist. And nowadays the banking world is full of particle theorists.

In the summer of 1979, I received an invitation from Ed Yao from the University of Michigan to spent a sabbatical year there. I knew Yao from his scientific work, and I immediately called my wife, asking if she was interested in a year in Michigan. For reasons that I have now forgotten we left in March 1980, to stay till December of that year. In Michigan we were asked to stay, but initially we answered rather firmly that I was not interested. In November we started wavering, and in fact Fermi lab (under the directorship of Leon Lederman, another experimental friend of mine) started to express interest as well. In December a nice house was auctioned near were we lived in Ann Arbor, and my wife told me: if you buy that house I will stay. I did not get the house, some richer medical person got it, but this somehow made us decide to stay in Ann Arbor. Part of it was a certain unhappiness with the situation in the Netherlands, and another part was the happiness of our sons with the American school system. My Utrecht colleagues were quite upset when I told them and they did sent me a telegram asking me to remain in Utrecht. But we decided to accept the offer of Ann Arbor, and when we came back in Utrecht I started preparations to leave, which we did in September of 1981.

I did my part in the scientific life in the United States, serving among others on the various committees that decide on experiments at the big laboratories, Fermilab near Chicago, SLAC at Stanford and Brookhaven National Laboratory, Long Island.

Shortly after we arrived I was offered a named chair, the John D. and Catherine T. MacArthur chair. Europeans think that this means an extra income, but that is not true. There are other things however. Apart from the prestige of this chair it had a really nice feature going with it: a yearly amount of $ 35000 that I could spend for scientific purposes. While it may seem a small amount, it nonetheless made quite a difference. I could pay certain deserving students during the summer, buy computer stuff, help the group with little things, visit conferences, invite collegues etc. It is wonderful that you can buy a new computer almost immediately if there is a need for it. In Europe that took often a long time, you had to put it on a budget and wait for approval etc. This is one aspect of the greater flexibility of the American system. If you want something done you either use discretionary funds at your disposal or you go around and try to find money (discretionary funds) from people or groups that feel it is also to their advantage to support the purpose. For example I remember supporting a Russian scientist partly from my MacArthur fund, partly from a fund from the astro-physics group.

In hindsight we do not regret this move to the USA, but it would take me to far to explain that. Too many strictly personal considerations are involved here. The University of Michigan has been good to me, and I feel loyal to that institution. Also the life in Ann Arbor was quite nice, and Anneke felt very much at home there, enjoying membership of a great many clubs. Among others she learned to make beautiful stained glass windows. Nonetheless, it was quite a step for a 50 year old man and his family to emigrate. Dutch people abroad have a saying: rather nostalgia than Holland. I would not go that far, we had certainly many friends in the Netherlands, and also most of our families (from my wife and me) live there.

My main tie with Europe during the US period (1981–1996) was with the University of Madrid (the Autonoma), Spain. There was a particle theory group headed by F. J. Yndurain. I would go there up to two months during the summer time, and conversely he would often come to Ann Arbor. This type of collaboration is usually very fruitful, not only for doing science, but also because it fosters the exchange of graduate students.

It was just in this period that Spain decided seriously to catch up with the rest of Europe, and that was an interesting and exciting thing to watch. While not everything went perfect, I would say that Spain made enormous strides forward in a relatively short time. Up to this day I have very good relations with Spanish physicists, both in Spain and at CERN. I should perhaps add that to them CERN is of crucial importance, as it has been to me.

On retirement we decided to return to the town of Bilthoven in the Netherlands that we had left in 1981, to find still many of our old friends there. That is were we now live happily. Our children however did not go back, they would really not fit anymore in the Dutch society. It rains too much.

FROM WEAK INTERACTIONS TO GRAVITATION

Nobel Lecture, December 8, 1999

by

MARTINUS J. G. VELTMAN

Emeritus MacArthur Professor of Physics, University of Michigan, Ann Arbor, MI 48109-1120, USA.

INTRODUCTION

This lecture is about my contribution to the proof of renormalizability of gauge theories. There is of course no perfectly clear separation between my contributions and those of my co-laureate 't Hooft, but I will limit myself to some brief comments on those publications that carry only his name. An extensive review on the subject including more detailed references to contemporary work can be found elsewhere [1].

As is well known, the work on the renormalizability of gauge theories caused a complete change of the landscape of particle physics. The work brought certain models to the foreground; neutral currents as required by those models were established and the discovery of the J/Ψ was quickly interpreted as the discovery of charm, part of those models as well. More precisely, we refer here to the model of Glashow [2], the extension to include quarks by Glashow, Iliopoulos and Maiani (GIM) [3] and the model of Weinberg-Salam [4] for leptons including a Higgs sector. The GIM paper contained discussions on the required neutral hadron currents and also the inclusion of charm as suggested first by Hara [5]. After an analysis by Bardeen, at a seminar in Orsay (see also [6]), the work of Bouchiat, Iliopoulos and Meyer [7] established the vanishing of anomalies for three color quarks. Without going into details, subsequently quantum chromodynamics came to be accepted. In this way the Standard Model was established in just a few years.

ESSENTIAL STEPS

Let me review here what I consider as my contribution to the subject. It can be described in three separate parts. I will try to simplify things as much as possible.

I. **The physics argument.** In 1965 Adler [8] and Weisberger [9] established what is now known as the Adler-Weisberger relation. This relation, numerically agreeing with experimental data, was interpreted by me as a consequence of a Ward identity of a non-Abelian gauge theory (also called Yang-Mills theories), and as such guided me to take up the study of such theories.

II. **The renormalizability argument.** Earlier calculations on the radiative corrections to the photon-vector-boson vertex showed a disappearance of many divergencies for a properly chosen vector-boson magnetic moment. In studying Yang-Mills theories I noted that those theories automatically produced this particular magnetic moment. I therefore concluded that Yang-Mills theories are probably the best one can have with respect to renormalizability. Thus I was led to the study of renormalizability of these theories.

III. **Technical progress.** Starting then on the study of diagrams in a Yang-Mills theory I established the vanishing of many divergencies provided the external legs of the diagrams were on the mass shell. That by itself is not enough with respect to renormalizability, because that requires diagrams and Feynman rules of a renormalizable type. I was thus led to search for a transformation of the theory such that new, renormalizable type Feynman rules were derived, without changing the S matrix. In this I succeeded up to one loop.

None of these points is trivial, as can be shown easily by considering work in that period. For example, Weinberg [10] in his 1979 Nobel lecture reports that he interpreted the success of the Adler-Weisberger relation as a property of strong interactions, namely, the validity of chiral SU2 × SU2. Consequently he continued working on things such as π–π scattering. Feynman is reputed to have exclaimed that he never thought of investigating the renormalizability aspect of Yang-Mills theories when he heard of that development. Finally, there existed several papers where the non-renormalizability of Yang-Mills theories was "proven", for example one by Salam [11].

In the following I will discuss these three points in detail as they developed historically.

THE PHYSICS ARGUMENT

In 1965 Adler and Weisberger derived their famous relation between the axial vector coupling constant of β decay in terms of a dispersion integral for pion-nucleon scattering. This relation, agreeing well with experiment, was based on Gell-Mann's current commutator rules [12]. Subsequently an extensive discussion developed in the literature concerning the so-called Schwinger terms that could invalidate the argument. I decided to try to derive these same results starting from another assumption, and as a starting point I took the well-known CVC and PCAC equations for the weak currents:

$$\partial_\mu J_\mu^V = 0$$

$$\partial_\mu J_\mu^A = ia\pi$$

These equations do not include higher-order electromagnetic (e.m.) or weak effects. As a first step I tried to include electromagnetic effects by using the well-known substitution $\partial_\mu \to \partial_\mu - iqA_\mu$, where q is the charge of the object on which ∂_μ operates. Since the currents were isovectors, that could be done easily using isospin notation. For the vector current the equation became

$$(\partial_\mu + ie\vec{A}_\mu \times)\vec{J}^V_\mu = 0,$$

treating the e.m. field as the third component of an isovector. Next I used the idea that the photon and the charged vector bosons may be seen as an isotriplet, thus generating what I called divergency conditions [13]. For the axial vector current this gave:

$$\partial_\mu \vec{J}^A_\mu = ia\vec{\pi} + ie\vec{A}_\mu \times \vec{J}^A_\mu + ig\vec{W}^V_\mu \times \vec{J}^A_\mu + ig\vec{W}^A_\mu \times \vec{J}^V_\mu .$$

As a matter of technical expedience two vector bosons were used to denote a vector-boson coupling to a vector current and a vector-boson coupling to an axial current. This equation turned out to be adequate to derive the Adler-Weisberger relation. An added benefit of this derivation was that no difficulties with respect to Schwinger terms arose, and the axial vector coupling constant was related directly to the pion-nucleon scattering length. The Adler-Weisberger relation evidently used an additional relation in which the pion-nucleon scattering length was given in terms of a dispersion integral.

In response, John Bell, then at CERN, became very interested in this derivation. He investigated what kind of field theory would generate such divergency conditions, and he found that this would happen in a gauge theory [14].

Subsequent developments were mainly about the consequences of those relations involving e.m. fields only. It is clear that specializing to the third component of the axial divergence condition there are no e.m. corrections. Following Adler, reading $\partial_\mu J^{A0}_\mu = a\pi^0$ in the opposite way imposed a condition on the pion field including e.m. effects. This was an extension of earlier work by Adler [15], known under the name of consistency conditions for processes involving pions. In this case one of the conclusions was that π^0 decay into two photons was forbidden, and without going into a detailed description this led to the work of Bell and Jackiw [16] on the anomaly. Simultaneously Adler [17] discovered the anomaly and in fact used precisely my (unpublished) derivation to connect this to π^0 decay. Later I became quite worried by this development, as I saw this anomaly as a diffculty with respect to renormalization.

THE RENORMALIZABILITY ARGUMENT

Here I must go back to 1962. In that year Lee and Yang [18] and later Lee [19] alone started a systematic investigation of vector bosons interacting with photons. The paper of Lee and Yang mainly concentrated on deriving the Feynman rules for vector bosons. The trouble at that time was that in doing the usual canonical derivation one encountered certain contact terms for the vector-boson propagator. I will not delve any further into this; later I found a simple way to circumvent these problems. In those days, however, these were considered serious problems.

Subsequently Lee started a complicated calculation, namely the lowest order radiative corrections to the vector-boson-photon coupling. The usual replacement $\partial_\mu \to \partial_\mu - ieA_\mu$ in the vector-boson Lagrangian is not suffcient to determine the vector-boson magnetic moment; it remains an arbitrary parameter. This is because of the occurrence of two derivatives such as $\partial_\mu\partial_\nu$; when making the minimal substitution it matters whether one writes $\partial_\mu\partial_\nu$ or $\partial_\nu\partial_\mu$, and this causes the arbitrariness in the magnetic moment. Anyway, Lee, concentrating on the electric quadrupole moment of the vector-boson calculated the appropriate triangle diagram using a cutoff procedure called the ξ-limiting process.

I was very interested in this calculation, because like many physicists I strongly believed in the existence of vector bosons as mediators in weak interactions. This belief was based on the success of the $V-A$ theory, suggesting a vector structure for the weak currents. Indeed, this led Glashow to his famous 1961 paper. I decided that Lee's work ought to be extended to other situations, but it was quite obvious that this was no mean task. Given the ξ method, and the occurrence of the magnetic moment as an arbitrary parameter, the triangle diagram if calculated fully (Lee limited himself to parts relevant to him) gives rise to a monstrous expression involving of the order of 50 000 terms in intermediate stages. There was simply no question of going beyond the triangle diagram.

At this point I decided to develop a computer program that could do this work. More specifically, I concentrated on the triangle graph, but I wrote the program in such a way that other processes could be investigated. In other words, I developed a general purpose symbolic manipulation program. Working furiously, I completed the first version of this program in about three months. I called the program *Schoonschip,* among others to annoy everybody not Dutch. The name means "clean ship," and it is a Dutch naval expression referring to clearing up completely a messy situation. In January 1964, visiting New York in connection with an American Physical Society meeting, I visited Lee and told him about the program. He barely reacted, but I heard later that after I left the office he immediately wanted one of the local physicists to develop an analogous program.

In toying with the calculation I tried to establish what would be the best value for the vector-boson magnetic moment with respect to the occurring divergencies. There was one value where almost all divergencies disappeared, but I did not know what to do with this result. It remained in my memory though, and it played a role as explained below.

TECHNICAL PROGRESS

To explain the development requires some backtracking. In 1959 I took up the study of the problem of unstable particles. The problem is of a nonperturbative character, because a particle is unstable no matter how small the coupling constant of the interaction that produces the decay. Thus the (unstable) particle will not appear in the in and out states of the S matrix. However, for zero value of the coupling constant the particle is stable and must be part of the in and out states. Thus the limit of zero coupling constant does not reproduce the zero-coupling-constant theory.

It was in principle well known at the time how to handle an unstable particle. Basically one did what is called a Dyson summation of the propagator, and that indeed removed the pole in the propagator. From the Källén-Lehmann representation of the propagator one knows that every pole corresponds to an in or out state, so the summation indeed seemed to correspond to removing the particle from the in and out states.

However, when performing the Dyson summation one found that the theory became explicitly nonperturbative, as self-energy diagrams and with them factors g (the coupling constant of the destabilizing interaction) appeared in the denominator of the propagator of the unstable particle. That propagator looked like this:

$$\frac{1}{k^2 + M^2 + g^2 F(k)}.$$

Obviously, this propagator cannot be expanded as a function of g in the neighbourhood of $k^2 + M^2 = 0$ if the imaginary part of $F(k)$ is nonzero in that point (the real part is made zero by mass renormalization). So, instead of a propagator with a pole the Dyson summation made it into a function with a cut in the complex k^2 plane. At this point it is no longer clear that the S matrix is unitary, because the usual equation for the S matrix $S = T[exp(iH)]$ is no longer valid. In other words, to establish unitarity one had to consider the diagrams by themselves.

Thus I attacked this problem, essentially finishing it in 1961. This was for my thesis, under the supervision of Leon van Hove. The article went its ponderous Dutch thesis way and was published in 1963 [20], in a somewhat unusual journal (*Physica*) for high-energy physics. Curiously, about the same time Feynman [21] considered the same problem, in connection with establishing unitarity for the massless Yang-Mills theory, a theory whose diagrams include ghosts. These ghosts make unitarity nonevident. Moreover the derivation by Feynman, done with path integrals, did not guarantee unitarity. I am quite sure that he never saw my article, and I never discussed it with him either. He tried to do it some other way, quite complicated, initially succeeding only up to one loop. Later, DeWitt [22] extended Feynman's proof to any number of loops, but my proof is much simpler and moreover connects quite directly to physical intuition. In fact, my proof had as a result that the imaginary part of a diagram equals the sum of all diagrams that can be obtained by cutting the initial diagram in all possible ways.

The importance of this work was twofold. Not only did unitarity become a transparent issue, I also learned to consider diagrams disregarding the way they were derived, for example using the canonical formalism. Given that it is not easy to derive Feynman rules for a Yang-Mills theory in the canonical way, that did gave me an advantage in studying that theory. For considering Yang-Mills theories the path-integral formalism is quite adequate; there is only one point, and that is that this formalism does not guarantee unitarity. In

1968 the path-integral formalism had all but disappeared from the literature, although students of Schwinger did still learn functional methods. I myself did not know the first thing about it.

In 1968, I was invited by Pais to spend a month at Rockefeller University. I happily accepted this invitation and decided to try to think through the present situation. For two weeks I did nothing but contemplate the whole of weak interactions as known at the time. I finally decided to take Bell's conclusion seriously and therefore assumed that the weak currents were those of a gauge theory. Thus I started to learn Yang-Mills theory and tried to find out how that would work in some simple weak processes. In the process of writing down the Feynman rules I noted that this theory gave precisely the "best" vertex (with respect to divergencies) as I had found out doing the work reported above on the photon-vector-boson interaction. This encouraged me to concentrate on the renormalizability aspect of the theory.

As far as I remember I started by considering the one-loop corrections to neutrino-electron scattering. Here the situation became quickly quite complicated. The vertices of the Yang-Mills theory were much more complicated than those that one was used to, and even the simplest diagrams gave rise to very involved expressions. In the end I decided to drop everything except the basic theory of vector bosons interacting with each other according to a Yang-Mills scheme. In addition, of course, I gave these vector bosons a mass, since the vector bosons of weak interactions were obviously massive. I started in blissful ignorance of whatever was published on the subject, which was just as well or I might have been convinced that Yang-Mills theories are non-renormalizable. As Feynman said in his Nobel lecture as presented at CERN, "Since nobody had solved the problem it was obviously not worthwhile to investigate whatever they had done." I want to mention here that at that time I started to get worried about the anomaly, but I decided to leave that problem aside for the moment.

Consider the propagator for a massive vector field:

$$\frac{\delta_{\mu\nu} + k_\mu k_\nu / M^2}{k^2 + M^2} .$$

The source of all trouble is of course the $k_\mu k_\nu$ term. So anyone starting at this problem tries to eliminate this term. In quantum electrodynamics that can indeed be done, but for Yang-Mills fields this is not possible. There is always some remnant. Now here a simple observation can be made: this bad term comes with a factor $1/M^2$. In fact, one can trace the worst divergencies in a diagram simply by counting factors $1/M^2$. But given that they will not ever completely cancel, as they indeed do not for a Yang-Mills-type theory, then one will never get rid of these divergencies unless somehow these factors $1/M^2$ cancel out. But where should the necessary factors M^2 come from? There is only one way, and that is through external momenta that are on the mass shell, meaning that the momentum p of such an external line satisfies $p^2 = -M^2$. And here is the problem.

Renormalization means that for a divergent graph one cannot take the external momentum on the mass shell and then do the necessary subtractions, because the graph may occur as part of a more complicated graph. For example, in a box diagram there may be a self-energy insertion in one of the internal lines. There momenta of the lines attached to the self-energy insertion are not on the mass shell, thus it is not suffcient to subtract only those divergencies that remain if those momenta are on the mass shell. You would still have to show that the extra divergencies arising when those lines are not on the mass shell actually cancel, a gruesome task. What to do?

Well, what I did was to reformulate the theory such that somehow all cancellations were implemented in the rules. In the first instance I took the Stueckelberg technique [23]: added a scalar field and made couplings involving derivatives such that it appeared together with the vector-boson propagator:

$$\frac{\delta_{\mu\nu} + k_\mu k_\nu / M^2}{k^2 + M^2} + \kappa \, \frac{k_\mu k_\nu / M^2}{k^2 + M^2} \, .$$

The second term would be due to the exchange of the scalar particle. The parameter κ was introduced to keep track of the counter term. Eventually κ was taken to be -1. Now this new field is physically undesirable, because to have $\kappa = -1$ is actually impossible. The scalar field would have to have indefinite metric, or some such horrible thing. In an Abelian theory it is easy to show that the field is noninteracting, but not in the non-Abelian case. Then I had an idea: introduce further interactions of this new scalar field in such a way that it becomes a free field. The result, hopefully, would be a new theory, involving a well-behaved vector-boson propagator and furthermore an interacting scalar field that would then be a ghost. Indeed, being a free field it could appear in the final state only if it was there in the initial state. At this point one would have new Feynman rules, presumably much less divergent because of the improved vector-boson propagator. It was all a matter of what Feynman rules would result for this scalar field. If they were those for a renormalizable field, then we would be in business!

So here is the important point: the theory must be formulated in terms of diagrams which would have to be of the renormalizable type. No matter that ghosts occur; those do not get in the way with respect to the renormalization program.

There is a bonus to this procedure: one can write an amplitude involving one such scalar field. Because the scalar field is a free field, that amplitude must be zero if all other external lines are on the mass shell. That then gives an identity. Using Schwinger's source technique one can extend this to the case in which one or more of the other external lines are off the mass shell. I later called the resulting identities generalized Ward identities.

There is another aspect to this procedure. Because the final diagrams contain a vector-boson propagator that has no $k_\mu k_\nu$ part, that theory is not evidently unitary. At that point one would have to use the cutting rules that I had obtained before and show, using Ward identities for the cut scalar lines, that the theory was unitary. All in all quite a complicated affair, but not that diffcult.

Here the miracle occurred. On the one-loop level almost all divergencies disappeared. It was not as straightforward as I write it here, because even with the new rules one needed to do some more work using Ward identities to get to the desired result. In any case, I arrived at Feynman rules for one-loop diagrams that were by ordinary power-counting rules renormalizable rules. For those who want to understand this in terms of the modern theory: instead of a Faddeev-Popov ghost (with a minus sign for every closed loop) and a Higgs ghost (no minus sign, but a symmetry factor) I had only one ghost, and on the one-loop level that was actually the difference of the two ghosts as we know them now.

No one will know the elation I felt when obtaining this result. I could not yet get things straight for two or more loops, but I was sure that that would work out all right. The result was for me a straight and simple proof that my ideas were correct. A paper presenting these results was published [24].

The methods in that paper were clumsy and far from transparent or elegant. The ideas, however, were clear. I cannot resist quoting the response of Glashow and Iliopoulos [25]. After my paper appeared they decided to work on that problem as well, and indeed, they showed that many divergencies cancelled, although not anywhere to the renormalizable level. For example, the one-loop box diagram is divergent like Λ^8 in the unitary gauge; their paper quoted a result of Λ^4. I of course obtained the degree of divergence of a renormalizable theory, i.e., $\log(\Lambda)$. Here then is a part of the footnote they devote to this point: "The divergencies found by M. Veltman go beyond the theorem proven in this paper, but they only apply to on-mass-shell amplitudes..." Indeed!

In present-day language one could say that I made a transformation from the unitary gauge to a "renormalizable" gauge. As I had no Higgs the result was not perfect. But the idea is there: there may be different sets of Feynman rules giving the same S matrix.

INTERIM

In the years 1969–1971 I expended considerable effort trying to go beyond the one-loop result. There were many open problems, and they had to be considered. In the beginning of 1970 I streamlined the derivation to the point that it became transparent. This was done by deriving Ward identities using Schwinger's source techniques [26]. This is really much like the way one derives Ward identities today, now called Slavnov-Taylor identities. The BRS transformation is a sophisticated form of the free-field technique (using anticommuting fields). I remember being upset when I first heard a lecture by Stora on what he called the Slavnov-Taylor identities. I told him that they were another variant of my generalized Ward identities. However, Stora is not a diagram man, and I am sure that he never understood my paper.

Another issue was the limit of zero mass of the massive Yang-Mills theory. In January 1969 there was a conference at CERN and I announced that two-loop diagrams for the massive Yang-Mills theory did not go over into those of the massless theory, in other words, the massless theory is not the limit of zero mass of the massive theory [27]. This argument was spelled out in an article with J. Reiff [28]. The main part of that article was to tie up another loose end: the Feynman rules for vector bosons in the unitary gauge. The argument is quite elegant and superseded the article by Lee and Yang mentioned before. This made it clear to me that spurious contact terms related to that part of the theory were not responsible for the two-loop problems.

Somewhere in the first half of 1970, I heard via Zumino that Faddeev (and Slavnov, as I learned later) [29] had established that already at the one-loop level the massless theory was not the limit of the massive theory. The difference was hiding in a symmetry factor for the one-loop ghost graphs: they have a factor of $\frac{1}{2}$ as compared to the Faddeev-Popov ghost loops of the massless case. In the summer of 1970, H. Van Dam and I reproduced and understood the argument and went further to consider gravitation [30]. Here we found one of the more astonishing facts in this domain: for gravity the limit from massive gravitons to zero mass is not the same as the massless theory (of Einstein). Thus a theory of gravitation with a massive spin 2 particle of exceedingly small mass (for example, of the order of an inverse galactic radius) would give a result for the bending of light by the sun that was discretely different (by a factor of $\frac{3}{4}$) from that of the massless theory. Thus by observing the bending of light in our solar system we can decide on the range of the gravitational field on a galactic scale and beyond. Many physicists (I may mention Kabir here) found this result hard to swallow. The discontinuity of the zero-mass limit as compared to the massless case has always been something contrary to physical intuition. Indeed, for photons there is no such effect. The work with Van Dam was actually my first exercise in the quantum theory of gravitation.

So, by the end of 1970, I was running out of options. I started to think of studying the difference between the massless and massive cases, more explicitly, to try to sort of subtract the massless theory from the massive theory, diagramwise. That would have produced a hint to the Higgs system. Indeed, the theory with a Higgs particle allows a continuous approach to the massless theory with, however, four extra particles. Furthermore, I was toying with the idea to see if the remaining infinities had a sign that would allow subtraction through some further interactions. Conceivably, all this could have resulted in the introduction of an extra particle, the Higgs, with interactions tuned to cancel unwanted divergencies, or to readjust the one-loop counter terms to be gauge invariant (in my paper the four-point counter terms were not). The result would have been in the worst kind of "Veltmanese" (a term used by Coleman to describe the style of 't Hooft's first article). However, that development never happened, and hindsight is always easy. That kind of work needed something else: a regularization procedure. Not only was the lack of a suitable method impeding further investigation or application of the results obtained so far to practical cases, but there was also the question of anomalies. It is at this point that 't Hooft entered into my program.

't Hooft became my student somewhere in the beginning of 1969. His first task was writing what was called a "scriptie", a sort of predoctoral thesis, or *thèse troisième cycle* (in France). The subject was the anomaly and the σ model. That being finished in the course of 1969 he then started on his PhD work. At the same time he took part in my path-integral enterprise, so let me describe that.

I spent the academic year 1968–1969 at the University at Orsay, near Paris. During the summer of 1968 I was already there for the most part and met Mandelstam, who had been working on Yang-Mills theories as well. He had his own formalism [31], and we compared his results with mine. We did not note the notorious factor of 2 mentioned above: Mandelstam had studied the massless case, while my results applied to the massive theory. Boulware was also there. As a student of Schwinger he knew about functional integrals, and he later applied his skills to the subject [32]. It became clear to me that there was no escape: I had to learn path integrals. At the end of my stay at Rockefeller University somebody had already told me that there was work by Feynman [21] and also Faddeev on the massless theory. The article by Faddeev and Popov [33] was, as far as I was concerned, written in Volapuk. It also contained path integrals, and although I had accepted this article in my function as editor of Physics Letters, I had no inkling what it was about at the time (summer 1967). I accepted it then because of my respect for Faddeev's work. Just as well!

My method of learning about path integrals was lecturing on it, in Orsay. Ben Lee happened to be there as well, and he was also interested. With some diffculty I obtained the book of Feynman and Hibbs. (This was not easy; the students were busy making revolution and had no time for such frivolous things as path integrals. I thus sent around a note asking them to return the book prior to making revolution, which indeed produced a copy. This gave me the reputation of an arch reactionary, which I considered a distinction, coming as it did from Maoists.) Somewhere during these lectures a Polish physicist (Richard Kerner, now in Paris) produced another article by Faddeev, in Russian, and I asked him to translate it. I have never read that article, Ben Lee took it with him. I was simply not up to it and I still felt that I did not understand path integrals. So, returning to the Netherlands I decided to do it once more, and in collaboration with Nico Van Kampen we did set up a course in path integrals (autumn 1969). My then student 't Hooft was asked to produce lecture notes, which he did. I would say that then I started to understand path integrals, although I have never felt comfortable with them. I distrust them. 't Hooft had no such emotional ballast, and he became an expert in the subject. So, by the end of 1969 't Hooft had been educated in the σ model, anomalies, and path integrals.

't HOOFT

At this point 't Hooft showed unhappiness with the provisional subject that I had suggested, namely, the double-resonance peak of Maglic. He wanted to enter into the Yang-Mills arena. I then suggested that he investigate the massless theory, with emphasis on finding a regulator method. This was so decided during a dinner, also attended by Van Kampen.

In studying the massless case 't Hooft used combinatorial methods (diagram manipulation) to establish various identities [34]. He could have used the Ward identities of my earlier paper, but I think he wanted to show that he could do better. Thus it came to pass that he never wrote the Slavnov-Taylor identities, an oversight that these two gentlemen quickly corrected [35,36]. 't Hooft derived mass shell identities, presumably enough for renormalization purposes.

Perhaps the main point that we argued about was the necessity of a gauge-invariant regularization scheme. He took the point of view that no matter what scheme one uses one simply adjusts the subtraction constants so that the Ward identities are satisfied, and that is all that is needed to renormalize the theory. Well, that is true provided there are no anomalies, and after some time he accepted the point. He developed a gauge-invariant method that worked up to one loop. A fifth dimension was used. Later, trying to go beyond one loop we developed the dimensional regularization scheme; in the summer of 1971 we had a rough understanding of that method. I should say that at all times I had an ulterior motive: I very much wanted an actually usable scheme. The existing methods (such as the Pauli-Villars scheme) are perhaps useful in doing quantum electrodynamics, but completely impractical for a Yang-Mills theory. I needed a good tool.

In an appendix to his paper 't Hooft presented, within the path-integral scheme, a gauge choosing method. I did not recognize this at all, but later, backtracking, I discovered that this was an evolved version of my original attempt at a change of gauge, including the "free field" technique. Russian physicists (Faddeev, Slavnov, Fradkin and Tyutin) took it over into the path-integral formalism, mangled, cleaned, and extended the method, mainly applying it to the massless case as well as (massless) gravitation. The actual scheme proposed by 't Hooft in the quoted appendix is the method that is mostly used today.

I am not going to describe the (substantial) Russian contributions here. This despite the fact that, like almost everywhere else, doing field theory was not very popular in the Soviet Union. I believe that in ref [1] a fair account has been given.

I am also skipping a description of the second paper of 't Hooft [37], introducing spontaneous symmetry breakdown and thus arriving at the renormalizable theories with massive vector bosons as known today. There are only two points that I would like to mention: I insisted that as much as possible the results should not depend on the path-integral formalism, i.e. that unitarity should be investigated separately, and secondly, that the issue of there being something in the vacuum not be made into a cornerstone. Indeed, once the Lagrangian including spontaneous symmetry breaking has been written down you do not have to know where it came from. That is how I wanted the paper to be formulated. I suspected that there might be trouble with this vacuum field, and I still think so, but that does not affect in any way 't Hooft's second paper. He sometimes formulates this as me opposing the cosmological constant, but at that time I did not know or realize that this had anything to do with the cosmological constant. That I realized for the first time during a seminar on gravitation at Orsay, at the beginning of 1974. See ref. [38].

So let me go on to the autumn of 1971. 't Hooft dived into massless Yang-Mills theory studying the issue of asymptotic freedom; I think that Symanzik put him on that track. I devoted much attention to the dimensional regularization scheme [39]. Again I refer the interested reader to ref. [1] for details, including the independent work of Bollini and Giambiagi.

After dimensional regularization was developed to an easily workable scheme I decided that it would be a good idea to write two papers: (i) a paper clearly showing how everything worked in an example, and (ii) a reasonably rigorous paper in which renormalizable gauge theories were given a sound basis, using diagram combinatorial techniques only. The result was two papers, one entitled "Example of a gauge field theory" [40], the other "Combinatorics of gauge field theories" [41]. The first was presented at the Marseille conference, summer 1972, where a preliminary version of the second paper was also presented. I have no idea how many physicists read the "Example" paper; I think it is a pity that we published it only in the conference proceedings and not in the regular literature. In that paper all one-loop infinities of the simple SU2 model with a two-parameter gauge choice were computed, and the informed reader may without any trouble use the counter Lagrangian given in that paper to deduce the β parameter for that theory (including a Higgs). That is the parameter relevant for asymptotic freedom. The calculations for this paper were fully automated and done by *Schoonschip.* When 't Hooft asked me about the divergencies of the massless theory as a check on his own calculations there was no problem doing that. I did not know about asymptotic freedom and did not understand the relevance of this particular calculation at that moment. He reported his result at the Marseille conference.

RADIATIVE CORRECTIONS

Much of my effort after 1972 were directed towards applying the theory, i.e., towards radiative corrections. In 1975 there was still considerable argument about neutral currents. Most people thought that the precise configuration contained in the Weinberg model was a must, not knowing that by choosing another Higgs sector one could adjust the Z_0 mass to any value. This was clearly a critical point, and Ross and I set out to investigate this issue [42]. This led to the introduction of a new parameter, now called the ρ parameter, that takes on the value of 1 for the simplest Higgs sector as chosen by Weinberg. The ρ parameter is essentially the square of the ratio of the charged vector-boson mass to the neutral vector-boson mass, with a correction related to weak mixing. This parameter has become an important part of today's physics, because it is the most sensitive location for radiative effects of heavy particles, quarks or Higgs. At the Paris conference on neutral currents of 1974 or 1975 I presented a very short contribution, consisting of, I believe, only two transparencies. All I said was this: the neutral vectorboson mass can be anything. Here is a convenient way to parametrize that. To this day I am flabbergasted that nobody, but nobody at that conference seemed to have gotten the message. They kept on thinking that finding the precise quantitative amount of neutral-current effects as predicted by the Weinberg model (extended to quarks according to Glashow, Iliopoulos, and Maiani) was crucial to the applicability of gauge theories. In reality, had they found deviant results, the only consequence would have been a different Higgs sector.

In 1976 it became reasonably clear that the Standard Model including the simplest possible Higgs sector was the right model. Meaningful calculations on radiative corrections were now possible, and I set out to do them. It appeared that there were at least three families. The following issues were of immediate importance to me:

(i) How many generations are there?
(ii) Is there an upper limit on the Higgs mass?

I will not enter into the argument on the number of neutrinos from astrophysics. Such arguments are less than airtight, because they build on the whole body of our understanding of the big bang and evolution of the Universe. Concerning the third generation, an interesting argument developed: what is the mass of the top quark? It would fill an amusing article to list all articles that made claims one way or the other, but I leave that to someone else. I realized that without a top quark the theory would be non-renormalizable, and therefore there ought to be observable effects becoming infinite as the top-quark mass goes to infinity. To my delight there was such a correction to the ρ parameter and furthermore it blows up proportional to the top-quark mass squared [43]. This is the first instance in particle physics of a radiative correction becoming larger as the mass of the virtual particles increases. That is our first window to the very-high-energy region. This radiative correction became experimentally better known and eventually produced a prediction for the top-quark mass of 175 GeV, which agrees with the result found when the top was discovered. This agreement also seems to indicate that there are no more generations, because there is little or no room for any quark mass differences in (hypothetical) new generations. Given the pattern of masses that we observe now, that appears unlikely, although it is strictly speaking not impossible.

From the beginning I was very interested in the Higgs sector of spontaneously broken theories. I started to look for a way to establish a limit on the Higgs mass; after all, if the Higgs is an essential ingredient of renormalization there must be terms in perturbation theory that cannot be renormalized away and that would be sensitive to the Higgs mass. It can easily be argued that the place to look for that is in the radiative corrections to the vector-boson masses, and the relevant parameter there is the ρ parameter, introduced in the paper with Ross mentioned before. As it happens, while there could have been an effect proportional to the square of the Higgs mass, it turned out that that piece cancels out and only a logarithmic dependence remains [44]. This makes it very diffcult to estimate the Higgs mass on the basis of radiative corrections, and I have introduced the name "screening theorem" in this connection. Nature seems to have been careful in hiding the Higgs from actual observation. This and other facts have led me to believe that something else is going on than the Higgs sector as normally part of the Standard Model.

After that I started to set up a systematic scheme for the calculation of radiative corrections, together with Passarino. As he, together with Bardin, has written a book that has just come out I refer the interested reader to that book [45].

There was another motive in doing the radiative corrections. I wanted ultimately to compute the radiative corrections to *W*-pair production at LEP, because it was clear to me that those corrections would be sensitive to the Higgs mass. This then would suggest a value for the LEP energy: it should be high enough that radiative corrections to *W*-pair production were sufficiently large and could be studied experimentally. One would either find the Higgs or see important radiative corrections. That calculation was done together with Lemoine [46] and finished in 1980. I did not succeed in making the case sufficiently strong: no one understood the importance of such considerations at that time. Thus the LEP energy came out at 200 GeV, too low for that purpose. As the vector bosons were still to be found, few were prepared to think beyond that. Also, I have no idea if a 250- to 300-GeV LEP would have been possible from an engineering point of view, let alone financially.

PRESENT STATUS

The Higgs sector of the Standard Model is essentially untested. Customarily one uses the simplest possible Higgs system, one that gives rise to only one physical Higgs particle. With that choice the Z_0 mass is fixed to be equal to the charged W mass divided by $\cos(\theta)$, where θ is the weak mixing angle. Let us first establish here a simple fact: by choosing the appropriate Higgs sector one can ensure that the Z_0 mass is unconstrained. Furthermore the photon mass need not be zero and can be given any value.

In the early days a considerable amount of verbosity was used to bridge the gap between the introduction of the models in 1967 and the later-demonstrated renormalizability. Two of the terms frequently used to this day are "electroweak unification" and "spontaneous symmetry breakdown." As I consider these terms highly misleading, I would like to discuss them in some detail.

To what extent are weak and electromagnetic interactions unified? The symmetry used to describe both is SU2 $\times$ U1, and that already shows that there is really no unification at all. True unification, as in Maxwell's theory, leads to a reduction of parameters; for example, in Maxwell's theory the propagation velocities of magnetic and electric fields are the same, equal also to the speed of light. In the electroweak theory there is no such reduction of parameters: the mixing angle can be whatever, and that makes the electric coupling constant $e = g\sin(\theta)$ a free parameter. If the Higgs sector is not specified, then the Z_0 mass and the photon mass are also free parameters. There is really no unification (apart from the fact that the isovector part of the photon is in the same multiplet as the vector bosons).

However, if one specifies the simplest possible Higgs system then the number of free parameters diminishes. The Z_0 mass is fixed if the weak mixing angle and the charged vector-boson mass are fixed, and the photon mass is necessarily zero. So here there seems to be some unification going on. It seems to me, however, utterly ridiculous to speak of "electroweak unification" when choosing the simplest possible Higgs system.

The question of spontaneous symmetry breakdown is more complicated. From my own perspective the situation is as follows. In 1968, I showed what I termed the one-loop renormalizability of the massive Yang-Mills theory. The precise meaning of that will become clear shortly. However, there is trouble at the two-loop level, so at the time I thought that there had to be some cutoff-mechanism that would control the (observable) divergencies occurring beyond the one-loop level. Actually, the Higgs system can be seen as such a cutoff mechanism. The Higgs mass becomes the cutoff parameter, and indeed this cutoff parameter is observable (which is the definition of non-renormalizability of the theory without a Higgs system). This parameter enters logarithmically in certain radiative corrections (the Z_0 mass, for example), and from the measurement of these corrections follows some rough estimate of the value of this parameter. However, part of the input is that the Higgs sector is the simplest possible; without that assumption there is no sensitivity at the one-loop level (because then the Z_0 mass is not known and the radiative correction becomes a renormalization of that mass). That is the meaning of one-loop renormalizable. Let us, however, assume that from a symmetry point of view things are as if the Higgs system were the simplest possible. Then from the radiative corrections the cutoff parameter (the Higgs mass) can be estimated.

From this point of view the question is to what extent we can be sure today that the cutoff system used by Nature is the Higgs system as advertised. Evidence would be if there is indeed a particle with a mass equal to the estimate found from the radiative corrections. But if there is none, that would simply mean that Nature uses some other scheme, to be investigated experimentally.

In all of this discussion the notion of spontaneous symmetry breakdown does not really enter. In the beginning this was a question that I kept on posing myself. Spontaneous symmetry breakdown usually implies a constant field in the vacuum. So I asked myself: is there any way one could observe the presence of such a field in the vacuum? This line of thought led me to the question of the cosmological constant [38]. Indeed, if nothing else, surely the gravitational interactions can see the presence of a field in the vacuum. And here we have the problem of the cosmological constant, as big a mystery today as 25 years ago. This hopefully also makes it clear that, with the introduction of spontaneous symmetry breakdown, the problem of the cosmological constant enters a new phase. I have argued that a solution to this problem may be found in a reconsideration of the fundamental reality of space-time versus momentum space [47], but this is clearly not the place to discuss that. Also, the argument has so far not led to any tangible consequences.

So, while theoretically the use of spontaneous symmetry breakdown leads to renormalizable Lagrangians, the question of whether this is really what happens in Nature is entirely open.

CONCLUSION

The mind-wrenching transition of field theory in the sixties to present-day gauge-field theory is not really visible anymore, and is surely hard to understand for the present generation of field theorists. They might ask: why did it take so long? Perhaps the above provides some answer to that question.

REFERENCES

1. M. Veltman, "The Path to Renormalizability", invited talk at the Third International Symposium on the History of Particle Physics, June 24–27, 1992. Printed in "The Rise of the Standard Model", Lillian Hoddeson, Laurie Brown, Michael Riordan and Max Dresden Editors. Cambridge University Press 1997.
2. S.L. Glashow, *Nucl. Phys.* 22 (1961) 579.
3. S.L. Glashow, J. Iliopoulos and L. Maiani, *Phys. Rev.* D2 (1970) 1285.
4. S. Weinberg, *Phys. Rev. Let.* 19 (1967) 1264; A. Salam, in "Elementary Particle Theory", Stockholm 1968, N. Svartholm editor.
5. Y. Hara, *Phys. Rev.* 134 (1964) B701.
6. W. Bardeen, *Phys. Rev.* 184 (1969) 1848.
7. C. Bouchiat, J. Iliopoulos and Ph. Meyer, *Phys. Let.* 38B (1972) 519.
8. S.L. Adler, *Phys. Rev. Let.* 14 (1965) 1051.
9. W.I. Weisberger, *Phys. Rev. Let.* 14 (1965) 1047.
10. S. Weinberg, Nobel lecture 1979.
11. A. Salam, *Phys. Rev.* 127 (1962) 331.
12. M. Gell-Mann, *Physics* 1 (1964) 63.
13. M. Veltman, *Phys. Rev. Let.* 17 (1966) 553.
14. J.S. Bell, *Il Nuovo Cim.* 50A (1967) 129.
15. S.L. Adler, *Phys. Rev.* 137B (1965) 1022.
16. M. Veltman, *Proc. Roy. Soc.* A301 (1967) 107.
    J.S. Bell and R. Jackiw, *Il Nuovo Cim.* A60 (1969) 47.
17. S.L. Adler, *Phys. Rev.* 177 (1969) 2426.
18. T.D. Lee and C.N. Yang, *Phys. Rev.* 128 (1962) 885.
19. T.D. Lee, *Phys. Rev.* 128 (1962) 899.
20. M. Veltman, *Physica* 29 (1963) 186.
21. R.P. Feynman, *Acta Phys. Pol.* 24 (1963) 697 (talk July 1962).
22. B.S. DeWitt, *Phys. Rev. Let.* 12 (1964) 742; Phys. Rev. 160 (1967) 1113; Phys. Rev. 162 (1967) 1195, 1239.
23. E.C.G. Stueckelberg, *Helv. Phys. Acta* 11 (1938) 299. See ref. [1] for other references to this subject.
24. M. Veltman, *Nucl. Phys.* B7 (1968) 637. See also M. Veltman, Copenhagen lectures, July 1968. Reprinted in "Gauge theory – past and future", R. Akhoury, B. De Wit, P. Van Nieuwenhuizen and H. Veltman, Editors. 1992, World Scientific, Singapore, p 293.
25. S.L. Glashow and J. Iliopoulos, *Phys. Rev.* D3 (1971) 1043.
26. M. Veltman, *Nucl. Phys.* B21 (1970) 288 [16 Apr 1970].
27. M. Veltman, Proc. Topical Conf. on Weak Interactions, CERN, Geneva, 14–17 Jan 1969, CERN yellow report 69-7, page 391.

28. J. Reiff and M. Veltman, *Nucl. Phys.* B13 (1969) 545.
29. A.A. Slavnov and L.D. Faddeev, *Teor. Mat. Fiz.* 3 (1970) 18.
30. H. Van Dam and M. Veltman, *Nucl. Phys.* B22 (1970) 397.
31. S. Mandelstam, *Phys. Rev.* 175 (1968) 1580.
32. D. Boulware, *Annals of Phys.* 56 (1970) 140.
33. L.D. Faddeev and V.N. Popov, *Phys. Let.* 25B (1967) 29.
34. G. 't Hooft, *Nucl. Phys.* B33 (1971) 173.
35. A.A. Slavnov, *Theoretical and Mathematical physics* 10 (1972) 153 (English translation page 99).
36. J.C. Taylor, *Nucl. Phys.* B33 (1971) 436.
37. G. 't Hooft, *Nucl. Phys.* B35 (1971) 167.
38. M. Veltman, Cosmology and the Higgs mechanism. Rockefeller University preprint May 1974. M. Veltman, *Phys. Rev. Let.* 34 (1975) 777.
39. G. 't Hooft and M. Veltman, *Nucl. Phys.* B44 (1972) 189.
40. Renormalization of Yang-Mills fields and applications to particle physics, Marseille Conference June 19-23, 1972. Proceedings edited by C.P. Korthals-Altes, p 37.
41. G. 't Hooft and M. Veltman, *Nucl. Phys.* B50 (1972) 318.
42. D.A. Ross and M. Veltman, *Nucl. Phys.* B95 (1975) 135.
43. M. Veltman, *Nucl. Phys.* B123 (1977) 89.
44. M. Veltman, *Acta Phys.* Pol. B8 (1977) 475.
45. D. Bardin and G. Passarino, The Standard Model in the Making. Clarendon Press, Oxford 1999.
46. M. Lemoine and M. Veltman, *Nucl. Phys* B164 (1980) 445.
47. M. Veltman, *Acta Phys.* Pol. B25 (1994) 1399.

Physics 2000

ZHORES I. ALFEROV, HERBERT KROEMER

““for developing semiconductor heterostructures used in high-speed and opto-electronics”

and

JACK S. KILBY

“for his part in the invention of the integrated circuit”

THE NOBEL PRIZE IN PHYSICS

Speech by Professor Tord Claeson of the Royal Swedish Academy of Sciences.
Translation of the Swedish text.

Your Majesties, Your Royal Highnesses, Ladies and Gentlemen,

Information technology (IT) influences our lives at many levels. We use it to collect, process, communicate and present information. IT controls high-tech processes as well as medical diagnostic instruments and everyday home appliances. Computers are linked in a global network and only a few years from now, they will number one billion. The performance of microelectronic circuits seems to be increasing one hundred-fold every ten years – at unchanged prices. IT is viewed as a prime mover in the economic upswing our society has experienced over the past decade.

This year's Nobel Prize in Physics rewards contributions to the early developments of microelectronics and photonics, focusing on the integrated circuit, or "chip," as well as semiconductor heterostructures for lasers and high-speed transistors.

The transistor was invented around Christmas 1947 and the discoverers of the transistor effect were awarded the 1956 Nobel Prize in Physics. Ten years after that discovery, transistors had replaced vacuum tubes on a large scale. Beaches were being flooded by pop music, and one of the inventors is said to have exclaimed: "If only I had never invented that transistor."

Discrete transistors were soldered on circuit boards together with other components. But the emerging computers required ten thousands of transistors on the same board, a time-consuming and error-prone task.

As a newly hired engineer, Jack Kilby did not get his two weeks of vacation in the summer of 1958. Instead he had the privilege of thinking undisturbed on working time. He designed a circuit out of components made from a single semiconductor material that had been processed in different ways. This had already been suggested, but it was in conflict with the prevailing industrial practice of producing parts in the cheapest available material. On September 12, he was able to demonstrate that an integrated circuit worked – the birth date of the integrated circuit is one of the most important birth dates in the history of technology. Since then, things have moved fast. Chips being made today contain nearly a billion bits of memories or logic gates in processors – the brains of computers.

What was needed was not only more, smaller and cheaper transistors, but also faster ones. Early transistors were relatively slow. Semiconductor heterojunctions were proposed as a way of increasing amplification and achieving higher frequencies and power. Such a heterostructure consists of two semiconductors whose atomic structures fit one another well, but which have different electronic properties. A carefully worked out proposal was published in 1957 by Herbert Kroemer. Today, high-speed transistors are found in mobile (cellular) phones and in their base stations, in satellite dishes and links. There they are part of devices that amplify weak signals from outer space or from a faraway mobile telephone without drowning in the noise of the receiver itself.

Semiconductor heterostructures have been at least equally important to the development of photonics – lasers, light emitting diodes, modulators and solar panels, to mention a few examples. The semiconductor laser is based upon the recombination of electrons and holes, emitting particles of light, photons. If the density of these photons becomes sufficiently high, they may begin to move in rhythm with each other and form a phase-coherent state, that is, laser light. The first semiconductor lasers had low efficiency and could only shine in short pulses.

Herbert Kroemer and Zhores Alferov suggested in 1963 that the concentration of electrons, holes and photons would become much higher if they were confined to a thin semiconductor layer between two others – a double heterojunction. Despite a lack of the most advanced equipment, Alferov and his co-workers in Leningrad (now St. Petersburg) managed to produce a laser that effectively operated continuously and that did not require troublesome cooling. This was in May 1970, a few weeks earlier than their American competitors.

Lasers and light emitting diodes (LEDs) have been further developed in many stages. Without the heterostructure laser, today we would not have had optical broadband links, CD players, laser printers, bar code readers, laser pointers and numerous scientific instruments. LEDs are used in displays of all kind, including traffic signals. Perhaps they will entirely replace light bulbs. In recent years, it has been possible to make LEDs and lasers that cover the full visible wavelength range, including blue light.

I have emphasized the technical consequences of these discoveries, since these are easier to explain than the spectacular scientific breakthroughs that they have also led to. Challenging problems and matching resources have led to large-scale basic research. The advanced materials and tools of microelectronics are being used for studies in nanoscience and of quantum effects. Scientific experiments and computations are, of course, highly computerized.

Semiconductor heterostructures can be regarded as laboratories of two-dimensional electron gases. The 1985 and 1998 Nobel Prizes in Physics for quantum Hall effects were based on such confined geometries. They can be reduced further to form one-dimensional quantum channels and zero-dimensional quantum dots for future studies.

Drs. Alferov, Kilby and Kroemer,

I have briefly described some consequences of your discoveries and inventions. Few have had such a beneficial impact on mankind as yours. I also predict that there will be continued development, as we may be only halfway through the information technology revolution. New effects may appear as a result of basic research. When, what and where we cannot say, but we can be sure they will come.

On behalf of the Royal Swedish Academy of Sciences, I would like to convey the warmest congratulations to you and ask you to step forward to receive the Nobel Prize from His Majesty the King.

ZHORES I. ALFEROV

Life goes on surprisingly fast. It seems to happen a short time ago that I would attend anniversary celebrations in honour of noted physicists, my teachers who to my mind looked quite old. But at the present time, I myself have recently marked the 70th birthday.

My parents, Ivan Karpovich, and Anna Vladimirovna, had been Byelorussia born and raised. At the age of eighteen my father arrived in Saint Petersburg, in the year 1912. In his early hard years, he had been a docker, an errand boy and consequently got a job as a worker at the "Lessner" plant (later the Karl Marx Plant).

During World War I, he was a brave hussar, a non-commissioned officer of the Life Guards, a holder of the St. George Order. In September 1917, my father joined the Bolshevik party and retained his adherence to the socialist and communist principles to the end of his life.

In childhood, my brother and I "with a sinking heart" used to listen to father's stories about the civil war and his military career. We learned how the formerly non-commissioned officer had been appointed to take command of a cavalry regiment in Red Army. Father also used to tell us about his meetings with revolutionary leaders: V. I. Lenin, L. D. Trotsky, B. E. Dumenko, "comrade Andrey" (A. Solts) who always put his apartment in the "Embankment House" at my father's disposal while we stayed in Moscow. Father graduated from the Industrial Academy in 1935 and since then destiny was throwing us all over the country: Stalingrad, Novosibirsk, Barnaul, Syas'stroy in the environs of Leningrad, Turinsk (Sverdlovsk region), where we lived throughout the war time, and eventually the Minsk-city lying in ruins after the war. Dad was given a new assignment as director of a factory, joint enterprises (corporation of enterprises), later director of a trust. Mother headed a public organization of housewives; worked as a librarian and always remained our close friend while bringing us up without discouraging words. As a result of being so-called "director's boys", my brother and I tried to behave ourselves and to act in the way that people thought was correct and proper both at school and in public.

Learning was easy to me, and dependable defender, my elder brother Marx, made my existence cloudless at school and outdoors as well. Marx had graduated high school on June 21, 1941 (next day the Nazi invasion started) in the town of Syas'stroy and shortly after that we left for the Urals to Turinsk city as Dad had been assigned there to a post of director of a newly-built gunpowder cellulose factory (at the time referred to as factory No. 3). My elder brother, who was seventeen years old then, joined the Urals Industrial Institute (the Energy Faculty). The young student considered the problem of energy to be of cardinal importance for the future. But not long did he study at the Institute. He decided to defend his Motherland and to fight against fascists at the front line.

He passed Stalingrad, Kharkov, the Kursk battle. Having recovered after heavy head injury he was sent to the Army in the Field again. That was so-called "another Stalingrad", i.e., the Korsun-Shevchenko battle, where in his 20 years was shot down a Guard junior lieutenant Marx Ivanovich Alferov, my elder brother who remained of 20 years forever.

In October 1943, on the way to front from a hospital he spent 3 days with us in Sverdlovsk. I often look back and reflect on those three days; on his description of the war, his youthful enthusiasm and faith in the power of science, technology and human intelligence.

In the post-war particular situation I attended an only boy's school in the destroyed Minsk-city, and was lucky in having an excellent physics teacher there Yakov Borisovich Meltserson. He delivered lectures on physics for us, rather naughty boys, and we were sitting quiet and listened attentively. The teacher loved physics devotedly and had a gift of making our imagination work. His explanation of the cathode oscilloscope operation and talk on radar systems greatly impressed me. When finishing the school I took his advice which institution to choose for education and that was a celebrated Ul'yanov Electrotechnical Institute in Leningrad (abbreviated to LETI).

Many of systematic studies in electronics and radio engineering that had been performed there made significant contributions into the electronics industry. As for me, it was my good fortune to meet my first supervisor there. Theoretical courses of studies were easy enough for me. It was the laboratory research that attracted me. Being a third-year student, I began to work in a laboratory of vacuum processes. My first investigations were directed by a research associate N. N. Sozina who studied semiconductor photodetectors. Since that time, half a century ago, semiconductors have become main objects of my scientific interests. A book "The Electroconductivity of Semiconductors" by F. F. Volkenshtein, which had been written in Leningrad (during the time of Leningrad's siege) was my Textbook then. My graduation thesis was devoted to the problem of obtaining the thin films and investigating the photoconductivity of bismuth telluride compounds.

In December 1952, I graduated from the Institute and was offered by my supervisor N. N. Sozina to stay in the LETI to continue my study. But I dreamed of working at the Physico-Technical Institute that had been founded by Abram Fedorovich Ioffe. His book "Fundamentals of Modern Physics" was a manual for me. Happily, three vacancies for graduates had been given to us by Ioffe's Institute. One of them fell to my lot. My joy was boundless. And may be it is this lucky distribution that has determined my happy scientific career.

In the letter to my parents, then residing in the Minsk-city, I wrote about my lucky chance. I did not know that Academician Ioffe was dismissed and left the Institute of which the director he had been for thirty years.

I recall my first day at the Physico-Technical Institute on January 30, 1953.

I was introduced to my new supervisor, V. M. Tuchkevich, head of a subdivision. It was a very important problem to be solved by our not very big team: creation of germanium diodes and triodes (transistors) on p-n junctions.

The Physico-Technical Institute, being regarded on today's scale, was not a big one. I was given an Institute pass No. 429, i.e., the total amount of employees was as high as the above mentioned number; most of famous physicists of the Physico-Technical Institute moved to Moscow (to I. V. Kurchatov's, and newly-built atomic centers). Semiconductors elite followed A. F. Ioffe in order to work under his supervision in a recently organized semiconductor laboratory belonging to the Presidium of the Academy of Sciences of the USSR. In the Physico-Technical Institute there retained only D. N. Nasledov, B. T. Kolomiets and V. M Tuchkevich as representatives of the old generation of physicists who formerly dealt with semiconductors.

Academician A. P. Komar was after A. F. Ioffe on charge of the Physico-Technical Institute. The new director's attitude to his predecessor was not quite correct but as to the restoration and development of the Institute, his strategy was O.K. Of utmost importance was the support of works on the creation of new semiconductor electronics, space investigations (gas dynamics of high velocities and high temperature protective coatings; development of the light isotope separation methods for the hydrogen weapon (under the guidance of B. P. Konstantinov).

Studies of fundamental problems of physics, both theoretical and experimental ones, were encouraged too: just in this time experimental discovery of exciton was done (E. F. Gross), it was formulated the principles of a kinetic theory of strength (S. N. Zhurkov), development of the pioneering works on physics of atomic collisions were initiated (V. M. Dukel'skii, N. V. Fedorenko).

Both the director of the Institute (A. P. Komar) and the deputy director (D. N. Nasledov) understood the importance of drawing the interests of young people to science. It was a practice then to welcome newcomers at the highest level. In this way many renowned Russian scientists started their work, among them were present members of the Academy of Sciences, B. P. Zakharchenya, A. A. Kaplyanskii, E. P. Mazets, V. V. Afrosimov and others.

I remember my first attendance of the seminar on semiconductors at the Physico-Technical Institute in February 1953 as one of the most impressive events I have ever experienced. That was a brilliant report delivered by E. F. Gross about the discovery of the exciton. The sensation I experienced then could not be compared to anything. I was stunned by the talk on the birth of a discovery in the area of science to which I myself had got the access.

Yet the main thing was everyday experimental work in the laboratory. Since that time I have been keeping, as a most precious thing, my laboratory daily report book that contains notes of mine about the creation of the first soviet p-n junction transistor on the 5th of March, 1953. And now, when recalling that time I cannot help feeling proud of what we had accomplished. We comprised a team of very young people. Under the guidance of V. M. Tuchkevich we succeeded in working out principles of the technology and the metrics of transistor electronics. Below are the names of researchers who had been working in our small laboratory: A. A. Lebedev, a Leningrad University graduate – the growth and doping of perfect germanium single crystals; Zh. I. Alferov – the preparation of transistors, their parameters being at the level of the best world samples; A. I. Uvarov and S. M. Ruvkin – the creation of a precise metrics of germanium single crystals and transistors; N. S. Yakovchuk, a graduate of the Faculty of Radio Engineering of the Leningrad Electrical Technical Institute – designing transistor-based circuits.

As early as in May 1953, the first Soviet transistor receivers were shown to the "top authorities". That work, of which the performers had been working with passion peculiar to their young hearts and with utmost sense of responsibility, exerted a great influence upon me. While quickly and effectively progressing as a scientist, I began to comprehend the significance of the technology not only for electronic devices, but in basic research work too, in regard with notorious "minor" details and sporadic results. And it is since then that I prefer to analyze experimental result proceeding from "simple" general laws prior to putting forward sophisticated explanations.

In subsequent years, our team of researchers at the Physico-Technical Institute expanded considerably and in a very short time the first Soviet germanium power rectifiers were created alongside with germanium photodiodes and silicon solar cells in V. M. Tuchkevich's laboratory. Works on studying the behavior of impurities in germanium and silicon also were being carried out then.

In the month of May 1958, Anatolii Petrovich Alexandrov (later the President of the Academy of Sciences of the USSR) asked our team of working out a special semiconductor device for the first Soviet atomic submarine. That required a perfectly new technology and in addition to – another construction of germanium rectifiers, which had been done in a record short space of time. In the month of October, these devices were mounted on a submarine. I was a junior research associate at the Institute then, and was somewhat surprised by a telephone call from the first Vice-Chairman of the Government of the USSR, Dmitrii Fedorovich Ustinov, who asked me of fortnight reduction of the term. There was no getting away from that: I directly moved in the laboratory premises and settled there but, of course, the request was fulfilled that was my first State Order, which I had been decorated with then and which I valued very much.

In 1961, I read my candidate degree thesis that had been mainly devoted to working out and investigating of power germanium and partially silicon rectifiers. Occurrence of Soviet power semiconductor electronics became possible as a result of those works. Of great importance there, in the sense of a scientific, purely physical standpoint, had been a conclusion drawn by me that in p-i-n, p-n-n semiconductor homostructures under working current densities (for most of semiconductor devices), the current had been determined by recombination in heavily doped p- and n (n^+)-regions while the recombination contribution in the middle i(n)-region of a homostructure was not the determining one: so, as soon as the first work on semiconductor lasers had appeared, it was natural for me to consider the advantages of employing in lasers the double heterostructure of p-i-n (p-n-n$^+$, n-n-p$^+$) type. The idea was formulated by us shortly after the appearance of the first work of R. Hall with co-workers, which described a semiconductor laser based on a GaAs homo-p-n-structure.

To realize principal advantages of heterostructures appeared to be possible only after obtaining of $Al_xGa_{1-x}As$ heterostructures. We did that and it turned out that we had been only one month ahead in relation to American researches from IBM.

When we began investigating heterostructures, I used to convince my young colleagues, that we were not one and only group of scientists in the world who understood the significance of the concept that the semiconductor physics and electronics would be developing on the basis of HETERO-, rather than HOMO-structures. Indeed, since 1968 we entered an era of a strong competition and the first of all were three laboratories of the biggest American companies: Bell Telephone, IBM and RCA.

In 1967, while on a short trip to U.K., I visited STL laboratories in Harlow. They were well equipped and the experimental base was excellent but English colleagues only discussed theoretical aspects of the heterostructures physics; they did not find experimental study of heterostructures to be promising then. In London I had some time for sightseeing and shopping. I bought there wedding gifts to my fiancee Tamara Darskaya. As soon as I returned to Leningrad, we celebrated our wedding in a splendid restaurant "Krysha" (the Roof) in the Grand Hotel "Europe".

Tamara was a daughter of a very popular actor of Voronezh Theater of Musical Comedy. Tamara worked then in the environs of Moscow at a big Space Enterprise under the guidance of Academician V. P. Glushko. She wonderfully combined incompatible beauty with cleverness and common sense and was always very kind toward her close friends. It was time of repeated weekly flights to Moscow. Holding a position of a Senior Research Associate at the Physico-Technical Institute, I could afford that. Leningrad-Moscow flight occurred in an hour time and the cost of a ticket to the TU-104 plane was as low as 11 rubles (about 15 US dollars). Nevertheless, after half a year shuttling between the two cities Tamara had moved to Leningrad.

In 1968–1969, we virtually realized all the ideas on control the electron and light fluxes in classical heterostructures based on the arsenid gallium-arsenid aluminum system. Apart from fundamental results that were quite new and important efficient one-side injection, the "superinjection" effect, diagonal tunneling, electron and optical confinement in a double heterostructure (which in a short while became the main element in studying the low-dimensional electron gas in semiconductors), we succeeded in employing principal benefits of heterostructure applications in devices, i.e., lasers, LEDs, solar cells, dynistors and transistors. Of utmost importance was, beyond doubts, the making of low threshold room temperature operating lasers on a double heterostructure (DHS) that had been suggested by us as far back as 1963. The approach developed by M. B. Panish and I. Hayashi (Bell Telephone) as well as by H. Kressel (RCA) was different from that of ours since they offered to use in lasers a single p-AlGaAs-p-GaAs heterostructure, which made their approach rather limited. A possibility of obtaining an efficient injection in the heterojunction seemed doubtful to them and, in spite of the fact that potential advantages of DHS had been recognized.

In August 1969, I first time visited the USA; my paper that I read there at **the International Conference on Luminescence in Newark (State of Delaware)** was devoted to AlGaAs-based DHS low threshold room temperature lasers and produced an impression of an exploded bomb on American colleagues. Professor Ya. Pankov from RCA, who just shortly before my reading the paper had explained to me that they had not got a permission for my visiting their laboratory, as soon as I concluded my speech told me that the permission had been received. I could not help enjoying my refusal explaining that now I had been invited by that moment to attend IBM and Bell Telephone laboratories.

My seminar in the Bell followed by the looking over the laboratories and discussions with researches clearly revealed to me our merits and demerits of our progress in my laboratory. I believe that the soon commenced emulation for being the first in getting the continuous wave operation of laser at the room temperature was at that time a rare example of an open and friendly competition between laboratories belonging to the antagonistic Great Powers. We won the competition overtaking by a month Panish's group in Bell Telephone. Significance of obtaining the continuous wave regime had the connection first and foremost with working out an optical fiber with low losses as well as the creation of our DHS lasers, which resulted in appearance and rapid development of optical fiber communication.

In the winter 1970–1971 and spring 1971, I spent six months in the USA working in laboratory of semiconductor devices at the University of Illinois together with Prof. Nick Holonyak. We met at the first time in 1967, when he visited my laboratory at the Physico-Technical Institute. Prof. Nick Holonyak, who is one of the founders of semiconductor optoelectronics, the inventor of the first visible semiconductor laser and LED became my closest friend. Now over 33 years we have discussed all semiconductor physics and electronics problems, political and life aspects and our interaction (visits, letters, seminars, telephone conversations) played very important role in our work and life.

In 1971, I became a recipient of the USA Franklin's Institute gold medal for DHS laser works. Being my first international award, it was of particular value to me. There are Soviet physicists besides me who have been given the Franklin's Institute gold medals too: Academician P. L. Kapitsa in 1944; Academician N. N. Bogolubov in 1974; Academician A. D. Sakharov in 1981. I consider it a big honour to belong to such a company!

An $Al_xGa_{1-x}As$ system of lattice-matched heterostructures, which in practice seemed to be a lucky exception, was infinitely expanded on the basis of multi-component solid solutions, first theoretically and later on experimentally (InGaAsP is the most convincing example).

Heterostructure-based solar cells were created by us as far back as 1970. And when American scientists published their early works, our solar batteries have been already mounted on the satellites (sputniks) and their industrial production was in full swing. The cells, when being employed in space, proved their efficiency. For many years they have been operating on the "MIR" skylab and in spite of the fact that forecasts of a substantial decrease of the value of one watt of the electrical power have not been justified so far, the most effective energy source in space is, nevertheless, a set of solar cells on heterostructures of III–V compounds.

In 1972, my pupils-colleagues and I were awarded the Lenin's Prize – the highest scientific Prize in the USSR. Our gladness regrettably was not cloudless. For some formal and obscure reasons we lost from the list of nominees R. F. Kazarinov and E. L. Portnoi.

On the day of the prize award I was in Moscow and called home, to Leningrad. But the telephone did not answer. Then, I called my parents (they have been living in Leningrad since 1963) and gladly told my father that I had been given the Lenin's Prize. But my father replied – And so what? Our grandson is born today! In my lucky 1972 year, in addition to the prestigious prize I was elected a member of the Academy of Sciences. But the happiest day was that of Vanya Alferov's birth.

Studies of superlattices and quantum wells were rapidly promoted in the West and afterwards in this country soon resulted in coming into being of a new area of the quantum physics of solid: the physics of low-dimensional electron systems. In this regard, studies of zero-dimensional structures – so-called "quantum dots" – form the summit of the above mentioned works. Gratifying is the circumstance that the Ioffe Institute today, while going through the hard times, remains the world leader in this area of physics. Works of the second and third generation of my students, those being well-known P. S. Kop'ev, N. N. Ledentsov, V. M. Ustinov, S. V. Ivanov have won general recognition nowadays. N. N. Ledentsov has become the youngest corresponding member of the Russian Academy of Sciences.

In 1987, I was elected director of the Ioffe Institute, in 1989, president of the Leningrad Scientific Center of the Academy of Sciences of the USSR; and in April 1990, Vice-President of the Academy of Sciences of the USSR. Afterwards, I was reelected and hold all these posts now within the Russian Academy of Sciences.

In the first years of my presidency and directorship we succeeded in remarkable scaling up research activity in our unique (for all the world) Academy of Sciences. We have also developed effective collaboration with Universities and Educational Institutions. The Physico-Technical Special Secondary School attached to Ioffe's Physico-Technical Institute had been opened at that time; ongoing was the process of creation of specialized University chairs: the first one, that of Optoelectronics was organized in the Electrotechnical University, (formerly the LETI) as far back as in 1973. On the basis of both then existing and newly organized chairs a Physicotechnical faculty was set up in the Polytechnical Institute in 1988.

A great contribution into the above mentioned system makes the Scientific-Educational Center that has been built by the Physico-Technical Institute and incorporates school boys, students and scientists in a magnificent edifice, which can be called "The Palace of Knowledge".

Still, throughout the years passed, of greatest importance has been so far the existence of our Academy of Sciences as a unique both scientific and educational structure in Russia. The Academy faced the menace of abolition in the twenties as "an inheritance from the tsarist regime". It faced the menace of abolition in the nineties as "an inheritance from the totalitarian Soviet regime". To insure its safety I gave my consent to be a member of the Russian Parliament (a deputy of State Duma) in 1995. President Yu. S. Osipov and Vice-Presidents, Academicians and Corresponding Members, doctors and candidates of sciences, senior and junior research associates, lab-assistants and mechanics took a firm stand on this kind of situation. For the saving of the Academy of Sciences, we made compromises with the power but never with the conscience.

All that had been made by human beings, in principle, was made due to Science. And if our country's choice is to be a Great Power, Russia will be the great power not because of the nuclear potential, not because of faith in God or president, or western investments but thanks to the labor of the nation, faith in Knowledge and Science and thanks to the maintenance and development of scientific potential and education.

When I was a little boy of ten, I have read a wonderful book "Two Captains" (by V. Kaverin). In essence, in my life I have been following the principle that was peculiar the main character of that book: "One should make efforts and search for. And having obtained whatever the purpose, to make efforts again".

Of great importance here is to know what you are struggling for.

THE DOUBLE HETEROSTRUCTURE: CONCEPT AND ITS APPLICATIONS IN PHYSICS, ELECTRONICS AND TECHNOLOGY

Nobel Lecture, December 8, 2000

by

ZHORES I. ALFEROV

A. F. Ioffe Physico-Technical Institute, Russian Academy of Sciences, 26 Politechnicheskaya, St. Petersburg 194021, Russian Federation.

1. INTRODUCTION

It is impossible to imagine now modern solid state physics without semiconductor heterostructures. Semiconductor heterostructures and, particularly, double heterostructures, including Quantum Wells, Wires and Dots are today the subject of research of 2/3 of the semiconductor physics community.

It can be said that if the possibility of controlling the type of conductivity of a semiconductor material by doping with various impurities and the idea of injecting nonequilibrium charge carriers were the seeds from which semiconductor electronics developed, heterostructures could make it possible to solve the considerably more general problem of controlling the fundamental parameters inside the semiconductor crystals and devices: band gaps, effective masses of the charge carriers and the mobilities, refractive indices, electrons energy spectrum etc.

Development of the physics and technology for the semiconductor heterostructures resulted in remarkable changes in our everyday life. Heterostructure electronics is widely used in many areas of human civilization. It is hardly possible to imagine our recent life without double heterostructure (DHS) laser-based telecommunication systems, without heterostructure-based light-emitting diodes, heterostructure bipolar transistors, without low-noise HEMT for high frequency applications including, for example, satellite television. The DHS laser enters now practically every house with CD-players. Heterostructure solar cells have been widely used for space and terrestrial applications.

Our interest in semiconductor heterostructures was not occasional. Systematic studies of semiconductors were started in the early 1930s at Physico-Technical Institute under the direct leadership of it's founder – Abraham Ioffe. V. P. Zhuze and B. V. Kurchatov studied the intrinsic and impurity conductivity of semiconductors in 1932 and the same year A. F. Ioffe and Ya. I. Frenkel created a theory of rectification in a metal-semiconductor contact based on the tunneling phenomenon [1]. In 1931 and 1936, Ya. I. Frenkel published his famous articles where he predicted, gave the name and developed the theory of excitons in semiconductors and E. F. Gross experimentally discovered excitons in 1951 [2]. The first diffusion theory of p-n heterojunction rectification, which became the base for W. Shockley's p-n junction theory, was published by B. I. Davydov in 1939 [3]. By A. F. Ioffe's initiative in the late 1940s at the Physico-Technical Institute the research into intermetallic compounds was started. Theoretical prediction and experimental discovery of semiconductor properties of A^3B^5 compounds were done independently by H. Welker and on the example of InSb by N. A. Gorunova and A. R. Regel at the Physico-Technical Institute [4]. We benefited a lot from the high theoretical, technological and experimental level in this area that had already existed at the Ioffe Institute at that time.

2. CLASSICAL HETEROSTRUCTURE

The idea of using heterojunctions in semiconductor electronics was put forward already at the very dawn of electronics. In the first patent concerned with p-n junction transistors W. Shockley [5] proposed a wide-gap emitter to obtain unidirectional injection. A. I. Gubanov at our Institute first theoretically analyzed volt-current characteristics of isotype and anisotype heterojunctions [6], but the important theoretical considerations at this early stage of heterostructure research have been done by H. Kroemer, who introduced the concept of quasi-electric and quasi-magnetic fields in a graded heterojunction and made an assumption that heterojunctions might exhibit extremely high injection efficiencies in comparison to homojunctions [7]. In the same period there were various suggestions about applying heterostructures in semiconductor solar cells.

The proposal of p-n junction semiconductor lasers [8], the experimental observation of effective radiative recombination in a GaAs p-n structure with a possible stimulated emission [9] and the creation of p-n junction lasers and LEDs [10] were the seeds from which semiconductor optoelectronics started to grow. However, lasers were not efficient because of high optical and electrical losses. The thresholds currents were very high and low temperature was necessary for lasing. The efficiency of LEDs was very low as well due to high internal losses.

The important step was made immediatly after the creation of p-n junction lasers when the concept of double heterostructure laser was formulated independently by us and H. Kroemer [11]. In his article H. Kroemer proposed to use the double heterostructures for carriers confinement in the active region. He proposed that "laser action should be obtainable in many of the indirect gap semiconductors and improved in the direct gap ones, if it is possible to supply them with a pair of heterojunctions injectors".

In our patent we also outlined the possibility to achieve high density of injected carriers and inverse population by "double" injection. We specially pointed out that homojunction lasers "do not provide CW at elevated temperatures" and as an additional advantage of DHS lasers we considered the possibility "to enlarge the emitting surface and to use new materials in various regions of the spectrum".

Initially the theoretical progress was much faster than the experimental realization. In 1966, [12] we predicted that the density of injected carriers could by several orders of magnitude exceed the carrier density in the wide-gap emitter ("superjunction" effect). At the same year in the paper [13] submitted to a new Soviet Journal "Fizika i Tekhnika Poluprovodnikov" (Sov. Phys. Semiconductors) I summarized our understanding of the main advantages of the DHS for different devices, especially for lasers and high power rectifiers: "The recombination, light emitting, and population inversion **zones coincide and are concentrated in the middle layer. Due to potential** barriers at the boundaries of semiconductors having forbidden bands of different width, the through currents of electrons and holes are completely absent, even under strong forward voltages, and there is no recombination in the emitters (in contrast to p-i-n, p-n-n$^+$, n-p-p$^+$ homostructures, in which the recombination plays the dominant role) ... Because of a considerable difference between the permittivities, the light is completely concentrated in the middle layer, which acts as a high-grade waveguide, and thus there are no light losses in the passive regions (emitters)".

Here are the most important peculiarities of semiconductor heterostructures we underlined at that time: (i) superinjection of carriers, (ii) optical confinement, (iii) electron confinement.

The realization of the wide-gap window effect was very important for photodetectors, solar cells and LEDs applications. It permitted to broaden considerably and to control precisely spectral region for solar cells and photodetectors and to improve drastically the efficiency for LEDs. Main physical phenomena in double and single classical heterostructures are shown in Fig. 1. Then it was only necessary to find heterostructures where these phenomena could be realized.

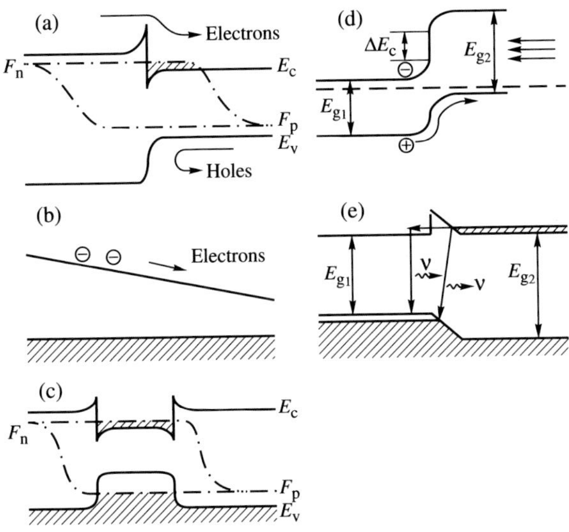

Figure 1. Main physical phenomena in classical heterostructures. (a) One-side injection and superinjection; (b) Diffusion in built-in quasi-electric field; (c) Electron and optical confinement; (d) Wide-gap window effect; (e) Diagonal tunneling through a heterostructure interface.

At that time general skepticism existed with respect to the possibility of creating the "ideal" heterojunction with a defect-free interface and first of all with theoretical injection properties. Even a very pioneering study of the first lattice-matched epitaxially grown single-crystal heterojunctions Ge-GaAs by R. L. Anderson [14] did not give any proof of the injection of noneqvilibrium carriers in heterostructures. Actual realization of the efficient wide-gap emitters was considered as simply close to the impossible and the double heterostructure laser patent was often referred to as a "paper patent".

Mostly due to this general skepticism there existed only a few groups trying to find out the "ideal couple", which was, naturally, a difficult problem. There should be met many conditions of compatibility between thermal, electrical, crystallochemical properties and between the crystal and the band structure of the contacting materials.

A lucky combination of a number of properties, i.e. small effective mass and wide energy gap, effective radiative recombination and a sharp optical absorption edge due to the "direct" band structure, a high mobility at the absolute minimum of the conduction band and its strong reduction of the nearest minimum at the (100) point ensured for GaAs even at that time a place of honor in semiconductor physics and electronics. Since the maximum effect is obtained by using heterojunctions between the semiconductor serving as the active region and more wide band material, the most promising systems looked at in that time were GaP-GaAs and AlAs-GaAs. To be "compatible", materials of the "couple" should have, as the first and the most important condition, close values of the lattice constants; therefore heterojunctions in the system AlAs-GaAs were preferable. However prior to starting work on preparation and study of these heterojunctions one had to overcome a certain psychological barrier. AlAs had been synthesized long ago [15], but many properties of this compound remained unstudied since AlAs was known to be chemically unstable and decompose in moist air. The possibility of preparing stable specimens adequate to applications of heterojunctions in this system seemed to be not very promising.

Initially, our attempts to create DHS were related to a lattice-mismatched GaAsP system. And we succeeded in fabricating by VPE the first DHS lasers in this system. However, due to lattice mismatch the lasing like that in homojunction lasers occurs only at liquid nitrogen temperature [16]. From a more curious point of view, I would like to mention that it was the first practical result obtained for a lattice mismatched, even partially relaxed, system.

Our experience, which we got from studying the GaAsP system, was very important for understanding many specific heterojunction physical properties and basics of heteroepitaxy. Development of a multichamber VPE method for the GaAsP system permitted us to create in 1970 superlattice structures with a 200 Å-period and to demonstrate the splitting of the conduction band [17].

But from the general point of view at the end of 1966, we came to a conclusion that even a small lattice mismatch in heterostructures such as $GaP_{0.15}As_{0.85}$–GaAs did not permitt to realize potential advantages of the DHS. At that time my co-worker D. N. Tret'yakov told me that small crystals of

$Al_xGa_{1-x}As$ solid solutions of different compositions, which had been prepared two years ago by cooling from a melt, were put in the desk drawer by Dr. A. S. Bortshevsky and nothing happened to them. It immediately became clear that $Al_xGa_{1-x}As$ solid solutions turned out to be chemically stable and suitable for preparation of durable heterostructures and devices. Studies of phase diagrams and the growth kinetics in this system and development of the LPE method especially for heterostructure growth soon resulted in fabricating the first lattice-matched AlGaAs heterostructures. When we published the first paper on this subject, we were lucky to be the first to find out a unique, practically an ideal lattice-matched system for GaAs, but as it frequently happened simultaneously and independently the same results were achieved by H. Rupprecht and J. Woodall at T. Watson IBM Research Center [18].

Then the progress in the semiconductor heterostructure area was very rapid. First of all we experimentally proved the unique injection properties of the wide-gap emitters and the superinjection effect [19], the stimulated emission in AlGaAs DHS [20], established the band-diagram of $Al_xGa_{1-x}As$–$GaAs_x$ heterojunction, carefully studied luminescence properties, diffusion of carriers in a graded heterostructure and very interesting peculiarities of the current flow through the heterojunction – that is similar, for instance, to diagonal tunneling-recombination transitions directly between holes of the narrow-band and electrons of the wide-band heterojunction components [21].

At the same time, we created the majority of the most important devices with realization of the main advantages of the heterostructure concepts:

– Low threshold at room temperature DHS lasers [22] (Fig. 2).

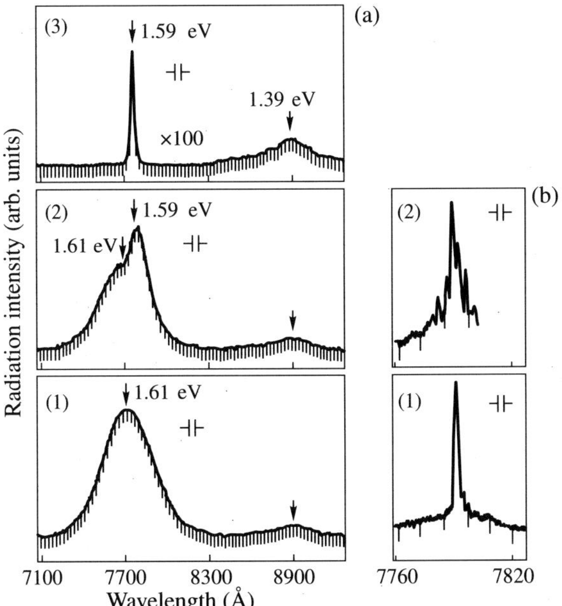

Figure 2. Emission spectrum of the first low-threshold Al_xGa_{1-x} DHS laser operating at room temperature (300 K), J_{th} = 4300A/ cm^2. The current rises: (a) from (1) 0.7 A to (2) 8.3 A and then to (3) 13.6 A; (b) from (1) 13.6 A to (2) 18 A; $s = 2.2 \times 10^{-3}$ cm^2.

– High effective SHS and DHS LED [23]
– Heterostructure solar cells [24]
– Heterostructure bipolar transistors [25]
– Heterostructure p-n-p-n switching devices [26].

One of the first successful applications in industrial scale production in our country was heterostructure solar cells in space research. We transferred our technology to the "Quant" company and since 1974, GaAlAs solar cells have been installed on many of our sputniks. Our space station "Mir" (Fig. 4) has been using them for 15 years.

Most of these results were achieved afterwards in other laboratories in 1–2 years and in some cases even later. But in 1970, the international competition became very strong. Later on one of our main competitors I. Hayashi, who was working together with M. Panish at Bell Telephone Lab in Murray Hill, wrote [27]: "In September 1969, Zhores Alferov of the Ioffe Institute in Leningrad visited our laboratory. We realized he was already getting a $J_{\rm th}^{(300)}$ of 4.3 kA/cm^2 with a DH. We had not realized that the competition was so close and redoubled our efforts ... Room temperature CW operation was reported in May 1970 ...". In our paper published in 1970, [28] the CW lasing was realized in stripe-geometry lasers formed by photolithography and mounted on copper plates covered by silver (Fig. 3). The lowest $J_{\rm th}$ density at 300 K was 940 A/cm^2 for broad area lasers and 2.7 kA/cm^2 for stripe lasers. Independently, CW operation in DHS lasers was reported by Izuo Hayashi and Morton Panish [29] (for broad area lasers with diamond heatsinks) in a paper submitted only one month later than our work. Achievement of CW at room temperature produced an explosion of interest in physics and technology of semiconductor heterostructures. If in 1969, AlGaAs heterostructures were studied just in a few laboratories mostly in USSR and US (A. F. Ioffe Institute, "Polyus" and "Quant" – industrial lab's where we transferred our technology for applications in USSR; Bell Telephone, D. Sarnoff RCA Research Center, T. Watson IBM Res.Center in US), in the beginning of 1971, many Universities, industrial labs in USA, USSR, United Kingdom, Japan and even in Brazil and Poland started investigations of III–V Heterostructures and Heterostructure Devices.

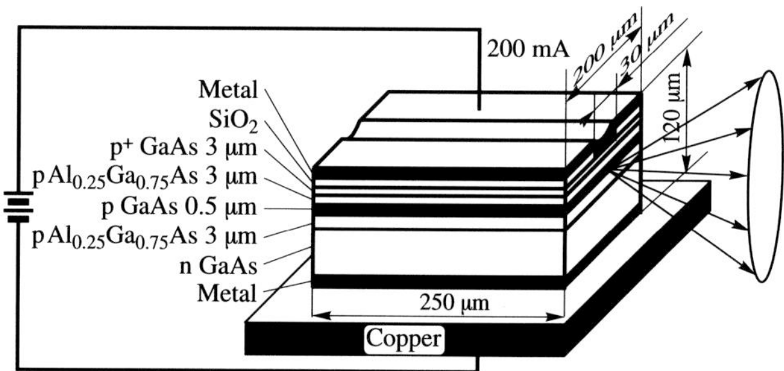

Figure 3. Schematic view of the structure of the first injection DHS laser operating in the CW regime at room temperature.

Figure 4. Space station "Mir" equipped with heterostructure solar cells.

At this early stage of the development of the heterostructure physics and technology it became clear that we needed to look for new lattice-matched heterostructures in order to cover a broad area of the energy spectrum. The first important step was done in our laboratory in 1970: in the paper [30] we reported that various lattice-matched heterojunctions based on quaternary III–V solid solutions were possible, which permitted independent variation between lattice constant and band gap. Later on G. Antipas with co-workers [31] came to the same conclusions. As a practical example utilizing this idea we considered different InGaAsP compositions and soon this material was recognized among the most important ones, for many different practical applications: photocathodes [32] and especially lasers in infra-red region for fiber optical communications [33] and the visible [34].

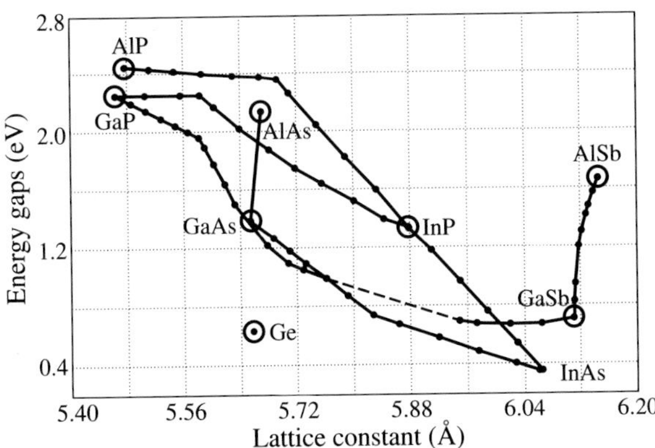

Figure 5. Energy gaps vs lattice constant for III–V Semiconductors.
Lattice matched heterojunctions: Ge-GaAs—1959 (R. L. Anderson);
AlGaAs—1967 (Zh. Alferov *et al.*, J. M. Woodall & H. S. Rupprecht);
Quaternary HS (InGaAsP & AlGaAsSb): Proposal—1970 (Zh. Alferov *et al.*);
First experiment—1972 (Antipas *et al.*)

In the early 1970s a "world map" of ideal lattice-matched heterostructures was shown (Fig. 5). Not before a decade later this "world map" was drastically changed (Fig. 6). Nowadays, it is necessary to add III-nitrides.

The main ideas of a semiconductor distributed-feed-back laser were formulated by us in the patent in 1971 [35]. The same year H. Kogelnik and C. V. Shank considered the possibility of replacing the Fabry–Perot or similar types of resonator in dye-lasers with volume periodical inhomogeneities [36]. It is necessary to note, that their approach is not applicable to semiconductor lasers and all laboratories that carried out research in DFB and DBR semiconductor lasers used the ideas formulated in [35]:

1. Diffraction grating created not in volume, but on a surface waveguide layer
2. Interaction of waveguide modes with a surface diffraction grating, giving not only distributed feedback but also highly collimated light output.

Detailed theoretical analyses of the semiconductor laser with surface diffraction grating was published in 1972 [37]. In this paper the authors established the way for the single-frequency generation. First semiconductor lasers with surface diffraction grating and distributed feedback were realized practically simultaneously at the Physico-Technical Institute [38], Caltech [39] and Xerox Lab in Palo Alto [40].

In the early 1980s H. Kroemer and G. Griffiths published a paper [41] which stimulated strong interest in staggered line-up heterostructures (type II heterojunction). Spatial separation of electrons and holes at the interface results in a tunability of their optical properties [21(c), 42]. Staggered band alignment gives a possibility to realize optical emission with a photon energy much smaller than the band-gap energy of each of the semiconductors form-

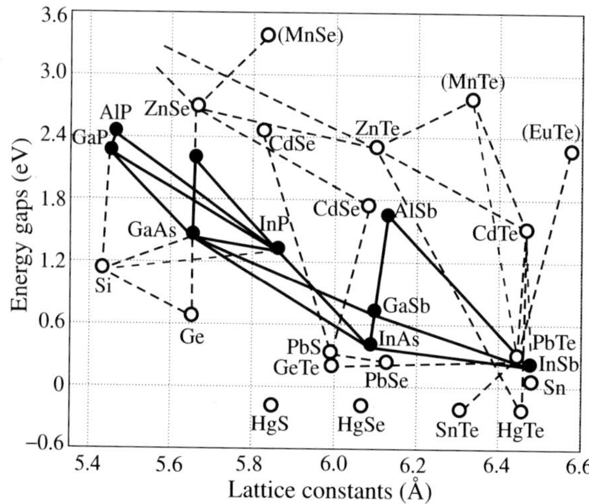

Figure 6. Energy gaps vs lattice constants for semiconductors of IV elements, III–V and II(IV)–VI compounds and magnetic materials in parentheses. Lines connecting the semiconductors, solid for the III–V's and dotted for the others, indicate quantum heterostructures that have been investigated.

ing a heterojunction. The demonstration of an injection laser based on a type-II GaInAsSb–GaSb heterojunction [42] opens good perspectives for the creation of effective coherent light sources in the infrared optical range. Radiation in such a device is due to the recombination of electrons and holes localized in self-consistent potential wells at different sides of the hetero-interface. Thus, type-II heterostructures open possibilities both for fundamental physics and for device applications, which cannot be realized with type-I heterostructures in the III–V material system. However, practical applications of these structures are still hampered by a poor understanding of their fundamental properties and the few actual systems, which have been studied experimentally up to now.

To summarize this brief review of the classical heterostructures development, it would be very convenient to classify the most important results in the following way:

Classical Heterostructures

I. Fundamental Physical Phenomena (Fig. 1)

- One-side injection
- Superinjection
- Diffusion in Built-in Quasi-Electric Fields
- Electron confinement
- Optical confinement
- Wide-gap window effect
- Diagonal tunneling through heterostructure interface

II. Important Applications in Electronics

- Semiconductor lasers – Low threshold and CW at room temperature, DFB and DBR Lasers, Vertical Surface Emitting Lasers, IR II-type Heterostructure Lasers
- High efficient LED
- Solar cells and photodetectors, based on the wide-gap window effect
- Semiconductor integrated optics, based on semiconductor DFB and DBR lasers
- Bipolar wide-gap transistors
- Transistors, thyristors, dynistors with photonic signal transmission
- High power diodes and thyristors
- Infra-red to visible converters
- Efficent cold cathodes.

III. Important Technological Peculiarities

- Lattice-matched structures are a necessity of principle
- Multi-component solid solutions are used for lattice-matching
- **Epitaxial growth technology is needed of principle.**

Concluding this concise summary of early development of bulk heterostructures one may say that the invention of an "ideal" heterojunction and the introduction of the heterostructure concept into semiconductor physics and technology have led to the discovery of new physical effects, cardinal improvement of characteristics of practically all known semiconductor devices, and the invention of new ones.

3. HETEROSTRUCTURE QUANTUM WELLS AND SUPERLATTICES

Owing to electron confinement in DHS, the double-heterostructure laser became an important precursor of a quantum well structure: when a middle-layer had a thickness of some hundred ångströms, the electron levels would split due to the quantum-size effect. The development of heterostructure growth techniques gave a possibility to fabricate high quality double heterostructures with ultrathin layers. Two main methods of growth with very precise control of thickness, planarity, compositions etc. were developed in the 70s. A modern molecular beam epitaxy method became practically important for III–V heterostructure technology first of all due to the pioneering work of A. Cho [43]. Metallo-organic chemical vapor deposition originated from an early work of H. Manasevit [44] and found broad application in III–V heterostructure research after R. Dupuis and P. Dapkus reported the room temperature injection of AlGaAs DHS lasers which had been grown by the MOCVD method [45].

Clear manifestation of the quantum-size effect in optical spectra of GaAs–AlGaAs semiconductor heterostructures with ultrathin GaAs layers (quantum wells) was demonstrated by Raymond Dingle *et al.* in 1974 [46]. The authors observed a characteristic step-like behavior in absorption spectra and systematic shifts of the characteristic energies with a quantum well width decrease.

Studies of superlattices were started by the work of L. Esaki and R. Tsu in 1970 [47] who considered the electron transport in a superlattice, i.e. an additional periodic potential created by doping or changing the composition of semiconductor materials with the period bigger, but comparable with a lattice constant of a crystal. In this, as Leo Esaki called it "man-made crystal", a parabolic band would break into mini-bands separated by small forbidden gaps and having Brillouin zones determined by this period. Similar ideas were described by L. V. Keldysh in 1962 [48] when considering the periodic potential produced on a semiconductor surface by an intense ultrasonic wave. At the Physico-Technical Institute R. Kazarinov and R. Suris theoretically considered the current flow in superlattice structures in the early 1970s [49]. It was shown that the current between wells is determined by tunneling through the potential barriers separating the wells and the authors predicted a very important phenomena: tunneling under electric field when the ground state of a well coincides with an excited state of the next well and stimulated emission resulting from photon-assisted tunneling between the ground state of one well and excited state of a neighboring well, which is lower by the energy due to applied electric field. At that time L. Esaki and R. Tsu independently considered resonant tunneling in superlattice structures [50].

The pioneering experimental studies of the superlattice structures were carried out by L. Esaki and R. Tsu: the superlattices were grown by VPE in the system GaP_xAs_{1-x}–GaAs. At the same time in our laboratory we developed the first multichamber apparatus and, as it was mentioned before, prepared a superlattice structure $GaP_{0.3}Al_{0.7}$–GaAs with the thickness of each layer equal to 100 Å and total number of the layers being 200 [17]. Observed peculiarities of the volt-current characteristics, their temperature dependence and photoconductivity were explained by the splitting of the conduction band due to the one-dimensional periodic potential of the superlattice. These first superlattices were also the first strained-layer superlattices. E. Blakeslee and J. Matthews who were working with L. Esaki and R. Tsu at IBM succeeded in the mid-70's in growing strained-layer superlattices with a very low concentration of defects. But many years later, after G. Osbourn's theoretical study [51] at Sandia lab and the first successful preparation of a high quality strained-layer superlattice GaAs–$In_{0.2}Ga_{0.8}As$ by M. Ludowise at Varian, N. Holonyak at the University of Illinois achieved the CW room-temperature laser action on those structures[52]. It became clear that in a strained-layer superlattice the lattice strain became an additional degree of freedom and by varying the layer thicknesses and compositions one can vary continuously and independently of one another the forbidden gap, lattice constant etc. of the overall superlattice.

In the early 1970s L.Esaki with co-workers passed to MBE technology in the AlGaAs system [53] and in March 1974, they submitted a paper on resonant tunneling [54]. It was the first experimental demonstration of quantum well heterostructure physics. They measured the tunneling current and conductance as a function of an applied voltage in GaAs–GaAlAs double barriers and found current maxima associated with resonant tunneling. Later in the same year L. Esaki and L. L. Chang observed resonant tunneling in a superlattice [55]. Strong interest in resonant tunneling obviously was connected with its potential applications in high-speed electronics. In the late 1980, picosecond operation has been achieved in a double resonant tunnel diode and oscillations up to 420 GHz were reported in a GaAs resonant tunnel diode at room temperature.

The restriction of the electron motion to two dimensions in field effect transistors were recognized long time ago [56] and for trapped electrons in inversion layers was first verified by the magneto-conductance experiment by A. B. Fowler *et al.* in 1966 [57]. Spectral effects due to spatial quantization were observed in thin bismuth films in 1968 by V. N. Lutskii and L. A. Kulik [58].

The pioneering work on modulation-doped superlattices [59] demonstrating a mobility enhancement with respect to the bulk crystal, stimulated research data on application of the high-mobility two-dimensional electron gas for microwave amplification. In France and Japan practically simultaneously new types of transistors based on a single nAlGaAs–nGaAs modulation-doped heterostructure were created that were labeled TEGFET (two dimensional electron gas FET) in France [60] and HEMT (high electron mobility transistor) in Japan [61].

The first quantum well laser operation was demonstrated by J. P. van der Ziel *et al.* [62] but parameters of the lasing were much worse than for average DHS lasers. In 1978, R. Dupuis and P. Dapkus in collaboration with N. Holonyak first reported about the quantum well laser with parameters comparable with conventional DHS lasers [63]. The name "quantum well" was used in that paper. Real advantages of quantum well lasers were demonstrated much later by W. T. Tsang at Bell Telephone Lab. Thanks to many improvements of MBE growth technology and introducing an optimized structure (GRIN SCH) he found threshold currents as low as 160 A/cm^2 [64].

We started to develop MBE and MOCVD methods of growth of III–V heterostructures only in the late 1970s. First of all, we stimulated the design and construction of the first Soviet MBE machine in our electronic industry. During a few years there were developed three generations of MBE machine and the last, which had the name "Cna" (the nice river not very far from Ryazan – the city where NITI – industrial laboratory of the Electronic Industry was located, this NITI carried out development of MBE machine) were good enough for our goals. In parallel, later on, we started to develop MBE system with NTO AN – scientific Instruments company of the Academy of Sciences in Leningrad. In the middle of the 80's we got a few systems of this version. Both types of MBE systems are still working at the Ioffe Institute and other laboratories.

The MOCVD systems we developed just in our Institute and later, in the 1980s, a Swedish company "Epiquip" specially designed with our participation a couple of systems for our Institute, which are still used in our research.

The strong interest in experimental study of low-dimensional structures and lack of equipment for MBE and MOCVD growth technology stimulated our research on the development of LPE suitable for quantum well heterostructures.

However, until the late 1970s it seemed impossible to grow III–V heterostructures with an active-region thickness of less than 500 Å by LPE because of the existence near the heterojunction of extended interface regions with varying chemical compositions.

The situation was changed due to work of N. Holonyak *et al.* [65] for superlattice like InGaAsP structures by using a rotating boat system. In our laboratory we developed a new LPE method with the usual translational motion in a standard horizontal system for InGaAsP heterostructures [66] and a low-temperature LPE method for AlGaAs heterostructures [67]. These methods permitted us to prepare practically any kind of excellent quality quantum well heterostructures with the thickness of the active region up to 20 Å and with the size of the interface regions comparable to one lattice constant. Of great practical importance for InGaAsP laser heterostructures was the creation of a record threshold current density for InGaAsP/InP (λ = 1.3 and 1.55 μm) and for InGaAsP/GaAs (λ = 0.65–0.9 μm) single quantum well separate confinement lasers [68]. For high power InGaAsP/GaAs (λ = 0.8 μm) lasers a total efficiency of 66 % with CW power 5 W for 100 μm width of the strips, a stripe-geometry laser was achieved [69]. In this lasers effective cooling of the semiconductor power device by recombination radiation was for the first time realized as it had been predicted much earlier [13]. Another important conclusion about InGaAsP heterostructure was its unusual resistance to multiplication of dislocations and defects (Fig. 7) [70]. It was this research that made a start to broad application of Al-free heterostructures.

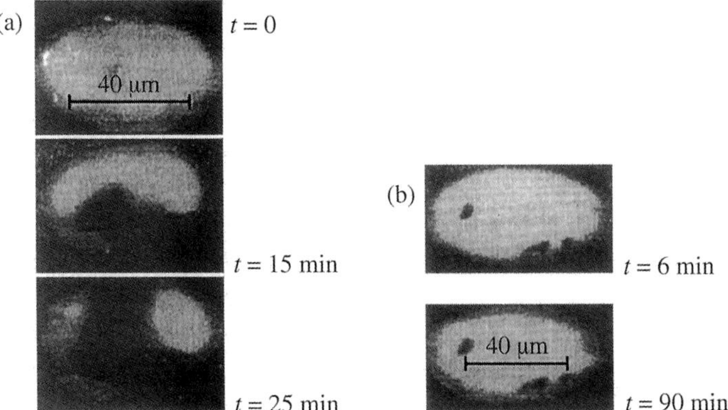

Figure 7. Time evolution of DHS active region under high level photoexcitation. AlGaAs/GaAs (a), InGaAsP/GaAs (b). Diameter of Kr^+-laser excitation beam – 40 μm. Excitation level (a) 10^4 W/cm^2, (b) 10^5 W/cm^2.

A most complicated quantum well laser structure, which combined a single quantum well with short period superlattices (SPSL), for the creation of GRIN SCH (the most favourable for the lowest J_{th}) was demonstrated in our laboratory in 1988 [71] (Fig. 8). Using SPSL we achieved not only the desirable profile of a graded wave-guide region, thus creating a barrier for dislocation movement to the active layer, but also got a possibility to grow different

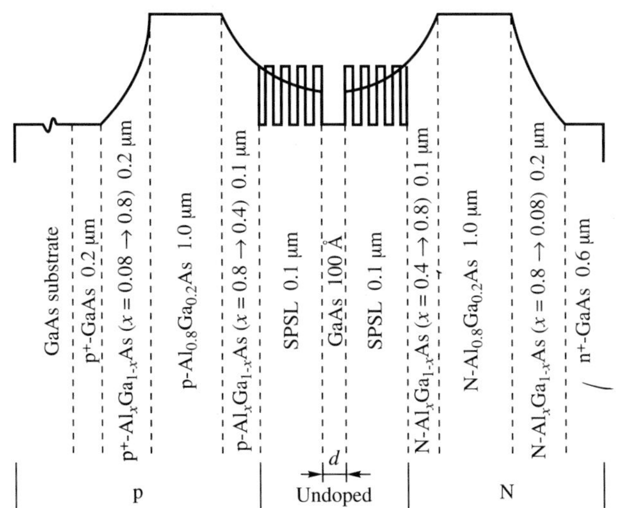

Figure 8. MBE grown SPSL QWSCH laser structure.

parts of the structure at sufficiently large differences of the temperature. In this way, we have obtained both an excellent surface morphology and a high internal quantum efficiency on a planar GaAs (100) surface. The lowest J_{th} = 52 A/cm^2 and, shortly after some small optimization, 40 A/cm^2 was for a long time a world record for semiconductor injection lasers and a good demonstration of the application of quantum wells and superlattices in electronic devices.

The idea of stimulated emission in superlattices that had been published by R. Kazarinov and R. Suris [49] was realized nearly a quarter of century after the proposal, by Federico Capasso [72]. The proposed structure was strongly improved and a cascade laser developed by F. Capasso gave rise to a new generation of unipolar lasers operating in the middle-infrared range.

The history of the semiconductor lasers is, from a certain point of view, the history of evolution of the semiconductor laser current threshold, which is shown in Fig. 9. The most dramatic changes happened just after the introduction of the DHS concept. Impact of SPSL QW led practically to a theoretical limit of this most important parameter. What can happen as a result of application of new quantum wires and quantum dots structures will be discussed in the next part of our paper.

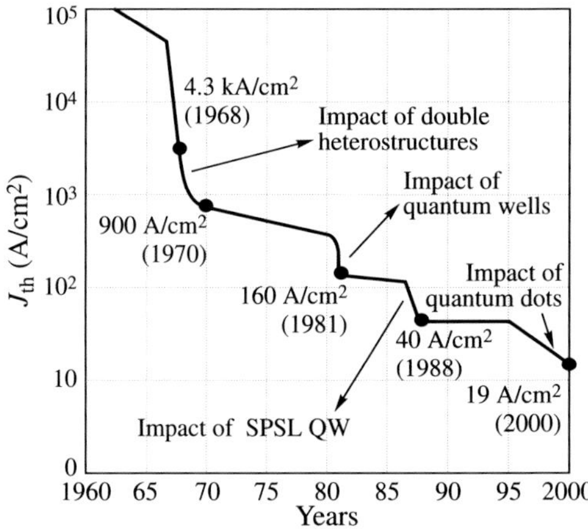

Figure 9. Evolution of the threshold current of semiconductor lasers.

Maybe the crown of the quantum well studies was the discovery of the Quantum Hall effect [73]. This discovery and its comprehensive studies in AlGaAs-GaAs heterostructures, which shortly led to the fractional quantum Hall effect discovery [74], had a principal effect on the whole physics of the solid state. Observation of the effect which deals only with fundamental quantities and does not rely on peculiarities of the band structure, carrier mobility and densities in a semiconductor, has shown that heterostructures can be used to model some very basic physical effects. Recently many studies in this area have been concentrated on understanding condensation of electrons and search of Wigner crystallization.

To summarize this part of our paper we present the summary in a way like that in the previous classical heterostructure part.

Heterostructure Quantum Wells and Superlattices

I. Fundamental Physical Phenomena
- 2D electron gas
- Step-like density-of-state function
- Quantum Hall effect
- Fractional Quantum Hall effect
- Excitons at room temperature
- Resonant tunneling in double-barrier structures and superlattices
- Energy spectrum in superlattices is determined by choice of potential and strain
- Stimulated emission at resonant tunneling in superlattices
- Pseudomorphic growth of strained structures.

II. Important Consequences for Applications
- Shorter emission wavelength, reduced threshold current, larger differential gain and reduced temperature dependence of the threshold current for semiconductor lasers
- IR quantum cascade laser
- SPSL QW laser
- Optimization of Electron and Light Confinement and Waveguiding for semiconductor lasers
- 2D electron gas transistors (HEMT)
- Resonant-tunneling diodes
- Precise resistance standards
- SEEDs and electro-optical modulators
- IR photodetectors based on quantum size level absorption.

III. Important Technological Peculiarities
- Lattice-match is unnecessary
- Low growth-rate technology (MBE, MOCVD) is needed of principle
- Submonolayer growth technique
- Blockading mismatch dislocations during epitaxial growth
- Sharp increase of the variety of heterostructure components.

4. HETEROSTRUCTURE QUANTUM WIRES AND QUANTUM DOTS

The principal advantage of application of quantum-size heterostructures for lasers originates from the noticeable increase of the density of states with reducing dimensionalities for the electron gas (Fig. 10).

During the 1980s, progress in 2D-quantum well heterostructures physics and its applications attracted many scientists to studying systems of far less dimensionality – quantum wires and quantum dots. In contrast to quantum

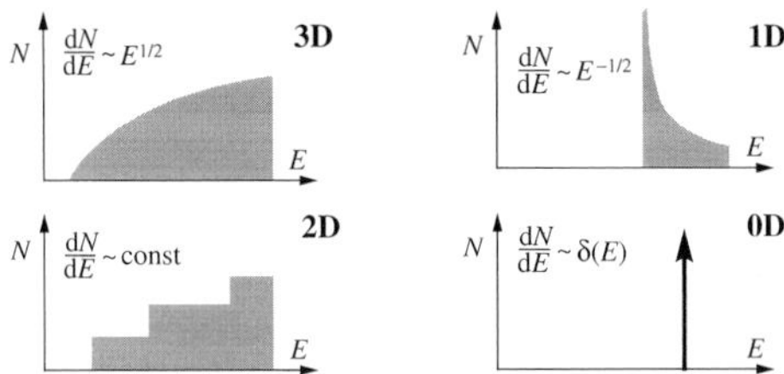

Figure 10. Density of states for charge carriers in structures with different dimensionalities.

"wells" where carriers are localized in the direction perpendicular to the layers but move freely in the layer plane, in quantum "wires" carriers are localized in two directions and move freely along the wire axis. And being confined in all three directions quantum "dots" – "artificial atoms" with a totally discrete energy spectrum are created (Fig. 11).

Experimental work on fabrication and investigation of quantum wire and dot structures began more than 15 years ago. In 1982, Y. Arakawa and H. Sakaki [75] theoretically considered some effects in lasers based on heterostructures with size quantization in one, two, and three directions. They wrote: "Most important, the threshold current of such a laser is reported to be far less sensitive than that of a conventional laser reflecting the reduced dimensionality of the electronic state." The authors performed experimental studies on a QW laser placed in high-magnetic fields directed perpendicular to the QW plane and demonstrated that the characteristic temperature (T_0) describing the exponential growth of the threshold current with temperature increases in a magnetic field from 144 to 313 °C. They pointed to a possibility to weaken the threshold current dependence on temperature for QWR lasers and full temperature stability for QD lasers (Fig. 12). By now there is a significant number of both theoretical and experimental papers in this field.

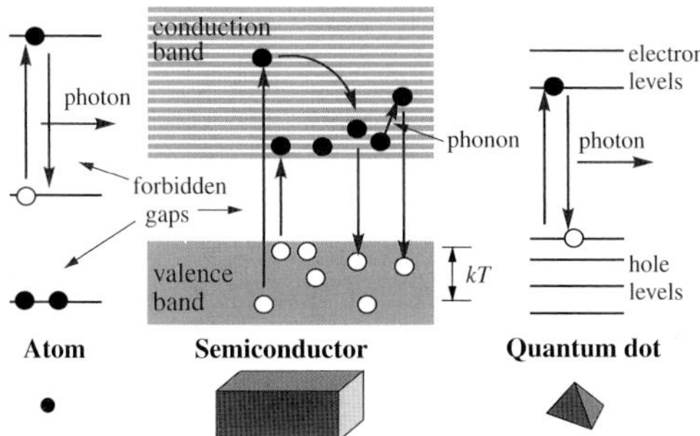

Figure 11. Schematic representation of energy diagrams in case of a single atom (left), a bulk crystal (center), and a quantum dot (right).

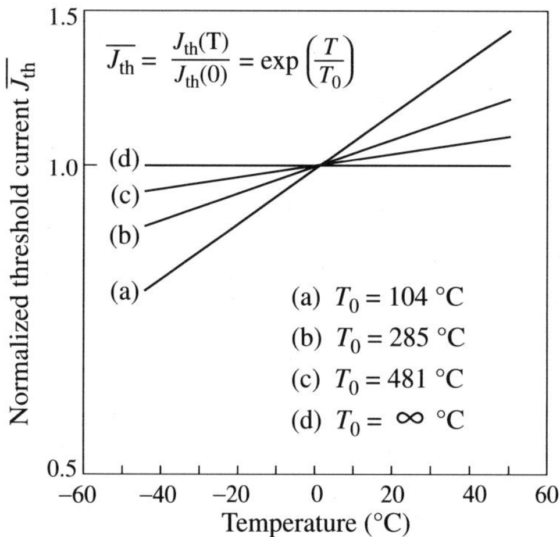

Figure 12. Normalized temperature dependence of the threshold current for various DHS lasers. (a) Bulk; (b) with QWs; (c) with QWRs, (d) with QDs.

The first semiconductor dots based on II–VI microcrystals in glass matrix were proposed and demonstrated by A.I. Ekimov and A.A. Onushchenko [76]. However, since the semiconductor quantum dots were introduced in an insulating glass matrix and the quality of the interface between glass and semiconductor dot was not high, both fundamental studies and device applications were limited. Much more exciting possibilities appeared after three dimensional coherent quantum dots had been fabricated in a semiconductor matrix [77].

Several methods were proposed for the fabrication of these structures. Indirect methods, such as the post-growth lateral patterning of 2D quantum wells suffer often from insufficient lateral resolution and interface damage caused by the patterning procedure. A more promising way is the fabrication by direct methods, i.e., growth in V-grooves and on corrugated surfaces which may result in formation of quantum wires and dots. The groups of the Ioffe Institute and Berlin Technical University – last years we carried out this research in close cooperation – contributed significantly to the last direction.

Finally we came to the conclusion that the most exciting method of the formation of ordered arrays of quantum wires and dots is the self-organization phenomena on crystal surfaces. Strain relaxation on step or facet edges may result in formation of ordered arrays of quantum wires and dots both for lattice-matched and lattice-mismatched growth.

The first very uniform arrays of three dimensional quantum dots exhibiting also lateral ordering were realized in the system InAs–GaAs both by MBE and MOCVD growth methods [78, 79].

Elastic strain relaxation on facet edges and island interaction via the strained substrate are driving forces for self-organization of ordered arrays of uniform, coherently strained islands on crystal surfaces [80].

In lattice-matched heteroepitaxial systems the growth mode is determined solely by the relation between the energies of two surfaces and the interface energy. If the sum of the surface energy of epitaxial layer γ_2 and energy of interface γ_{12} is lower than the substrate surface energy, $\gamma_2 + \gamma_{12} < \gamma_1$, i.e., if the material 2 being deposited wets the substrate, then we have the Frank–van der Merve growth. Changing the $\gamma_2 + \gamma_{12}$ value may result in a transition from the Frank–van der Merve mode to on Volmer–Weber one where 3D islands are formed on a bare substate.

In a heteroepitaxial system with lattice mismatch between the material being deposited and the substrate the growth may initially proceed in a layer-by-layer mode.

However, a thicker layer has a higher elastic energy, and the elastic energy tends to be reduced via formation of isolated islands. In these islands the elastic strains relax and, correspondingly, the elastic energy decreases. This results in a Stranski–Krastanow growth mode (Fig. 13). The characteristic size of islands is determined by the minimum in the energy of an array of 3D coherently strained islands per unit surface area as a function of the island size (Fig. 14) [80]. Interaction between islands via elastically strained substrate would results in lateral island ordering typical of the square lattice.

Experiments show in most cases rather narrow size distribution of the islands, and on top of that coherent islands of InAs form under certain conditions a quasi-periodic square lattice (Fig. 15) The shape of quantum dots can be significantly modified during regrowth or post-growth annealing, or by applying complex growth sequences. Short period alternating deposition of strained materials leads to a splitting of QDs and to formation of vertically coupled quantum dot superlattice structures (Fig. 15) [81]. Ground state QD emission, absorption and lasing energies are found to coincide [78].

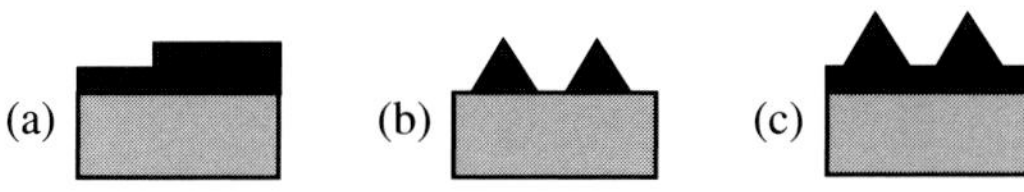

Figure 13. (a) Frank–van der Merve, (b) Volmer–Weber, (c) Stranski–Krastanow growth modes.

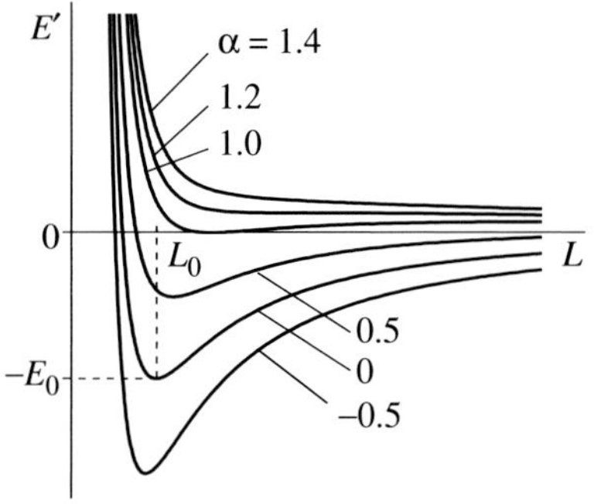

Figure 14. Energy of a sparse array of 3D coherently strained islands per unit surface area as a function of island size. The parameter α is the ratio between the change in the surface energy upon island formation and the contribution from island edges to the elastic relaxation energy. When $\alpha > 1$, the system tends thermodynamically toward island coalescence. When $\alpha < 1$, there exists an optimal island size and the system of islands is stable against coalescence.

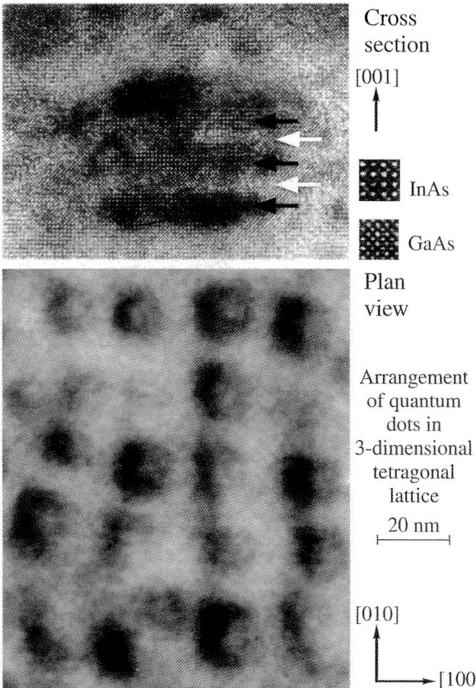

Figure 15. Vertical and transverse ordering of coupled QDs in the system InAs–GaAs.

Observation of ultranarrow (< 0.15 meV) luminescence lines from single quantum dots [78], which do not exhibit broadening with temperature, is the proof of the formation of an electronic quantum dot (Fig. 16).

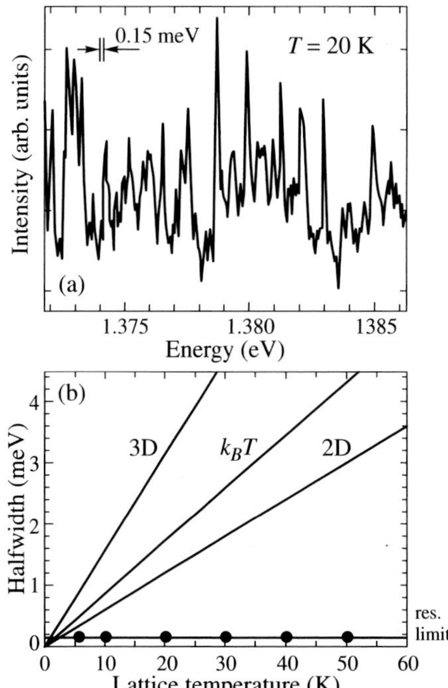

Figure 16. (a) High-resolution cathodoluminescence (CL) spectrum of InAs-GaAs QD structures. (b) Temperature dependense of the full width at half-maximum of the cathodoluminescense peak.

Quantum dot lasers are expected to have superior properties with respect to conventional QW lasers. High differential gain, ultralow threshold current density and high temperature stability of the threshold current density are expected to occur simultaneously. Additionally ordered arrays of scatterers formed in an optical waveguide region may result in distributed feedback and (or) in stabilization of single-mode lasing. Intrinsically buried quantum dot structures spatially localize carriers and prevent them from recombining nonradiatively at resonator faces. Overheating of facets, being one of the most important problems for high-power and high efficiency operation of AlGaAs–GaAs and AlGaAs–InGaAs lasers, may thus be avoidable.

Since the first realization of QD lasers [82], it has become clear that the QD size uniformity was sufficient to achieve good device performance. But even at that time, it was recognized that the main obstacle for QDHS laser operation at room and elevated temperatures was connected to temperature-induced evaporation of carriers from QD's. Different methods were developed to improve the laser performance: (i) the increase of the density of QD's by stacking of QD's (Fig. 17); (ii) the insertion of QD's into a QW sheet; (iii) the use of a matrix material with a higher bandgap energy. As a result, we got many parameters of QDHS lasers better than ones for QWHS lasers based on the same materials. As an example, the world-record threshold-current density of 19 A/cm^2 has been recently achieved [83]. Further, the cw-output power up to 3.5–4.0 W (CW) for a 100-μm strip width, the quantum efficiency of 95 % and the wall-plug efficiency of 50 % were obtained [84].

Significant activities in theoretical understanding of QD lasers with realistic parameters have been performed. For a QD size dispersion of about 10 % and other practical structure parameters, the theory [85] predicts typical threshold-current densities of 5 A/cm^2 at room temperature. The value of 10 A/cm^2 at 77 K [86] and even 5 A/cm^2 at 4 K [87] have been experimentally observed.

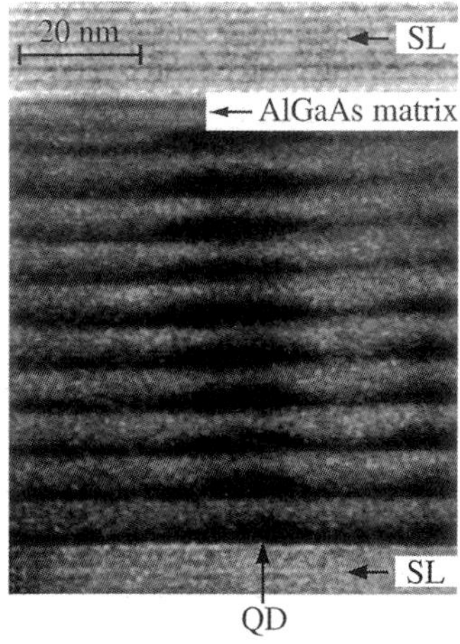

Figure 17. Transmission electron microscopy image of the active region of high-power QDHS laser.

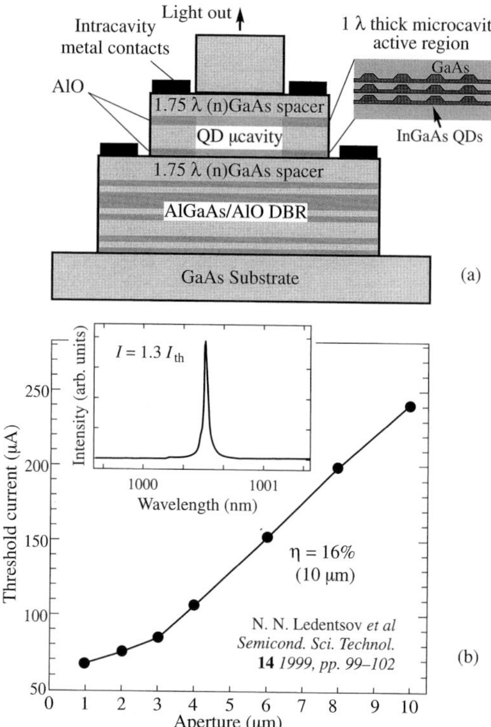

Figure 18. Schematic view of the QD VCSEL structure (a).
Basic advantages of quantum dots: (i) no interface recombination at oxide-defined apertures; (ii) reduced lateral spreading of carriers out of the aperture region. Single QD laser at ultralow threshold current is possible.
Dependence of the threshold current on the aperture size in a QD VCSEL (b).
(i) low threshold current densities (170 A/cm^2 at 300 K); (ii) low threshold currents at ultrasmall apertures; (iii) 1.3 μm range on GaAs substrate?

In view of the advanced device applications of QD's, the incorporation of QD's in vertical-cavity surface-emitting lasers (VCSEL's) seems to be very important. QD VCSEL's with parameters, which fit to the best values for QW devices of the similar geometry, have been demonstrated (Fig. 18) [88]. Recently, very promising results for 1.3-μm QD VCSEL's on a GaAs substrate to use in fiber optical communications have been obtained (Fig. 19) [89].

In a free-standing 3D island formed on a lattice-mismatched substrate, the strains can relax elastically, without the formation of dislocations. Thus, sufficiently large volume of a coherent narrow-gap QD material can be realized. This makes it possible to cover a spectral range of 1.3–1.5 μm using a GaAs substrate and to develop wavelength-multiplexing systems on the basis of QD VCSEL's in the future.

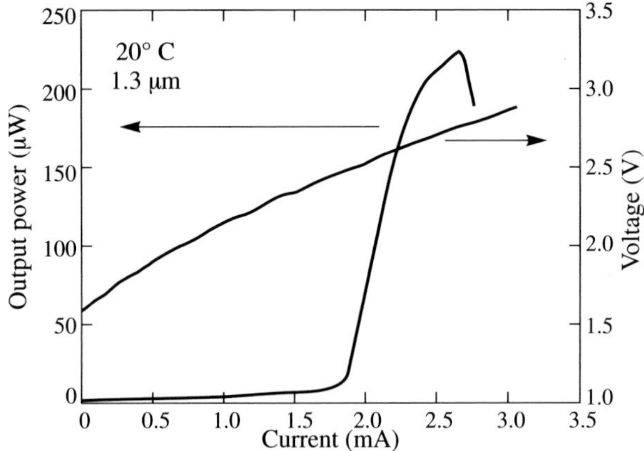

Figure 19. GaAs-based QD VCSEL emitting at 1.3 μm.

It is very important to emphasize that we always realized the DHS concept for QWR and QD structures because in both cases we have a narrow gap-band material in a wide-gap matrix.

Let us summarize this part again by the same way as it had been done for other parts.

Heterostructure Quantum Wires and Dots

I. Fundamental Physical Phenomena

- 1D electron gas (wires)
- Density-of-state function with sharp maximums (wires)
- 0D-electron gas (dots)
- delta-function type of density-of-state function (dots)
- Increasing binding energy of excitons.

II. Important Applications in Electronics

- Reduced lasing threshold current and larger differential gain
- Reduced temperature dependence of threshold current (wires)
- Temperature stability of the threshold current (dots)
- Discrete amplification spectrum and a possibility of obtaining performance characteristics similar to those of solid-state or gas lasers (dots)
- Higher modulation factor in electro-optical modulators
- A possibility of creation "single-electron" devices
- A new possibility for the development of FET transistors.

III. Important Technological Peculiarities

- The application of self-organization effects for growth
- Epitaxial growth in V-grooves (wires)
- High resolution lithography of HSQWL.

5. FUTURE TRENDS

Recently very impressive results for short wave-length light sources have been achieved on the basis of II–VI selenides and III–V nitrides. The success in this research was mostly determined by application of heterostructure concepts and methods of growth which had been developed for III–V quantum wells and superlattices. The natural and most predictable trend is the application of the heterostructure concepts as well as technological methods and peculiarities to new materials. Different III–V, II–VI and IV–VI heterostructures, developed in recent time, are good examples of this statement.

But from a general and more deep point of view, heterostructures (it concerns all of them: the classical, QWs and SLs, QWRs and QDs) are a way of the creation of new types of materials – Heterosemiconductors. By using Leo Esaki words – instead of "God made crystals" we create by ourselves – "Man made crystals".

The classical heterostructures, quantum wells and superlattices are quite mature and we exploit many of their unique properties. Quantum wires and dots structures are still very young: exciting discoveries and new unexpected applications are awaiting us on this way. Even now we can say that ordered equilibrium arrays of quantum dots may be used in many devices: lasers, light modulators, far-infrared detectors and emitters, etc. Resonant tunneling via semiconductor atoms introduced in larger band-gap layers may lead to significant improvement in device characteristics. More generally speaking, QD structures will be developed both "in width" and "in depth".

In width means new material systems to cover new energy spectrum. The life-time problems of the green and blue semiconductor lasers and even more general problems of the creation of defect-free structures based on wide-gap II–VI and III–V (nitrides) would be solved by using QDs structures in these systems.

As to *in depth* it is necessary to mention that the degree of ordering depends on very complicated growth conditions, materials constants, concrete values of the surface free energy. The way to resonant tunneling and "single" electron devices including optical ones is a deep detailed investigation and evaluation of these parameters in order to achieve the maximal possible degree of ordering. In general, it is necessary to find out more strong self-organization mechanisms for ordered arrays of QDs creation.

In the early 1980s I was invited to deliver a lecture about heterostructures and applications at the Amoco Photonic Center near Chicago.

The summary of my lecture was as follows:

1. Heterostructures – a new kind of semiconductor materials:
   - expensive, complicated chemically & technologically but most efficient
2. Modern optoelectronics is based on heterostructure applications
   - DHS laser – key device of the modern optoelectronics
   - HS PD – the most efficient & high speed photo diode
   - OEIC – only solves problem of high information density of optical communication systems

3. Future high speed microelectronics will mostly use heterostructures
4. High temperature, high speed power electronics – a new broad field of heterostructure applications
5. Heterostructures in solar energy conversion: the most expensive photocells and the cheapest solar electricity producer
6. In the 21st century heterostructures in electronics will reserve only 1 % for homojunctions.

And 20 years later I do not like to change any word there.

It is hardly possible to describe even the main directions of the modern physics and technology of semiconductor heterostructures. There is much more than was mentioned. Many scientists contributed to this tremendous progress which not only defines to a great extent the future prospects of condensed matter physics and semiconductor laser and communication technology but, in a sense also, the future of the human society. I would like also to emphasize the impact of scientists of previous generations who prepared our way. I am very happy that I had a chance to work in this field from the very beginning. I am even more happy that we can continue to contribute to the progress in this area now.

REFERENCES

[1] V. P. Zhuze and I. V. Kurchatov, "Two questions about conductivity of cuprous oxide". *Zh. Eksp. Teor. Fiz. (J.E.T.P.)* **2**, p. 309, 1932.
V. P. Zhuze and B. V. Kurchatov, "Zur electrischen Leitfähigkeit von Kupferoxydul". *Phyz. Zs. SU* **2**, No 6, p. 453, 1932.
Ya. I. Frenkel, A. Ioffe, "On the electrical and photoelectric properties of contacts between a metal and semiconductor". *Phys. Z. SU 1*, No 1, p. 60, 1932.

[2] Ya. I. Frenkel, "On the transformation of light into heat in solids". *Phys. Rev.* **37,** p. 17, 1276, 1931.
Ya. I. Frenkel, "Light absorption and electrons and holes sticking in dielectric crystals". *Zh. Eksp. Teor. Fiz.* **6**, p. 647, 1936.
E. F. Gross, N. A. Karryev, "Light absorption by cuprous oxide crystal in infrared and visible spectrum". *DAN SSSR (Doklady Academii Nauk)* **84**, p. 261, 1952.
E. F. Gross, N. A. Karryev, "Exciton optical spectrum". *DAN SSSR,* **84**, p. 471, 1952.

[3] B. I. Davydov, "Contact resistance of semiconductors". *Zh. Eksp. Teor. Fiz.* **9**, p. 451, 1939.

[4] N. H. Welker, "Über neue halblestende Verbindungen (New semiconductor compounds)". *Zs. Naturforsch.*, **7a**, p. 744, 1939; *Zs. Naturforsch.*, **8a**, p. 248, 1953.
N. A. Goryunova, "Seroe olovo (Gray tin)", Thesis, Leningrad State University-Physico-Technical Inst., 1951.
A. I. Blum, N. P. Mokrovsky, A. R. Regel, "The study of the conductivity of the semiconductors and intermetallic compounds in solid and liquid state". Proc. of the VII Conference on semiconductor properties. Kiev, 1950. *Izv. AN USSR, Ser. Fiz.* **XVI**, p. 139, 1952.

[5] W. Shockley, "Circuit Element Utilizing Semiconductor Material", U.S. Patent 2269347, September 25, 1951.

[6] A. I. Gubanov, "Theory of the contact between two semiconductors with different types of conduction". *Zh. Tekh. Fiz.* **20**, p. 1287, 1950.
A. I. Gubanov, "Theory of the contact of two semiconductors of the same type of conductivity". *Zh. Tekh. Fiz.* **21**, p. 304, 1951.

[7] H. Kroemer, "Theory of a wide-gap emitter for transistors", *Proc. IRE*, **45**, p. 1535, 1957; "Quasi-electric and quasi-magnetic fields in a non-uniform semiconductor," *RCA Rev.*, **28**, p. 332, 1957.

[8] N. G. Basov, O. N. Krokhin and Yu. M. Popov, "The possibility of use of indirect transitions to obtain negative temperature in semiconductors", *Soviet Physics -JETP (USA)*, **12**, p. 1033, (May, 1961).

[9] D. N. Nasledov, A. A. Rogachev, S. M. Ryvkin and B. V. Tsarenkov, "Recombination radiation of galium arsenic", *Fiz. Tverd. Tela*, **4**, p. 1062–1065, 1962 (Soviet Physics- Solid State (USA), **4**, pp. 782–784, 1962.)

[10] R. H. Hall, G. E. Fenner, J. D. Kingsley, T. J. Soltys and R. O. Carlson, "Coherent light emission from GaAs junction", *Phys. Rev. Lett.*, **9**, pp. 366–378, 1962.
M. I. Nathan, W. P. Dumke, G. Burns, F. H. Dill Jr and G. I. Lasher, "Stimulated emission of radiation from GaAs p-n junctions". *Appl. Phys. Lett.*, **1**, pp. 62–64, 1962.
N. Holonyak Jr. and S. F. Bevacgua, "Coherent (visible) light emission from Ga $(As_{1-x}P_x)$ junctions", *Appl. Phys. Lett.*, **1**, pp. 83–83, 1962.

[11] Zh. I. Alferov and R. F. Kazarinov, "Semiconductor laser with electric pumping," Inventor's Certificate No. 181737 [in Russian], Application No. 950840, priority as of March 30, 1963; H. Kroemer, "A proposed class of heterojunction injection lasers". *Proc. IEEE*, **51**, p. 1782, 1963.

[12] Zh. I. Alferov, V. B. Khalfin, and R. F. Kazarinov, "A characteristic feature of injection into heterojunctions," *Fiz. Tverd. Tela*, **8**, pp. 3102–3105, 1966 [*Sov. Phys. Solid State*, **8**, p. 2480, 1967].

[13] Zh. I. Alferov, "Possible development of a rectifier for very high current densities on the bases of a p-i-n (p-n-n^+, n-p-p^+) structure with heterojunctions," *Fiz. Tekh. Poluprovodn.*, **1,** pp. 436–438, 1966 [*Sov. Phys. Semicond.* **1**, pp. 358–361, 1967.]

[14] R. L. Anderson, "Gemanium-gallium arsenide heterojunctions". *IBM J. Res. Develop.* **4**, p. 283, 1960.
P. L. Anderson, "Experiments on Ge-GaAs heterojunctions". *Solid State Electron.* **5**, p. 341, 1962.

[15] G. Natta, L. Passerini, *Gazz. Chim. Ital.* **58**, p. 458, 1928.
V. M. Goldschmidt, "Crystal structure and chemical constitution". *Trans. Farad. Soc.* **25**, p. 253, 1929.

[16] Zh. I. Alferov, Garbuzov D. Z., Grigor'eva V. S., Zhilyaev Yu. V., Kradinova L. V., Korol'kov V. I., Morozov E. P., Ninua O. A., Portnoy E. L., Prochukhan V. D., Trukan M. K. "Injection luminescence of epitaxial heterojunctions in the GaP-GaAs system". *Sov. Phys. Sol. State.* **9**, p. 208–210, 1967.

[17] Zh. I. Alferov, Yu. V. Zhilyaev and Yu. V. Shmartsev, "The splitting of the conduction band in `superlattice' on the base GaP_xAs_{1-x}". *Fiz. Tekh. Polupr.* **5**, p. 196, 1971 [*Sov. Phys.- Semicond.* **5**, p. 174, 1971).

[18] Zh. I. Alferov, V. M. Andreev, V. I. Korol'kov, D. N. Tret'yakov, and V. M. Tuchkevich, "High-voltage p-n junctions in $Ga_xAl_{1-x}As$ crystalls," *Fiz. Tekh. Poluprovodn.*, **1**, pp. 1579–1581, 1967 [*Sov. Phys. Semicond.* **1**, pp. 1313–1314, 1968];
H. S. Rupprecht, J. M. Woodall, and G. D. Pettit, "Efficient visible electroluminescence at 300 K from $Ga_{1-x}Al_xAs$ p-n junctions grown by liquid-phase epitaxy," *Appl. Phys. Lett.*, **11**, p. 81, 1967.

[19] Zh. I. Alferov, V. M. Andreev, V. I. Korol'kov, E. L. Portnoi, and D. N. Tret'yakov, "Injection properties of n-$Al_xGa_{1-x}As$-p-GaAs heterojunctions," *Fiz. Tekh. Poluprovodn.*, **2**, pp. 1016–1017, 1968 [*Sov. Phys.-Semicond.* **2**, pp. 843–844, 1969.]

[20] Zh. I. Alferov, V. M. Andreev, V. I. Korol'kov, E. L. Portnoy, and D. N. Tret'yakov, "Coherent radiation of epitaxial heterojunction structures in the AlAs-GaAs system," *Fiz. Tekh. Poluprovodn.* **2,** pp. 1545–1547, 1968 [*Sov. Phys.-Semicond.* **2**, pp. 1289–1291, 1969].

[21] (a) Zh. I. Alferov, V. M. Andreev, V. I. Korol'kov, E. L. Portnoi, and D. N. Tret'yakov, "Recombination radiation in epitaxial structures in the system AlAs-GaAs," in *Proc. Ninth Int. Conf. on Semiconductor Structures*, Moscow, July 23–29, 1968, **1** [in Russian] (Nauka, Leningrad, 1969), pp. 534–540;

(b) Zh. I. Alferov, "Electroluminescence of heavily-doped heterojunctions p Al_xGa_{1-x}-nGaAS," in *Proc. Int. Conf. on Luminescence*, Newark, Delaware, August 25–29, 1969, *J. Lumin.*, **1**, pp. 869–884, 1970;

(c) Zh. I. Alferov, D. Z. Garbuzov, E. P. Morozov, and E. L. Portnoi, "Diagonal tunneling and polarization of radiation in $Al_xGa_{1-x}As$-GaAs heterojunctions and in GaAs p-n junctions," Fiz. Tekh. Poluprovodn., **3**, pp. 1054–1057, 1969 [*Sov. Phys.-Semicond.* **3**, pp. 885–887, 1970];

(d) Zh. I. Alferov, V. M. Andreev, V. I. Korol'kov, E. L. Portnoi, and A. A. Yakovenko, "Recombination radiation in $Al_xGa_{1-x}As$ solid solutions with variable forbidden gap" *Fiz. Tekh. Poluprovodn.* **3**, pp. 541–545, 1969 [*Sov. Phys.-Semicond.* **3**, pp. 460–463, 1970].

[22] Zh. I. Alferov, V. M. Andreev, E. L. Portnoy, and M. K. Trukan, "AlAs-GaAs heterojunction injection lasers with a low room-temperature threshold," *Fiz. Tekh. Poluprovodn.*, **3**, pp. 1328–1332, 1969 [*Sov. Phys. Semicond.* **3**, pp. 1107–1110, 1970].

[23] Zh. I. Alferov, V. M. Andreev, V. I. Korol'kov, E. L. Portnoi, and A. A. Yakovenko, "Spontaneous radiation sources based on structures with AlAs-GaAs heterojunctions," *Fiz. Tekh. Poluprovodn.*, **3**, pp. 930–933, 1969 [*Sov.-Phys. Semicond.* **3**, pp. 785–787, 1970].

[24] Zh. I. Alferov, V. M. Andreev, M. B. Kagan, I. I. Protasov, and V. G. Trofim, "Solar-energy converters based on p-n $Al_xGa_{1-x}As$-GaAs heterojunctions," *Fiz. Tekh. Poluprovodn.*, **4**, pp. 2378–2379, 1970 [*Sov. Phys.-Semicond.* **4**, pp. 2047–2048, 1971].

[25] Zh. I. Alferov, F. A. Ahmedov, V. I. Korol'kov, and V. G. Nikitin, "Phototransistor utilizing a GaAs-AlAs heterojunction," *Fiz. Tekn. Poluprovodn.*, **7**, pp. 1159–1163, 1973 [*Sov. Phys.-Semicond.*, **7**, pp. 780–782, 1973].

[26] Zh. I. Alferov, V. M. Andreev, V. I. Korol'kov, V. G. Nikitin, and A. A. Yakovenko, "p-n-p-n structures based on GaAs and on $Al_xGa_{1-x}As$ solid solutions," *Fiz. Tekn. Poluprovodn.*, **4**, pp. 578–581, 1970 [*Sov. Phys.-Semicond.*, **4**, pp. 481–483, 1971].

[27] I. Hayashi, "Heterostructure lasers," *IEEE Trans. Electron Devices*, **ED-31**, pp. 1630–1645, 1984.

[28] Zh. I. Alferov, V. M. Andreev, D. Z. Garbuzov, Yu. V. Zhilyaev, E. P. Morozov, E. L. Portnoi, and V. G. Trofim, "Investigation of the influence of the AlAs-GaAs heterostructure parameters on the laser threshold current and the realization of continuous emission at the room temperature," *Fiz. Tekh. Poluprovodn.*, **4**, pp. 1826–1829, 1970 [*Sov. Phys. Semicond.* **4**, pp. 1573–1575, 1971].

[29] I. Hayashi, M. B. Panish, P. W. Foy, and S. Sumski, "Junction lasers which operate continuously at room temperature," *Appl. Phys. Lett.*, **17**, pp. 109–111, 1970.

[30] Zh. I. Alferov, V. M. Andreev, S. G. Konnikov, V. G. Nikitin, and D. N. Tret'yakov, "Heterojunctions on the base of $A^{III}B^{V}$ semiconducting and of their solid solutions," in *Proc. Int. Conf. Phys. Chem. Semicond. Heterojunctions and Layer Structures*, Budapest, October, 1970, **1**, G. Szigeti, Ed. (Academiai Kiado, Budapest, 1971), pp. 93–106.

[31] G. A. Antipas, R. L. Moon, L. W. James, J. Edgecumbe and R. L. Bell, "In gallium arsenid and related compounds". Conf.Ser. IOP, **17**, p. 48, 1973.

[32] L. James, G. Antipas, R. Moon, J. Edecumbe, R. L. Bell, "Photoemission from cesium-oxide-activated InGaAsP". *Appl. Phys. Lett.* **22**, p. 270, 1973.

[33] A. P. Bogatov, L. M. Dolginov, L. V. Druzhinina, P. G. Eliseev, L. N. Sverdlova, and E. G. Shevchenko, "Heterolasers on the base of solid solutions $Ga_xIn_{1-x}As_4P_{1-y}$ and $Al_xGa_{1-x}Sb_yAs_{1-y}$," *Kvantovaya Electron.* **1**, p. 2294, 1974 [*Sov. J. Quantum Electron,* **1**, p. 1281, 1974];

J. J. Hsieh, "Room-temperature operation GaInAsP/InP double-heterostructure diode lasers emitting at 1.1 μm," *Appl. Phys. Lett.*, **28**, p. 283, 1976.

[34] Zh. I. Alferov, I. N. Arsent'ev, D. Z. Garbuzov, S. G. Konnikov, and V. D. Rumyantsev, "Generation of coherent radiation in $nGa_{0.5}In_{0.5}P$-$pGa_{x\sim0.55}In_{1-x}As_{y\sim0.10}P_{1-y}$-$nGa_{0.5}In_{0.5}P$" *Pisma Zh. Tech. Fiz.*, **1**, pp. 305–310, 1975 [*Sov. Phys.-Tech. Phys. Lett.*, **1**, pp. 147–148, 1975];

"Red injection heterolasers in the Ga-In-As-P system," *Pisma Zh. Tech. Fiz.*, **1**, pp. 406–408, 1975 [*Sov. Phys.-Techn. Phys. Lett.*, **1**, pp. 191–192, 1975];

W. R. Hitchens, N. Holonyak Yr., P. D. Wright, and J. J. Coleman, "Low-theshold LPE $In_{1-x'}Ga_{x'}P_{1-z'}As_{z'}/In_{1-x}Ga_{x}P_{1-z}As_{z}/In_{1-x'}Ga_{x'}P_{1-z'}As_{z'}$ yellow double-heterojunction laser diodes ($J < 10^4$ A/cm^2, I = 5850 A, 77 K)," *Appl. Phys. Lett.*, **27**, p. 245, 1975.

[35] Zh. I. Alferov, V. M. Andreev, R. F. Kazarinov, E. L. Portnoi, and R. A. Suris, "Semiconductor optical quantum generator," Inventor's Certificate No. 392875 [in Russian], Application No. 1677436, priority as of July 19, 1971.

[36] H. Kogelnik and C. V. Shank, "Stimulated emission in a periodic structure," *Appl. Phys. Lett.*, **18,** p. 152, 1971.

[37] R. F. Kazarinov and R. A. Suris, "Injection heterolaser with diffraction grating on contact surface". *Fiz. Tekh. Polupr.* **6,** p. 1359, 1972 (Sov. Phys.-Semicond. **6**, p. 1184, 1973).

[38] Zh. I. Alferov *et al.* "Laser with supersmall divergens of radiation". *Fiz. Tekh. Polupr.* **8**, p. 832, 1974 (Sov.Phys.-Semicond. **8,** 541, 1974).
Zh. I. Alferov et al. "Semiconductor laser with distributed feedback in second order". *Pisma Zh. Tekh. Fiz.* **1**, p. 645, 1975 (Tech. Phys. Lett. **1**, p. 286, 1975).

[39] M. Nakamura, A. Yariv, H. W. Yen, S. Somekh and H. L. Garvin, "Optically pumped GaAs surface laser with corrugation feedback". *Appl. Phys. Lett.* **22**, p. 315, 1973.

[40] D. R. Scifres, R. D. Burnham and W. Streifer, "Distributed-feedback single heterojunction GaAs diode laser". *Appl. Phys. Lett.* **25**, p. 203, 1974.

[41] H. Kroemer and G. Griffiths, "Staggered-lineup heterojunctions as sources of tunable bellow-gap radiation: Operating principle and semiconductor selection". *IEEE Electron Device Lett.* **EDL-4**, p. 1, 20, 1983.

[42] A. N. Baranov et al. "Generation of the coherent radiation in quantum-sizeed structure of the single heterojunction". *Fiz. Tekh. Polupr.* **20**, p. 2217, 1986 (Sov.Phys.-Semicond. 20, 1385, 1986).

[43] A. Y. Cho, "Film deposition by molecular beam techniques," *J. Vac. Sci. Technol.*, **8**, p. 31, 1971.
A. Y. Cho, "Growth of periodic structures by the molecular-beam method". *Appl. Phys. Lett.* **19**, p. 467, 1971.

[44] H. M. Manasevit, "Single crystal GaAs on insulating substrates," *Appl. Phys. Lett.*, **12**, p. 156, 1968.

[45] R. D. Dupuis and P. D. Dapkus, "Room temperature operation of $Ga_{1-x}Al_xAs$/GaAs double-heterostructure lasers grown by metalloorganic chemical vapor deposition," *Appl. Phys. Lett.* **31**, p. 466, 1977.

[46] R. Dingle, W. Wiegmann, and C. H. Henry, "Quantized states of confined carriers in very thin $Al_xGa_{1-x}As$-GaAs-$Al_xGa_{1-x}As$ heterostructures," *Phys. Rev. Lett.*, **33**, p. 827, 1974.

[47] L. Esaki and R. Tsu, "Superlattice and negative differential conductivity". *IBM J. Res. Dev.* **14**, p. 61, 1970.

[48] L. V. Keldysh, "Effect of ultrasonics on the electron spectrum of crystals". Fiz. Tverd. Tela **4,** p. 2265, 1962 (Sov.Phys.-Sol.State **4,** 1658, 1963).

[49] R. F. Kazarinov and R. A. Suris, "Possibility of the amplification of electromagnetic waves in a semiconductor with a superlattice," *Fiz. Tekh. Poluprovodn.*, **5**, pp. 797–800, 1971 [*Sov. Phys. Semicond.*, **5**, pp. 707–709, 1971]; "Electric and electromagnetic properties of semiconductors with a superlattice," *Fiz. Tekh. Poluprovodn.*, **6**, pp. 148–162, 1972 [*Sov. Phys. Semicond.*, **6**, pp. 120–131, 1972]; "Theory of electrical properties of semiconductors with superlattices," *Fiz. Tekh. Poluprovodn.*, **7**, pp. 488–498, 1973 [*Sov. Phys. Semicond.*, **7**, pp. 347–352, 1973].

[50] R. Tsu and L. Esaki, "Tunneling in finite superlattice". *Appl. Phys. Lett.* **22**, p. 562, 1973.

[51] G. Osbourn, "Strained-layer superlattices from lattice mismatched materials". *J. Appl. Phys.* **53**, p. 1586, 1982.

[52] M. Ludowise *et al.* "Continuous 300 K laser operation of strained superlattices". *Appl. Phys. Lett.* **42**, p. 487, 1983.

[53] L. L. Chang, L. Esaki, W. E. Howard and R. Ludke, "Growth of GaAs-GaAlAs superlattices". *J. Vac. Soc. Technol.* **10**, p. 11, 1973.

[54] L. L. Chang, L. Esaki, R. Tsu, "Resonant tunneling in semiconductor double barriers". *Appl. Phys. Lett.* **24**, p. 593, 1974.

[55] L. Esaki and L. L. Chang, *New Transport phenomenon in a semiconductor 'Superlattice'"*. *Phys. Rev. Lett.* **33**, p. 686, 1974.

[56] J. R. Shriffer, "Semiconductor surface physics" (R. H. Kingston; ed.), p.68, University of Pennsylvania Press, Philadelphia.

[57] A. B. Fowler, F. F. Fang, W. E. Howard and P. J. Stiles, "Magnetooscillatory conductance in silicon surfaces". *Phys. Rev. Lett.* **16**, p. 901, 1966.

[58] V. N. Lutskii, "Quantum-size effect – present state and perspective on experimental investigations". *Phys. Stat. Sol. (a)* **1**, p. 199, 1970.

[59] R. Dingle, H. L. Stormer, H. L. Gossard and W. Wiegmann, "Electron mobilities in modulation-doped semiconductor heterojunction superlatticies". *Appl. Phys. Lett.* **33**, p. 665, 1978.

[60] D. Delagebeaudeuf *et al.* "Two-dimensional electron gas MESFET structure". *Electron Lett.* **16**, p. 667, 1980.

[61] T. Mimura, S. Hiyamizu, T. Fuji and K. Nanbu, "A new field-effect transistor with selectively doped GaAs/n-$Al_xGa_{1-x}As$ heterojunctions". *Jpn. J. Appl. Phys.* **19**, L225, 1980.

[62] J. P. van der Ziel, R. Dingle, R. C. Miller, W. Wiegmann, and W. A. Nordland, Jr., "Laser oscillations from quantum states in very thin GaAs-$Al_{0.2}Ga_{0.8}As$ multilayer structures," *Appl. Phys. Lett.*, **26**, pp. 463–465, 1975.

[63] R. D. Dupuis, P. D. Dapkus, N. Holonyak, Yr., E. A. Rezek and R. Chin, "Room temperature operation of quantum-well $Ga_{1-x}Al_xAs$-GaAs laser diodes grown by metalorganic chemical vapor deposition," *Appl. Phys. Lett.*, **32**, pp. 295–297, 1978.

[64] W. T. Tsang, "Extremely low threshold (AlGa)As graded-index waveguide separate-confinement heterostructure lasers grown by molecular-beam epitaxy," *Appl. Phys. Lett.*, **40**, pp. 217–219, 1982.

[65] E. Rezek, H. Shichijo, B. A. Vojak and N. Holonyak, "Confined-carrier luminescence of a thin $In_{1-x}Ga_xP_{1-z}As_z$ well ($x_0 \sim 0.13$, $z \sim 0.29$, ~ 400 A) in an InP p-n-junction". *Appl. Phys. Lett.* **31**, p. 534, 1977.

[66] Zh. I. Alferov *et al.* "Auger-profiles and luminescence investigations LPE grown InGaAsP-heterostructures with active regions (1.5–5) × 10^{-6} cm". *Fiz. Tekh. Polupr.* **19**, p. 1108, 1985 (Sov. Phys.-Semicond. **19,** p. 679, 1985).

[67] Zh. I. Alferov *et al.* "AlGaAs-heterostructure quantum wells grown by low temperature LPE". *Pisma Zh. Tekh. Fiz.* **12**, p. 1080, 1986 (Sov. Phys.-Techn. Phys. Lett. **12**, p. 450,1986).

[68] Zh. I. Alferov *et al.* "Low threshold InGaAsP/InP separate confinement lasers λ =1.3 μm and λ = 1.55 μm (j_{th} = 600–700 A/cm^2)". *Pisma Zh. Tekhn. Fiz.* **12**, p. 210, 1986 (Sov. Phys.-Techn. Phys. Lett. **12**, p. 87, 1986). Zh. I. Alferov *et al.* "Low threshold quantum well InGaAsP/GaAs separate confinement double heterostructures lasers formed by the LPE (λ = 0.86 μm, j_{th} = 90 A/cm^2, $L = \infty$; j_{th} = 165 A/cm^2, L = 1150 μm, 300 K)". *Fiz. Tekh. Polupr.* **21**, p. 914, 1987 (Sov.Phys.-Semicond. **21**, p. 914, 1987).

[69] Zh. I. Alferov *et al.* "Quantum-well InGaAsP/GaAs (λ = 0.86: 0.78 μm) separate confinement lasers (j_{th} = 100 A/cm^2, efficiency = 59 %)". *Fiz. Tekh. Polupr.* **22**, p. 1031, 1988 (Sov. Phys.-Semicond. **22**, p. 650, 1988). D. Z. Garbuzov *et al.* "Technical Digest CLEO", paper THU44, p. 396, 1988.

[70] D. Z. Garbuzov *et al. Conf. Digest 12th Intern. Semicond. Laser Conf.*, Davos, Switzerland, paper L-33, 238, 1990.

[71] Zh. I. Alferov *et al.* "Reducing of the threshold current in GaAs-AlGaAs DHS SCH quantum well lasers (j_{th} = 52 A/cm^2, T = 300 K) with quantum well restriction by short period superlattice of variable period". *Pisma Zh. Tekh. Fiz.* **14**, p. 1803, 1988 (Sov. Phys.-Technical Physics Lett. **14**, p. 782, 1988).

[72] J. Faist, F. Capasso, D. L. Sivco, C. Sirtori, A. L. Hutchinson, A. Y. Cho "Quantum cascade laser," *Science*, **264**, pp. 553–556, 1994.

[73] K. v. Klitzing, G. Dorda and M. Pepper, "New method for high-accuracy determina-

tion of the fine-structure constant based on quantized Hall resistance". *Phys. Rev. Lett.* **45**, p. 494, 1980.

[74] D. C. Tsui, H. L. Stormer and A. C. Gossard, "Two dimensional magnetotransport in the extreme quantum limit". *Phys. Rev. Lett.* **48**, p. 1559, 1982.

[75] Y. Arakawa and H. Sakaki, "Multidimensional quantum well laser and temperature dependence of its threshold current," *Appl. Phys. Lett.*, **40**, pp. 939–941, 1982.

[76] A. I. Ekimov and A. A. Onushchenko, "Quantum size effect in three dimensional microscopic semiconductor crystals," *JETP Lett.*, **34**, p. 345, 1981.

[77] L. Goldstein, F. Glas, J. Y. Marzin, M. N. Charasse and G. Le Roux, "Growth by molecular beam epitaxy and characterization of InAs/GaAs strained-layer superlattices," *Appl. Phys. Lett.*, **47**, pp. 1099–1101, 1985.

[78] N. N. Ledentsov, M. Grundmann, N. Kirstaedter, J. Christen, R. Heitz, J. Bohrer, F. Heinrichsdorff, D. Bimberg, S. S. Ruvimov, P. Werner, U. Richter, U. Gösele, J. Heydenreich, V. M. Ustinov, A. Yu. Egorov, M. V. Maximov, P. S. Kop'ev and Zh. I. Alferov, "Luminescence and structural properties of (In,Ga)As-GaAs quantum dots," in *Proc. 22nd Int. Conf. on the Physics of Semiconductors*, Vancouver, Canada, 1994, D. J. Lockwood, Ed. (World Scientific, Singapore, 1995), **3**, p. 1855.

[79] Zh. I. Alferov, N. Yu. Gordeev, S. V. Zaitsev, P. S. Kop'ev, I. V. Kochnev, V. V. Khomin, I. L. Krestnikov, N. N. Ledentsov, A. V. Lunev, M. V. Maksimov, S. S. Ruvimov, A. V. Sakharnov, A. F. Tsatsul'nikov, Yu. M. Shernyakov, and D. Bimberg, "A low-threshold injection heterojunction laser based on quantum dots, produced by gas-phase epitaxy from organometallic compounds," *Fiz. Tekh. Poluprovodn.*, **30**, pp. 357–363, 1996 [*Semiconductors*, **30**, pp. 197–200, 1996].

[80] V. A. Shchukin, N. N. Ledentsov, P. S. Kop'ev and D. Bimberg, "Spontaneous ordering of arrays of coherent strained islands," *Phys. Rev. Lett.*, **75**, pp. 2968–2972, 1995.

[81] Zh. I. Alferov, N. A. Bert, A. Yu. Egorov, A. E. Zhukov, P. S. Kop'ev, A. O. Kosogov, I. L. Krestnikov, N. N. Ledentsov, A. V. Lunev, M. V. Maksimov, A. V. Sakharov, V. M. Ustinov, A. F. Tsatsul'nikov, Yu. M. Shernyakov, and D. Bimberg, "An injection heterojunction laser based on arrays of vertically coupled InAs quantum dots in a GaAs matrix," *Fiz. Tekh. Poluprovodn.*, **30**, pp. 351–356, 1996 [*Semiconductors*, **30**, pp. 194–196, 1996].

[82] N. Kirstaedter *et al.*, "Low threshold, large T_0 injection laser emission from (InGa)As quantum dots," *Electron. Lett.*, **30**, pp. 1416–1418, 1994.

[83] G. Park, O. B. Shchekin, D. L. Huffaker, and D. G. Deppe, "Low threshold oxide-confined 1.3 μm quantum dot laser," *IEEE Photon. Technol. Lett.* **33**, p. 230, 2000.

[84] A. E. Zhukov, A. R. Kovsh, S. S. Mikhrin, N. A. Maleev, V. M. Ustinov, D. A. Lifshits, I. S. Tarasov, D. A. Bedarev, M. V. Maximov, A. F. Tsatsul'nikov, I. P. Soshnikov, P. S. Kop'ev, Zh. I. Alferov, N. N. Ledentsov, and D. Bimberg, "3.9 W CW power from submonolayer quantum dot diode laser," *Electron. Lett.*, **35**, pp. 1845–1846, 1999.

[85] L. V. Asryan and R. A. Suris, "Inhomogeneous line broadening and the threshold current density of a semiconductor quantum dot laser," *Semicond. Sci. Tehnol.* **11**, pp. 554–569, 1996.

[86] A. E. Zhukov, V. M. Ustinov, A. Yu. Egorov, A. R. Kovsh, A. F. Tsatsul'nikov, N. N. Ledentsov, S. V. Zaitsev, N. Yu. Gordeev, P. S. Kop'ev, and Zh. I. Alferov, "Negative characteristic temperature of InGaAs quantum dot injection laser," *Jpn. J. Appl. Phys.*, pt. 1, **36**, pp. 4216–4218, 1997.

[87] G. Park, O. B. Shchekin, S. Csutak, D. L. Huffaker, and D. Deppe, "Room-temperature continuous-wave operation of a single-layered 1.3 μm quantum dot laser," *Appl. Phys. Lett.*, **75**, pp. 3267–3269, 1999.

[88] J. A. Lott, N. N. Ledentsov, V. M. Ustinov, A. Yu. Egorov, A. E. Zhukov, P. S. Kop'ev, Zh. I. Alferov, and D. Bimberg, "Vertical cavity lasers based on vertically coupled quantum dots," *Electron. Lett.*, **33**, pp. 1150–1151, 1997.

[89] J. A. Lott, N. N. Ledentsov, V. M. Ustinov, N. A. Maleev, A. E. Zhukov, A. R. Kovsh, M. V. Maximov, B. V. Volovik, Zh. I. Alferov, and D. Bimberg, "InAs-InGaAs quantum dot VCSELs on GaAs substrates emitting at 1.3 μm," *Electron. Lett.*, **36**, pp. 1384–1385, 2000.

HERBERT KROEMER

I was born on August 25, 1928 in Weimar, Germany. My father was a civil servant working for the city administration of my home town; my mother was a classical German "Hausfrau." Both came from simple skilled-craftsmen families. Neither had a high-school education, but there was never any doubt that they wanted to have their children obtain the best education they could afford. My mother, in particular, pushed relentlessly for top performance in school: simply doing well was not enough. Fortunately, I breezed through 12 years of school almost effortlessly, not once requiring help with homework from my parents.

Despite their insistence on excellence, my parents never pushed me in any particular academic direction; I was completely free to follow my inclinations, which ran towards math, physics, and chemistry. When I finally told my parents that I wanted to study physics, my father merely wondered what that is, and whether I could make a living with it. I certainly could become a physics teacher at a High School, or "Gymnasium," a thoroughly respectable profession.

I did have one major problem in school, though: Discipline! I was often bored, and entertained myself in various disruptive ways. A frequent punishment was an entry into the "Klassenbuch," the daily class ledger. These entries were considered a very serious matter, and if I had not been excellent academically, I would have risked being expelled. Once, after I had again been entered as having disturbed the class, the teacher who had overall responsibility for the class – Dr. Edith Richter, whom I adored – asked me in great exasperation: "Why again?" I told her that I had been bored, whereupon she exploded: "Mr. Kroemer, one of the purposes of a higher education is that you learn to be bored gracefully." I will never forget that outburst – nor have I ever really learned to be bored gracefully.

Another teacher – Willibald Wimmer – had his own clever way of handling me. Before the end of the war, he had been an instructor at a local engineering college, ending up teaching math and physics at our high school. He was used to dealing with more mature students, and he treated us as adults. I was way ahead of the curriculum in math, and kept showing off. Worse, I taught some of my classmates math "tricks," that were not part of the curriculum. So, Mr. Wimmer made a "treaty" with me: While he could not excuse me from attending class, I was guaranteed a top grade without being required to turn in the homework assignments, and was permitted to do whatever I wanted to do during the hour, *provided* I kept absolutely quiet – except when explicitly asked to speak up. Both of us kept that treaty.

Mr. Wimmer also became our physics teacher, a subject about which he clearly knew little more than what was in the textbook. Realizing that I was deeply into physics, he simply enlisted me and one other student to help him in lecture preparations, like setting up what apparatus had survived the war. Once I even was asked to present the lecture myself, with him sitting in the front row and enjoying the show. It was a wonderful experience.

Having graduated from the gymnasium in 1947, I was accepted as a physics student at the University of Jena, where I fell under the spell of the great Friedrich Hund, the most brilliant lecturer I ever encountered. The joy did not last long. In early-1948 the political suppression in East Germany became very severe, especially at rebellious universities like Jena. Every week, some of my fellow students had suddenly disappeared, and you never knew whether they had fled to the West, or had ended up in the German branch of Stalin's Gulag, like the uranium mines near the Czech border. During the Berlin airlift, I was in Berlin as a summer student at the Siemens company, and I decided to go West via one of the empty airlift return flights.

From Berlin, I had written to several west German universities for admission, including Göttingen, but did not receive a reply before leaving Berlin (they had turned me down). I followed the advice of one of my Jena professors "why don't you give my greetings to Professor König in Göttingen." König told me that physics admissions were closed, but he passed me on for what was ostensibly just a friendly chat to Professor Richard Becker and his alter-ego assistant, Dr. Günther Leibfried. They in turn passed me on to Wolfgang Paul (Nobel 1989), and I think also to Robert Pohl. It soon dawned on me that this was not just a friendly social chat with people who had nothing better to do, but a thorough examination. I remember one of the questions Paul asked me: "You know that a mirror interchanges left and right? – Then why doesn't it interchange top and bottom?" In the end, I was returned to Becker, who told me that two of the students who had been admitted were not coming, and a meeting was scheduled for the next day to select who would get the two openings. A few days later I received a postcard that I had been accepted.

Post-war Göttingen was – intellectually – a wonderfully stimulating place. I was attracted to one of the younger instructors – "Privatdozent" Dr. Hellwege – who offered a so-called Proseminar, where pre-research students would present papers assigned to them, and I participated in this for several semesters in a row. Once, the famous Fritz Houtermans visited Hellwege, and sat in on several of the presentations, including mine. I presented someone's data that yielded a reasonable straight line on a double-log plot, and proudly claimed a power law for the data. Houtermans was not impressed: "On a double-log plot, my grandmother fits on a straight line." I keep quoting Houtermans' grandmother to my own students. Eventually, I signed up with Hellwege for a Diploma Thesis, which would probably have led to an experimental study of the optical spectra of some rare-earth salts. But Hellwege had a long waiting list, and in the meantime, Professor Fritz Sauter – a refugee who had found a temporary home as a guest in Becker's Institute for Theoretical Physics – offered me a theoretical Diploma Thesis, based on a talk that I had given in one of his seminars. Hellwege suggested that I accept Sauter's offer: "You will be finished with him before you can start with me." So I became a theorist.

The diploma thesis was an extension of a 1939 paper by Shockley on the nature of surface states in one-dimensional potentials. As one of the elaborations, I looked at the interface between two different periodic potentials, which confronted me for the first time with what we would today call the band offsets at heterojunctions.

There was another early encounter with heterojunctions while working under Sauter. We made a field trip to the AEG research laboratories in Belecke, a small town in Westphalia. There, a Dr. Poganski gave a beautiful demonstration that the selenium rectifier was not a Schottky barrier, but a p-n junction between p-type selenium and n-type CdSe, a true heterojunction – although that term did not exist yet. This must have had an at least sub-conscious influence on me: when I later started thinking about heterojunctions in earnest, the question whether such things could actually exist as real devices had an obvious answer: Of course!

While working on my diploma, I gave another colloquium talk under Sauter, reporting on the famous Bardeen/Brattain paper "Physical Principles Involved in Transistor Action" (or some title like that). At the end I made some suggestion about some open questions raised by the authors. Sauter was intrigued and suggested that as a possible Ph.D. topic. Sometime later, he came into my office and told me to stop further work on my Diploma thesis, and to simply write up what I had done so far. When I protested, he insisted that it was time to move on to the real thing, the Ph.D. dissertation.

I had thus come into contact with one of Sauter's strong beliefs, apparently dating back to the tradition of the 20s: that degrees should not be awarded on the basis of having "served time," but were basically certificates that the recipient had proven capable of executing creative work independently, and no longer required supervision. In fact, he clearly preferred quick dissertations. As a result, I received my Ph.D. before my 24th birthday, fast even for a theorist: Wonderful!

The Ph.D. dissertation involved what we would today call *hot-electron effects*, in the collector space-charge layer of the then-new transistor. The idea was simple. Almost nothing was known about the energy band structure of Ge, but someone's theoretical estimates suggested – quite incorrectly – very narrow bands, especially for the valence band. In this case, if the field was strong enough, any holes in the valence might undergo what we now call Bloch oscillations. A few lines of algebra suggested that, for a given current density, the traveling hole concentration would increase with increasing field ("Staueffekt"), leading to strong space charge effects. The influence of these space charges on the current-voltage characteristics of point contact diodes and transistors formed the main body of the dissertation.

My algebra also implied a decrease of electron drift velocity with increasing field, implying a negative differential conductivity. Knowing nothing about electrical circuit theory, I was unaware how useful such a phenomenon could be, until Shockley pointed it out to me in a personal discussion two years later.

But it became clear soon that my dissertation was unrelated to reality. My assumptions about the band structure and about an energy-independent mean free path had been invalid, and after the discovery of avalanche breakdown it became obvious that the huge fields required for Bloch oscillations in a *bulk* semiconductor could never be reached. Twenty years later, after the pathbreaking work of Esaki and Tsu on negative differential conductivity in superlattices, I realized that I had in fact anticipated their basic physics, albeit in a more primitive form: What was not possible in bulk semiconductors, appeared to become possible in superlattices with their much longer period.

Back to Sauter. He was not interested in closely supervising his students; he simply watched what they were doing on their own initiative. Still, he had a tremendous influence on me in matters of methodology. Whenever I came to him with a pure physics idea, he would invariably say, with slight sarcasm: "But Mr. Kroemer, you ought to be able to formulate this mathematically!" If I came to him with a math formulation, I would get, in a similar tone: "But Mr. Kroemer, that is just math, what is the physics?" After a few encounters of this kind, you got the idea: You had to be able to go back and forth with ease. Yet, in the last analysis, concepts took priority over formalism, the latter was simply an (indispensable) means to an end.

This set of priorities clearly showed, and it had a profound influence on me. As a student of Sommerfeld, Sauter was a superb mathematician himself. But he detested it when people were showing off their math skills by using math that was more advanced than necessary for the problem at hand. To the contrary: You were expected to show how simple you could make it. Because he was a great expert on Bessel functions, I once felt compelled to put, into the draft of my dissertation, an ad-hoc problem that required Bessel functions. He was not amused: "This has no business here; you just put it in to impress me. Take it out!"

Richard Becker had exactly the same attitude (the two were close friends), and I later encountered it again in Shockley. Under influences such as these, I never developed into a "hard-core Theorist with a capital T," but became basically a conceptualist who remained acutely aware of his limitations as a formalist, and whose personal role model was Niels Bohr more than anybody else amongst the Greats of Physics.

The German 1952 job market for theoretical physicists was all but nonexistent. New university positions were not created, and there were plenty of more senior people waiting to occupy any vacancies that might open up. So I never even considered a university career. The situation in industry was hardly any better. As luck would have it, the small semiconductor research group at the Central Telecommunications Laboratory (FTZ) of the German postal service was looking for a "house theorist" who knew semiconductor theory, and I got the job. My duties were simple. I had to be available for whatever theoretical questions anybody had, and also take an active role by poking my nose into the work of my experimentalist and technologist colleagues, to look on my own for topics to which I could contribute – provided I would never touch any equipment. Every week or two, I had to give a talk of 1 to 2 hours to the group, on any subject of my choosing of which I thought that the group should be taught about it. Other than that, I was left completely free to pick whatever problems I felt were worth tackling. So I had become a "professor" of sorts after all, teaching a small but highly motivated "class." From day-1 I was forced to learn to communicate, not with other theorists, but with experimentalists and technologists. It was a fascinating challenge, with a range of topics far beyond what I myself had learned in Göttingen, very often going beyond physics, into metallurgy, chemistry, and electrical engineering.

Of course I ceased to be a "real" theoretical physicist – if I ever was one. Call me an Applied Theorist if you want. However, the awareness of doing something truly useful helped overcome the uneasy feelings over ending a theorist career as soon as it had begun. By hindsight, maybe it wasn't such a bad career move after all!

As my research topic at the FTZ, I picked the problem of the severe frequency limitations of the new transistors – and what one might be able to do about them. It was this problem that led directly to heterostructure ideas. In a 1954 publication of mine there are a couple of paragraphs outlining in a rudimentary form the first ideas for what was later to be called the heterostructure bipolar transistor, or HBT. I proposed both a transistor with a graded gap throughout the base, and the simpler form of just a wide-gap emitter. The rest is history. This history is described in some detail in my Nobel Lecture, so I will give here only the highlights.

Some time after joining RCA Laboratories in Princeton, NJ, in 1954, I returned to heterojunctions. I actually tried – unsuccessfully – to build some HBTs with a Ge/Si alloy emitter on a Ge base. But my principal contributions to the field were two theoretical papers. One of these, in the *RCA Review*, is essentially unknown to this day, but it clearly spelled out the concept of *quasi-electric fields*, which I considered the fundamental design principle for all heterostructures.

The final step came in 1963, while I worked at Varian Associates in Palo Alto, CA. A colleague – Dr. Sol Miller – gave a research colloquium on the new semiconductor diode laser. He reported that experts had concluded that it was fundamentally impossible to achieve a steady-state population inversion at room temperature, because the injected carriers would diffuse out at the opposite side of the junction too rapidly. I immediately protested: "But that's a pile of ... ; all you have to do is give the outer regions a wider energy gap." I wrote up the idea and submitted the paper to *Applied Physics Letters*, where it was rejected. I was talked into not fighting the rejection, but to submit it to the *Proceedings of the IEEE*, where it was published, but ignored. I also wrote a patent, which is probably a better paper than the one in *Proc. IEEE.*

Then came the final irony: I was refused resources to work on the new kind of laser, on the grounds that there could not possibly be any applications for it. By a coincidence, the Gunn effect had just been discovered, and having a long-standing interest in hot-electron negative-resistance effects, I worked on the Gunn effect for the next ten years, and did not participate in the final technological realization of the laser.

I left Varian in 1966, and in 1968 joined the University of Colorado. There I eventually returned to heterostructures, and in the early-70s tackled the theory of band offsets together with my student Bill Frensley – now at UT Dallas – who worked out the first ab-initio theory of the band offsets. Shortly afterwards – now at UCSB – I developed a powerful method to determine band offsets experimentally, by capacitance-voltage profiling *through* the hetero-interface.

In the late-70s, I returned to the device that had started it all, the HBT. The technology developments that had made possible the DH laser offered great promise also for the HBT, and I became a strong advocate of developing the full potential of that device.

In addition to heterostructures, I have worked on numerous other semiconductor topics, be it in physics, materials, devices, or technology. Second only to heterostructures has been a continuing interest in hot-electron negative-resistance effects, dating back to my Ph.D. dissertation. I already mentioned the work on the Gunn effect, but there was more. During my RCA years, I had come up with a crazy scheme to obtain a negative resistance *perpendicular* to a strong bias field, by drawing on the fact that some of the heavy holes in Ge have negative transverse effective masses – that is, perpendicular to their velocity. Experimentally, it was another failure, but conceptually, I found it extraordinarily stimulating. So did others, and it earned me a great deal of early notoriety. Today, I am back to one of the sins of my youth: to the superlattice Bloch oscillator, an exciting combination of heterostructures and hot electron physics.

At the opposite end from hot electrons has been recent work on superconducting weak links in which a degenerately modulation-doped InAs/AlSb quantum well acts as a ballistic coupling medium between superconducting Nb electrodes. They exhibit some utterly delightful large discrepancies between experiment and accepted theory.

There are numerous additional topics scattered throughout my career. I have basically been an opportunist – and not at all ashamed of it.

QUASI-ELECTRIC FIELDS AND BAND OFFSETS: TEACHING ELECTRONS NEW TRICKS

Nobel Lecture, December 8, 2000

by

HERBERT KROEMER

ECE Department, University of California, Santa Barbara, CA 93106, USA.

I. INTRODUCTION

Heterostructures, as I use the word here, may be defined as heterogeneous semiconductor structures built from two or more different semiconductors, in such a way that the transition region or interface between the different materials plays an essential role in any device action. Often, it may be said that *the interface is the device.*

The participating semiconductors all involve elements from the central portion of the periodic table of the elements (Table I). In the center is silicon, the backbone of modern electronics. Below Si is germanium. Although Ge is rarely used by itself, Ge-Si alloys with a composition-dependent position play an increasingly important role in today's heterostructure technology. In fact, historically this was the first heterostructure device system proposed, although it was also the system that took longest to bring to practical maturity, largely because of the 4 % mismatch between the lattice constants of Si and Ge.

Table I. Central portion of the periodic table of the elements, showing the element from columns II through VI actively used in current heterostructure technology.

II	III	IV	V	VI
	Al	Si	P	S
Zn	Ga	Ge	As	Se
Cd	In		Sb	Te
Hg				

Silicon plays the same central role in electronic metallurgy that steel plays in structural metallurgy. But just as modern structural metallurgy draws on metals other than steel, electronics draws on semiconductors other than silicon, namely, the compound semiconductors. Every element in column III may be combined with every element in column V to form a so-called III–V compound. From the elements shown, twelve different discrete III–V compounds may be formed. The most widely used compound is GaAs – gallium arsenide – but all of them are used in heterostructures, the specific choice depending on the application. In fact, today the III-V compounds are almost always used in heterostructures, rather than in isolation.

Two or more discrete compounds may be used to form alloys. A common example is aluminum-gallium arsenide, $Al_xGa_{1-x}As$, where x is the fraction of column-III sites in the crystal occupied by Al atoms, $1-x$ is occupied by Ga atoms. Hence we have not just 12 discrete compounds, but a continuous range of materials. As a result, it becomes possible to make compositionally graded heterostructures, in which the composition varies continuously rather than abruptly throughout the device structure.

Similar to the III–V compounds, every element shown in column II may be used together with every element in column VI to create II–VI compounds, and again alloying is possible to create a continuous range of the latter.

II. BAND DIAGRAMS AND QUASI-ELECTRIC FORCES

Whenever I teach my semiconductor device physics course, one of the central messages I try to get across early is the importance of energy band diagrams. I often put this in the form of "Kroemer's Lemma of Proven Ignorance":

> If, in discussing a semiconductor problem, you cannot draw an **Energy Band Diagram**, this shows that you don't know what you are talking about,

with the corollary

> If you can draw one, but don't, then your audience won't know what you are talking about.

Nowhere is this more true than in the discussion of heterostructures, and much of the understanding of the latter is based on one's ability to draw their band diagrams – and knowing what they mean.

To illustrate the idea, consider first a homogenous piece of semiconductor, say, a piece of uniformly doped silicon, but with an electric field applied. The band diagram then looks like the top diagram in Fig. 1, consisting simply of two parallel tilted lines representing the conduction and valence band edges. The separation between the two lines is the energy gap of the semiconductor; the slope of the two band edges is the elementary charge e multiplied by the electric field E. When an electron or a hole is placed into this structure, a force $-eE$ is acting on the electron, $+eE$ on the hole; the two forces are equal in magnitude and opposite in direction, their magnitude is the slope of the bands, just the signs differ.

In a heterostructure, the energy gap becomes position-dependent, and the two band edge slopes are no longer equal, hence the two forces are no longer equal in magnitude. It would, for example, be possible to have a force acting only upon one kind of the carriers (Fig. 1b), or to have forces that act in the same direction for both types of carriers (Fig. 1c). Purely electrical forces in homogeneous crystals can never do this. This is why I call these forces "quasi-electric." ***They present a new degree of freedom for the device designer to enable him to obtain effects that are basically impossible to obtain using only "real" electric fields.***

This is the underlying **general design principle** of all heterostructure de-

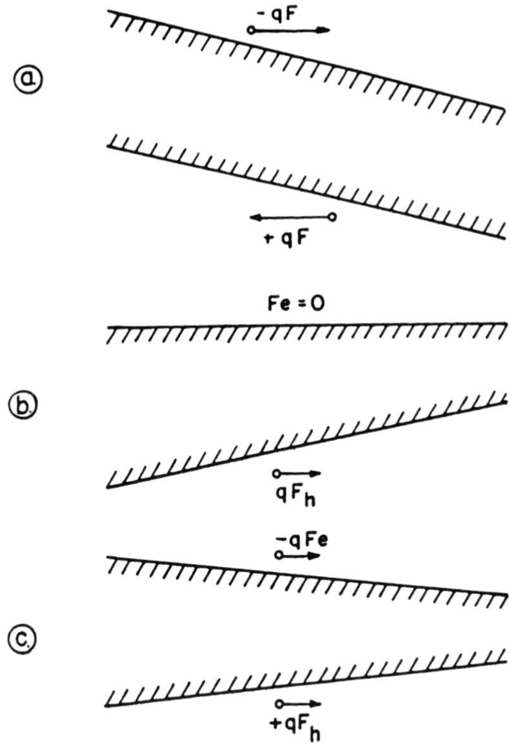

Figure 1. Quasi-Electric Fields: (a) A true electric field simply tilts the bands; (b) quasi-electric fields, with no force on electrons, but a force on holes; (c) quasi-electric fields forcing electrons and holes in the same direction. From Kroemer (1957a).

vices, first spelled out in a 1957 paper of mine (Kroemer, 1957a). In fact, the preceding paragraph is an only slightly edited version of a key paragraph in that paper.

When I wrote those lines, I did not know about Shockley's famous 1951 patent (Shockley), where the possibility of a bipolar transistor with an emitter of wider energy gap is explicitly mentioned. However, the wide-gap emitter idea appears to have been presented principally to cover alternative design possibilities, a procedure typical in patents. The patent gives no indication why such a design would have distinct advantages over a homostructure design, much less a general design principle extending to other kinds of devices. My own formulation might be viewed as a broad generalization of the idea in Shockley's patent. But my point of departure was different: not an *abrupt* energy gap change with accompanying band offset steps, but explicitly a *continuous* energy gap variation of "designable" width, of which the abrupt gap change is simply a limiting case.

Returning to Fig. 1b, it should be emphasized that the zero conduction band slope shown there does not imply a zero electric field. A true electric field is of course present, and it can in principle be determined by the integration of Poisson's equation, provided the local space charge densities are known, often a non-trivial task. But this true field is not part of the band diagram. Nor do the electrons care: The band edge slopes are what matters, not the true electric field. The difference between the two becomes even more drastic in Fig. 1c, where we could not guess even the *direction* of the true field, much less its magnitude.

III. HETEROSTRUCTURE BIPOLAR TRANSISTORS

A. Graded-gap transistor

I had been led to the 1957 principle by a very practical question dating back to 1953/54, when I was working at the telecommunications research laboratory (Fernmeldetechnisches Zentralamt; FTZ) of the German Postal Service: The early bipolar junction transistors were far too slow for practical applications in telecommunications, and I set myself the task of understanding the frequency limitations theoretically – and what to do about them. One approach – not the only one – was to speed up the flow of the minority carriers from the emitter to the collector by incorporating an electric field into the base region. This could be done by using, not a uniform doping in the base, but one that decreased exponentially from the emitter end to the collector end – the so-called *drift transistor* (Krömer, 1953). While working out the details, I realized that

> "... a drift field may also be generated through a variation of the energy gap itself, by making the base region from a non-stoichiometric mixed crystal of different semiconductors with different energy gaps (for example, Ge-Si), with a composition that varies continuously through the base." [Translated from Krömer (1954)]

This was not yet the full general design principle, but it constituted the original conception of what has become known as the heterostructure bipolar transistor (HBT), and ultimately of the heterostructure device field in general.

The appropriate band diagram (Fig. 2) followed in the 1957 paper mentioned earlier, where I gave the 1954 idea as one example of the general design principle. Note that Fig. 2 shows a flat conduction band, as would be the case for a sufficiently heavy uniform doping; the band diagram of Fig. 1b represents essentially the base region of that early concept. The case of Fig. 1c illustrates the generality of the design principle.

Note that the original proposal explicitly gave the Ge-Si system as an example, rather than a III/V compound system. It was to take some four decades until Ge-Si HBTs were finally becoming commercially available, long after devices based on III/V compounds had done so.

B. Wide-gap emitter

The proposed graded-gap base structure was far beyond the technologies then available, a situation that was to remain unchanged for decades. The only possibility one of my colleagues – Mr. Alfons Hähnlein – could envisage

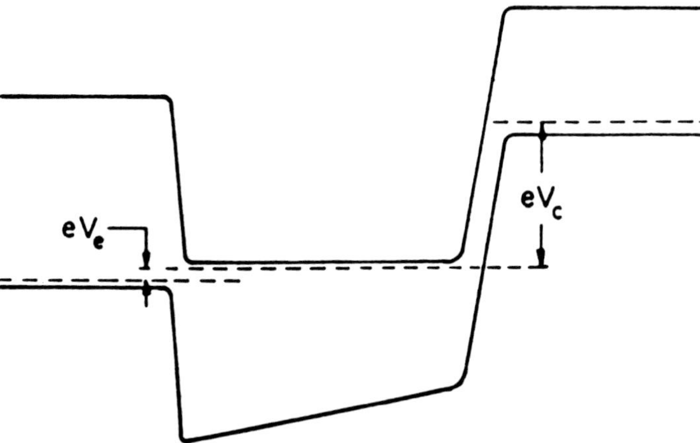

Figure 2. P-n-p transistor with a base region with a graded gap, to speed up minority carrier flow from emitter to collector [from Kroemer (1957a)]. P-n-p transistors were the preferred design for the Ge-based transistors of the mid-50's.

was a design in which the emitter was made from a wider-gap semiconductor than the base, with a quasi-abrupt transition at the interface between the two, leading to a band diagram as in Fig. 3, in essence – but unknowingly – re-inventing Shockley's design.

It was of course obvious that the objective of putting a drift field into the base of the transistor could not be achieved in this way. But on reflecting about what exactly might be the properties of such a structure, I realized that a wide-gap emitter has advantages of its own (Kroemer, 1957b; 1982): One of the problems with all bipolar transistors is minimizing the highly undesirable back-injection of majority carriers from the base (electrons in a p-n-p transis-

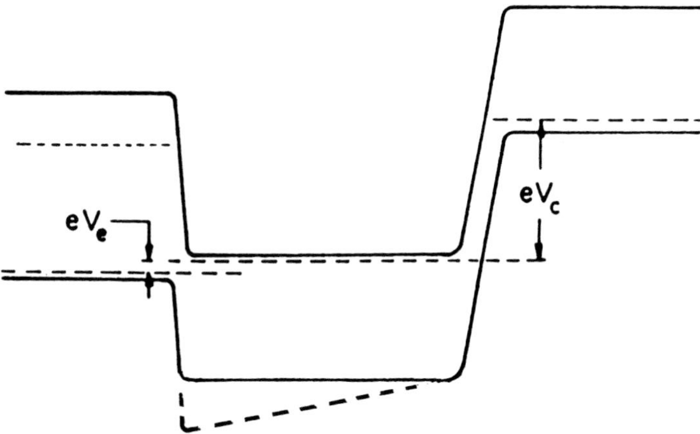

Figure 3. Wide-gap emitter. The energy gap variation has been compressed into a quasi-abrupt transition at the emitter-to-base interface. The base region still has a uniform energy gap without the transport-aiding quasi-field, but there is now a potential barrier for the escape of electrons from the base into the emitter that is larger than the barrier for holes entering the base from the emitter.

tor) into the emitter. In a homojunction transistor, this requirement sharply limits the base doping, which has other undesirable consequences, like a large base access resistance. A wide-gap emitter greatly suppresses this back-injection current: Expressed in terms, not of the quasi-electric forces, but of the associated potentials, any electrons escaping from the base into the emitter must overcome a higher potential barrier than the holes entering the base from the emitter. As a result, the electron escape current density is reduced roughly by a factor $\exp(-\Delta E_G/kT)$, where ΔE_G is the difference in energy gaps. This is very effective: An easily achieved energy gap difference of 0.2eV ($\approx 8kT$) implies a reduction by a factor $e^{-8} \approx 1/3000$.

Given this reduction, it now becomes possible to dope the base much more heavily, to reduce the base resistance. But in the presence of the inevitable junction capacitances, a reduction of base resistance reduces the *RC* time constants of the device, and thereby enhances its speed .

Because of the much greater technological simplicity of the wide-gap emitter design over the graded-base design, it was the wide-gap emitter design that dominated HBT technology until recently, but the highest-performance HBTs now use both approaches (Kroemer, 1983).

C. Follow-up

Because of the absence of any credible technology, I did not follow up the above 1954 ideas until three years later, after I had joined RCA Laboratories in Princeton, NJ. I realized the generality of the design principle outlined above, and wrote the *RCA Review* paper referred to earlier (Kroemer, 1957a). The paper was almost totally ignored, not only because the *RCA Review* was a somewhat obscure journal, but probably even more because I myself somehow never explicitly referred to the paper (nor to its 1954 precursor) in my own subsequent work until about 40 years later (Kroemer, 1996). The general design principle itself was extensively discussed in a 1982 HBT review (Kroemer, 1982), but without reference to the 1954 paper and the 1957 *RCA Review* paper.

The 1957 paper of mine that *is* widely cited was a second paper in that year, which gives a detailed analysis of the wide-gap emitter version of the HBT (Kroemer, 1957b). Having been published in a more visible journal, it drew considerable attention, and stimulated several attempts by others to realize the wide-gap emitter version of the HBT during the '60s. Unfortunately, technology was still not ready, and none of these early attempts led to anything useful. By 1970, people seemed to have largely given up.

While at RCA, I also made an unsuccessful attempt to build a Ge transistor with a Ge-Si alloy emitter, which might be sufficiently amusing (and characteristic of the primitive state of 1957 technology) to be told here (Kroemer, 1957c). The idea was to utilize the fact that the Au-Si phase diagram exhibits a low-melting (370 °C) eutectic. I prepared such a eutectic, smashed the fairly brittle material with a hammer into a coarse powder, placed small grains of the powder onto a Ge chip, and alloyed the combination at a temperature somewhere between 500 °C and 600 °C. The Au-Si alloy would then melt and penetrate into the Ge chip, dissolving some Ge. Upon cooling, a Ge-Si alloy emitter would re-crystallize (Fig. 4). I actually got one or two transistors to work, but as a rule, the large thermal strains generated during the solidification of the eutectic caused the Ge chip to crack. The attempt was sufficiently unsuccessful that I never published the work. It was followed up by Diedrich and Jötten (1961), who knew about my work, but the technology clearly was unpromising, and Si-Ge HBTs had to wait several decades for their practical realization.

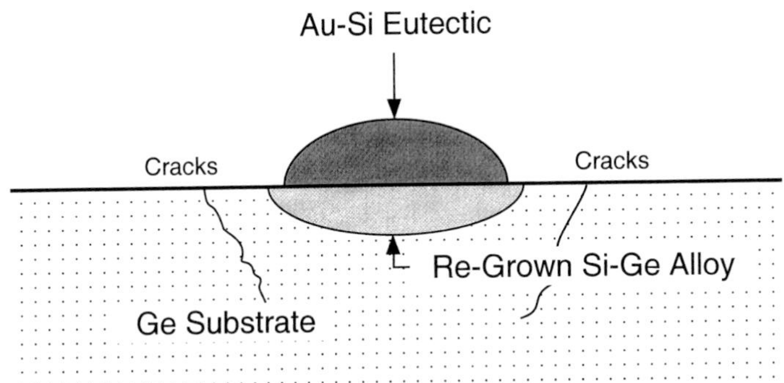

Figure 4. Attempt to realize a Ge transistor with a Ge-Si alloy emitter. A piece of Au-Si eutectic was alloyed into a Ge base, forming a Si-Ge alloy emitter upon cooling. From Kroemer (1957c).

IV. DOUBLE-HETEROSTRUCTURE LASER

Neither the graded-gap HBT nor the wide-gap emitter HBT draw on the full power of the idea expressed in the general design principle that the quasi-electric fields '***enable the device designer to obtain effects that are basically impossible to obtain using only "real" electric fields***.' They represent major improvements, alright, but do they represent something *basically impossible* otherwise?

An example of something that was indeed truly impossible to achieve otherwise emerged abruptly in March 1963. I was working at Varian Associates in Palo Alto at the time, and a colleague of mine – Dr. Sol Miller – had taken a strong interest in the new semiconductor junction lasers that had emerged in 1962, a topic then outside my own range of interests. In a colloquium on the topic he gave a beautiful review of what had been achieved, not failing to point out that successful laser action required either low temperatures or short low-duty-cycle pulses, usually both. Asked what the chances were to achieve continuous operation at room temperature, Miller replied that certain experts had concluded that this was fundamentally impossible.

It is instructive to review this argument here. Consider the (highly oversimplified) energy band diagram of a GaAs p-n junction, heavily doped on both sides, and forward-biased to the point that flatband conditions were reached (Fig. 5). Electrons then diffuse from the n-type side to the p-type side, and holes diffuse in the opposite direction, creating a certain concentration of electron-hole pairs in the junction region proper; their recombination would cause light emission. But in order to obtain *laser* action, a popula-

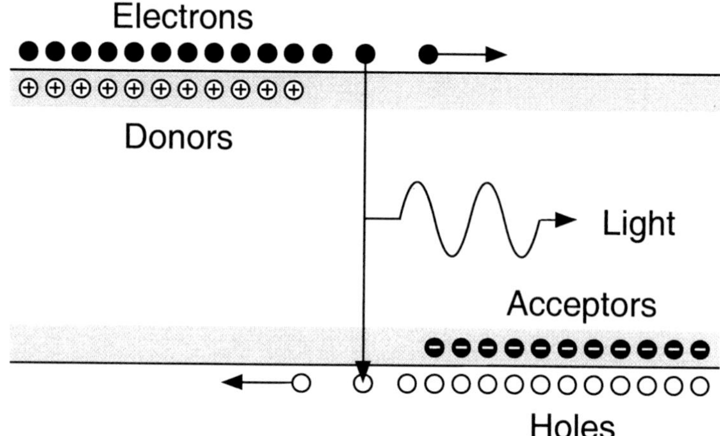

Figure 5. Schematic energy band diagram of a p-n homojunction forward-biased to flatband conditions, creating a high concentration of electron-hole pairs in the vicinity of the junction plane, leading to emission of recombination radiation.

tion *inversion* has to be achieved, which means that, in the active region, the occupation probability of the lowest states in the conduction band has to be higher than that of the highest states in the valence band. A *necessary* condition for such a population inversion is a forward bias larger than the energy gap. But even then, a population inversion is hard to achieve in an ordinary p-n junction. First of all, the electron concentration in the active region will always be lower than in the n-type doped region, with an analogous limitation for the holes. Inversion, therefore, requires degenerate doping on both sides. But even with degenerate doping, both the electrons and holes would diffuse out of the active region immediately into the adjacent oppositely doped region, preventing a population inversion from building up. Increasing the forward bias would not help much, because it would increase the rate of outflow just as much as the rate of injection.

I immediately protested against this argument with words somewhat like "but that is a pile of ..., all one has to do is give the injector regions a wider energy gap ." As is shown in Fig. 6, such a change would cause an electron-repelling quasi-electric field to be present on the p^+ side, and a similar hole-repelling barrier on the n^+ side. Carrier confinement would thus be achieved.

By increasing the forward bias further, potential wells develop for both the electrons and the holes (Fig. 7), with quasi-electric forces on *both* sides pushing *both* electrons and holes towards the active region. As a result, electron and hole concentrations can become much larger than the doping levels in the contact regions, and it becomes readily possible to create the population inversion necessary for laser action. This double-heterostructure (DH) laser finally represented a device truly impossible with only the real electric fields available in homostructures; note that the idea for it arose essentially at the instant I had been made aware that there was a problem.

I wrote up a paper describing the DH idea, along with a patent application.

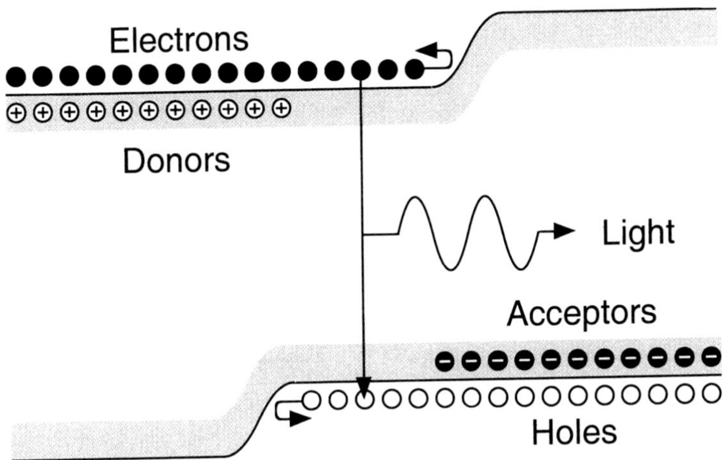

Figure 6. Carrier confinement in a double heterostructure, due to the presence of quasi-electric potential barriers at the ends of the light-emitting active region, preventing the outflow of injected electrons and holes, without interfering with the flow of majority carriers from the injector regions.

The paper was submitted to *Applied Physics Letters*, where it was rejected. I was persuaded not to fight the rejection, but to submit the paper to the *Proceedings of the IEEE* instead, where it was published (Kroemer, 1963) – but largely ignored. Fig. 8 shows the band diagram actually published.

The patent was issued in 1967 (Kroemer, 1967). It is probably a better paper than the *Proc. IEEE* letter. It expired in 1985.

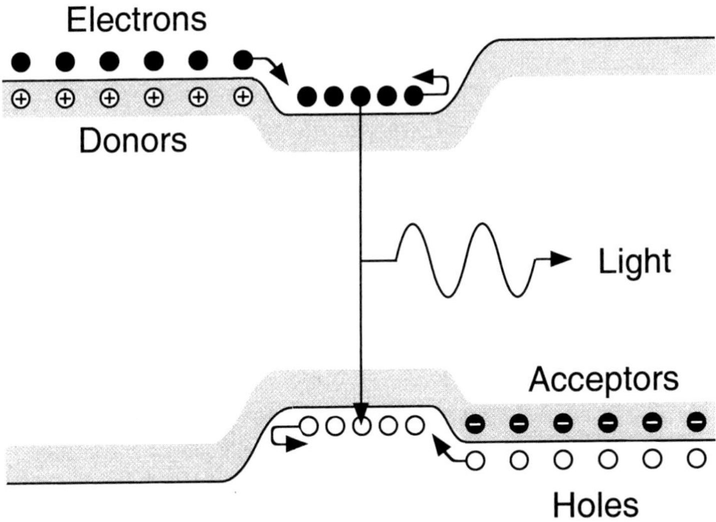

Figure 7. With a further increase of the forward bias, potential wells form for both electrons and holes, which permit the accumulation of the injected carriers to degenerate concentrations much higher than the values in the injector regions.

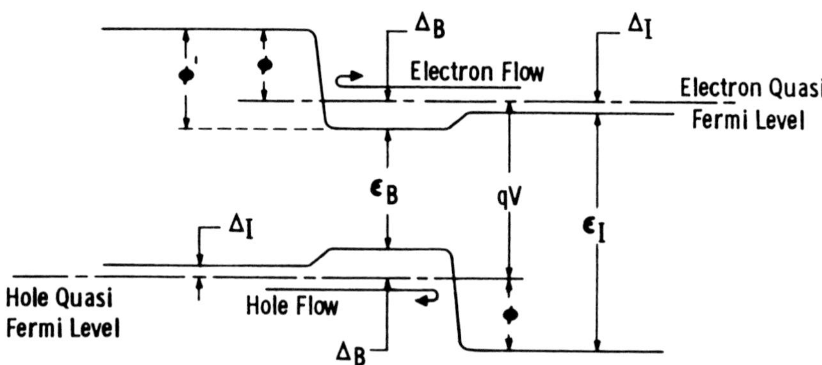

Figure 8. Band diagram of the double-heterostructure laser, as originally published (Kroemer, 1963).

Once again, here was an idea far ahead of any technology to realize it. DH lasers operating continuously at room temperature were finally demonstrated in 1970, first by Alferov et al. (1970), and shortly afterwards by Hayashi *et al.* (1970). For the history of the experimental work, see Alferov (2001); Alferov (1996); Casey and Panish (1978).

For reasons discussed below, I myself was not able to be a participant in the technological realization of the idea. For the next 10 years I worked on research on the Gunn effect, to return to heterostructures in the mid-70s.

V. ON HOW NOT TO JUDGE NEW TECHNOLOGY

When I proposed to develop the technology for the DH laser, I was refused the resources to do so, on the grounds that "this device could not possibly have any practical applications," or words to that effect. By hindsight, it is of course obvious just how wrong this assessment was.

It was really a classical case of judging a fundamentally new technology, not by what new applications it might *create*, but merely by what it might do for already-existing applications. This is extraordinarily short-sighted, but the problem is pervasive, as old as technology itself. The DH laser was simply another example in a long chain of similar examples. Nor will it be the last. I therefore believe it is worthwhile to say a few words about this kind of argument here.

Any detailed look at history provides staggering evidence for what I have called, on another occasion (Kroemer, 1995), the *Lemma of New Technology*:

> The principal applications of any
> sufficiently new and innovative technology always have been
> – and will continue to be –
> applications *created* by that technology.

As a rule, such applications have indeed arisen – the DH laser is just a good recent example – although usually not immediately.

But this means that we must take a long-term look when judging the applications potential of any new technology: It must *not* be judged simply by how it might fit into already existing applications, where the new discovery may have little chance to be used in the face of competition with already-entrenched technology. Dismissing it on the grounds that it has no known applications will only stifle progress towards those applications that *will* grow out of that technology.

I do not think we can realistically predict which new devices and applications may emerge, but I believe we can create an environment encouraging progress, by not always asking immediately what any new science might be good for (and cutting off the funds if no answer full of fanciful promises is forthcoming). In particular, we must educate our funding agencies about this historical fact. This may not be easy, but it is necessary. We must make it an acceptable answer to the quest for applications to defer that answer, and that at the very least a search for applications should be considered a part of the research itself, rather than a result to be promised in advance. Nobody has expressed this last point better than David Mermin in his recent put-down of so-called "strategic research" (Mermin, 1999):

> "I am awaiting the day when people remember the fact that discovery does not work by deciding what you want and then discovering it."

What is *never* acceptable – and what we must refrain from doing – is an attempt to justify the research by promising credibility-stretching mythical improvements in *existing* applications. Most such claims are not likely to be realistic and are easily refuted; they only trigger criticism of just how unrealistic the promises are, thereby discrediting the whole work.

Ultimately, progress in applications is not *deterministic*, but *opportunistic*, exploiting for new applications whatever new science and technology happen to be coming along.

VI. CONSTRAINTS

1. Lattice Matching

Let me now turn to some of the problems in implementing heterostructures.

When two materials with significantly different lattice parameters are grown upon each other, whether graded or not, huge strains rapidly build up with increasing thickness, and eventually misfit dislocations will form, a defect without any redeeming features. As a result, the need for lattice matching is all but obvious. The problem is somewhat less severe in modern structures calling for very thin layers (see below); but even there, the lattice-matched case serves as the conceptual point of departure.

Historically, the importance of lattice matching was recognized almost from the beginning, especially for bipolar devices such as lasers. In my 1967 DH laser patent (Kroemer, 1967), I gave a table listing numerous semiconductors in the order of increasing lattice parameter (see Table II); the accompanying text in the patent called for semiconductor pairs with a lattice mismatch below 0.01Å (≈ 0.2 %) as the most promising ones, indicating a recognition of the stringency of the lattice matching demand. The possibility to achieve lattice matching by alloying was explicitly recognized, though.

Table II. Partial copy of the 1963 table of semiconductors ordered by lattice constant (second column) from ref. (Kroemer, 1967). The third column gives the increase in lattice constant relative to the preceding material. Note that no distinction is made between column-IV elements, the III-V compounds, and the II-VI compounds. Also, the 1963 lattice constant of AlAs was significantly in error: The correct room-temperature value (5.661Å) is actually 0.02Å *larger* than the GaAs value, and the difference is much less at typical crystal growth temperatures. [Only the semiconductors up to ZnSe are shown here; the complete 1963 table can be found in Kroemer (1996)].

Semiconductor	a [Å]	Δa [Å]
ZnS	5.406	
Si	5.428	.022
GaP	5.450	.022
AlP	5.46	.01
AlAs	5.63	.17
GaAs	5.653	.02
Ge	5.658	.005
ZnSe	5.667	.009
...	...	...

Ironically, the 1963 literature value for the lattice constant of AlAs was incorrect. As a result, the GaAs-AlAs pair initially did not seem to meet the proposed stringent criterion, and the known poor stability of (binary) AlAs against oxygen did not help. It took some time to recognize its promise, not so much as a binary material, but as an alloy with GaAs, which greatly reduced the oxidation problems, and reduced the lattice mismatch to a completely negligible level.

A more instructive way to represent the information of Table II, including energy gaps as well, is in terms of what some of us call *The Map of the World,* a display of the energy gaps of semiconductors of interest vs. their lattice constants (Fig. 9), with interconnect lines shown to represent binary alloys.

Much of the reason for the continued dominance of the (Al,Ga)As alloy system in heterostructure studies is precisely the "Great Crystallographic Accident" that AlAs and GaAs have essentially the same lattice parameter. This natural lattice matching means, in particular, that an ideal substrate is readily available for the growth of such heterostructures, namely bulk GaAs, obtainable as high-quality single crystals with low dislocation densities, especially in semi-insulating form. If there remains *one* bad aspect to the (Al,Ga)As system, it is the obnoxious chemical affinity of aluminum to oxygen, the source of many residual defects in (Al,Ga)As. Following a 1983 suggestion by myself (Kroemer, 1983), the use of (Ga,In)P lattice-matched to GaAs has recently drawn some attention as an alternative to (Al,Ga)As, especially in HBTs, for which the band lineups at the (Ga,In)P-GaAs interface are more favorable than those of (Al,Ga)As-GaAs.

A second natural substrate is InP, widely used for both optoelectronic and

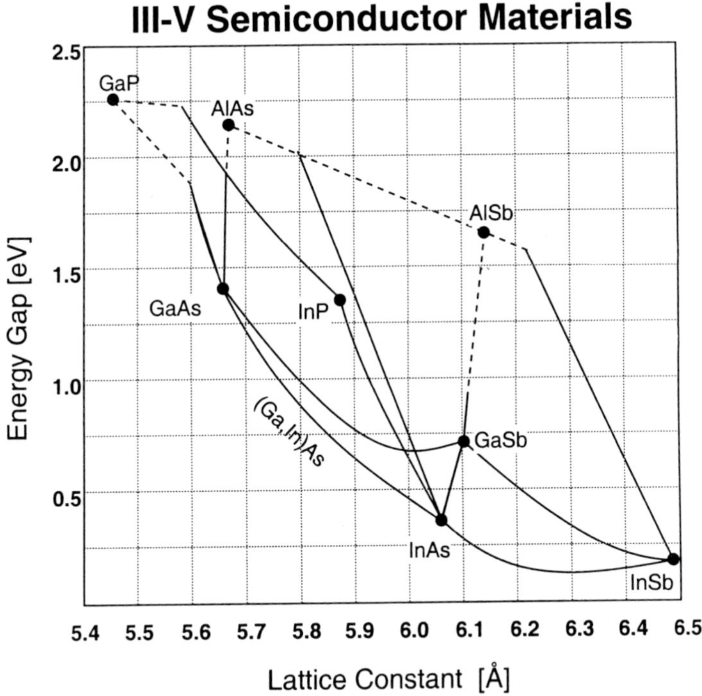

Figure 9. Partial "Map of the World," plotting the energy gap of various III-V compounds vs. lattice constant. The map omits the "Old-World Continents" of the column-IV and the II-VI semiconductors, and the "New World" of the nitrides.

high-speed device applications that call for energy gaps less than that of GaAs. There is no binary III-V compound lattice-matched to InP, but InP is widely used in devices, combined with a wide variety of alloys ranging from (Ga,In)As to Al(As,Sb).

With the emergence of quantum wells, superlattices, and other structures calling for very thin layers, the issue of strain induced by lattice mismatch has lost some of its tyrannical dominance. In sufficiently thin structures, remarkably large strains can be accommodated without dislocation formation, to the point that the modification of the energy band structure of a heterostructure by *deliberate* introduction of strain has become an important device design principle in its own right. The recent evolution of successful Si-Ge HBTs is perhaps the most dramatic triumph of this idea (see, for example, Abstreiter (1996); König (1996), but other examples are close behind, both in field-effect transistors (FETs) and in photonic devices. Some of the recent developments in self-assembling quantum dots are explicitly based on utilizing strain already during the crystal growth process.

2. Valence Matching

If lattice matching were the only constraint, the Ge-GaAs system would be the ideal hetero-system, as was in fact believed by some of us – including myself – in the early-'60s. At that time, the most successful heterojunctions that had been demonstrated were the Ge-on-GaAs heterojunctions studied by Anderson (1960), suggesting a bright future for this system (the term *heterojunction* seems to have appeared first in Anderson's papers). Table II reflects this idea, in the form of combining III–V compounds, II–VI compounds, and group-IV semiconductors into a common table, making the GaAs-Ge system appear to be the most promising candidate It took a few years to realize that this was a blind alley – and why.

It is not a questions of chemical incompatibility, or even of cross-doping effects. Covalent bonds between Ge on the one hand, and Ga or As on the other are readily formed, but they are what I would like to call *valence-mismatched*, meaning that the number of electrons provided by the atoms is not equal to the canonical number of exactly two electrons per covalent bond. Hence the bonds themselves are not electrically neutral, as first pointed out in a 1978 "must-read paper" by Harrison *et al.* (1978).

Consider a hypothetical idealized (001)-oriented interface between Ge and GaAs, with Ge to the left of a mathematical plane, and GaAs to the right (Fig. 10). In GaAs, an As atom brings along 5 electrons (= 5/4 electrons per bond), and expects to be surrounded by 4 Ga atoms, each of which brings along 3 electrons (3/4 per bond), adding up to the correct number of 8/4 = 2 electrons per Ga-As covalent bond. But when, at a (001) interface, an As atom has two Ge atoms as bonding partners, each Ge atom brings along 1 electron per bond, which is one-half electron too many. Loosely speaking, the As atom "does not know" whether it is a constituent of GaAs, or a donor in Ge.

As a result, each Ge-As bond acts as a donor with a fractional charge, and each Ge-Ga bond as an acceptor with the opposite fractional charge. To be electrically neutral, a Ge-GaAs interface would have to have equal numbers of both charges, not only averaged over large distances, but locally. Given chemical bonding preferences, such an arrangement will not occur naturally during epitaxial growth. If only one kind of bonds were present, as in Fig. 10, the interface charge would support an electric field of 4×10^7 V/cm. Such a huge field would force atomic re-arrangements during growth, trying to equalize the number of Ge-As and Ge-Ga bonds. However, these re-arrangements will never go to completion, but will leave behind ill-defined locally fluctuating residual charges, with deleterious consequences for any device application. Interfaces with perfect bond charge cancellation are readily drawn on paper; but in practice there are always going to remain some local deviations from the perfect charge compensation, leading to performance-degrading random potential fluctuations along the interface.

Although Harrison *et al.* discuss only the GaAs-Ge interface, their argument applies to other interfaces combining semiconductors from different columns of the periodic table. In the specific case of compound semiconductor growth on a column-IV elemental semiconductor, the additional problem of antiphase domains on the compound side arises (see, for example, Kroemer (1987)).

The above discussion pertained to the most-widely used (001)-oriented interface. The interface charge at a valence-mismatched interface actually de-

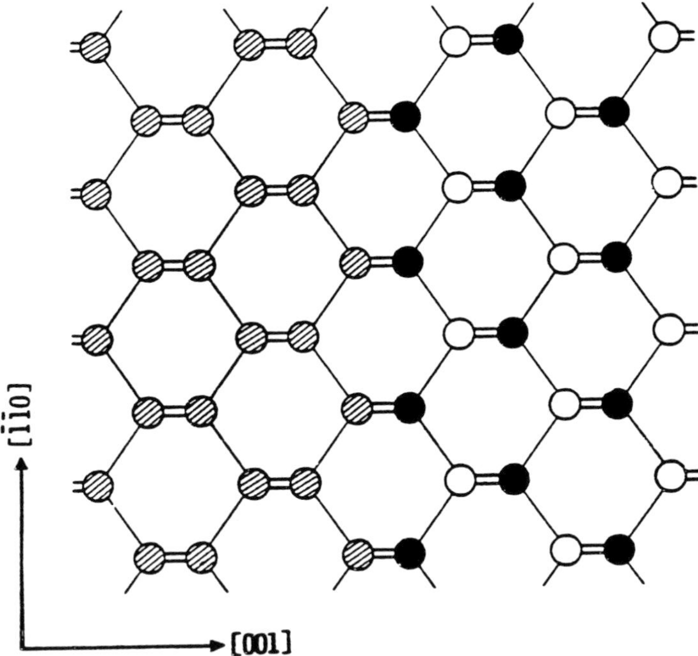

Figure 10. Departure from electrical neutrality at a "mathematically planar" (001)-oriented Ge/GaAs interface. The different atomic species – Ga or As atoms (white and black circles) and Ge atoms (shaded circles) – do not bring along the correct number of electrons to form electrically neutral Ga-Ge or As-Ge covalent bonds of 2 electrons per bond. From Harrison *et al.* (1978).

pends on the crystallographic orientation. It has been shown by Wright *et al.* that an ideal (112) interface exhibits neither an interface charge, nor antiphase domains, and it was in fact possible to demonstrate GaP-on-Si interfaces that had a sufficiently low defect density that they operated as emitters in a GaP-on-Si HBT (Wright *et al.*, 1982; 1984). However, the performance was still sufficiently poor that the approach was not pursued further.

VII. MOLECULAR BEAM EPITAXY AND ABRUPT HETEROSTRUCTURES

The 1970 DH laser demonstration was accomplished by liquid-phase epitaxy (LPE), a beautifully simple technology, but with severe limitations. The big technological breakthrough for heterostructures came only with the emergence of molecular beam epitaxy (MBE) as a practical crystal growth technology, largely pioneered by Al Cho (followed later by organometallic vapor phase epitaxy). In contrast to LPE, MBE permitted combining a wide range of semiconductors, even such hetero-valent combinations as GaP and GaAs on Si. Moreover, it offered a very high degree of control over the local composition, almost on an atomic layer scale. Suddenly, we could realize experimentally almost any band diagram we could draw, at least in the growth direction (lateral control on a similar scale remains an elusive goal to this day). By 1980, the progress in heterostructures had been so large, that I was able to give an invited paper the provocative title "Heterostructures for Everything: Device Principle of the 1980's?" (Kroemer, 1981). It turned out to be an accurate prediction.

In particular, it had become possible to grow almost atomically abrupt heterojunctions. This also meant that two heterojunctions could be placed sufficiently closely together that quantum effects in the space between them became important, and could be utilized for new kinds of devices. The most obvious development was that of quantum wells (QWs), especially for laser applications, which soon became dominated by QW lasers. But we also saw an increasing use of heterostructures in non-bipolar applications, in effect applying the general quasi-electric field design principle outside its range of origin.

One such example is the use of pairs of tunneling barriers in resonant-tunneling diodes, for application as high-frequency sources up into the sub-terahertz frequency range. Another is the idea of Esaki and Tsu to use a periodic heterostructure superlattice as a quasi-bulk negative-resistance medium with an even higher frequency limit (Esaki and Tsu, 1970). It has so far remained an elusive goal, but it continues to be a very active field of research (including by myself).

I would like to single out here a less obvious new concept, that of *modulation doping*, due to Dingle *et al.* (1978). Consider a heterojunction in which only the side with the higher conduction band is doped (Fig. 11). The downward quasi-electric potential step at the interface will cause electrons to drain into the lower conduction band on the other side. Once they are past the range of the quasi-electric potential step associated with the abrupt hetero-interface itself, the electrons still see the ordinary electric field associate with

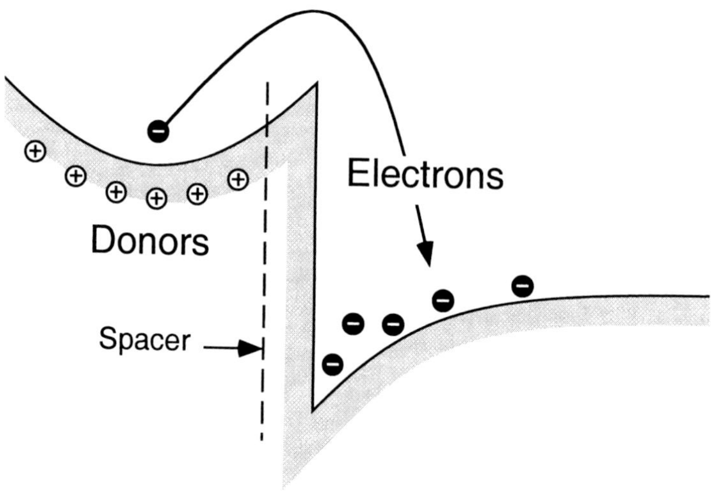

Figure 11. Modulation doping. At an abrupt heterojunction, electrons contributed by donors on the higher-energy side drain onto the lower-energy side, creating a quasi-two-dimensional electron gas there. Because the electrons are now spatially separated from the donors, impurity scattering is reduced, especially if an undoped spacer is inserted on the higher-energy side. The band curvature shown is due to the space charges on the two sides of the interface.

the Coulomb attraction by the donors left behind on the other side. It pulls the electrons towards the interface, creating a 2-dimensional electron gas (2DEG) inside a roughly triangular quantum well. Moreover – and most importantly – because the electrons have been spatially separated from "their" donors, impurity scattering is reduced, and the electron mobility is enhanced. To maximize these benefits, an undoped spacer region is left adjacent to the interface.

The idea had extremely far-reaching consequences, both for devices, and in basic solid-state physics. In devices, it formed the basis of a new class of field effect transistors (FETs), commonly referred to as HEMTs, meaning *H*igh-*E*lectron-*M*obility *T*ransistors (Mimura *et al.*, 1980; Delagebeaudeuf *et al.*, 1980). Their properties are superior to those of earlier classes of FETs. Because of their low noise, they are now used as the sensitive input stage in cellular phones, and thus have contributed to the explosive growth of this aspect of modern information technology.

In basic physics, the suppression of impurity scattering by modulation doping with optimized spacers has permitted the achievement of huge low-temperature mobilities. There is a direct path from the idea of modulation doping to the discovery of the fractional quantum Hall effect, by Tsui, Störmer, and Gossard (Tsui *et al.*, 1982; Stormer, 1999), in 2DEG samples of unprecedented structural perfection grown by Gossard. The subsequent theoretical interpretation of the effect by Laughlin (1999) revealed it as a true fundamental breakthrough in solid-state physics, for which Tsui, Störmer, and Laughlin received the 1998 Nobel Prize in Physics. Unfortunately, the Nobel statute prohibition against dividing the prize amongst more than three individuals excluded Gossard from sharing in the award.

VIII. BAND OFFSETS

In wake of the emergence of MBE technology in the early-70s, my own research returned to heterostructure problems, especially to the problem of band offsets at abrupt heterojunctions. In that limit, the energy band structure makes a discontinuous transition, and exactly how the bands on the two sides are lined up becomes a central question, both experimentally and theoretically. One of the reasons all my early device band diagrams show graded transitions was to sidestep this question of band lineups, of which I was actually well aware.

A. Offset Types

Given two semiconductors, there are evidently three different band lineups possible (Fig. 12)

1. Straddling Lineups

The most common lineup is the straddling one, with conduction and valence band offsets of opposite sign. It is, in essence, the abrupt limit of the graded band structure of Fig. 1c. In quantum wells and superlattices made from such

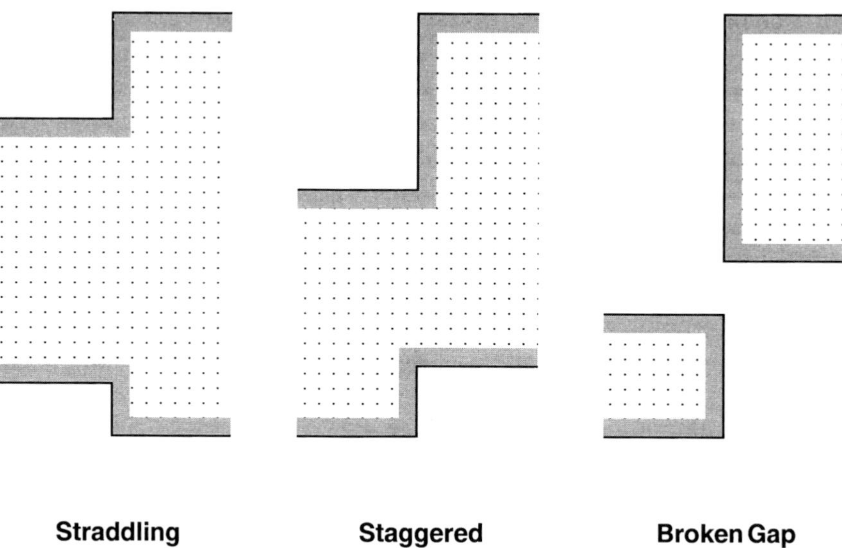

Figure 12. Straddling, staggered, and broken-gap band lineups.

pairs, the lowest conduction band states occur in the same part of the structure as the highest valence band states, which makes these pairs of particular interest for opto-electronic applications, like lasers, which are bipolar kinds of devices, with both electrons and holes involved in the device operation. The two kinds of carriers then occur in the same layers; hence such structures are sometimes referred to as *spatially direct.* Many of today's opto-electronic devices, such as quantum well lasers, are based on such a lineup. The most-widely studied heterojunction system, GaAs-(Al,Ga)As, is of this kind, as are a number of other systems, for example, (Ga,In)As lattice-matched to InP, and (Ga,In)P lattice-matched to GaAs.

2. Staggered Lineups

For some materials pairs, the two bands are shifted in the same direction, leading to a band structure in which the lowest conduction band minimum occurs on one of the sides, the highest valence band maximum on the other, with an energy separation between the two less than the lower of the two bulk gaps. The combination of AlAs-$Al_xGa_{1-x}As$ for $x > 0.3$ is of this kind, as is (Al,In)As lattice-matched to InP; there are several others. In bipolar structures with this lineup, the electrons and holes are confined to *different* layers, hence these structures are *spatially indirect.* Nevertheless, the wave functions overlap at the interface, making radiative recombination possible, with a photon energy less than the narrower of the two gaps (Kroemer and Griffiths, 1983; Caine *et al.*, 1984).

Staggered lineups imply large band offsets in either the conduction or the valence band, and for some applications this property is more important than the spatial indirectness. For example, the conduction band lineup at the InAs-AlSb interface, 1.35eV (Nakagawa *et al.*, 1989), is the highest that has been reported for any III-V system, and several applications are based on this property, along with the low electron effective mass in InAs. The fastest resonant tunneling diode reported in the literature (Brown *et al.*, 1991), oscillating up to 712 GHz, was based on this system.

The high barriers also offer superb electron confinement in FETs, and the possibility of achieving extremely high levels of electron concentration (approaching $10^{13}cm^{-2}$) by modulation doping (i. e., putting the donors into the barriers rather than into the wells), while retaining high mobilities. This combination makes the InAs-AlSb system ideal for investigating the properties of quantum wells in the metallic limit, for example as coupling medium in a new class of superconducting weak links (Kroemer *et al.*, 1994).

3. Broken-Gap Lineup

If a staggered lineup is carried to its extreme, the result is a broken-gap lineup, in which the bottom of the conduction band on one side drops below the top of the valence band on the other. There exists at least one nearly-lattice-matched pair of this kind, InAs-GaSb, with a break in the forbidden gap at the interface on the order of 150 meV (Sakaki *et al.*, 1977).

The broken-gap InAs-GaSb lineup by itself is an exotic lineup, of interest especially to research physicist. To the theorist interested in understanding band offsets, the ability to predict such an offset, at least approximately, is one of the litmus tests of any lineup theory, and recent lineup theories pass this test with flying colors.

B. Theory

It should be self-evident from the above that the question as to the exact values of the band offsets at the various semiconductor pairs of interest is a central one, both theoretically and experimentally. I tried to contribute to both.

At the end of the '60s, the only rule for estimating band offsets theoretically was the *electron affinity rule* (Anderson, 1960), according to which the conduction band offset should be equal to the difference in electron affinity at the two free semiconductor surfaces. In a 1975 paper (Kroemer, 1975), I pointed out that this is an extraordinarily unsatisfactory rule. Even if good electron affinity data were available, the validity of the rule depended on hidden assumptions about the relations between the properties of the interface between two semiconductors, and those of the much more drastic vacuum-to-semiconductor interfaces, assumptions that almost certainly were invalid. Harrison aptly characterized the rule by saying that it "replaces one simple problem by two very difficult problems." (Harrison, 1977)

I called for a theory that would determine the band offsets from the *bulk* properties of the participating semiconductors, and I suggested it as a Ph. D. topic to Bill Frensley (now at the University of Texas in Dallas). One of the specific question I asked Bill to look into was whether broken-gap lineups might in fact occur. The resulting theory (Frensley and Kroemer, 1976; 1977), based on pseudopotentials, was the first to give a semi-quantitative derivation, from bulk properties, not only of band offsets that were already known, like

GaAs/AlAs; it also had a considerable predictive value. In particular, the theory predicted that the InAs/GaSb heterojunction either had a broken-gap lineup, or came very close to it.

The Frensley-Kroemer theory has since then been followed by the work of others based on different principles; see Harrison (1977) and Christensen (1988).

C. Band Offsets by C-V Profiling

Sometime in 1979, Jim Harris (then at the Rockwell Science Center, now at Stanford) showed me some capacitance-voltage (*C-V*) profiling data on an LPE-grown (Al,Ga)As/GaAs heterojunction. *C-V* profiling is a common technique to determine electron concentrations in semiconductors by measuring the capacitance of a reverse-biased Schottky barrier placed upon the surface of the semiconductor. By varying the bias, one can explore the depth distribution of the electrons over some distance. Near the hetero-interface, Harris' data showed a clear indication of an electron accumulation on the GaAs side, and an electron depletion on the (Al,Ga)As side, as one would expect from an appropriate band diagram. However, the apparent electron concentration was strongly smeared out by averaging over a Debye length. When I tried to understand the averaging process quantitatively, I realized that the dipole moment associated with the accumulation/depletion pair should be preserved during the averaging, and that its measurement should permit a determination of the conduction band offset (Kroemer *et al.*, 1980; Kroemer and Chien, 1981; Kroemer, 1985). The analysis yielded a band offset of approximately 66 % of the energy gap difference (Kroemer *et al.*, 1980), not far from today's generally accepted value of 62%.

The *C-V* technique has since then been used by many others and has provided some of the best data for band offsets for many heterojunction pairs.

IX. EPILOGUE

Throughout this paper, I have concentrated on my own work towards heterostructures, especially on the early parts of it, through 1963, which were dominated by bipolar device concepts. But today's heterostructure field would not be what it is without the subsequent contributions – technological or conceptual – by numerous others, especially on non-bipolar structures. It was only through this work of numerous others, on topics that went beyond my own contributions, that the significance of the latter eventually emerged. For this I owe all of them my thanks.

X. REFERENCES

Abstreiter, G., 1996, Physica Scripta T68, 68.

Alferov, Z. I., V. M. Andreev, D. Z. Garbuzov, Y. V. Zhilyaev, E. P. Morozov, E. L. Portnoi and V. G. Trofim, 1970, Fiz. Tekh. Poluprovodn. **4**, 1826. [Sov. Phys. - Semicond. 4, 1573-1575 (1971)].

Alferov, Z. I., 1996, Physica Scripta **T68**, 32.

Alferov, Z. I., 2001, this volume.
Anderson, R. L., 1960, IBM J. Res. Dev. **4**, 283.
Brown, E. R., J. R. Söderström, C. D. Parker, L. J. Mahoney, K. M. Molvar and T. C. McGill, 1991, Appl. Phys. Lett. **58**, 2291.
Caine, E. J., S. Subbanna, H. Kroemer, J. L. Merz and A. Y. Cho, 1984, Appl. Phys. Lett. **45**, 1123.
Casey, C. and M. Panish, 1978, *Heterostructure Lasers – Part A: Fundamental Principles* (Academic Press, New York). See Sec. 1.2.
Christensen, N. E., 1988, Phys. Rev. B **38**, 12687.
Delagebeaudeuf, D., P. Delescluse, P. Etienne, M. Laviron, J. Chaplart and N. T. Linh, 1980, Electron. Lett. **16**, 667.
Diedrich, H. and K. Jötten, 1961, Procs. of *Colloque international sur les dispositifs à semiconducteurs*, Paris (Editions Chiron, Paris) p. 330.
Dingle, R., H. L. Störmer, A. C. Gossard and W. Wiegmann, 1978, Appl. Phys. Lett. **33**, 665.
Esaki, L. and R. Tsu, 1970, IBM J. Res. Dev. **14**, 61.
Frensley, W. R. and H. Kroemer, 1976, J. Vac. Sci. Technol. **13**, 810.
Frensley, W. R. and H. Kroemer, 1977, Phys. Rev. B **16**, 2642.
Harrison, W. A., 1977, J. Vac. Sci. Technol. **14**, 1016.
Harrison, W. A., E. A. Kraut, J. R. Waldrop and R. W. Grant, 1978, Phys. Rev. B **18**, 4402.
Hayashi, I., M. B. Panish, P. W. Foyt and S. Sumski, 1970, Appl. Phys. Lett. **17**, 109.
König, U., 1996, Physica Scripta **T68**, 90.
Kroemer, H., 1957a, RCA Review **18**, 332. (Re-printed from the Proceedings of the Symposium *"The Role of Solid State Phenomena in Electric Circuits,"* Polytechnic Institute of Brooklyn, April 1957, p. 143)
Kroemer, H., 1957b, Proc. IRE **45**, 1535.
Kroemer, H., 1957c, unpublished.
Kroemer, H., 1963, Proc. IEEE **51**, 1782.
Kroemer, H., 1967, US patent 3,309,553 (filed Aug. 16, 1963).
Kroemer, H., 1975, Crit. Revs. Solid State Sci. **5**, 555.
Kroemer, H., W.-Y. Chien, J. S. Harris and D. D. Edwall, 1980, Appl. Phys. Lett. **36**, 295.
Kroemer, H., 1981, Jpn. J. Appl. Phys. Supplem. **20-1**, 9.
Kroemer, H. and W.-Y. Chien, 1981, Solid-State Electron. **24**, 655.
Kroemer, H., 1982, Proc. IEEE **70**, 13.
Kroemer, H., 1983, J. Vac. Sci. Technol. B **1**, 126.
Kroemer, H. and G. Griffiths, 1983, IEEE Elect. Dev. Lett. **4**, 20.
Kroemer, H., 1985, Appl. Phys. Lett. **46**, 494.
Kroemer, H., 1987, J. Cryst. Growth **81**, 193.
Kroemer, H., C. Nguyen, E. L. Hu, E. L. Yuh, M. Thomas and K. C. Wong, 1994, Physica B **203**, 298.
Kroemer, H., 1995, Procs. of *NATO Adv. Res. Wkshp. on Future Trends in Microelectronics*, Ile de Bendor, France, edited by S. Luryi et al., NATO ASI Series E **323** (Kluwer, Dordrecht) p. 1.
Kroemer, H., 1996, Physica Scripta **T68**, 10.
Krömer, H., 1953, Naturwissensch. **40**, 578.
Krömer, H., 1954, Archiv d. Elekt. Übertragung **8**, 499.
Laughlin, R. B., 1999, Revs. Mod. Phys. **71**, 863.
Mermin, D., 1999 (Aug.), Physics Today **52** (8), 11.
Mimura, T., S. Hiyamizu, T. Fujii and K. Nanbu, 1980, Jpn. J. Appl. Phys. **19**, L225.
Nakagawa, A., H. Kroemer and J. H. English, 1989, Appl. Phys. Lett. **54**, 1893.
Sakaki, H., L. L. Chang, R. Ludeke, C. A. Chang, G. A. Sai-Halasz and L. Esaki, 1977, Appl. Phys. Lett. **31**, 211.
Shockley, W., 1951, US patent 2,569,347 (filed 26 June 1948).
Stormer, H. L., 1999, Revs. Mod. Phys. **71**, 875.
Tsui, D. C., H. L. Störmer and A. C. Gossard, 1982, Phys. Rev. Lett. **48**, 1559.
Wright, S. L., M. Inada and H. Kroemer, 1982, J. Vac. Sci. Technol. **21**, 534.
Wright, S. L., H. Kroemer and M. Inada, 1984, J. Appl. Phys. **55**, 2916.

Jack S Kilby

JACK S. KILBY

The Nobel Committee has asked me to discuss my life story, so I guess I should begin at the beginning.

I was born in 1923 in Great Bend, Kansas, which got its name because the town was built at the spot where the Arkansas River bends in the middle of the state. I grew up among the industrious descendents of the western settlers of the American Great Plains.

My father ran a small electric company that had customers scattered across the rural western part of Kansas. While I was in high school, a huge ice storm knocked down most of the poles that carried the telephone and electric power lines. My father worked with amateur radio operators to communicate with areas where customers had lost their power and phone service.

My dad's goal was to do whatever it took to run his business and to help people, but I thought that amateur radio was a fascinating subject. It sparked my interest in electronics, and that's when I decided that this field was something I wanted to pursue.

I also was a fan of broadcast radio. In the 1940s, I especially enjoyed listening to Big Band music. Even today, there's a radio station in Dallas that plays this kind of music, and with a little luck, I don't have to listen to much else.

After high school, I studied electrical engineering at the University of Illinois. Most of my classes were in electrical power, but because of my childhood interest in electronics, I also took some vacuum tube and engineering physics classes.

I graduated in 1947, just one year before Bell Labs announced the invention of the transistor. It meant that my vacuum-tube classes were about to become obsolete, but it offered great opportunities to put my physics studies to good use.

In line with the interests that occupied my thoughts in Great Bend, I hired on with an electronics manufacturer in Milwaukee, Wisconsin, that made parts for radios, televisions and hearing aids.

While in Milwaukee, I took evening classes at the University of Wisconsin towards a master's degree in electrical engineering. Working and going to school at the same time presents some challenges, but it can be done, and its well worth the effort.

In 1958, my wife and I moved to Dallas, Texas, when I took a job with Texas Instruments. TI was the only company that agreed to let me work on electronic component miniaturization more or less full time, and it turned out to be a great fit.

After proving that integrated circuits were possible, I headed teams that built the first military systems and the first computer incorporating integrated circuits. I also worked on teams that invented the handheld calculator and the thermal printer, which was used in portable data terminals.

In 1970, I took a leave of absence from TI to do some independent work. While on leave, one of the things I worked on was how to apply silicon technology to help generate electrical power from sunlight.

From 1978 to 1984, I spent much of my time as a Distinguished Professor of Electrical Engineering at Texas A&M University. The "distinguished" part is in the eye of the beholder, and I really didn't do much "professing." However, I did have a rewarding time doing research and working with students and faculty on various projects.

I officially retired from TI in the 1980s, but I have maintained a significant involvement with the company that continues to this day.

Along the way, I've been honored to receive awards such as the National Medal of Science and to be inducted into the National Inventors Hall of Fame. Seeing your name alongside the likes of Henry Ford, Thomas Edison, and the Wright Brothers is a very humbling experience, and I appreciate these and the other honors very much.

Receiving the Nobel Prize in Physics was a completely unexpected, yet very pleasant surprise. I had to start my pot of coffee very early the morning I received the news that I had been chosen.

It's gratifying to see the committee recognize applied physics, since the award is typically given for basic research. I do think there's a symbiosis as the application of basic research often provides tools that then enhance the process of basic research. Certainly, the integrated circuit is a good example of that. Whether the research is applied or basic, we all "stand upon the shoulders of giants," as Isaac Newton said. I'm grateful to the innovative thinkers who came before me, and I admire the innovators who have followed.

Four decades of hindsight is perhaps a unique experience among those who have been awarded the Nobel Prize in Physics. As I noted in my lecture, there were various efforts to solve the electronic miniaturization problem at the time I invented the integrated circuit. Humankind eventually would have solved the matter, but I had the fortunate experience of being the first person with the right idea and the right resources available at the right time in history.

I would like to mention another right person at the right time, namely Robert Noyce, a contemporary of mine who worked at Fairchild Semiconductor. While Robert and I followed our own paths, we worked hard together to achieve commercial acceptance for integrated circuits. If he were still living, I have no doubt we would have shared this prize.

Now that I am retired, I still occasionally consult on various industry and government projects, mostly in the area of semiconductors. I also serve on the board of directors of a company or two.

People often ask me what I'm proud of, and, of course, the integrated circuit is at the top of the list. I'm also proud of my wonderful family. I have two daughters and five granddaughters, so you could say that the Kilbys specialized in girls.

I've reached the age where young people frequently ask for my advice. All I can really say is that electronics is a fascinating field that I continue to find fulfilling. The field is still growing rapidly, and the opportunities that are ahead are at least as great as they were when I graduated from college. My advice is to get involved and get started.

TURNING POTENTIAL INTO REALITIES: THE INVENTION OF THE INTEGRATED CIRCUIT

Nobel Lecture, December 8, 2000

by

JACK S. KILBY

Texas Instruments Incorporated, 12500 TI Boulevard, Dallas, TX 75243-4136, USA.

Thank you for that kind introduction and welcome. It is an honor to share the Nobel Prize in Physics and to speak with you today. My title for this lecture is "Turning Potential into Realities: The Invention of the Integrated Circuit."

The field of electronics had strong potential when I invented the integrated circuit in 1958. The reality of what people have done with integrated circuits has gone far beyond what anyone – including myself – imagined possible at the time.

Charles Townes won the Nobel for his work with laser technology, and he summed up how I feel. Townes said, "It's like the beaver told the rabbit as they stared at the Hoover Dam. 'No, I didn't build it myself. But it's based on an idea of mine!'"

At its most basic level, the integrated circuit manipulates the characteristics of electricity. In fact, the entire field of electronics is about manipulating the flow of electrons – in other words, making electricity perform advanced kinds of work. A long line of people have been involved in harnessing electricity, from William Gilbert, who first used the word "electricity" 400 years ago, to Thomas Edison, who built one of the first practical electrical generators in the 1880s.

The invention of the vacuum tube launched the electronics industry. Sometimes called "electron tubes," these devices controlled the flow of electrons in a vacuum. They initially were used to amplify signals for radio and other audio devices. This allowed broadcast radio to reach the masses in the 1920s. Vacuum tubes steadily spread into other devices, and the first tube was used as a switch in calculating machines in 1939.

By the end of World War II, it was obvious that the cost, bulk and reliability of vacuum tubes would limit commercial and military electronic systems. A naval destroyer built in 1937 had 60 vacuum tubes on board. Even the B-29, probably the most complex equipment used in the war, had only around 300 vacuum tubes. Most of those were single-function tubes.

Integration was a low priority at that point. However, electronic systems rapidly grew more complex. The post-war destroyers of 1952 had more than 3,200 vacuum tubes. And, of course, the ENIAC computer of 1946 was a vacuum tube monster. It was the world's fastest computer for years, but it con-

Figure 1. Vacuum tube.

tained more than 17,000 vacuum tubes, weighed 60,000 pounds, occupied 16,200 cubic feet, and consumed 174 kilowatts of electricity – equivalent to 233 horsepower.

Vacuum tubes clearly could not support any significant evolution of computers. People were able to visualize and design systems that – if realized with the prevailing technology – would be too big, too heavy, consume too much power and simply get too hot to work. And because these components would have to be assembled from tens or hundreds of thousands of individual parts, they would have been unreliable and unaffordable[1].

By the mid-1950s, these problems were visible to many in the industry, and many researchers began seeking a solution. Military applications played a major role in the development of electronics. I think the Nobel Peace Prize is important, because it honors people who work to bring peace. And peace means we can use electronics to benefit mankind rather than wage war.

Having said that, the early proponents of miniaturized electronic systems were the U.S. military and space agencies. Most of the approaches tended to make individual components smaller, so collectively these were called miniaturization programs. Each had a common objective – to build complete electronic circuits.

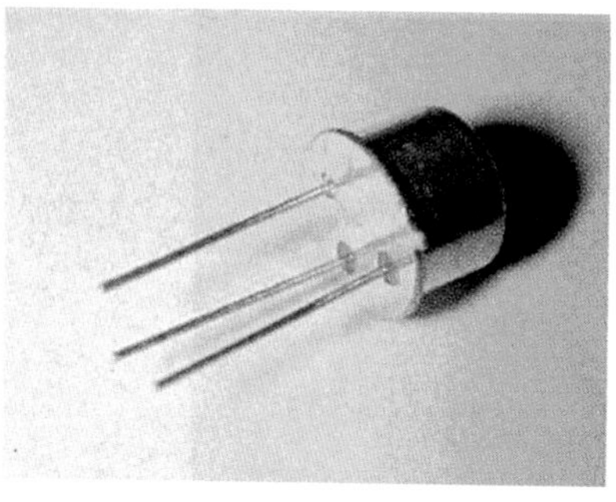
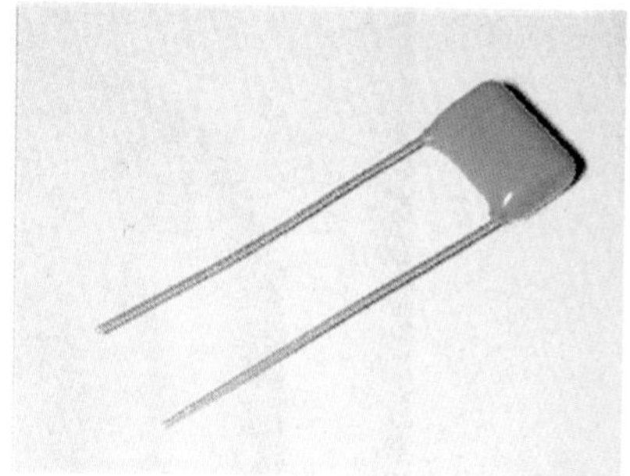
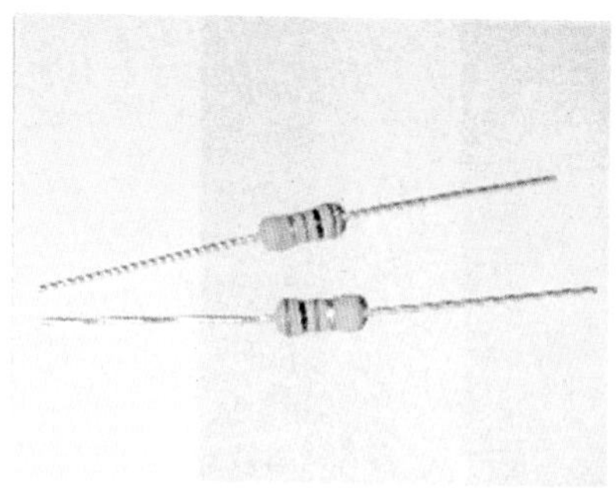
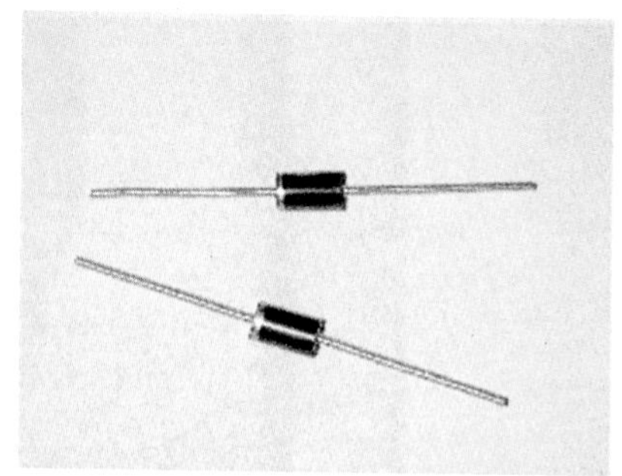

Figure 2. Circuit components.

T.R. Reid has pointed out that building a circuit is like building a sentence[2]. There are certain standard components – nouns, verbs and adjectives in a sentence, and resistors, capacitors, transistors and diodes in a circuit. Each has its own function By connecting the components in different ways, you can get sentences, or circuits, that perform in different ways. Over the years, each component had developed a specialized set of manufacturing processes.

A huge step forward came in 1948, when Bell Labs unveiled the transistor. And in creating solid-state components, the race for miniaturized electronic circuits intensified.

Transistors rely on the nearly free travel of electrons through crystalline solids known as semiconductors. Semiconductors, such as silicon or germanium, have electrical properties that are between conductors and insulators. Not surprisingly, that's why they're called "semi" conductors.

We can change the electrical characteristics of semiconductors through a process called "doping," which carefully adds impurities to predetermined areas[3]. Starting with pure silicon, one dopant will add electrons to a particular area. A different dopant will take electrons away to create holes in an area. In either case, we increase the conductivity of the basic material.

Semiconductor material that conducts by free electrons is known as n-type material. Semiconductors that conduct through electron deficiencies are called p-type material. At its simplest level, a transistor is a "sandwich" of these materials, and the components are frequently encased in ceramic.

As with most technologies, the techniques for making transistors made

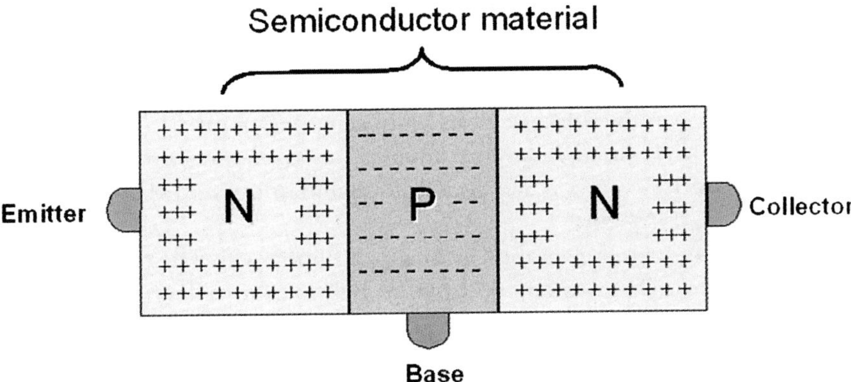

Figure 3. Creating a transistor.

steady progress. The first transistors were known as point contact. They were very primitive by subsequent standards, but they did provide the initial breakthrough.

Later techniques included the grown-junction transistor – pictured here, and which will crop up again later in the lecture. In these transistors, a single crystal of germanium was grown and doped at the same time.

The crystal was pulled from a melt containing n-type impurities, and then p-type impurities were added to the molten germanium. Later, more n-type was added. This created an n-doped crystal with a thin p-type layer within it. The crystal was then cut into small blocks, and each block formed a single NPN transistor.

Subsequent developments included grown-diffused, alloy, surface barriers and eventually the mesa and planar types.

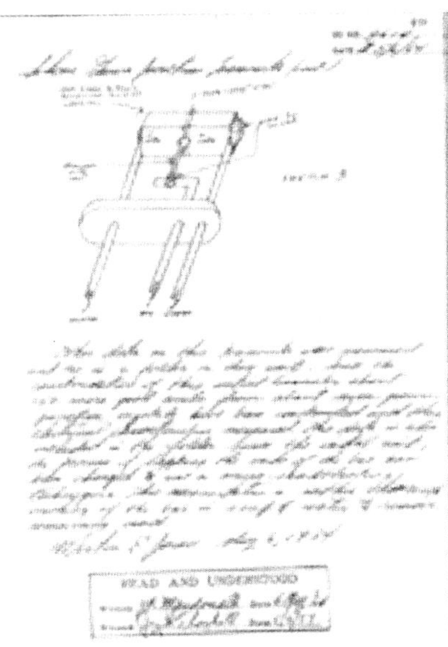

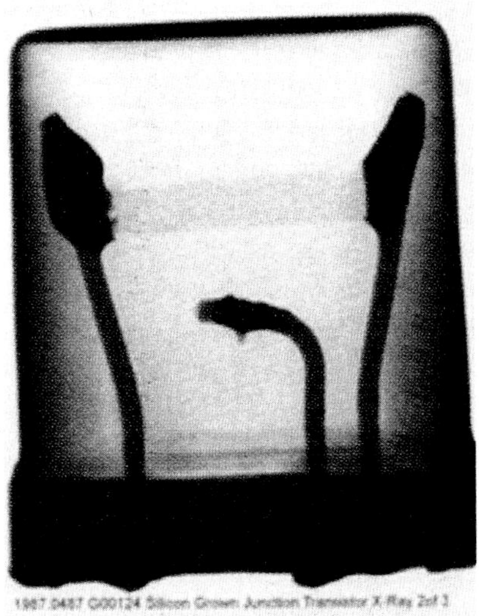

Figure 4. Grown-junction transistor.

The earliest transistors were made from germanium. But silicon offered some big advantages in terms of much higher operating temperatures, breakdown voltage and power-handling ability. The problem was that silicon was more difficult to refine because of its higher melting point.

Shortly before I joined Texas Instruments, that company solved the puzzle, making silicon transistors commercially available in the mid-1950s. But, I'm getting ahead of my story.

In 1947, I graduated from the University of Illinois with a degree in electrical engineering. That was one year before Bell Labs unveiled the transistor. That meant that I started my career at a very exciting time. But my formal training was in electrical power, though I had taken some vacuum-tube classes.

Fortunately, I also had studied engineering physics, because I felt it might be more useful than knowing how to connect three-phase transformers! Looking back, I'm glad I took those physics classes.

I wanted to work in electronics, in part because of a childhood interest. My dad ran a power company that served a wide area in rural Kansas, and he used amateur radio in his work. I found it very interesting.

In fact, it was during an ice storm during my teens, when customers throughout his area lost power, that I first saw how radio – and by extension, electronics – could really impact people's lives, keeping them informed and connected, giving them hope.

After graduation, I was hired by Centralab, a Milwaukee-based electronics manufacturer. They made parts for radios, televisions and hearing aids. Centralab was a fortunate choice for me, because they worked with hybrid circuits – an early form of miniaturization.

Centralab also developed what would become known as thick-film hybrid circuits. In this process, silver paint was deposited on a ceramic substrate – or base layer – to form conductors. Carbon-based inks formed resistors. Small capacitors were formed in the substrate, and larger ones were attached. The necessary vacuum tubes could then be attached with sockets or soldered directly to the substrate[4].

Centralab was ideal for me in another way. The group I worked in was small, so I saw the entire process, from engineering through sales and production. My initial duties included the design and product engineering work on hearing aid amplifiers and resistor-capacitor networks for television sets.

Since it was a new field, making inventions was easy. Almost anything that departed from previous designs was novel and probably could be patented. In this period, I received about a dozen patents, the most notable being a capacitor design using the first reduced titanates and a technique for automatically adjusting resistors using sandblasting.

In 1951, Bell Labs held their first transistor symposium and began licensing transistor technology for a $25,000 fee. At this time, Centralab became interested in making transistors and acquired a license.

I felt the transistor pointed the way to the future, and I wanted to be there. After some home study and some formal training that included Bell Labs symposia, I became the leader of a three-man project to build transistors and incorporate them into Centralab products.

We built a reduction furnace, crystal puller and zone refiner. This was relatively complex at the time, but growing semiconductor crystals is similar in principle to the crystal-growing kit you might buy for a child today.

We used our new equipment to make germanium alloy devices. We mounted unprotected transistors in a plastic carrier, which was a novel design. Environmental protection was achieved by using the ceramic substrate as part of the hermetic seal.

By 1957, Centralab had established a small production facility that sold small quantities of amplifiers for hearing aids and some other applications. They were even marginally profitable. It was clear, however, that major expenditures would soon be required, especially for the military market that was becoming a major opportunity. The military required silicon devices, which added to the necessary capital expense.

By 1958, I had decided to leave and began looking for another job. I talked with several companies and found that IBM had committed to technology similar to the thick-film work I had done at Centralab. I felt they had made a basic mistake by choosing a substrate that was much too small. Motorola was quite interested and thought I could work part-time on my own ideas about miniaturization.

Texas Instruments was more enthusiastic than the others and offered me a job working in miniaturization full-time.

When I started at TI in May of 1958, I had no vacation coming that year. So I worked through the period when about 90 percent of the workforce took what we called "mass vacation." I was left with my thoughts and imagination[5].

I already had reviewed the earlier attempts at miniaturization. The vast majority of existing transistors held no potential for integration since their electrodes were on different surfaces – offering no realistic possibility of interconnecting them.

In the early 1950s, the Englishman Geoff Dummer of the Royal Radar Establishment suggested that all electronics could be made as a single block. He mentioned the use of amplifying layers, resistive layers and things of that sort. The electrical functions would be connected directly by cutting out areas of the various layers[6]. It was a remarkable statement. But Dummer gave no proposal for how to do it.

In 1956, Dummer gave a small contract to a British manufacturer to build such a device. They were completely unsuccessful, partly because they worked with the grown-junction technology I mentioned earlier and tried to connect the various layers.

While his device failed, Dummer was on target in general. The body resistance of the semiconductor itself – and the capacitance of the junctions between the positive and negative regions that could be created within it – could be combined with transistors to make a complete circuit out of the same material.

My contribution was taking this idea and making it a practical reality.

While my first duties at TI were not precisely defined, I was free to choose an approach in the general area of miniaturization. The company made transistors, resistors and capacitors. Since the company could make semiconductors cost effectively, I thought it would be worthwhile to try and make everything from semiconductors.

This was contrary to most other major efforts at the time, which fell into three basic categories.

One group felt the main problem was the assembly of individual parts, and that by making all parts the same size and shape, they could automate the assembly process. A second group thought that thin-films were the way to go – a more modern form of the thick-film technology used at Centralab. Both of these approaches would have used conventional transistors and assembled them to the other components.

A third group felt a more radical approach must be taken. They felt that our knowledge of material was now complete enough so that entirely new structures could be invented to perform circuit functions. A quartz crystal, which performed the functions of an inductor and capacitor, was the favorite example[7].

It was clear to me that one of the major problems with all of the existing approaches to microminiaturization was that they involved different materials and fabrication processes. I began to consider an approach that would reduce the number of both.

It was obvious that transistors and diodes could be made of semiconductor materials. Discrete – or standalone – resistors and capacitors could also be made from semiconductors. However, that would be relatively expensive, and neither would perform as well as the best ones made with more conventional techniques and materials. For example, titanium nitride was better for resistors, and Teflon made better capacitors.

Even so, since all components *could* be made with a single kind of material, it was possible to consider making them all within a single *piece* of material. By connecting them properly, complete circuits could be formed.

On July 24, 1958, I described in my lab notebook what would come to be known as "The Monolithic Idea." It stated that circuit elements such as resis-

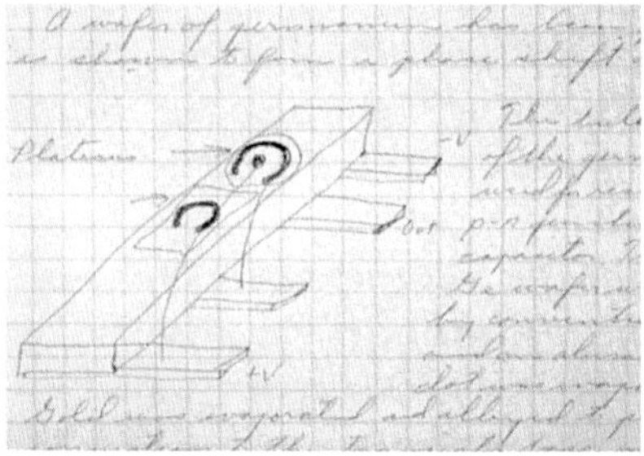

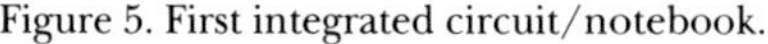

Figure 5. First integrated circuit/notebook.

tors, capacitors, distributed capacitors and transistors – if all made of the same material – could be included in a single chip.

I quickly sketched a proposed design for a flip-flop circuit using components all made from silicon. Resistors were provided by the bulk resistance in the silicon, and capacitors were formed at the p-n junctions[8].

I showed the design to my direct supervisor, Willis Adcock, upon his return from vacation. He was enthused, but skeptical. He asked for proof that circuits made entirely from semiconductors would work.

Therefore, I built up a circuit using discrete – or separate – silicon elements, starting with packaged grown-junction transistors. I formed resistors by cutting small bars of silicon and etching them to the required values. Meanwhile, capacitors were cut from diffused silicon power transistor wafers, which had been metallized on both sides.

Once assembled, the unit was demonstrated to Adcock on August 28, 1958. Fortunately, the test proved that all circuit elements could be built of semiconductor materials.

But since I had used discrete components, it was not, of course, an integrated circuit. I immediately set out to build a truly integrated structure.

At the time, TI was very strong in grown-junction devices and had just begun to work seriously on diffused structures. One silicon transistor, a power device with an alloyed emitter, was in production. Several small-signal germanium devices also were in production. The emitter and base contacts were evaporated through metal masks. Mesas were etched after hand masking with black wax.

I obtained several of these wafers, already diffused and with the contacts in place. By choosing the circuit, I was able to lay out two structures that would use the existing contacts on the wafers.

The first circuit attempted was a phase-shift oscillator, a favorite demonstration vehicle for linear circuits at the time. Technicians Pat Harbrecht and Tom Yeargan cut the wafers into bars about 0.12 inches by 0.4 inches. Metal tabs were alloyed to the back of the bar to provide contacts to the bulk resistors.

Black wax was applied by hand to mask the mesas, one for the transistor, and a larger one for a diffused region forming a distributed resistor-capacitor network.

On September 12, 1958, the first three oscillators of this type were completed. When power was applied, the first unit oscillated at about 1.3 megahertz.

To show that digital circuits could be built, the same techniques were used to build a flip-flop. That unit was completed on September 19.

By early October, we had begun designing a new germanium flip-flop integrated circuit. This unit was the first to be built entirely from scratch. It used bulk resistors, junction capacitors and mesa transistors. The first working units were completed early in 1959 and were later used to publicly announce the "Solid Circuit" concept in March of 1959.

Shortly thereafter, Robert Noyce of Fairchild Semiconductor showed the desirability of using the planar process with metal leads over the oxide, and several others, such as folks from Westinghouse, contributed to the integrated circuit field.

Today, it is a little hard to believe, but in 1959 as we began to announce this idea – and Fairchild publicly discussed Noyce's innovations – there was tremendous criticism.

At the time, it was not obvious that the monolithic semiconductor approach was going to succeed over the others I outlined earlier. Gordon Moore, Noyce, a few others and myself provided the technical entertainment at professional meetings for the next five years as we described and debated the merit of the various miniaturization systems.

There were three basic objections[9]. The first was a belief that production yields would always be too low to be profitable. At the time, less than 10 percent of all transistors manufactured actually worked properly. Another group of people felt it did not make very good use of materials, since the best resistors and transistors were not made with semiconductors. Also, the true transistor people did not want to see their elegant devices messed up with all the other stuff on the chip. These arguments were difficult to counter, since they were basically true.

And finally, many people working in the large companies thought if semiconductor technology succeeded, all of the world's circuit designers would be out of work. Of course, employment of circuit designers has actually increased over the years, but they are doing their work differently than in the transistor era.

The turning point came from two highly visible military programs in the 1960s – the Apollo moon mission and the Minuteman missile. The adoption of integrated circuit technology by these programs was a strong endorsement.

By 1964, a few adventurous companies had begun docking integrated circuits into their commercial equipment. To help popularize integrated circuits, I participated in a TI team that developed the first handheld electronic calculator.

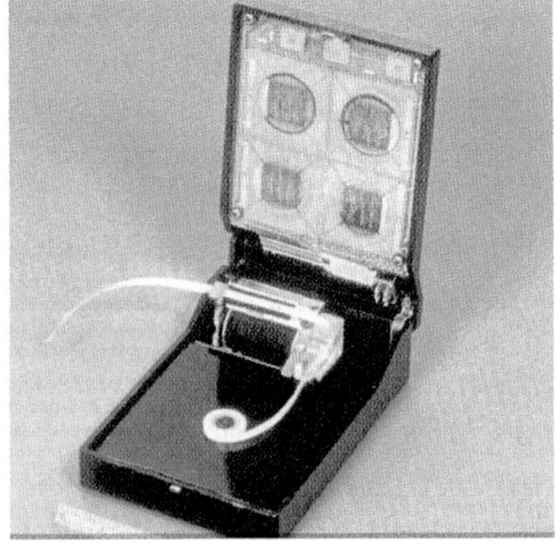

Figure 6. Handheld calculator.

Digital Equipment Corporation also was a very early user, and by the late 1960s, most engineers had accepted the fact that integrated circuits were here to stay.

As the concept was accepted, hundreds and later thousands of the best engineers in the world began to work on it.

My lecture has focused on the environment immediately preceding and following the invention of the integrated circuit.

The innovation and development that has followed in the past 40 years has been more remarkable and far more rapid than all the developments in the prior 400 years after William Gilbert first coined the term "electricity."

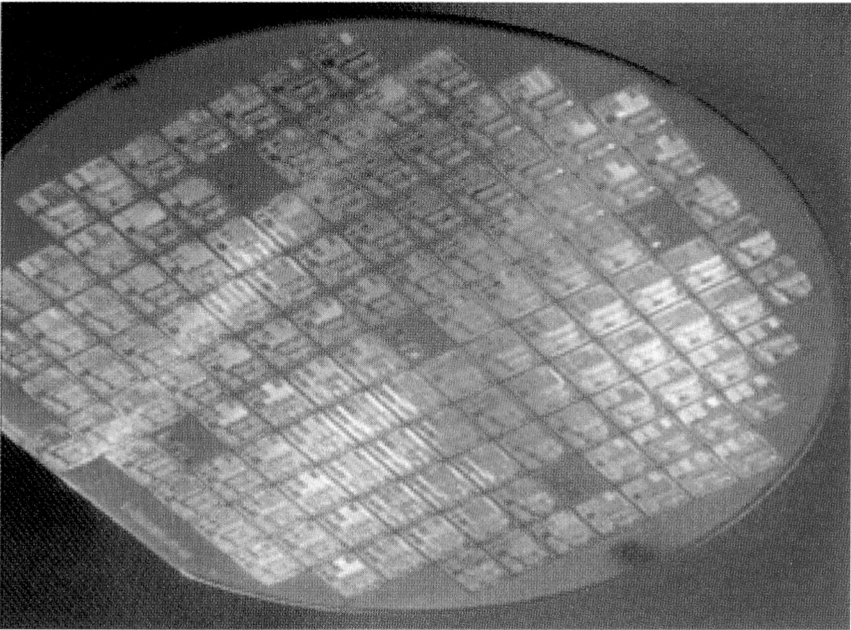

Figure 7. Increasing ICs per wafer.

Thanks to the work by hundreds of thousands of the world's best engineers, we've not only created new applications for integrated circuits, we've also gotten much better at making them. New manufacturing processes have been devised, better transistors have been invented, and sophisticated techniques for computer-aided design have been developed. Consequently, progress in the field was rapid.

The early simple chips with a dozen components grew to chips with 10,000 components by 1970 and more than a hundred million components today. This progress has been accompanied by a rapid decrease in the cost of electronic circuitry[10].

In 1958, a single transistor cost about $10. Today, you can buy a chip with over 100 million transistors for about that price. Costs are almost certain to continue declining in the future. This decrease in cost of 100 million to one has greatly expanded the field of electronics.

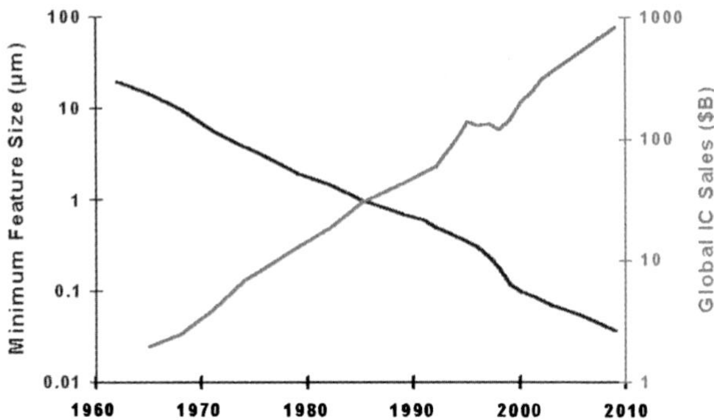

Figure 8. Smaller features, lower cost, larger market.

Today, powerful personal computers sell for less than $1,000. And these are far more capable than the $10 million versions of the 1960s.

While integrated circuits are used for military applications, many more are used to improve the quality of life for everyone[11]. Automobiles are safer and emit fewer pollutants because of their integrated circuit systems. Radio and TV have become nearly universal, and hundreds of millions of people are united by the networking power of the Internet. Wireless communications keep people in contact with information and other people anywhere they go on the planet.

I believe the best is yet to come.

Today, approximately 1,000 electrons are necessary to turn an individual transistor on or off. By 2010, it's estimated this will be accomplished by only 100 electrons. The 2010 projection assumes that higher dielectric constant materials will be introduced.

If they were not, then the continuation of geometrical scaling would extrapolate the reduction to a mere 10 electrons per transistor by 2010 and just one electron by 2020.

That, of course, would present a fundamental physical limitation.

Some proposed approaches around this obstacle include quantum-cellular automata and molecular switches, among others[12]. When we reach this nanometer-length scale, many people think chemically assembled configurations will begin replacing today's patterned and etched structures.

I don't really know how all that will play out. I do know that engineers in all corners of the world continue to refine integrated circuits while others are working on what might come next.

I know how they feel. In 1958, my goals were simple: to lower the cost, simplify the assembly, and make things smaller and more reliable. Although I do not consider myself responsible for all of the activity that has followed, it has been very satisfying to witness the integrated circuit's evolution.

I am pleased to have had even a small part in helping turn the potential of human creativity into practical reality.

REFERENCES

[1] Murphy, Bernard T., Haggan, Douglas E., Troutman, William W., *From Circuit Miniaturization to the Scalable IC*, Proceedings of the IEEE, Vol. 88, No. 5, May 2000.

[2] Reid, T.R., *The Chip: How Two Americans Invented the Microchip and Launched a Revolution*, Simon and Schuster, New York (1984).

[3] Kilby, J.S., *Silicon FEB Techniques*, Solid/State Design, July 1964.

[4] Roup, R.R., and Kilby, J.S., *US Patent 2,841,508*, issued July 1958.

[5] Kilby, J.S., *Invention of the Integrated Circuit*, IEEE Transactions on Electron Devices, Vol. Ed-23, No. 7, July 1976.

[6] Kilby, J.S., *Invention of the Integrated Circuit*, IEEE Transactions on Electron Devices, Vol. Ed-23, No. 7, July 1976.

[7] WADD Technology Notes, *Molecular Dendritic Approach*, Westinghouse Electric Corp. Reports, February 1960 and July 1960.

[8] Kilby, J.S., *Semiconductor Solid Circuits*, presented at American Rocket Society 14th Annual Meeting, November 1959.

[9] Kilby, J.S., *Invention of the Integrated Circuit*, IEEE Transactions on Electron Devices, Vol. Ed-23, No. 7, July 1976.

[10] Nishi, Yoshio and Doering, Robert, editors, *Handbook of Semiconductor Manufacturing Technology*, Marcel Dekker, New York (2000).

[11] Doering, Robert, *Implications of Scaling to Nanoelectronics*, Workshop on Societal Implications of Nanoscience and Nanotechnology.

[12] Jasinksi, J., Petroff, P., *Applications Nanodevices, Nanoelectronics, and Nanosensors as reported in Nanotechnology Research Directions: IWGN Workshop Report, Vision for Nanotechnology R&D in the Next Decade*, 1999.